MICROSCALE ORGANIC LABORATORY

MICROSCALE ORGANIC LABORATORY

DANA W. MAYO
Professor of Chemistry
Bowdoin College

RONALD M. PIKE
Professor of Chemistry
Merrimack College

SAMUEL S. BUTCHER
Professor of Chemistry
Bowdoin College

JOHN WILEY & SONS
New York   Chichester   Brisbane   Toronto   Singapore

All experiments contained herein have been performed several times by students in college laboratories under the supervision of the authors. If performed with the materials and equipment specified in this text, in accordance with the methods developed in this text, the authors believe the experiments to be a safe, valuable educational experience. However, all duplication or performance of these experiments are conducted at one's own risk. The authors do not warrant or guarantee the safety of individuals performing these experiments. The authors hereby disclaim any liability for any loss or damage claimed to have resulted from or related in any way to the experiments, regardless of the form of action.

Production Supervisor: Pamela A. Pelton
Cover design: Dawn L. Stanley
Cover photo by Jeffrey Krein Studio
Manuscript Editor: Vivi Danser
    under the supervision of Martha Cooley

Library of Congress Cataloging-in-Publication Data

Mayo, Dana W.
    Microscale organic laboratory

    Includes index.
    1. Chemistry, Organic—Laboratory manuals.
I. Pike, Ronald M.   II. Butcher, Samuel S.,
1936–   III. Title.

QD261.M38   1986   547′.007′8   85-16872
ISBN 0-471-82448-8

Printed in the United States of America

10 9 8 7 6 5 4 3

To Jeanne d'Arc, Marilyn, and Sally

Foreword

Even a casual glance through this book will convince the reader that it represents a striking new departure from lab manuals. But the real question is, is it workable? Can students work successfully at this scale? If not, this text will be a useless *tour de force*, a piece of music playable only by its composer. Novelty alone is not sufficient.

So, ever the empiricist, I used myself as a test animal. I am not an organic chemist by training and have not seriously done any organic chemistry in twenty-odd years. Yet virtually *every* one of the experiments I tried (I have tested about 30% of them) worked well if the directions were followed, and often even if inadvertent variations were made! This alone would put this text in the front rank of lab books.

But more important is my discovery that the microscale adds immensely to the fun of working in organic chemistry. Gone are the revolting odors, the long waits while things heat up and cool down, the difficulties and hazards of cleaning up massive amounts of tar, and the brooding sense of impending doom if something goes wrong. Doing organic chemistry becomes a pleasant way to spend an afternoon.

Educational research studies have shown that the appeal of the laboratory ranks very high among the reasons why students choose to major in chemistry. And how many professional chemists were not originally drawn into the field by seeing magical changes of color, and the startling transformations of matter? The microscale gives students a lot fewer reasons to "hate lab." Things happen faster, so lab is less boring. The reduced hazard level means that the experience is no longer one of wading through a swamp infested with very real crocodiles. And most of all, the primitive joy of learning a craft may well sing to some students at a deep level.

Many colleges, held hostage by their macro glassware, will be unable to change over to microscale immediately. But whether or not this book is popular or a commercial success, it has an assured place as a seminal book, a turning point, in the history of laboratory education. New tools have been placed in the hands of laboratory curriculum designers, and a rebirth of interest in the teaching laboratory may result.

Miles Pickering
Lecturer and Director
Undergraduate Laboratories
Princeton University

September 1985

MICROSCALE ORGANIC LABORATORY describes a well thought-through, thoroughly tested, and entirely novel approach to the standard organic chemistry laboratory. It merits the careful attention of the entire academic chemistry community. It is the first genuinely innovative approach to the teaching of laboratory aspects of organic chemistry in many years.

Whatever their political persuasions, chemists in academia are profoundly conservative in terms of didactic approaches to their subject matter. New insights into chemistry are eagerly incorporated into the substance of what is taught; in contrast, the structure and methodology of courses are altered slowly and reluctantly. Incremental changes are the order of the day and have been for decades, and these are accepted only following careful and critical review. One need look no further than the exceptional longevity of Morrison and Boyd as the teaching standard for the theoretical side of organic chemistry to appreciate this point. The pace of changes in the acquisition of knowledge in organic chemistry stands in sharp contrast to the pace of innovations in teaching its subject matter.

This is by no means an inappropriate state of affairs. The way that organic chemistry and its laboratory practice are taught is the product of lengthy experience. The success of organic chemistry in the United States, both at the academic and industrial levels, provides a cogent argument that what is being done in teaching is effective and productive. A conservative approach to change is insurance against foolishly abandoning that which has proved successful over the years. At the same time, it may blind us to the opportunity for making a quantum jump in the teaching of organic chemistry.

MICROSCALE ORGANIC LABORATORY offers a qualitative quantum jump. It describes an entirely new way of doing the undergraduate lab in organic chemistry: everything on a very small scale. The laboratory also offers a strong instrumental focus; the individual student is responsible for generating most of his or her own data. There are several compelling theoretical arguments for believing that this approach offers highly significant advantages over the standard laboratory. In no particular order of priority they include: (a) the development of students' self-reliance; (b) cost-effectiveness, in terms of both capital for equipment and the cost of reagents; (c) safety benefits deriving from small-scale experiments; (d) the lack of a toxic chemical disposal issue; and (e) *better laboratory experience, since organic chemistry in both academia and industry is typically carried out on a very small scale.* It will prove easier to scale up than scale down.

Theoretical arguments aside, there are two good reasons to believe that MICROSCALE ORGANIC LABORATORY represents an important advance in laboratory teaching. The first and more important is that the approach has been carefully and thoroughly tested on several generations of Bowdoin students. The experiments described can be done successfully (and willingly) by undergraduate students. They learn the techniques, profit from them, and do well in subsequent chemistry courses and in chemistry beyond the undergraduate level. In short, the teaching process has been demonstrated to be successful at the only level that really counts. Theoretical arguments faithfully translate into practical benefits.

The second reason is more personal: my daughter Jennifer and son Matthew, both Bowdoin undergraduates, have participated at various levels in this organic laboratory experience. Both are critical observers and both have emerged supportive of the program and well educated in lab techniques in organic chemistry.

Selective undergraduate institutions such as Bowdoin College are insulated to some extent from the "publish or perish" ethic and devote thoughtful consideration to the teaching function. They thus have the potential to lead research-intensive institutions in the teaching of chemistry, and the best of them do. In my view, MICROSCALE ORGANIC LABORATORY does just that. New ground has been broken.

No thoughtful person will conclude that MICROSCALE ORGANIC LABO-
RATORY is the last word in teaching the organic laboratory. It promises to be
the new standard upon which future incremental improvements will be based.
All thoughtful teachers of organic chemistry laboratory courses owe what follows
in this book their most careful consideration.

E. H. Cordes
Vice President
Biochemistry
Merck Institute for Therapeutic Research

September 1985

Preface

This introductory organic laboratory textbook represents a major departure from all other modern texts dealing with this subject. We have chosen to introduce experimental organic chemistry at the microscale level. Currently, beginning students perform most of their experimental work at a level that is at least two orders of magnitude above that described in this text. Although contraction from the multigram to the milligram scale is the most obvious difference between this and other books, *Microscale Organic Laboratory* has a number of other unique aspects.

1. The laboratory environment has been made a distinct part of the experimental process. The student is given the means to easily determine his or her exposure to all volatile substances employed in the experiments. These calculations can be carried out for any laboratory by using instructor-supplied ventilation rates.
2. Chemical instrumentation is given high priority *in the laboratory*. The text avoids emphasizing data not directly determined by the student. Product characterization by infrared spectroscopy is routine, and a detailed development of the interpretation of such data is provided.
3. Modern separation and purification techniques, including preparative gas chromatography, thin-layer chromatography, and column chromatography, are extensively utilized in product workups.
4. Over one-third of the 82 reaction products are new to the undergraduate laboratory. Many of the reactions involve reagents or substrates that would present potential safety problems or exorbitant costs in a macroscale laboratory program; nevertheless, the use of these materials becomes safe and practical for experiments at the microscale level. For example, reagents such as 100% nitric acid, diborane, chloroplatinic acid, "instant ylids," silver persulfate, chromium trioxide resin, tetrabutylammonium bromide, and triflic acid are representative of materials that the student will encounter during the year.

Why are we committed to the goal of attempting to significantly reduce the scale of starting materials in the introductory organic laboratory? The academic community has become increasingly aware of the necessity of improving air quality in instructional laboratories. Health and safety concerns have resulted in a detailed examination of the laboratory program at our institution. The standard solution to the problem of poor air quality—a costly upgrading of the ventilation systems—has generally been considered the most reasonable answer.

Our study of the problem led us to the following conclusions. First, although current organic laboratory texts are filled with details of product characterization employing the latest spectroscopic methods, the descriptions of techniques for preparing compounds have changed very little from those of a century ago. In particular, the scale of synthesis has changed very little during this period. [Indeed, the quantities of materials employed have decreased only modestly, and in some cases they have actually increased!] Clearly, the strategy of introducing the student to organic chemistry laboratory techniques (originally and today) is centered on the multigram level (see Table 1).

We question the wisdom of continuing to introduce laboratory work at conventional levels, especially when one considers the quantities of costly substances commonly employed in natural product, pharmaceutical, biochemical, and other fields of modern research.

Second, most instructors agree that there has been a serious decrease in

Table 1 Starting Material Employed in Classic Organic Laboratory Syntheses 1902–1980

Date	Author	Acetanilide	4-Bromoacetanilide	Benzoin
			Starting Materials Required (grams)	
		Aniline	Acetanilide	Benzaldehyde
1902	Levy, 4th ed.	46.2	—	50.0
1915	Cohen, 3rd ed.	25.0	5.0	25.0
1933	Adkins	25.0	13.5	10.0
1941	Fieser, 2nd ed.	18.2	13.5	25.0
1963	Adams	20.0	13.5	16.0
1980	Durst	10.0	5.2	10.0

undergraduate laboratory contact time over the past two decades. It seems to us that an increasingly important question to consider in the evaluation of the introductory organic chemistry program is this: At what scale of introductory laboratory work can the student gain the greatest ability to handle organic materials within the shortest period of time?

We now firmly believe that the microscale laboratory approach resolves both of the concerns stated above, and that it affords a number of significant bonuses. The immediate result of "going micro" is that there is a change in the laboratory air quality that can only be described as spectacular. Of importance (and a point we have come to appreciate in retrospect) is the fact that many reactions and operations carried out at the micro level require far less time to reach completion. Indeed, *many of the time-consuming aspects of current instructional experiments are dramatically shortened.* We believe that this facet of the microscale approach will have a major pedagogic impact because the number of manipulations per laboratory period will substantially increase. Herein lies the significant advantage of this approach.

We also see a further pedagogic advantage in operating at the microscale level which requires the increased use of chemical instrumentation. Routine use of gas chromatography may be avoided in macro experiments; however, without increased student use of this instrumentation, successful experimentation with liquid substances at the microscale level becomes difficult or impossible. The capital investment required to allow for routine use of this equipment in many undergraduate laboratories might seem to put a microscale program out of the reach of institutions with limited budgets. Yet, if carefully managed, the substantial savings in chemical costs (75 to 90% of current budget) can offer a short payback period for expanded GC capacity and/or other equipment. The end result of conversion to microscale is a far more effective integration of modern instrumentation into the organic laboratory program.

Once the microscale approach is initiated, the advantages are endless. They include (1) major cost reduction in chemicals; (2) elimination of fire or explosion danger; (3) elimination of chemical waste disposal costs; (4) expansion in variety and sophistication of experiments; (5) elimination of dependence on commercially available starting materials; and (6) more durable and less expensive glassware.

These advantages represent compelling arguments for the microscale approach. Initially, however, we considered this approach with reservations. First, it appeared to involve a sophisticated set of techniques and manipulations too advanced for sophomore undergraduates to master. Second, we thought that with experience only at the microscale level, a student might encounter problems when larger-scale preparations are undertaken. Third, significantly less organic chemistry would be covered during the year because increased attention would have to be paid to the development of microtechniques. Fourth, certain classical procedures, such as fractional distillation, would have to be abandoned.

As of the completion of this text, we have conducted eight semesters of microscale organic laboratory assessment. Our experience with the test laboratory groups (composed of a cross section of volunteers) has been a revelation. It is clear that the entire range of the class achieves significantly better results on microexperiments. Better yields are realized and, in particular, the class appears to master experimental details and procedures more effectively. We feel that these results stem from several causes, one of which is the increased attention to detail required in the laboratory. We also sense a synergistic interaction between the microscale and the analytical chemistry laboratory, which often takes place in the sophomore year concurrently with the organic course. Analytical chemistry is carried out at a scale not unlike that employed in the microscale organic laboratory.

At present there is no indication that learning microtechniques in an introduction to organic chemistry causes any adverse effects when scale-up work is introduced in advanced laboratory courses or in research areas. To the contrary, our students appear to be performing significantly better in upper-level work.

The development of the experimental section of the text, which contains eighty reactions, has been a major effort. The conditions for most of these reactions have been optimized to maximize yields at the micro level. We have chosen to describe the experimental work in language similar to formal journal style. At first, this impersonal construction would appear to be ''user unfriendly''; in field testing, however, we have found that students quickly adapt to the style and soon come to appreciate the use of precise, routine terminology. This introduction to formal style also pays substantial dividends in upper-level courses, where students are expected to consult the original literature.

The philosophy of this text is to focus the student to a large extent on the *experimental* aspects of organic chemistry. We have purposely attempted to keep to a minimum the theoretical discussions, supplying only sufficient background material to cover potential discontinuities between lecture and laboratory. We want the student to become comfortable, and to develop a substantial degree of independence with the use of chemical instrumentation. To meet this objective, we have focused attention specifically on infrared spectroscopy and gas chromatography, these basic instrumental techniques are more generally utilized by sophomores than the more expensive techniques.

We recognize the dominant role of nuclear magnetic resonance in modern organic chemistry. The current situation however, is, that only in a limited number of cases do second-year students ever have the opportunity to generate their own data with this type of instrumentation. As we are primarily concerned with raising students' excitement and interest in gathering actual laboratory data, we feel that artificially incorporating material from outside the laboratory as a means of including the nmr or mass spectroscopy experiment may not enhance, but instead divert, attention from the laboratory experience. We are, however, exploring ways to overcome this particular problem and plan to address it in future editions.

The six-year road to completion of this text has been a long but rewarding one. Many individuals have made vital contributions. Arnold Brossi, David Brooks, Miles Pickering, Lea Clapp, Eugene Cordes, and Henry Horner provided encouragement and sound advice at crucial points along the way.

We are grateful to acknowledge that most of the infrared and nmr spectra not recorded at Bowdoin College were obtained from Aldrich Laboratories of FT-IR and NMR Spectra through the courtesy of Mr. Charles Pouchert and the Aldrich Chemical Company.

John Ryan, Larry Riley, Don Sellar, and Hugh Bowe of Ace Glass, Inc., are primarily responsible for the development of the novel microglassware described in this manual. They exhibited considerable patience throughout the ordeal. Ed Hollenbach and Lyle Phifer of Chem Service saw the advantage of making available small quantities of high-purity reagents and starting materials as a way to ensure success at the microscale level.

The vast amount of experimental development was shared by Teaching Research Fellows Janet Hotham, David Butcher, Paulette Fickett, and Caroline Foote, along with several Bowdoin College students (Mark Bowie, Sandy Hebert, Rob Hinkle, Marcia Meredith, and Gregory Merklin). The experimental ground covered by this group, much of which required ingenious solutions, is remarkable. The breadth of experiments available is a tribute to their dedication. Janet deserves a special thanks. A Merrimack College graduate adopted by Bowdoin College, she has been with the program almost from the beginning. Her thoughtful suggestions, based on experience in the trenches, and her willing contributions in any area of need are most gratefully remembered. The enthusiasm of the sophomore organic students who volunteered for the initial pilot sections at Bowdoin and the field testing sections at Merrimack, played a key role in encouraging us to continue our efforts.

Judy Foster is responsible for the majority of the illustrations and reaction schemes. Her efforts have greatly enhanced our descriptions of the techniques involved in microscale work. Her ingenuity also lead to the pictorial keying of the equipment setups. The patient understanding and thoughtful advice of Dennis Sawicki, Chemistry Editor at John Wiley, has been particularly valuable.

We wish to thank Dean Alfred Fuchs of Bowdoin for his constant encouragement of this program during its development. The initial exploratory work was supported by a grant from Bowdoin and a Department Grant from the du Pont-Nemours Company. A semester leave granted to DWM (Spring 1984) was funded by a grant from the ARCO Corporation, and a semester sabbatical was granted to RMP by Merrimack College (Spring 1985), as was an appointment as Visiting Charles Weston Pickard Professor of Chemistry at Bowdoin College (1980–1981, and Spring 1984). The Surdna Foundation awarded two major grants that allowed the complete development and implementation of the program at Bowdoin College and field testing of experiments at Merrimack College. These institutions' support of and faith in this educational concept is gratefully acknowledged.

Dana W. Mayo
Ronald M. Pike
Samuel S. Butcher
March, 1985
Brunswick, Maine

Contents

NOTE: A separate Contents for Chapters 5 and 6 is provided at the end of the following Contents. It classifies the experiments in those Chapters by mechanism.

H. Aromatic Nucleophilic Substitution

I. Aromatic Free Radical Substitution

J. Oxidation and Reduction Reactions

K. Other Experimental Transformations

L. Rearrangements

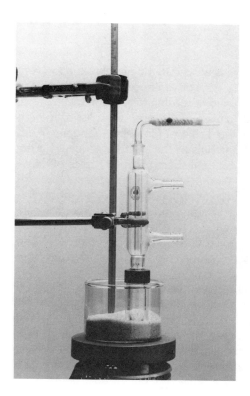

Chapter 1

Introduction

You are breaking new ground in the organic chemistry laboratory!

Your course is going to be quite different from the conventional manner in which this laboratory has been taught. You will be learning the experimental side of organic chemistry from the microscale level. Surprising as it may seem, because you will be working with very small amounts of materials, you will be able to observe and learn more organic chemistry than many of your predecessors did in nearly two years of laboratory work. You will find this laboratory an exciting and interesting place to be. While we cannot guarantee it for you individually, the majority of students who have been through the program during its development have found the microscale organic laboratory to be a pleasant adventure.

At the very beginning we want to acquaint you with the organization and contents of the text. We will then give you a few words of advice, which, if they are heeded, will allow you to avoid many of the sand traps as you work your way through the course. Finally, we will turn philosophical and attempt to describe what we think you should derive from this experience.

Following this brief introduction, the second chapter is concerned with safety in the laboratory. This chapter is unique. It provides the background necessary for you to effectively utilize the environmental data given in each experiment. For the first time in a laboratory text, information is supplied that will allow you to easily calculate your maximum possible exposure to any volatile chemical employed in any of the experiments. The theoretical basis for these calculations is discussed in detail in Appendix A. Chapter 2 also discusses general safety protocol for the laboratory. It is vitally important that you become familiar with the details of the material contained in this chapter; your health and safety depend on this knowledge.

The next section of the text consists of three chapters that are primarily concerned with the development of experimental technique. Chapter 3 describes in detail the glassware employed in microorganic chemistry: the logic behind its construction, tips on its usage, the common arrangements of equipment, and various other laboratory manipulations, including techniques for transferring microquantities of materials. Suggestions for the organization of your laboratory notebook are considered at the end of this chapter.

Chapter 4 deals with equipment and techniques for determining a number of physical properties of microscale samples. Chapter 5 is divided into 11 experimental sections. These exercises, and the detailed discussions attached to them, develop the major areas of experimental technique to be used in the microscale organic laboratory. A number of specific reactions incorporated in this section illustrate the use of the techniques being introduced. At the discretion of your instructor, these particular examples may be easily replaced to correspond more closely to local development of the subject, so do not be surprised if you find an experiment from Chapter 6 mated to a discussion in Chapter 5.

Chapters 6 and 7 encompass the major experimental section of the text. Chapter 6 is focused on preparative organic chemistry and consists of 41 experiments. Many of these exercises involve of a number of optional parts. Experiments 47 to 52 are made up of four sets of sequentially arranged experiments in which the product from one experiment is utilized in the next (a situation very similar to the real-life research laboratory). For convenience in organization, the reactions contained in Chapter 6 are grouped according to reaction mechanism, and the sequential experiments by mechanism and group. (The list of Experiments Classified by Mechanism follows the Table of Contents.) Chapter 7 develops the characterization of organic materials at the microscale level by the use of classical organic reactions and vibrational spectroscopy. Tables of derivative data for use in compound identification by these techniques are given in Appendix D. A detailed discussion of the interpretation of infrared group frequencies and a brief introduction to the interpretation of NMR spectral data make up a significant portion of Chapter 7. The theoretical basis for these spectroscopic sections is considered in Appendices B and C. Appendix B also contains a convenient collection of infrared group frequency data derived from discussions in Chapter 7.

The organization of the experimental procedures given in Chapters 5 and 6 is arranged in the following fashion. A short opening statement describing the reaction to be studied is followed by the reaction scheme. Generally, a brief discussion of the reaction follows, including a mechanistic interpretation. In a few cases of particularly important reactions, or where the experiment is likely to preceed consideration in lecture, a more detailed description is given. The estimated time to complete the work and a table of reactant data come next. For ease in organizing your laboratory time, the experimental section is divided into four subsections: reagents and equipment, reaction conditions, isolation of product, and purification and characterization. A table of the relevant environmental data follows the experimental procedure.

We then introduce a series of questions and problems designed to enhance and focus your understanding of the chemistry and the experimental procedures involved in a particular laboratory exercise. Finally, a list of literature references is given. Although this list comes at the end of the experimental section, we view it as a very important part of the text. The discussion of the chemistry involved in each experiment is necessarily brief. We hope that you will take time to read and expand your knowledge about the particular experiment that you are conducting. (You may, in fact, find that some of these references become assigned reading.) The formal language used in the experimental sections has been intentionally introduced to ease your transition into the chemical literature.

The experimental apparatus and materials involved at important stages of an experiment are indicated by a prompt sign (■) in the text and are shown in the margin. Important comments are italicized in the text, and **Warnings** and **Cautions** are given in boxes.

GENERAL RULES FOR THE MICROSCALE LABORATORY

1. *Study the experiment before you come to lab.* This rule is an historical plea from all laboratory instructors. In the microscale laboratory it takes on a more important meaning. You will not survive if you do not prepare ahead of time. In microscale experiments, operations happen much more quickly than in the macroscale laboratory. Your laboratory time will be filled to overflowing with many more events. If you are not familiar with the sequences of steps you are to follow, you will be in deep trouble. While the techniques employed at the microscale level are not particularly difficult to acquire, they do demand a significant amount of attention. In order for you to reach a successful and happy conclusion, you cannot afford to have the focus of your concentration broken by having to constantly refer to the text during the experiment. Disaster is ever present for the unprepared.

2. *Always work with clean equipment.* You must take the time to scrupulously clean your equipment before you start any experiment. Contaminated glassware ultimately will cost you additional time, and you will face the frustration of experiencing inconsistent results and lower yields. Dirty equipment is the primary cause of reaction failure at the microscale level.

3. *Carefully measure the quantities of materials to be used in the experiments.* A little extra time at the beginning of the laboratory can speed you on your way at the end of the day. A great deal of time has been spent optimizing the conditions employed in these experiments to maximize yields. Many organic reactions are very sensitive to the relative quantities of substrate (the material on which the reaction is taking place) and reagent (the reactive substance or substances that bring about the change in the substrate). The second largest cause of failed reactions, after equipment contamination, is attempting to run a reaction with incorrect quantities of the reactants present. Do not be hurried or careless at the balance.

4. *Clean means DRY.* Water or cleaning solution can be as detrimental to the success of a reaction as dirt or sludge in the system. You often will be working with very small quantities of moisture-sensitive reagents. The glass surface areas with which these reagents come in contact, however, are relatively large. A slightly damp piece of glassware can rapidly deactivate a critical reagent and result in reaction failure. *This rule must be strictly followed.*

5. *ALWAYS work on a clean laboratory bench surface,* preferably glass!

6. *ALWAYS protect the reaction product* that you are working with from a disastrous spill by carrying out all solution or solvent transfers over a crystallizing dish.

7. *ALWAYS place reaction vials or flasks in a clean beaker* when standing them on the laboratory bench.

8. *NEVER use cork rings to support round-bottom flasks,* particularly if they contain liquids. You are inviting disaster to be a guest at your laboratory bench.

9. *ALWAYS think through the next step* you are going to perform *before* starting it. Once you have added the wrong reagent, it's back to square one.

10. *ALWAYS save everything* you have generated in an experiment until it is successfully completed. You can retrieve a mislabeled chromatographic fraction from your locker, but not from the waste container!

THE ORGANIC CHEMISTRY LABORATORY

This laboratory experience will give you a brief glimpse of how organic chemistry operates as an experimental science. Historically, the organic lab has had a reputation of being smelly, long, tedious, and pockmarked with fires and explosions along the way. Modern organic chemistry still has trouble shaking this image, but the fact is that present-day organic chemistry is undergoing a revolution at the laboratory bench. New techniques are sweeping away many of the old complaints, as an increasing fraction of industrial and academic research is being carried on at the microscale level.

This text allows the interested beginning student to rapidly develop the necessary skills to slice more deeply into organic chemistry, as a sophomore, than ever before. The attendant benefits are more confidence and independence in acquired laboratory techniques. The happy result is that in the microscale organic chemistry laboratory, you are more likely to have a satisfying first encounter with the experimental side of this fascinating field of knowledge.

Chapter 2

Safety and the Environment of the Laboratory

The responsibilities that one assumes while working in an organic chemistry laboratory are not unlike those taken on when driving a car. We are clearly responsible for injuries to ourselves or others that may arise from improper use of the car. In another sense, as we drive we also add pollution to the environment in which everyone must live and work. Although we cannot reduce this pollution to zero, we are responsible for keeping our car in proper operating condition in order to minimize adverse effects.

There are similar risks associated with working in an organic chemistry laboratory (or any chemistry laboratory). The focus in this chapter and the text as a whole is upon the nature of these risks and the ways in which one may minimize the risks while still learning something about chemistry. In the laboratory, as in a car, we cannot reduce to zero the risk of an acute injury that might be associated with a fire or an explosion—nor can we reduce to zero the risk of chronic effects resulting from exposure to certain toxic chemicals. What we can do is to minimize these risks by *thinking safety* and working at maintaining a reasonably safe laboratory for each of us individually and for others in the lab.

NATURE OF HAZARDS

There is a wide range of risks in the chemistry laboratory. They are outlined briefly below so that you can begin to see the kinds of steps necessary to reduce the risks.

1. *Physical hazards.* Injuries resulting from flames, explosions, and injuries from equipment (cuts from glass, electrical shocks from faulty instrumentation, or improper use of instruments).

2. *External exposure to chemicals.* Injuries to skin and eyes resulting from contact with chemicals that have splashed or have been left on the bench top or on equipment.

3. *Internal exposure.* Longer term (usually) health effects resulting from breathing hazardous vapors or ingesting chemicals.

Reduction of Risks

Your instructor may have a more specific list of rules to be observed in your laboratory section. The following rules can serve as a starting point.

1. *Stick to the procedures described by your instructor and to the materials available in the laboratory.*

2. *Wear safety glasses.* We often recover quickly from injuries affecting only a few square millimeters on our bodies, *unless* that area happens to be on our eyes.

3. *Do not put anything into your mouth while in the laboratory.* This includes food, drinks, and pipets. There are countless ways in which surfaces can become contaminated in the lab. Since there are substances that must *never* be pipetted by mouth, one should get into the habit of *not* mouth pipetting *anything*.

4. *Be cautious with flames and flammable solvents.* Use them only under conditions specified by your instructor. Remember that the flame at one end of the bench may ignite the flammable liquid at the other end in the event of a spill or improper disposal. Flames must never be used when certain liquids are present in the laboratory, and they must always be used with care.

5. *Minimize the loss of chemicals to air or water and dispose of waste properly.* Some water-soluble materials may be safely disposed of in the water drains. Special receptacles will be provided for other wastes—sometimes in the hoods. Sodium metal and other active metals must be disposed of properly.

6. *Minimize skin contact with any chemicals.* Use impermeable gloves, when directed, and wash any chemical off your body promptly. If you have to wash something off with water, use lots of it.

7. *Tie back or confine long hair and loose items of clothing.* You don't want things to be falling into reagents or getting near flames.

8. *Don't work alone.* There are too many things that can happen to a person working alone that might leave that person unable to obtain assistance. In the rare case that you have permission to work alone, be sure that someone checks on you at regular intervals.

9. *Exercise care in assembling glass apparatus.* Separating standard taper glassware, pushing thermometers through rubber stoppers, and other operations with glassware all involve the risk that the glass may break and that lacerations or punctures may result. Make sure that thermometers and glassware are lubricated so that a great deal of force need not be used. Seek advice when unsure or for difficult cases.

10. *Report* any *injury or accident to your instructor.* This is important so that medical assistance can be requested, if necessary. It is also essential for the instructional staff to be made aware of any safety problems. Many of these are correctable.

11. *Keep things clean.* Put unused apparatus away. Wipe or care for spills on the bench top or on the floor immediately.

Precautionary Measures
Locate the nearest

- Fire extinguisher (understand uses of different types)
- Eye wash
- Spray shower
- Fire blanket
- Exit

MINIMIZING EXPOSURES TO VAPORS   The reduction in quantities used in the microscale laboratory carries with it a reduction in hazards caused by fires and explosions. Hazards associated with skin contact are also reduced; however, care must be exercised when working in close proximity to the small quantities involved.

There is a great potential for reducing the exposure to chemical vapors, but these reductions will be realized only if careful practices are observed by *every* person in the laboratory. One of the characteristics of vapors is that they mix rather rapidly throughout the lab and quickly reach the nose of the student on the other side of the room. For this reason some operations will have to be performed in hoods. Some institutions may have the majority of the experiments carried out in hoods. In the open laboratory each experimenter becomes a "polluter" whose emissions affect the people nearby the most, but become added to the lab and to the burden each of us must bear.

The hazard imposed by a given vapor is measured by its toxicity and by its concentration in the air. Toxicities for vapors are often reflected in the time-weighted averages (TWA) promulgated as regulations by the Occupational Safety and Health Administration (OSHA). The OSHA figures are described in the *Code of Federal Regulations* and are usually reprinted in the *Chemical Rubber Co. Handbook.* The American Conference of Governmental Industrial Hygienists adopts toxicity-related figures known as threshold limit values (TLV). The TWA and TLV values are intended to protect the health of a worker exposed to a chemical during five eight-hour workdays each week. While these figures are not directly applicable to the academic laboratory, we will use them here for guidance.

Other things (availability, suitability) being equal, one would, of course, choose the least toxic chemical for a given experiment. There are chemicals that play very important roles in synthetic organic chemistry that are quite toxic, and the toxicities of chemicals in regular use in the laboratory vary greatly. Bromine and benzene have TWA values of 0.7 and 30 mg/m^3, respectively, and are at the more toxic end of chemicals used, while acetone has a TWA of 2400 mg/m^3. These representative figures do not mean that acetone is "harmless" or that bromine cannot be used. In general one should exercise care at all times (make a habit of good laboratory practice) and should take special precautions when working with highly toxic materials.

The concentration of the vapor in the laboratory will depend on the vapor pressure of the liquid, the area of liquid or solid exposed, the nature of air currents near the liquid, and ventilation characteristics of the laboratory. One factor over which each individual has control is the evaporation process, which can be affected by the following procedures:

- Certain liquids must remain in hoods.
- Reagent bottles must be recapped when not in use.
- Spills must be quickly cleaned up and the waste properly discarded.

ESTIMATION OF CONCENTRATIONS IN THE LAB

A simple method is available for the estimation of the vapor concentrations of substances used in the lab. The average concentration of a substance in the laboratory is the ratio of the mass of a substance emitted into the laboratory to the volume of air in which the substance is diluted. The volume of air is *not* the volume of the laboratory, but rather the total volume of air moved through the laboratory by the ventilation system during the laboratory period. Since the material emitted is easily expressed in per-student terms, it is also convenient to consider the volume of air per student. The average concentration is then given by the following expression.

$$c = \frac{m}{V'} = \frac{m}{\dfrac{kQt_0}{n}} \tag{2.1}$$

The concentration in milligrams per cubic meter is given by c. This value may be compared with the TLV for each compound used. The numerator, m, is the milligrams of the material in question emitted by each student. A value for m is given with each experiment. $V' = kQt_0/n$, is the effective volume of air per student. It has dimensions in cubic meters per student. The factor k is known as the mixing factor. It accounts for imperfect mixing of the emissions before they are exhausted from the lab.

In many cases more than one chemical is emitted during an experiment. For these cases an equivalent exposure may be defined as E.

$$E = \frac{c_1}{L_1} + \frac{c_2}{L_2} + \cdots \frac{c_n}{L_n} \tag{2.2}$$

c_k is the concentration of the kth vapor and the L_k is the corresponding TLV. The value of E should be less than unity. Equation 2.1 may be combined with this expression to obtain a new relationship.

$$\frac{kQt_0}{n} > \frac{m_1}{L_1} + \frac{m_2}{L_2} + \cdots \tag{2.3}$$

The left-hand side of the inequality describes the total volume of ventilation air available to dilute each student's emissions during the laboratory period. Each term on the right-hand side describes the volume of air required to dilute emissions m_k to the TLV defined by L_k.

The factor (kQt_0/n) should be a constant for your laboratory section throughout the term. The terms m_k, L_k, and m_k/L_k are given for each experiment.

Compounds with m/L values comparable to or greater than kQt_0/n must be restricted to hoods, or other precautions must be taken to limit emissions. Compounds with smaller m/L values may be used with a normal amount of care in the open laboratory. The assumptions made in developing these quantities are described in Appendix A. Generally, concentrations will be much less than the TLV values in these experiments. Do not take this as a license for sloppiness. There is something to be said for minimizing our exposure to most of the chemicals used in the laboratory, and there are good reasons for not being sloppy to the point of losing product.

One should avoid skin contact with any of the chemicals used in these experiments. The term *skin* in the environmental section of each experiment indicates that there are special hazards from skin contact. Where emissions have not been characterized for a given process, the term *Max.* is used. In these cases the assumption is made that all of the material handled is vaporized. In many cases these values will greatly overstate the emissions and serve only as a rough upper limit to actual emissions.

EMISSIONS AND ENVIRONMENTAL DATA

Emissions in the laboratory result from a wide range of processes involving many chemicals. Emissions from some of these processes have been estimated from measurements of surrogate procedures. In other cases estimates have been made by interpolation, or maximum losses have been assumed. The environmental data presented with each experiment have been developed in the following way.

Amount Used

This figure is obtained directly from the description of the experiment. In many cases where the same material is used more than once in an experiment (i.e., methylene chloride as a reaction solvent and as the mobile phase for column chromatography), the uses are listed separately. This is done to permit an evaluation of the effects of an experiment when experiments are modified by eliminating some steps.

TLV

This term is a registered trademark of the American Conference of Governmental Industrial Hygienists. TLV figures have been taken from the 1984–1985 tabulation of the ACGIH. The booklet listing TLVs is available from the ACGIH at 6500 Glenway Avenue, Building D-5, Cincinnati, OH 45211.

Emissions

The estimated emissions are given in milligrams. The assumptions used in making these estimates and other data for the compounds are given in the line under the numbers. Some notes on the abbreviations follow.

Max.

Max. indicates that all of the material used is assumed to be emitted. This assumption is generally used for experiments in which the solvent is evaporated to recover a solute. Emissions of many processes have not been evaluated, and the maximum assumption has been used where there is reason to believe that a substantial fraction of the material will be emitted. The emissions for many of these processes will be much less if adequate hood space is available.

Open Container

The following losses from 50-ml beakers have been measured.

Compound	Loss (mg/hour)
Acetone	920
2-Butanone	390
3-Pentanone	140

Reflux

The loss of acetone is about 75 mg/hour for a water-cooled condenser. In cases involving other compounds the 75 mg/hour figure is multiplied by the ratio of open container losses for appropriate compounds (those having volatilities similar to the ketones listed above).

Transfer

Losses from liquid transfer are 30 mg and 75 mg for 3-pentanone (bp = 103°C) and acetone (bp = 56°C), respectively, when the transfer is performed at the macro level with a graduated cylinder. Losses from liquid transfer are 1 mg and 9.4 mg for the same two compounds when the transfers are done with a micropipet.

Treated as Hexane

Petroleum ether and ligroin are hydrocarbon mixtures for which TLVs have not been assigned. As a conservative step, these solvents have been treated as hexane, the substance with the lowest TLV likely to be present.

Volume

This is the figure for emissions divided by TLV

Abbreviations

Col. Chromatog. – purification by column chromatography

Craig – recrystallization with a Craig tube

Evap. – evaporation of nearly all of the solvent

Extract – liquid–liquid extraction

React. – substance used as a reactant

Recryst. – recrystallization

Solv. – substance used as a solvent

Vac. Filtr. – vacuum filtration

MTDS – mutation dose

IRDS – irritation dose	rat – rat
TXDS – toxic dose	ham – hamster
LDLo – lethal dose low	ipr – intraperitoneal
LD50 – lethal dose for 50%	hmn – human
TFX:Car – carcinogenic	inv – intravenous
ihl – inhalation	skn – skin
orl – oral	rbt – rabbit
unk – unknown	ims – intramuscular
mam – mammal	scu – subcutaneous
mus – mouse	

QUESTIONS   **2-1.** Think about what you would do in the following cases. (Note that you may need more information for some of these problems.)

    a. A hot solution "bumps," splashing your face.

    b. A beaker of solvent catches fire.

    c. A reagent bottle falls, spilling concentrated sulfuric acid.

    d. You hear a sizzle as you pick up a hot test tube.

2-2. A laboratory has four hoods each of which is 39 in. wide. When the hood door is open to a height of 8 in. and the hoods are operating, the average air velocity through the hood face is 170 ft/minute.

    a. Evaluate the total ventilation rate for this room assuming that there are no other exhausts.

    b. The laboratory is designed for use by 30 students. Evaluate the air available per student if the mixing factor is 0.3, and experiments last three hours.

    c. An experiment is considered in which each student would be required to evaporate 7 mL of methylene chloride. Estimate the average concentration of methylene chloride. Look up the TLV or the TWA for methylene chloride and consider how the evaporation might be performed.

2-3. A laboratory has a ventilation system that provides 20 m³ for each student during the laboratory period. (This figure includes the mixing factor.) An experiment is considered for this program that uses the following quantities of materials A, B, and C. The TLV is also listed for each compound.

Substance	Quantity (mg)	TLV (mg/m³)
A	400	1200
B	500	200
C	200	5

Assess the relative risks of these three compounds. Is there a likely need for operations to be conducted in a hood if the compounds are assumed to be entirely evaporated?

2-4. An experiment is considered in which 1 mL of diethylamine would be used by each student. The ventilation rate for the laboratory is 5 m³/minute. Look up the TLV (or TWA) for diethylamine. What restrictions might be placed on the laboratory to keep the average concentration over a three-hour period less than one-third of the TWA? Assume a mixing factor of 0.3.

Chapter 3

Introduction to Microscale Organic Laboratory Equipment and Technique

We will first describe a number of special pieces of glassware that you will use to perform the experiments described in this text. Most of this equipment will be found in your laboratory locker. You will find the following descriptions helpful when checking your equipment list at the beginning of the semester. Next, we will consider a series of standard experimental apparatus setups that utilize this equipment. At the end of these short discussions, you will find a series of references. The reference refers to an experiment in which the use of this particular setup is described.

Most of the equipment listed below is glassware that has been specially developed for use in the microscale organic laboratory program. We feel that this particular collection offers significant advantages for student use at this scale of laboratory operation. You may find, however, that similar but not identical components are present in your locker. Consult your laboratory instructor for further details on those pieces of equipment that are not obvious extensions of the following list (see Figs. 3.1–3.6).

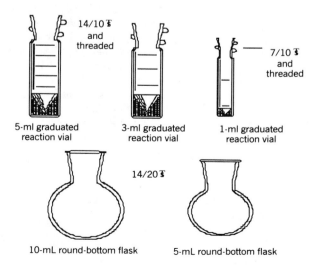

14/10 ℥ and threaded

7/10 ℥ and threaded

5-ml graduated reaction vial

3-ml graduated reaction vial

1-ml graduated reaction vial

14/20 ℥

10-mL round-bottom flask

5-mL round-bottom flask

Fig. 3.1 *Reaction flasks.*

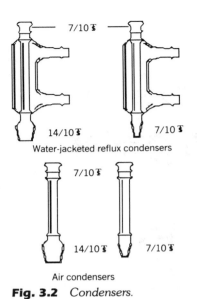

Water-jacketed reflux condensers

7/10 ℥

14/10 ℥   7/10 ℥

Air condensers

7/10 ℥

14/10 ℥   7/10 ℥

Fig. 3.2 *Condensers.*

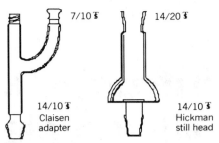

7/10 ℥   14/20 ℥

14/10 ℥
Claisen
adapter

14/10 ℥
Hickman
still head

Fig. 3.3 *Distillation heads.*

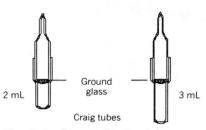

2 mL

Ground
glass

3 mL

Craig tubes

Fig. 3.4 *Recrystallization tubes.*

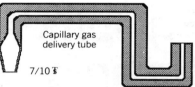

Capillary gas
delivery tube

7/10 ℥

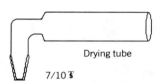

Drying tube

7/10 ℥

Fig. 3.5 *Miscellaneous items.*

5/5 ℥
and
threaded

5/5 ℥

GC collection tube   5/5 ℥

Product vial, 0.1 mL

Fig. 3.6 *Gas chromatographic collection items.*

**STUDENT
MICROGLASSWARE
EQUIPMENT**

No.	Glassware Items
1	conical vial, 5-mL, 14/10℥ and threaded
2	conical vial, 3-mL, 14/10℥ and threaded
2	conical vial, 1-mL, 7/10℥ and threaded
1	round-bottom flask, 10-mL, 14/20℥
1	round-bottom flask, 5-mL, 14/20℥
1	water-jacketed reflux condenser, male 14/10℥, female 7/10℥
1	water-jacketed reflux condenser, male 7/10℥, female 7/10℥
1	air condenser, male 14/10℥, female 7/10℥
1	air condenser, male 7/10℥, female 7/10℥
1	Hickman still head, male 14/10℥, female 14/20℥
1	Claisen head, male 14/10℥, threaded, female 7/10℥
1	Craig tube, 3-mL
1	Craig tube, 2-mL
2	drying tube, male 7/10℥
1	capillary gas delivery tube, male 7/10℥

No.	Gas Chromatographic Collection Items
2	GC collection tube, male 5/5℥
2	conical vial, 0.1-mL, 5/5℥ and threaded

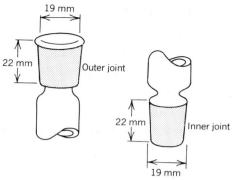

Fig. 3.7 *Standard taper joints [⊤ 19/22]. From Zubric, J. W. "The Organic Chem Lab Survival Manual"; Wiley: New York, 1984. (Reprinted by permission of John Wiley & Sons, New York, NY.)*

Fig. 3.8 *Threaded female joint.*

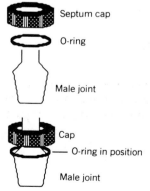

Fig. 3.9 *Male joint with septum cap and O-ring.*

Pieces of glassware are most efficiently assembled into a functional apparatus using standard taper ground glass joints (see Fig. 3.7). The symbol ⊤ is commonly used to indicate the presence of this type of connector. Normally, ⊤ is either followed or preceded by #/#. The first # refers to the maximum inside diameter of a female (outer) joint or the maximum outside diameter of a male (inner) joint, measured in millimeters. The second number corresponds to the total length of the ground surface of the joint. The advantage of this type of connection is that if the joint surfaces are lightly greased, a vacuum seal is achieved. One of the drawbacks in employing these joints, however, is that contamination of the reacting system readily occurs through extraction of the grease by the solvents present in the reaction vessel. In carrying out microscale reactions, this is particularly troublesome. An alternative to using greased joints as connectors is to employ either outside or inside screw-threaded glass systems. This type of joint utilizes plastic Teflon-lined screwcap connectors. These threaded glass joints are commercially available; however, they suffer from the same problem as the ungreased standard taper joint, that is, they are not gas tight (at least not in the hands of most undergraduates). The glassware joints used in the microscale experimental organic laboratory program have the ease and physical integrity of standard taper ground glass joints. The dimensions usually are either ⊤ 14/10 or 7/10, male or female. The conical vials in which the large majority of reactions are carried out, however, employ an alternative type of sealing system. You will note that in addition to being ground to a standard taper on the inside surface, these vials also possess a screw thread on the outside surface (Fig. 3.8). This arrangement allows a standard taper male joint to be sealed to the reaction flask by a septum type (open) plastic screwcap. The screwcap applies compression to a retaining silicone rubber O-ring positioned just above the male joint (Fig. 3.9). The compression of the O-ring thereby achieves a greaseless gas-tight seal while at the same time clamping the two pieces of equipment together. The ground joint provides protection from intimate solvent contact with the O-ring. This type of joint connection provides a quick, easy, and reliable mechanism for assembling the glassware required for carrying out the microscale reactions. It should be emphasized that the ground joint surfaces are grease free; therefore, it is important to disconnect joints soon after use or they may become locked or "frozen" together. Joints of the size employed in these experiments seldom are a problem to separate, if given proper care (keep them clean!).

A complete set of glassware would involve 18 different components and a total of 22 items grouped according to function as follows (see Figs. 3.1–3.6):

7 Reaction Flasks

1 conical vial, 5-mL, 14/10$ and threaded

2 conical vials, 3-mL, 14/10$ and threaded

2 conical vials, 1-mL, 7/10$ and threaded

1 round-bottom flask, 10-mL, 14/20$

1 round-bottom flask, 5-mL, 14/20$

The five conical vials are designed to be connected via an O-ring compression cap installed on the male joint of the adjacent part of the system. The standard round-bottom flasks are assembled with copper wire hooks and rubber bands or plastic spring clamps (Fig. 3.1).

4 Condensers

1 water-jacketed reflux condenser, male 14/10$, female 7/10$

1 water-jacketed reflux condenser, male 7/10$, female 7/10$

1 air condenser, male 14/10$, female 7/10$

1 air condenser, male 7/10$, female 7/10$

These items form two sets of condensers for use with 14/10$ or 7/10$ jointed reaction flasks. The female joints allow connection of the condenser to the 7/10$ drying tube or 7/10$ capillary gas delivery tube (Fig. 3.2).

2 Distillation Heads

1 Hickman still head, male 14/10$, female 14/10$

1 Claisen head, male 14/10$, female 7/10$, threaded septum connector

The Hickman still is used with an O-ring compression cap to carry out semimicro simple or fractional distillations. The Claisen head, which contains a threaded septum connector, often is employed to facilitate the syringe addition of reagents to sealed or moisture-sensitive systems (such as Grignard reactions). See Fig. 3.3.

2 Recrystallization Tubes

1 Craig tube, 2-mL

1 Craig tube, 3-mL

Craig tubes are a particularly effective method for recrystallizing small quantities of reaction products in an all-glass system. These tubes possess a nonuniform ground joint. Thus, all 22 items of the microglassware equipment involve some type of ground surface (Fig. 3.4).

3 Miscellaneous Items

2 drying tubes, male 7/10$

1 capillary gas delivery tube, male 7/10$

The drying tubes are used to protect moisture-sensitive reaction components from atmospheric water vapor while allowing a reacting system to be kept unsealed. The capillary gas delivery tube is employed in transferring gases formed during reactions to storage containers (Fig. 3.5).

4 Gas Chromatographic Collection Items

2 GC collection tubes, male 5/5$

2 conical vials, 0.1-mL, 5/5$ and threaded

The collection tube is connected directly to the gas chromatographic column for fraction collection followed by transfer of the sample to a 0.1-mL conical vial for storage. The system is employed in the resolution and isolation of two component mixtures (Fig. 3.6).

Heating and Stirring Reactions

It is extremely important to be able to carry out microscale experiments at accurately determined temperatures. Very often successful transformations require precise temperature control. In addition, many reactions require intimately mixed reactants in order to obtain a substantial yield of product. Therefore, the majority of the reactions you perform in this laboratory will be conducted with rapid stirring of the reaction mixture.

The most convenient piece of equipment for heating or stirring or for performing both operations simultaneously on a microscale level is the hot plate stirrer. Heat transfer from the hot surface to the reaction flask is generally accomplished by employing a crystallizing dish containing a shallow layer of sand that conforms to the size and shape of the particular vessel. The temperature of the system may be monitored by embedding a thermometer in the sand near the reaction vessel. The high sides of the crystallizing dish act to protect the apparatus from air drafts, and act somewhat as a hot air bath. An increase in the uniformity of heating can be obtained by covering the crystallizing dish with aluminum foil (Fig. 3.10).

Stirring of the reaction mixture in a conical vial is carried out with Teflon-coated magnetic spin vanes, and in round-bottom flasks with Teflon-coated magnetic stirring bars (see Figs. 3.10 and 3.11).

It is important that you become acquainted with the settings on the controls

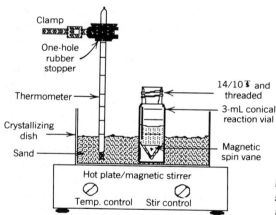

Fig. 3.10 *Hot plate/magnetic stirrer with sand bath and reaction vial.*

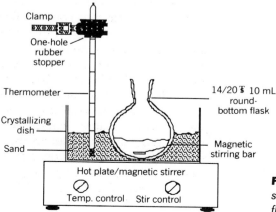

Fig. 3.11 *Hot plate/magnetic stirrer with sand bath and reaction flask.*

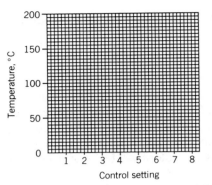

Fig. 3.12 *Graph of temperature (°C) vs. hot plate control setting.*

that adjust the current to the heating element and the motor that spins the magnet. You can make a rough graph of heat control setting versus temperature (Fig. 3.12) for your particular hot-plate system. These data will save considerable time when you bring a reaction apparatus to operating temperature. As your first step in the laboratory, it is advisable to adjust the temperature setting on the hot plate stirrer with the sand bath in place. The setting is determined from your control-temperature calibration curve. This procedure will allow the heated bath to reach a relatively constant temperature by the time it is required. You will then readily be able to make small final adjustments.

It should be emphasized that heavy layers of sand act as an insulator on the hot plate surface, which can result in damage to the heating element at high temperature settings. Therefore, when temperatures over 150°C are required, remember to use the minimum amount of sand.

It is important to adjust the reaction flask close to the bottom surface of the crystallizing dish in those experiments that depend on magnetic stirring. This arrangement is a good practice, in general, as it leads to the use of the minimum amount of sand.

If the reaction does not require elevated temperatures, but only needs to be stirred, the system is assembled without the sand bath. Some stirred reactions, on the other hand, require cooling. In this case the sand bath may be replaced with a crystallizing dish filled with ice water or ice water-salt, if lower temperatures are required.

Reflux Apparatus

We often find that in order to bring about a successful reaction between two substances, it is necessary to intimately mix the materials together and to maintain a specific temperature. The mixing operation is conveniently achieved by dissolution of the materials in a solvent in which they are mutually soluble. If the reaction is carried out in solution under reflux conditions, the choice of solvent can be used to control the temperature of the reaction. The very large majority of all the organic reactions described in this text involve the use of a reflux apparatus in one arrangement or another. What do we mean by "reflux"? The term means to "return" or "run back." This is exactly how the reflux apparatus functions. When the temperature of the reaction system is raised to the boiling point (constant temperature), *all* solvent vapors are condensed and returned to the reaction flask or vial (this operation is not a distillation). In the microscale reactions described, we employ two basic types of reflux condensers: the *air condenser* and the *water-jacketed condenser*. The air condenser operates, as its name implies, by condensing solvent vapors on the cool vertical wall of an extended glass tube that dissipates the heat by contact with laboratory room air. This simple arrangement functions quite effectively with liquids boiling above 150°C. Air condensers can occasionally be used with lower boiling systems. The water-jacketed condenser employs cold water to remove heat from the vertical column, thus facilitating condensation. This latter apparatus is highly efficient at condensing vapor from low-boiling liquids. In order to accommodate various size reaction flasks, both styles of condensers are supplied in two different male standard taper joint sizes: 7/10$ and 14/10$. The tops of both condenser columns possess the female 7/10$ joint.

In refluxing systems that do not require agitation, the stirrer (magnetic spin vane or bar) usually is replaced by a "boiling stone." These sharp-edged stones possess highly fractured surfaces that are very efficient at initiating bubble formation as the reacting medium approaches the boiling point. The boiling stone acts to protect the system from disastrous boilovers and also reduces "bumping." (They should be used only once and *never* should a boiling stone be added to a hot solution. Why?)

Below are listed eight different arrangements of reflux apparatus followed by references to the experiments in which they are utilized.

Various Arrangements of Reflux Apparatus (see Figs. 3.13–3.20)

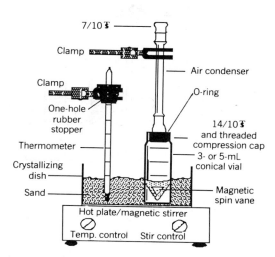

Fig. 3.13 *Air condenser with 3- or 5-mL conical vial, arranged for heating and magnetic stirring.*

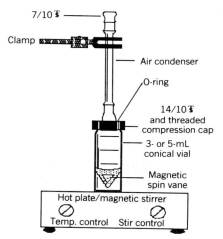

Fig. 3.14 *Air condenser with 3- or 5-mL conical vial, arranged for magnetic stirring.*

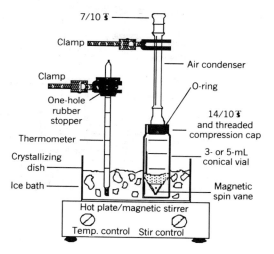

Fig. 3.15 *Air condenser with 3- or 5-mL conical vial, arranged for cooling and magnetic stirring.*

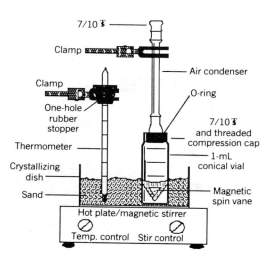

Fig. 3.16 *Air condenser with 1-mL conical vial arranged for heating and magnetic stirring.*

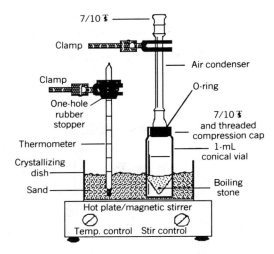

Fig. 3.17 Air condenser with 1-mL conical vial, arranged for heating.

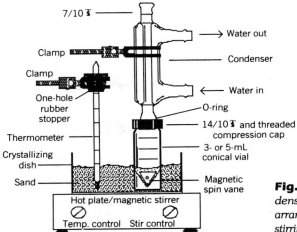

Fig. 3.18 Water-jacketed condenser with 3- or 5-mL conical vial, arranged for heating and magnetic stirring.

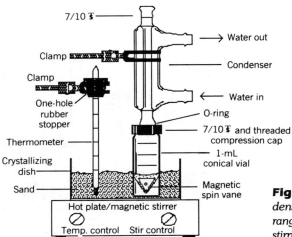

Fig. 3.19 Water-jacketed condenser with 1-mL conical vial, arranged for heating and magnetic stirring.

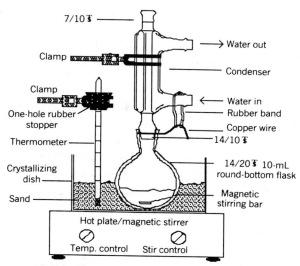

Fig. 3.20 *Water-jacketed condenser with 10-mL round-bottom flask, arranged for heating and magnetic stirring.*

Air condenser with 3- or 5-mL conical vial, arranged for heating and magnetic stirring (Experiments 19; 20; 34B; 45; 48C; 52A).

Air condenser with 3- or 5-mL conical vial, arranged for magnetic stirring (Experiments 21A, B, C, D; 22; 26A, B; 27; 34A, C, D; 43A, B; 50A).

Air condenser with 3- or 5-mL conical vial, arranged for cooling and magnetic stirring (Experiment 31).

Air condenser with 1-mL conical vial, arranged for heating and magnetic stirring (Experiment 49).

Air condenser with 1-mL conical vial, arranged for heating (Experiments 14; 15; 50B).

Water-jacketed condenser with 3- or 5-mL conical vial, arranged for heating and magnetic stirring (Experiments 6; 7A, B; 13; 46; 48D).

Water-jacketed condenser with 1-mL conical vial, arranged for heating and magnetic stirring (Experiments 25A, B, C; 37).

Water-jacketed condenser with 10-mL round-bottom flask, arranged for heating and stirring (Experiments 40A; 42; 47A, C).

Distillation Apparatus

Distillation is a laboratory operation used to separate substances that have different boiling points. The mixture is heated, vaporized, and then condensed, with the more volatile component being enriched in the early fractions of condensate. Unlike the reflux operation, in distillations, none or only a portion of the condensate is returned to the flask where vaporization is taking place. There are a large number of distillation apparatus designed to carry out this basic operation. They differ mainly in small features used to solve particular types of separation problems. In the present microscale experiments, a number of simple semi-microscale distillations are required. For these distillations the Hickman still head is ideally suited. This system has a 14/10 male joint for connection to either the 3- or 5-mL conical vials or the 5- or 10-mL round-bottom flasks. The

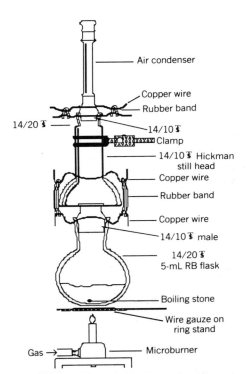

Fig. 3.21 *Hickman still head and air condenser with 5-mL round-bottom flask, arranged for microburner heating.*

still head functions both as an air condenser and a condensate trap. For a detailed discussion of this piece of equipment see Experiments 1A and 1B.

Three experimental arrangements of the Hickman still head, and references to the experiments where these setups are employed, are discussed below.

Various Distillation Apparatus (see Figs. 3.21–3.23)

Hickman still head and air condenser with 5-mL round-bottom flask, arranged for microburner heating (Experiment 34A, B, C, D).

Hickman still head with 3-, 5-mL conical vial or 10-mL round-bottom flask, arranged for heating and magnetic stirring (Experiments 1A, B; 30; 39; 48A).

Hickman still head with 5-mL conical vial, arranged for cooling and magnetic stirring (Experiment 39).

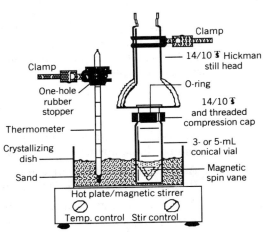

Fig. 3.22 *Hickman still head with 3- or 5-mL conical vial, arranged for heating and magnetic stirring.*

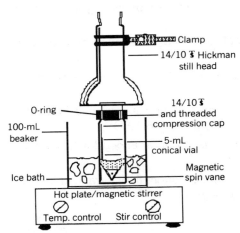

Fig. 3.23 *Hickman still head with 5-ml conical vial, arranged for cooling and magnetic stirring.*

Moisture-Protected Reaction Apparatus

Many organic reagents react rapidly and preferentially with water. *The success or failure of many of the following experiments depends to a large degree on how well you can exclude atmospheric moisture from your reaction system.* The "drying tube," which is packed with a desiccant such as anhydrous calcium chloride, is a handy way to carry out a reaction in apparatus that is not closed to the atmosphere, but which is reasonably well protected from water vapor. The microscale apparatus described here are designed to be used mainly with the 7/10 ⑀ drying tube. The reflux condensers discussed above are constructed with female 7/10⑀ joints at the top of the column, which allows convenient connection of the drying tube if the refluxing system is moisture-sensitive.

A number of the reactions to be studied are highly sensitive to moisture, thus successful operation at the microscale level can be rather challenging. If reagents are to be added after an apparatus has been assembled and dried, it is important to be able to add these reagents without exposing the system to the atmosphere, particularly when operating under humid conditions. In reactions not requiring reflux conditions and conducted at room temperature, this addition procedure is best accomplished by use of the microscale Claisen head adapter. The adapter has a screw-threaded joint that will accept a septum cap. The septum seal allows syringe addition of reagents and avoids the necessity of opening the apparatus to the laboratory atmosphere.

In a few instances reactions are unusually moisture-sensitive. In this situation, reactions are best carried out in completely sealed systems that are scrupulously dry. The use of the Claisen head adapter with a balloon substituted for the drying tube provides a satisfactory solution to the problem. Occasionally, it becomes important to maintain dry conditions during a distillation. The Hickman still is constructed with a 14/10𝔗 joint at the top of the head. Moisture protection in this case is best achieved by employing the 14/10𝔗 air condenser, which can act as a drying tube. Care should be exercised in packing the drying agent in this piece of equipment. Since the drying tube will be placed in a vertical position, contamination of the reaction by small particles of the desiccant is difficult to avoid.

Below are listed six experimental arrangements that involve protection of the reaction medium from contact with moisture. They are followed by references to the particular experiments where these apparatus are used.

Various Moisture-Protected Reaction Apparatus (see Figs. 3.24–3.29)

Moisture-protected air condenser with 1-mL conical vial, arranged for heating and magnetic stirring (Experiments 8B, C, E; 27).

Moisture-protected water-jacketed condenser with 3- or 5-mL conical vial, arranged for heating and stirring (Experiments 8A, D; 24; 47B).

Moisture-protected water-jacketed condenser with 1-mL conical vial, arranged for heating and stirring (Experiments 28A, B; 29; 38).

Moisture-protected Claisen head with 3- or 5-mL conical vials, arranged for syringe addition and magnetic stirring (Experiments 13; 17; 18; 23; 32A, B; 35; 48B).

Sealed Claisen head with 3- or 5-mL conical vials, arranged for N_2 flushing, heating, and magnetic stirring (Experiment 16A, B).

Moisture-protected Hickman still head with 10-mL round-bottom flask, arranged for heating and stirring (Experiment 30).

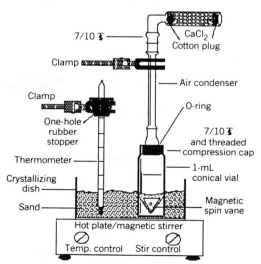

Fig. **3.24** *Moisture-protected air condenser with 1-mL conical vial, arranged for heating and magnetic stirring.*

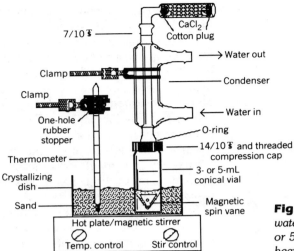

Fig. 3.25 *Moisture-protected water-jacketed condenser with 3- or 5-mL conical vial, arranged for heating and magnetic stirring.*

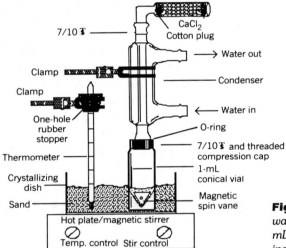

Fig. 3.26 *Moisture-protected water-jacketed condenser with 1-mL conical vial, arranged for heating and magnetic stirring.*

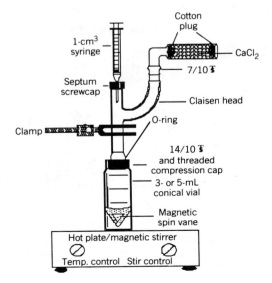

Fig. 3.27 *Moisture-protected Claisen head with 3- or 5-mL conical vial, arranged for syringe addition and magnetic stirring.*

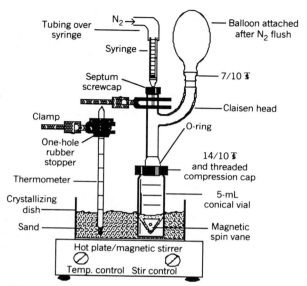

Fig. 3.28 *Sealed Claisen head with 3- or 5-mL conical vial, arranged for N₂ flushing, heating, and magnetic stirring.*

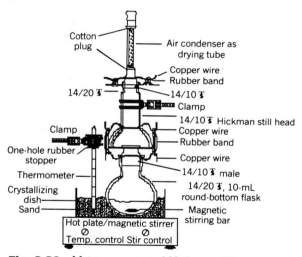

Fig. 3.29 *Moisture-protected Hickman still head with 10-mL round-bottom flask, arranged for heating and magnetic stirring.*

Specialized Pieces of Equipment

In a few experiments the reaction leads to gaseous products. The collection or trapping of these products is conveniently carried out using the capillary gas delivery tube. This item is designed to be attached directly to a 1-mL conical vial or to the female 7/10ℑ joint of a condenser connected to a reaction flask or vial. The tube leads to the collection system, which is a simple inverted graduated cylinder. The 0.1-mm capillary bore considerably reduces dead volume and increases efficiency of product transfer.

The trapping and collection of gas chromatographic liquid fractions becomes particularly important at the microscale level of experimentation. A number of the experiments depend upon this type of substance isolation. The ease and efficiency of carrying out this operation are greatly facilitated by employing the 5/5 ꝋ collection tube and 0.1-mL 5/5 ꝋ conical collection vial.

The operation of these systems is discussed in detail in the references listed below, along with the description of the experimental arrangements.

Various Sample-Collection Apparatus (see Figs. 3.30–3.32)

Conical vial (1 mL) and capillary gas delivery tube, arranged for heating and stirring (Experiment 9).

Water-jacketed condenser with 3-mL conical vial and capillary gas delivery tube, arranged for heating and stirring (Experiment 41).

Gas chromatographic collection tube and 0.1-mL conical vial (Experiments 2; 4A, B; 18).

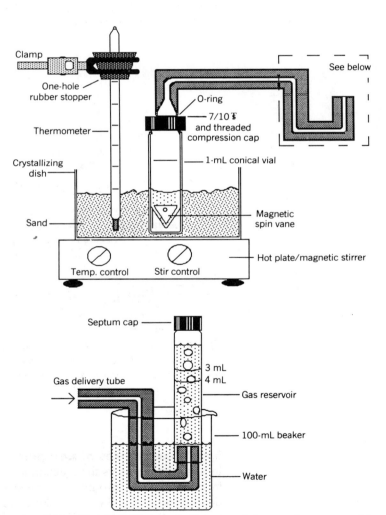

Fig. 3.30 *Vial (1-mL) and capillary gas delivery tube, arranged for heating and magnetic stirring.*

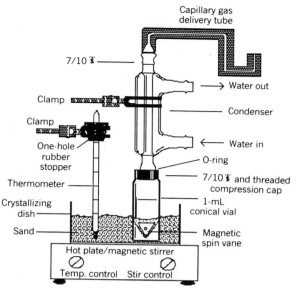

Fig. 3.31 *Water-jacketed condenser with 3-mL conical vial and capillary gas delivery tube, arranged for heating and magnetic stirring.*

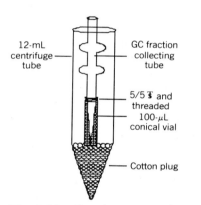

Fig. 3.32 *Gas chromatographic collection tube and 0.1-mL conical vial.*

Rules of the Trade for Handling Organic Materials at the Microscale Level

Now that we have had a brief look at the kind of equipment that we will be using to carry out the microscale organic reactions, let us examine specific techniques used to deal with the small quantities of material involved. These reactions by our definition start with 15 to 150 mg of the limiting reagent. These quantities sound small, and they are. Although 150 mg of a light powdery material will half fill a 1-mL conical vial, you will have a hard time observing 15 mg of a clear liquid in the same container even with magnification. On the other hand, this volume of liquid is reasonably easy to observe when placed in a 0.1-mL conical vial. A vital part of the game of working with small amounts of materials is to become familiar with microscale techniques and to practice them as much as possible in the laboratory.

Rules to Follow When Working with Liquids at the Microscale Level

1. *Liquids are never poured at the microscale level.* The transfer of liquid substances is carried out with pipets or syringes. We employ three different types of pipets and two standard syringes for most experiments where liquids are involved. This equipment is a prime source of contamination. *Be very careful to thoroughly clean the equipment after each use.*

   a. *Pasteur pipet.* This is a piece of simple glass tubing with the end drawn to a fine capillary (they are often called capillary pipets). These pipets are filled using a small rubber bulb and can hold several milliliters of liquid (Fig. 3.33a). You will do many transfers using this device. It is suggested that you calibrate several of them for approximate delivery of 0.5, 1.0, 1.5, and 2.0 mL of liquid. This is easily done by drawing the measured amount of a liquid from a 10 mL graduated cylinder and marking the level of the liquid in the pipet. This can be done with transparent tape or by scratching with a file. Indicate the level with a marking pen before trying to tape or file the pipet.

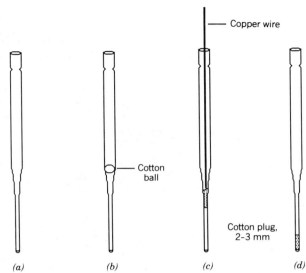

Fig. 3.33(a-d)   *Preparation of Pasteur filter pipet.*

b. *Pasteur filter pipet.* This is a very handy adaptation of the Pasteur pipet. This pipet is constructed by taking a small cotton ball and placing it in the large open end of the standard Pasteur pipet. Hold the pipet vertically and tap gently to position the cotton ball in the drawn section of the tube (Fig. 3.33*b*). Now form a plug in the capillary section by pushing the cotton ball down the pipet with a piece of copper wire (Fig. 3.33*c*). Finish by seating the plug flush with the end of the capillary (Fig. 3.33*d*). The optimum size plug will allow easy movement along the capillary while it is being positioned by the copper wire. The compression of the cotton will build enough pressure against the walls of the capillary (once the plug is in position) to prevent plug slippage while filling the pipet with liquid. If the ball is too big, it will wedge in the capillary before the end is reached, and wall pressure will be so great that liquid flow will be shut off. Even some plugs that are loose enough to be positioned at the end of the capillary still will have developed sufficient lateral pressure to make the filling rate unacceptably slow. With a little practice, however, these plugs can be quickly and easily inserted. Once in place the plug is rinsed with 1 mL of methanol and 1 mL of hexane and dried before use.

The reason for placing the cotton plug in the pipet is two-fold: First, a particular problem with the transfer of volatile liquids via the standard Pasteur pipet is the rapid buildup of back-pressure from solvent vapors in the rubber bulb. This pressure quickly tends to force the liquid back out of the pipet. The result can be valuable product dripping on the bench top. The cotton plug tends to resist this back-pressure and allows much easier control of the solution once it is in the pipet. The time delay factor becomes particularly important when the Pasteur filter pipet is employed as a microseparatory funnel (see discussion on extraction techniques, Experiment 3).

Second, each time a transfer of material is made, the material is automatically filtered. This process effectively removes dust and lint, which are a constant problem when working at the microscale level with unfiltered room air.

c. *Automatic pipet (considered the Cadillac of pipets).* These pipets quickly, safely, and reproducibly measure and dispense specific volumes of liquids. They are particularly valuable when operating at the microscale level, as they generate the precise, accurate liquid measurements that are absolutely necessary when handling microliter volumes of reagent. The automatic pipet adds considerable

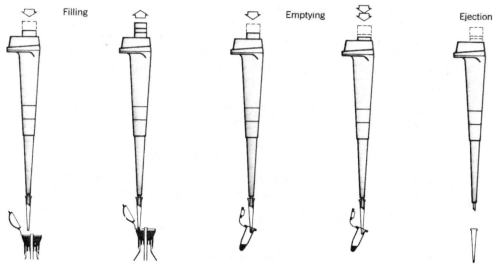

Filling   Emptying   Ejection

Fig. 3.34 *Operation of automatic delivery pipet. (Courtesy of Brinkmann Instruments Co., Westbury, NY.)*

insurance for the success of an experiment, as any liquid can be efficiently measured, transferred, and delivered to the reaction flask. They become almost an essential instrument in laboratory sections with large numbers of students.

The automatic pipet system consists of a calibrated piston pipet with a specially designed disposable plastic tip. You may encounter any one of three pipet styles: (i) single volume; (ii) multirange; (iii) continuously adjustable (see Fig. 3.34). The first type is calibrated to deliver only a single volume. The second type is adjustable to two or three predetermined delivery volumes. The third type is the most versatile and can be user set to deliver any volume within the range of the pipet. Obviously, the price of these valuable laboratory tools goes up with increasing attributes. They are expensive and their use must be shared in the laboratory. Treat them with respect!

The automatic pipet is designed so that the liquid comes in contact with the special tip only: (i) Never load the pipet without the tip in place. (ii) Never immerse the tip completely in the liquid that is being pipetted. (iii) Always keep the pipet vertical when the tip is attached. Follow these three rules and most automatic pipets will give many years of reliable service. The following suggestions are a few general rules for improving reproducibility with an automatic pipet:

(i) Try to effect the same uptake and delivery motion for all samples. Smooth depression and release of the piston will give the most consistent results. Never allow the piston to snap back.

(ii) *Always* depress the piston to the first stop before inserting the tip into the liquid. If the piston is depressed following submersion, formation of an air bubble in the tip becomes likely. Bubble formation will result in a filling error.

(iii) *Never* insert the tip more than 5 mm into the liquid. It is good practice not to allow the body of the pipet to contact any surface or bottleneck that might be wet with a corrosive chemical.

(iv) If an air bubble forms in the tip during uptake, return the fluid, discard the tip, and repeat the sampling process.

d. *Syringes.* These are particularly helpful pieces of equipment when transferring liquid reagents or solutions to sealed reaction systems. They can be inserted through a septum, which avoids opening the apparatus to the atmosphere (see Experiment 13). They are also routinely employed in the determination of ultramicro boiling points. It is critically important to clean the syringe needle after each use. To effectively clean a syringe requires as many as a dozen flushes. The microscale laboratory utilizes a low-cost glass 1-mL insulin syringe in which the rubber plunger seal is replaced with a Teflon seal. For gas chromatographic separation work, the standard 50- or 100-μL syringes are preferred (see Experiment 2).

2. *Liquid volumes may be converted easily to weight measure by the following relationship.*

$$\text{Volume (mL)} = \frac{\text{weight (g)}}{\text{density (g/mL)}}$$

3. *Work with liquids in conical vials* and work in vials that are approximately double the volume of the material. The trick here is to reduce the surface area of the flask in contact with the sample to an absolute minimum. Conical systems are far superior to the spherical surface of the conventional round-bottom flask.

Rules to Follow When Working with Solids at the Microscale Level

1. *General considerations.* Working with a crystalline solid is much easier than working with the equivalent quantity of a liquid. Unless the solid is in solution, a spill on a clean glass working surface usually can be recovered quickly and efficiently. *Be careful, however, when working in solution. ALWAYS use the same precautions that you would use if you were handling a pure liquid.*

2. *The transfer of solids.* This operation normally is carried out with microspatulas and is not a difficult technique to develop.

3. *Weighing solids at the milligram level.* The current generation of single-pan electronic balances have removed much of the drudgery from this operation. These systems can automatically tare an empty vial. Once the vial is tared, the reagent is added in small portions. The weight of each addition is instantly registered; material is added until the desired quantity has been transferred.

THE LABORATORY NOTEBOOK

Written communication is the most important method by which a chemist transmits his or her work to the scientific community. It begins with the record kept in a laboratory notebook. The reduction to practice of an experiment recorded in the laboratory notebook is the source of information used to prepare scientific papers published in journals or papers presented at scientific meetings. For the industrial chemist it is especially critical in obtaining patent coverage.

It is important that potential scientists, in whatever field, learn the art of keeping a detailed account of their work. A laboratory notebook has several key ingredients, outlined below. Note how each is incorporated in the illustration shown in the example.

Key Parts to a Laboratory Experimental Writeup

1. Date experiment was conducted.
2. Title of experiment.
3. Purpose for running the reaction.
4. Reaction scheme.
5. Table of reagents and products.
6. Details of procedure used.
7. Characteristics of the product(s).
8. References to product or procedure (if any).
9. Analytical and spectral data.
10. Signature of person performing the experiment and that of a witness, if required.

In reference to point No. 6, it is the obligation of the person doing the work to list the equipment, the amounts of reagents, experimental conditions, and the method used to isolate the product. Any color or temperature changes should be carefully noted and recorded.

There are several additional points to be made with respect to the proper maintenance of a laboratory record.

1. A hardbound, permanent notebook is essential.
2. Each page of the notebook should be numbered in consecutive order. For convenience, an index at the beginning or end of the book is recommended and blank pages should be retained for this purpose.
3. If a page is not completely filled, an X should be used to show that no further entry was made.
4. Always record your data in ink. If a mistake is made, draw a neat line through the word or words so that they remain legible. Data are always recorded directly into the notebook, *never* on scrap paper!
5. Make the record clear and unambiguous. Pay attention to grammar and spelling.
6. In industrial research laboratories, your signature as well as that of a witness is required. This is because the notebook may be used as a legal document.
7. Always write and organize your work so that someone else could come into the laboratory and repeat your directions without confusion or uncertainty. *Completeness* and *legibility* are key factors to remember.

Since the reactions you will be carrying out in this laboratory have been worked out and checked in detail, your instructor may not require you to keep your notebook in such a meticulous fashion. For example, in describing the procedure (item No. 6), it may be acceptable to make a clear reference to the material in the laboratory manual and to note any modifications or deviations from the prescribed procedure. In some cases, it may be more practical to use an outline method. In any event, the illustration should be studied carefully. It may be used as a reference when detailed records are important in your work.

Note. *Due to the length of the example, it is presented typed. Normally the entry in the notebook is handwritten. However, with the use of computers gaining wide acceptance in laboratory work, many chemists are using this means to record their data.*

EXAMPLE OF A LABORATORY NOTEBOOK ENTRY[1]

① ② ④

16 AUG. 1985

PREPARATION OF DIPHENYL SUCCINATE.

③

⑧

$$\begin{array}{c} CH_2COOH \\ | \\ CH_2COOH \end{array} + 2\ C_6H_5OH + POCl_3 \longrightarrow \begin{array}{c} CH_2COOC_6H_5 \\ | \\ CH_2COOC_6H_5 \end{array} + HPO_3 + 3HCl$$

Diphenyl succinate is being prepared as one of a series of dicarboxylic acid esters, which are to be investigated as growth stimulants for selected fungi species.

This procedure was adapted from that reported by Daub, G. H.; Johnson, W. S. "Organic Syntheses"; Wiley: New York, 1963; Collect. Vol. IV, p 390.

[1]Circled numbers refer to list on p. 28.

(5) ———————— Physical Properties of Reactants and Products

Compound	MW	Wt/Vol	mmol	mp(°C)	bp(°C)
Succinic acid	118.09	118 mg	1	182	
Phenol	94.4	188 mg	2		182
Phosphorous oxychloride	153.33	84 μL	0.9		105.3
Diphenyl succinate	270.29			121	

HOOD

(6) ————————

In a 3.0-mL conical vial containing a magnetic stirrer and equipped with a reflux condenser protected by a calcium chloride drying tube was placed succinic acid (118 mg, 1 mmol), phenol (188 mg, 2 mmol), and phosphorous oxychloride (84 μL, 0.9 mmol). The reaction mixture was heated with stirring at a temperature of 115°C using a sand bath [**HOOD**] for a period of 1.25 hours. It was necessary to conduct the reaction in the **HOOD** since HCl gas evolved during the course of the reaction. The drying tube was removed, toluene (0.5 mL) added through the top of the condenser using a Pasteur pipet, and the drying tube replaced. The mixture was then heated for an additional hour at 115°C.

The hot toluene solution was separated from the red syrupy residue of phosphoric acid using a Pasteur pipet. The toluene extract was filtered by gravity, using a fast grade filter paper, and the filtrate collected in a 10-mL Erlenmeyer flask. The phosphoric acid residue was then extracted with two additional 1.0-mL portions of hot toluene. These extracts were also separated using the Pasteur pipet, and were filtered, and the filtrate was collected in the same Erlenmeyer flask. The combined toluene solutions were concentrated to a volume of approximately 0.6 mL by warming in a sand bath under a gentle stream of nitrogen gas [**HOOD**]. The pale-yellow liquid residue was then allowed to cool to room temperature. Diphenyl succinate separated as colorless crystals. The solid was collected by vacuum filtration using a Hirsch funnel and the filter cake washed with three 0.5-mL portions of cold diethyl ether. The product was placed on a porous clay plate to dry overnight.

HOOD

(7) ————————

(7) ————————

(8) ————————

There was obtained 181 mg (67%) of diphenyl succinate having m.p. 120–121°C (Lit. value 121°C; "CRC Handbook of Chemistry and Physics", 60th ed.; CRC Press: Boca Raton, FL, 1979; # S197, p C-501).

(9) ————————

The IR spectrum exhibits the expected peaks for the compound. [*At this point the data may be listed or the spectrum pasted on a separate page of the notebook.*]

(10) ————————   *Marilyn C. Waris*

witnessed by
D. Jeanne d'Arc Mailhiot   16/aug./1985

CALCULATION OF YIELDS   Almost without exception, in each of the experiments presented in this text, you are requested to calculate the percentage yield. For any reaction, it is always important for the chemist to know how much of a product is *actually* produced (experimental) in relation to the *theoretical* amount (maximum) that could have been formed. The percentage yield is calculated based on the relationship

$$\% \text{ yield} = \frac{\text{actual yield (experimental)}}{\text{theoretical (calculated maximum)}} \times 100$$

The percentage yield is generally calculated on a weight (gram or milligram) or on a mole basis. In the present text, the calculations are made using milligrams.

Several steps are involved in the calculation of the percentage yield. They are as follows:

1. Write a *balanced* equation for the reaction. For example, consider Experiment 8A, the esterification of lauric acid to yield ethyl laurate.

$$CH_3(CH_2)_{10}C{\overset{\overset{\displaystyle \cdot\cdot}{O}:}{\underset{OH}{\diagup}}} \;+\; CH_3CH_2OH \; \underset{}{\overset{CH_3C\overset{\overset{\cdot\cdot}{O}:}{\diagdown}_{Cl}}{\rightleftharpoons}} \; CH_3(CH_2)_{10}C{\overset{\overset{\displaystyle \cdot\cdot}{O}:}{\underset{OCH_2CH_3}{\diagup}}} \;+\; HCl \;+\; CH_3COOH$$

2. Identify the *limiting* reactant. The ratio of reactants is calculated on a millimole (or mole) basis. In the example, 0.35 mmol of lauric acid is used compared to 17 mmol of ethanol. Thus, lauric acid is the limiting reactant. Note that the acetyl chloride is used as a reagent (0.35 mmol) but *does not* appear in the product, ethyl laurate. Therefore, it is not considered in the calculations.

3. Calculate the *theoretical* (maximum) amount of the product that could be obtained for the conversion, based on the limiting reactant. In the present case, referring to the balanced equation, one mole of lauric acid affords one mole of ethyl laurate. Therefore, if we start with 0.35 mmol of the acid, the maximum amount of ethyl laurate that can be produced is 0.35 mmol, or 80 mg.

4. Determine the *actual* (experimental) yield (milligrams) of product isolated in the reaction. This amount is invariably less than the theoretical quantity. For example, student yields for the preparation of ethyl laurate average 59 mg.

5. Calculate the *percentage yield* using the weights determined in steps 3 and 4. The percentage yield is then

$$\% \text{ yield} = \frac{59 \text{ mg (actual)}}{80 \text{ mg (theoretical)}} \times 100 = 74\%$$

As you carry out each reaction in the laboratory, strive to obtain as high a percentage yield of product as possible. The reaction conditions have been carefully developed and, therefore, it is essential that you master the microscale techniques concerned with transfer of reagents and isolation of products as soon as possible.

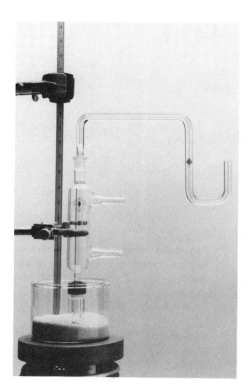

Chapter 4

Determination of Physical Properties

The physical properties of a compound are those measurable characteristics of a material that are independent of the influence of any other substance. Determination of these properties is important for (1) substance identification and (2) an indication of material purity. Historically, the physical constants of prime interest for determination in liquids include boiling point, density, and refractive index; for solids, melting point. In special cases, optical rotation and molecular weight values may be determined. Today, with the widespread availability of spectroscopic instrumentation, particularly infrared spectrometers, the absorption characteristics of a substance can be quickly recorded. The infrared spectrum obtained with one data point per wavenumber can add over four-thousand transmission measurements to the few classically determined properties. Indeed, even with development of high-resolution mass spectrometry, and proton and ^{13}C NMR spectroscopy, the infrared spectrum of a material is still the most powerful set of physical properties (transmission elements) available to the organic chemist for the *identification* of an unknown compound.

Determination of simple physical constants, at the present time, is carried out mainly to assist in establishing the purity of *known* compounds. The boiling point or the melting point of a material is very sensitive to small quantities of impurities. These data can be very helpful in determining whether or not a starting material needs further purification, or whether a product has been isolated in acceptable quality. When a new compound has been synthesized, an elemental (combustion) analysis normally would be included in the above list of determinations if sufficient material is available. In the case of a new substance, one is interested in establishing, not only the identity, but also the molecular structure of the substance. In this situation, other modern techniques such as proton and ^{13}C NMR, high-resolution mass spectrometry, and single crystal X-ray diffraction can provide powerful structural information.

When making comparisons between experimental data and those values obtained from the literature, it is essential that the latter information be obtained from the most reliable sources available. Certainly, judgment, which improves with experience, must be exercised in accepting any value as a standard. The known classical properties of the large majority of the compounds prepared in this text are taken from the *CRC Handbook of Chemistry and Physics*. This reference work is a valuable source that lists the physical constants of a very large number of inorganic, organic, and organometallic compounds. The handbook is kept current and a new edition is published each year.

ULTRAMICRO BOILING POINT

The vapor pressure of a liquid increases in a nonlinear fashion on heating. When the pressure reaches a point where it matches the local atmospheric or applied pressure, the liquid boils. That is, internally it spontaneously begins to form large vapor bubbles that rapidly rise to the surface. If heating is continued, both the

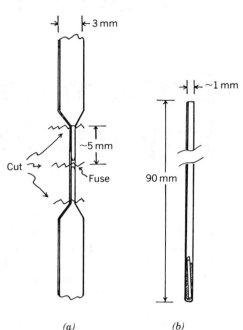

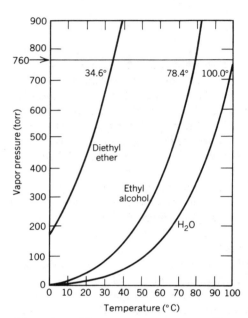

Fig. 4.1 *Vapor-pressure curves. From Brady, J. E.; Humiston, G. E. "General Chemistry," 3rd ed.; Wiley: New York, 1982. (Reprinted by permission of John Wiley & Sons, New York, NY.)*

Fig. 4.2a *Preparation of small glass bell for ultramicro boiling point determination. Mayo, D. W.; Pike, R. M.; Butcher, S. S.; Meredith, M. L. J. Chem. Educ. 1985, 62, Dec., 1985.*

Fig. 4.2b *Ultramicro boiling point assembly. Mayo, D. W., Pike, R. M.; Butcher, S. S.; Meredith, M. L. J. Chem. Educ. 1985, 62, Dec. 1985.*

vapor pressure and the temperature of the liquid will remain constant until the substance has been completely vaporzied (Fig. 4.1).

Since microscale preparations generally yield quantities of liquid products in the 30–70 μL range, the allocation of 5 μL or less to boiling point measurements becomes highly desirable. The modification of the Wiegand ultramicro boiling point procedure described here has established that reproducible and reasonably accurate (± 1 °C) boiling points can be observed on 3–4 μL of many liquids.[1]

Procedure

Ultramicro boiling points can be conveniently determined in standard (90-mm length) Pyrex glass capillary melting-point tubes. The melting-point tube replaces the conventional 3–4 mm (o.d.) tubing used in the Siwoloboff[2] procedure. The sample (3–4 μL) is loaded into the m.p. capillary via a 10-μL syringe and centrifuged to the bottom.

A small glass bell (which replaces the conventional melting-point tube as the bubble generator in micro boiling point determinations) is formed by heating 3-mm (o.d.) Pyrex tubing with a microburner and drawing it out to a diameter small enough to be accepted by the melting-point capillary. A section of the drawn capillary is fused and then cut to yield two small glass bells approximately 5 mm long (Fig. 4.2a).

One of the glass bells is inserted into the loaded melting-point capillary, open end first (down), and allowed to fall to the bottom. The assembled system (Fig. 4.2b) is then inserted into the stage of a Thomas-Hoover Uni-Melt Capillary Melting Point Apparatus[3] or similar system (Fig. 4.3).

[1]Wiegand, C. Angew. Chem. **1955**, *67*, 77. Mayo, D.W.; Pike, R.M.; Butcher, S.S.; Meredith, M.L. *J. Chem. Educ.* **1985**, *62*, No. 12 (Dec.).

[2]Siwoloboff, A. Ber. **1886**, *19*, 795.

[3]Thomas Scientific, 99 High Hill Road, P.O. Box 99, Swedesboro, NJ 08085.

The temperature is rapidly raised to 15–20° below the expected boiling point (the temperature should be monitored carefully in the case of unknown substances) and then adjusted to a 2°/minute rise rate until a fine stream of bubbles is emitted from the glass bell. The heat control is then adjusted to drop the temperature. The boiling point is taken at the point where the last escaping bubble collapses (i.e., when vapor pressure of the substance equals the atmospheric pressure). The heater is then rapidly adjusted to again raise the temperature at 2°/minute and induce a second stream of bubbles. The above procedure may then be repeated several times. *It should be emphasized that the precise and sensitive temperature control provided by the Thomas-Hoover system is essential to the successful application of this technique.*

The utilization of the conventional melting-point capillary as the "boiler" tube has the particular advantage that the system is ideally suited for observation in a conventional melting-point apparatus. The illumination and magnification available make the observation of rate changes in the bubble stream readily apparent. Economical gas chromatographic syringes (10 µL) appear to be the most successful instrument for dealing with the small quantities of liquids involved in these transfers. The 3-in. needles normally supplied with the 10-µL barrels will not reach the bottom of the capillary; however, liquid samples deposited on the walls of the tube are easily and efficiently moved to the bottom by centrifugation. After packing the sample in the bottom of the capillary tube, the glass bell is introduced. The use of the glass bell is necessitated by the fact that if a conventional Siwoloboff fused capillary insert is employed (it would extend beyond the top of the boiling tube), capillary action between the "boiler" tube wall and the capillary insert draws the majority of the sample from the bottom of the tube up onto the walls. This effect often precludes the formation of the requisite bubble stream.

Little loss of low-boiling liquids occurs (see Table 4.1). If the boiling point is overrun and the sample flashed from the bottom section of the "boiler" capillary, it rapidly condenses on the upper, cooler sections of the tube, which extend above the heat-transfer liquid. The sample can easily be recentrifuged to the

Fig. 4.3 *Thomas-Hoover melting point determination device. (Courtesy of Arthur Thomas Scientific, Swedesboro, NJ.*

Table 4.1 Observed Boiling Points (°C)[a]

Compound	Observed	Literature Value	Reference
Methyl iodide	42.5	42.4	b
Isopropyl alcohol	82.3	82.4	c
2,2-Dimethoxypropane	80.0	83.0	d
2-Heptanone	149–150	151.4	e
Cumene	151–153	152.4	f
Mesitylene	163	164.7	g
p-Cymene	175–178	177.1	h
Benzyl alcohol	203	205.3	i
Diphenylmethane	263–265	264.3	j

[a]Observed values uncorrected for changes in atmospheric pressure (corrections all estimated to be less than ±0.5°C).
[b]"CRC Handbook of Chemistry and Physics", 62nd ed.; CRC Press: Boca Raton, FL, 1981; #9082, p C-373.
[c]Ibid. #11971, p C-470.
[d]"Dictionary of Organic Compounds", 4th ed.; Oxford University Press: London, 1965; Vol. I, p 11.
[e]"CRC Handbook of Chemistry and Physics", 62nd ed.; CRC Press: Boca Raton, FL, 1981; #7627, p C-321.
[f]Ibid., #5394, p C-244.
[g]Ibid., #8987, p C-370.
[h]Ibid., #2192, p C-138.
[i]Ibid., #3160, p C-169.
[j]Ibid., #6282, p C-274.

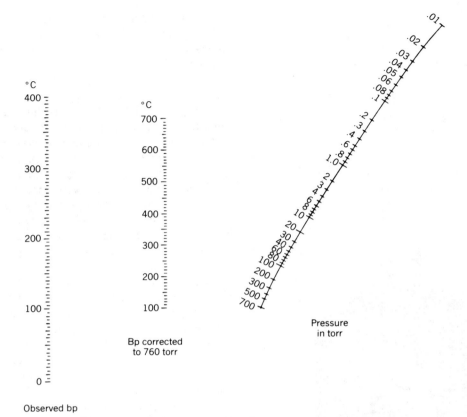

°C
400
300
200
100
0

Observed bp

°C
700
600
500
400
300
200
100

Bp corrected
to 760 torr

.01
.02
.03
.04
.05
.06
.08
.1
.2
.3
.4
.6
.8
1.0
2
3
4
6
8
10
20
30
40
60
80
100
200
300
500
700

Pressure
in torr

Fig. 4.4 *Pressure–temperature nomograph.*

bottom of the tube and a new determination of the boiling point commenced. Indeed, if the bell cavity fills completely during the cooling point of a cycle, it is often difficult to reinitiate the bubble stream without first emptying the entire cavity by overrunning the boiling point.

Observed boiling points for a series of compounds that boil over a wide range of temperatures are summarized in Table 4.1.

Materials that are thermally stable at their boiling point will give identical values on repeat determinations. Substances that begin to decompose will give values that slowly drift after the first few measurements. The observation of color and/or viscosity changes plus a variable boiling point all signal the need for caution in making extended repeat measurements.

The comparison of boiling points obtained experimentally at various atmospheric pressures with reference boiling points at 760 torr is greatly facilitated by the use of pressure–temperature nomographs such as shown in Fig. 4.4. A straight line from the observed boiling point to the observed pressure will pass through the corrected boiling point.

DENSITY Density, defined as mass per unit volume, is generally expressed as grams per milliliter (g/mL) or grams per cubic centimeter (g/cm^3) for liquids. Accurate procedures have been developed for the measurement of this physical constant at the microlevel. A micropycnometer (density meter), developed by Clemo and McQuillen,[4] requires approximately 2 μL (Fig. 4.5).

[4]Clemo, G.R. and McQuillen, A. *J. Chem. Soc.* **1935,** 1220.

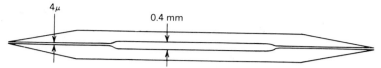

Fig. 4.5 *Pycnometer of Clemo and McQuillen. From Schneider, F. L. "Qualitative Organic Microanalysis," in Vol. II of Monographien aus dem Gebiete der Qualitativen Mikroanalyse; Benedetti-Picher, A. A., Ed.; Springer-Verlag: Wien, Austria; 1964.*

This very accurate device gives the density to three significant figures. The system is self-filling and the fine capillaries' ends do not need to be capped either while coming to temperature equilibrium or while weighing. The apparatus, however, must first be tared, filled, and then reweighed on an analytical balance. Knowing the volume of the pycnometer and the weight of liquid, the density is calculated. Since density changes with temperature, these measurements are made at a constant temperature. In the table of data given for each experiment, no indication of temperature reflects the value measured at 20°C.

REFRACTIVE INDEX

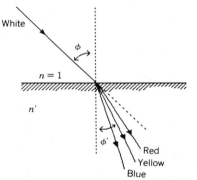

Fig. 4.6 *Upon refraction white light is spread out into a spectrum. This is called dispersion.*

It is commonly observed that a beam of light "bends" as it passes from one medium to another. For example, an oar looks bent as one views the portion under the water. This effect is a consequence of the refraction of light. It results from the change in velocity of the radiation at the media interface, and the angle of refraction is related to the velocity change as follows (see Fig. 4.6).

$$\frac{\sin \phi}{\sin \phi'} = \frac{\text{velocity in vacuum}}{\text{velocity in sample}} = n \text{ (refractive index)}$$

Since the velocity of light in a medium must be less than that in a vacuum, the index of refraction n will always be greater than one. In practice n is taken as the ratio of the velocity of light in air relative to the medium being measured. The refractive index is wavelength dependent.

The wavelength dependence gives rise to the effect of dispersion or the spreading of white light into its component colors. When we measure n, therefore, we must specify the wavelength at which the measurement is made. The standard wavelength for refractive index determinations has become the bright yellow sodium 589 nm emission, the sodium D line. Sodium is a poor choice of wavelength for these measurements with organic substances, but as the sodium lamp represents one of the easiest monochromatic sources of radiation to obtain experimentally, it is widely used. Because the density of the medium is sensitive to temperature, the velocity of radiation also changes with temperature, and therefore refractive index measurements must be made at constant temperatures. Many values are reported in the literature at 20°C. In the table of data given for each experiment, no indication of temperature reflects the value measured at 20°C. The index can be measured optically quite accurately to four decimal places. Since it is particularly sensitive to the presence of impurities, this measurement can be a valuable physical constant for tracking the purification of liquid samples.

The measurement (for example) is reported as:

$$n_D^{20} = 1.4628$$

Procedure

Fig. 4.7 *Abbe-3L refractometer. (Courtesy of Milton Roy Co., Rochester, NY.)*

In the Abbe-3L refractometer (Fig. 4.7), white light is used as the source, but compensating prisms give indices for the D line. This refractometer is commonly used in many undergraduate organic laboratories. Samples (approximately

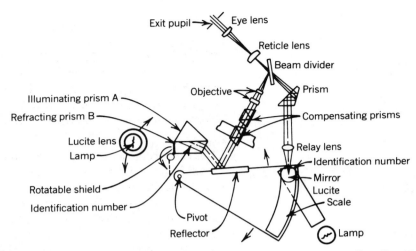

Exit pupil → Eye lens
Reticle lens
Beam divider
Illuminating prism A
Objective
Prism
Refracting prism B
Compensating prisms
Lucite lens
Relay lens
Lamp
Identification number
Rotatable shield
Mirror
Lucite
Identification number
Scale
Pivot
Reflector
Lamp

Fig. 4.8 *Diagram of a typical refractometer. (Courtesy of Milton Roy Co., Rochester, NY.)*

10 μL) are applied between the horizontal surfaces of a pair of hinged prisms (Fig. 4.8).

The refractometer is adjusted so that the field of view has a well-defined light and dark split image (see your instructor for the correct routine for making instrument adjustments on your particular refractometer).

> **CAUTION:** *Do not touch the prisms with a Pasteur pipet as they may be easily marred and will then give erroneous results.*

When using the refractometer, always clean the prisms with alcohol and lens paper before and after use. Record the temperature at which the reading is taken. A reasonably good extrapolation of temperature effects can be obtained by assuming that the index of refraction changes 0.0004 units/°C and varies inversely with the temperature.

MELTING POINTS

In general, the crystalline lattice forces holding organic solids together are distributed over a relatively narrow energy range. The melting points of organic compounds, are usually relatively sharp, that is, less than two degrees. The range and maximum temperature of the melt, however, are very sensitive to impurities. Small amounts of sample contamination by soluble impurities nearly always will result in melting-point depressions. A useful application of this effect involving inorganic compounds is the addition of calcium chloride to the sand used to clear ice-covered roads.

The drop in melting point is also accompanied by an expansion of the melt range. Thus, while the melting point is a useful guide in identification, it also is a particularly effective indication of sample purity.

Procedure

In the microscale laboratory program, two different types of melting-point determinations are carried out: (1) simple capillary melting points, and (2) evacuated melting points.

Simple Capillary Melting Point

Because the microscale laboratory utilizes the Thomas-Hoover Uni-Melt Apparatus or similar system for determining boiling points, melting points are conveniently obtained on the same apparatus. The Uni-Melt system utilizes an electrically heated and stirred silicone oil bath. The temperature readings require no correction in this case since only the mercury reservoir of the thermometer is immersed in the heat-transfer liquid. Melting points are determined in the same capillaries as boiling points. The capillary is loaded by introducing ~1 mg of material into the open end. The sample is then tightly packed (~2 mm) into the closed end by dropping the capillary down a length of glass tubing held vertically to the bench top. The melting-point tube is then ready for mounting in the metal stage, which is immersed in the silicone oil bath of the apparatus. If the melting point of the substance is expected to occur in a certain range, the temperature can be rapidly raised to ~20° below the expected value. At that point the temperature rise should be adjusted to 2°/minute, which is the standard rate of change at which the reference determinations are obtained. The melting-point range is recorded from the temperature when the first drop of liquid forms and ends when the last crystal melts.

Evacuated Melting Points

Many organic compounds begin to decompose at their melting points. This decomposition often begins as the melting point is approached and may adversely affect the values measured. The decomposition can be invariably traced to reaction with oxygen at elevated temperatures. If the melting point is obtained in an evacuated tube, much more accurate melting points can be obtained. These more reliable values arise not only from increased sample stability but because several repeat determinations can often be made on the same sample. The multiple measurements then may be averaged to provide more accurate data.

Evacuated melting points are quickly and easily obtained with a little practice. The procedure is as follows: Shorten the capillary portion of a Pasteur pipet to approximately the same length as a normal melting-point tube (see Fig. 4.9a). Seal the capillary end by rotating in a microburner flame. Touch the pipet only to the very edge of the flame, and keep the large end at an angle below the end being sealed (see Fig. 4.9b). This will prevent water from the flame being carried into the tube where it will condense in the cooler sections. Then load 1–2 mg of sample into the drawn section of the pipet with a microspatula (see Fig. 4.9c). Tap the pipet gently to seat the solid powder as far down the capillary as it can be worked (see Fig. 4.9d). Then push the majority of the sample part way down the capillary with the same diameter copper wire that you use to seat the cotton plug in constructing the Pasteur filter pipet (see Fig. 4.9e). Next, connect the pipet to a mechanical high-vacuum pump with a piece of vacuum tubing. Turn on the pump and evacuate the pipet for 30 seconds (see Fig. 4.9f). Then with a microburner gently warm the surface of the capillary tubing just below the drawn section. On warming, the remaining fragments of the sample (the majority of which has been forced further down in the tube) will sublime in either direction away from the hot section. Once the traces of sample have been "chased" away, the heating is increased and the capillary tube collapsed, fused, and separated from the shank, which remains connected to the vacuum system (see Fig. 4.9g). The pump is then turned off and the shank discarded. The sample is tightly packed into the initially sealed end of the evacuated capillary by dropping it down a section of glass tubing, as in the case of packing open melting-point samples. After packing the sample (~2 mm), a section of the evacuated capillary is once more quickly heated and collapsed by the microburner flame about 10–15 mm above the sample (see Fig. 4.9h).

This is required in order to trap the sample and to avoid sublimation up the tube during the melting-point measurement. This operation is a little tricky and should be practiced a few times. It is very important that a complete fusion take place. Now the sample is ready to be placed in the melting-point apparatus. The procedure is the same as in the open capillary case, except that after melting

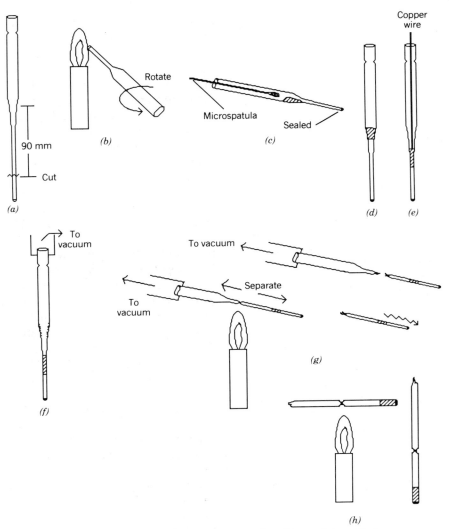

Fig. 4.9a-h *Procedure for obtaining evacuated melting point capillaries.*

the sample can be cooled, crystallized, and remelted several times and the average value of the range reported. If these latter values begin to drift downward, the sample can be considered to be decomposing even under evacuated, deoxygenated conditions. In this case the first value observed should be recorded as the melting point.

MIXED MELTING POINTS   Advantage can be taken of the sensitivity of the melting point to impurities. In the case where two different substances possess identical melting points, it would be impossible to identify an unknown sample as either material based on the melting point alone. If reference standards of the two compounds are available, however, then mixtures of the unknown and the two standards can be prepared. It is important to prepare several mixtures of varying concentrations for melting-point comparisons, as the point of maximum depression need not occur on the phase diagram at the 50:50 point. The melting points of the unknown and the mixed samples are conveniently obtained simultaneously (the Uni-Melt stage will accept up to seven capillaries at one time). The unknown sample and the mixture of the unknown with the correct reference will have identical values, but the mixture with the reference of a different substance will give a depressed melting point. This procedure is the classical route to positive identification of a crystalline solid.

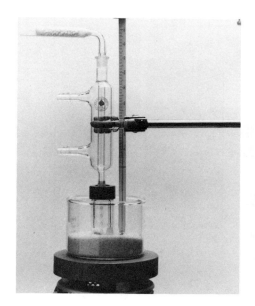

Chapter 5

Development of Experimental Technique

TECHNIQUE 1: DISTILLATION

Distillation is the process of heating a liquid to the boiling point, condensing the heated vapor by cooling it, and returning only a portion or none of the condensed vapors to the distillation flask. Distillation varies from the process of reflux (*see* p. 16) only in that a fraction of the condensate is diverted from the boiling system. Distillations in which a fraction of the condensed vapors is returned to the boiler are often referred to as being under "partial reflux."

Theory

Distillation techniques often can be used for separating two or more components on the basis of their differences in vapor pressure. Separation can be accomplished by taking advantage of the fact that the vapor phase is generally richer in the more volatile component (lower boiling component) of the liquid mixture. Molecules in a liquid are in constant motion and possess a spectrum of kinetic energies. Those with higher energies (a larger fraction of the lower boiling component) moving near the surface have a greater tendency to escape into the vapor (gas) phase. If a pure liquid (hexane, for example) is in a closed container, eventually hexane molecules in the vapor phase will reach equilibrium with hexane molecules in the liquid phase. The pressure exerted by the hexane vapor molecules at a given temperature is called the *vapor pressure,* and given by the symbol P_H^o, where the superscript o indicates a pure component. For any pure component A, the vapor pressure would be P_A^o.

Suppose a second component (toluene, for example) is added to the hexane. The total vapor pressure (P_{total}) is then the sum of the individual component *partial vapor pressures* (P_H, P_T) as given by *Dalton's law:*

$$P_{total} = P_H + P_T$$

or in general

$$P_{total} = P_A + P_B + P_C + \cdots P_n$$

It is important to realize that the vapor pressure (P_A^o) and the partial vapor pressure (P_A) are not equivalent since the presence of a second component in the system has an effect on the vapor pressure of the first component. The partial vapor pressure of any one volatile component can be obtained using *Raoult's law:*

$$P_A = P_A^o X_A$$

where X_A is the mole fraction of component A in the liquid system.

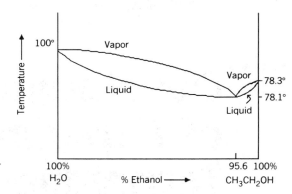

Fig. 5.1 *Ethanol-water minimum-boiling-point phase diagram.*

Azeotropic Mixtures

The major deviation from Raoult's law occurs with mixtures that form azeotropes. The term comes from the Greek meaning "not to boil with change." Particular combinations of substances undergo intermolecular interactions that perturb the normal vapor pressure relationship. The majority of these interactions result in two types of vapor–liquid composition behavior: (1) the minimum-boiling-point diagram, and (2) the maximum-boiling-point diagram. In the case of a two-component minimum-boiling azeotrope, the vapor pressure at a specific composition of the mixture is higher than the predicted value. The higher vapor pressure leads to a lower boiling point for the mixture at that particular composition than for either of the two components. Thus, a "constant-boiling mixture" results that can never be separated by distillation techniques. The most well-known binary minimum-boiling azeotrope is the ethanol-water mixture (Fig. 5.1).

As the phase diagram indicates, distillation of this system will result in a constant-boiling mixture at 78.1°C and 95.6% ethanol. While azeotrope formation can frustrate attempts to separate a mixture by distillation, azeotropes can be effective tools in the laboratory. For example, in Experiment 48A toluene is intentionally used as a solvent in order to remove water by azeotropic distillation (84°C). When a lower than predicted combined vapor pressure occurs, a maximum-boiling azeotrope results. This type of system will exhibit a similar type of phase diagram as the minimum-boiling azeotrope, differing only in that the constant-boiling mixture has a temperature maximum rather than a minimum.

SIMPLE DISTILLATION AT THE SEMI-MICROSCALE LEVEL

Simple distillation involves the use of the distillation process to separate a liquid from minor components that are nonvolatile or that have boiling points substantially (>30–40°C) above the major component. A typical setup for a macroscale distillation of this type is shown in Fig. 5.2.

At the microscale level when working with volumes smaller than 100 μL, gas chromatographic techniques (see Experiment 2) have replaced conventional microdistillation processes.[1] Semi-microscale simple distillation in the volume range of 0.1 to 2 mL remains an effective separation technique. Apparatus have been developed that achieve good separation of mixture samples smaller than

[1]Schneider, F. L. "Qualitative Organic Microanalysis"; Vol. II of Monographien aus dem Gebiete der Qualitativen Mikroanalyse, Benedetti-Pichler, Ed.; Springer-Verlag: Wien, Austria, 1964; p 31.

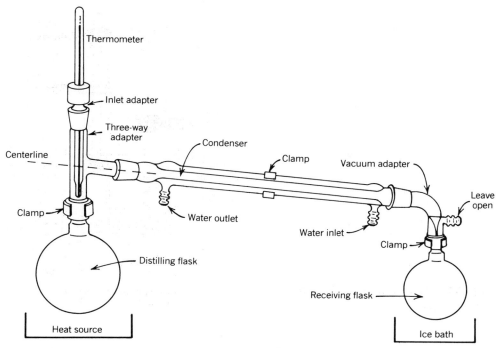

Fig. 5.2 A complete, simple distillation setup. From James W. Zubrick, The Organic Chem Lab Survival Manual (New York: Wiley, 1984). (Reprinted by permission of John Wiley & Sons, New York, NY.)

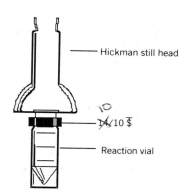

Fig. 5.3 Hickman still.

2.0 mL in volume. One of the most useful of these still designs is the Hickman still, shown in Fig. 5.3.

This still is employed in several of the microscale experiments to purify solvents, carry out reactions, and concentrate solutions for recrystallization. An introduction to the use of the Hickman still is given in Experiments 1A and 1B.

Experiment 1A
Simple Distillation at the Semi-microscale Level: The Separation of Ethyl Acetate from *trans*-1,2-Dibenzoylethylene

Physical Properties of Reactants and Products

Compound	MW	Wt/Vol	mp(°C)	bp(°C)	Density	n_D
Ethyl acetate	88.12	1.0 mL		77	0.90	1.3723
trans-1,2-Dibenzoylethylene	236.27	50 mg	111			

Transfer 1-mL of the yellow stock solution (*trans*-1,2-dibenzoylethylene/ethyl acetate, 50 mg/mL) to a 3-mL conical vial by automatic delivery pipet (remember to place the vial in a small beaker to prevent tipping during the transfer). Place

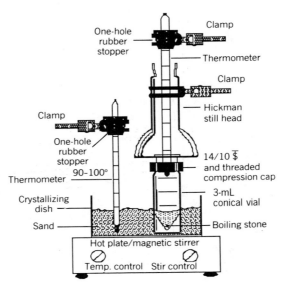

One-hole rubber stopper
Clamp
Thermometer
Clamp
Hickman still head
Clamp
One-hole rubber stopper
90–100°
Thermometer
14/10 ₮ and threaded compression cap
3-mL conical vial
Crystallizing dish
Sand
Boiling stone
Hot plate/magnetic stirrer
Temp. control   Stir control

Fig. 5.4 *Hickman still (14/10₮) with conical vial (3 mL).*

trans-1,2-Dibenzoylethylene/ethyl acetate (50 mg/mL), 1 mL

a boiling stone in the vial and assemble the Hickman still head. The still assembly is mounted in a sand bath and placed on a hot plate (see Fig. 5.4).

The temperature of the bath is raised to 90–100°C at a rate of 5°C/minute using a hot plate.

> **CAUTION:** *Do not let the temperature of the still rise too rapidly.*

Once boiling commences, the rate of heating should be lowered to 2–3°C/minute. A slow distillation rate is very important in establishing equilibrium between the vapor and liquid components in the mixture. Follow the course of the distillation by the rise of condensate on the sides of the Hickman column. When the condensate reaches the trap, adjust the bath temperature so that liquid is removed from the column slowly (approximately $100\mu L$/minute). A smooth slow distillation will provide a cleaner separation of the components and will also avoid mechanical transfer of nonvolatile components via splattering to the condensate trap (if the condensate appears yellow, mechanical contamination has occurred).

Collect approximately 50–150 μL of the ester in the collar of the still (the first fraction collected is often referred to as the forerun; give the temperature range). As the distillation continues, remove the forerun with a Pasteur pipet having a slightly bent tip (microburner). Place the fraction in a clean, dry 1-dram screwcapped vial (use an aluminum foil liner to avoid cap contamination). Number the fraction with a marking pen. Collect a second fraction of ester (400–500 μL; give the temperature range), which should be clear and colorless. Remove and store as before. Discontinue the distillation. Allow the distilling flask to cool slowly by leaving it in the warm sand bath while measuring the physical properties of the distillate fractions. The measurement of three physical properties of the ester will be carried out to establish the identity and purity of the compound by comparison with known literature values.

Determine the refractive index (see p. 37) of the two fractions collected. Compare the experimental values to those found in the literature for ethyl acetate. If the values are within 0.0010 units of each other, the fractions are considered to have the same constitution. Are the values for the two fractions the same? If not, which one deviates the most from the reference data? Attempt to explain the result.

Determine the density (see p. 36) of the ester using material contained in the second fraction. Conduct this density measurement by the simple procedure of transferring the entire fraction by calibrated pipet (use this value as the volume) to a previously tared screwcap vial (remember to include the aluminum foil in the tare), and weigh on a balance to ±1 mg. This measurement is nondestructive. The material employed may be recovered for use in further tests. Compare your results with those values found in the literature.

Determine the boiling point of the second fraction by the ultramicro boiling-point procedure (see p. 33). Compare your result with the literature value. Does this fraction appear to be pure ethyl acetate?

In the next step, disconnect and cool the 3-mL conical vial in an ice water bath for 10 minutes. *trans*-1,2-Dibenzoylethylene will crystallize from the concentrated solution. Remove the remaining solvent from the distillation vial with a Pasteur filter pipet and place the crystals on a porous clay plate to air dry. The melting point of the crystalline material is obtained by the open capillary method and compared to the literature value.

Reference values of the physical constants are available in the *CRC Handbook of Chemistry and Physics*. Submit a copy of the table prepared in your laboratory notebook to the instructor after tabulating the experimentally measured values of the physical properties in addition to those reported in the literature for ethyl acetate (see "acetic acid, ethyl ester").

FRACTIONAL SEMI-MICROSCALE DISTILLATION

Process

Fractional distillation is the application of the distillation process to the separation of liquid mixtures in which the boiling points of the components differ by less than 30–40°C. In this situation a fractionating column is required to increase the efficiency of the separation. As discussed earlier, it may be seen from a liquid–vapor composition curve (Fig. 5.5) that the lower boiling component of a binary mixture makes a larger contribution to the vapor composition than does the higher boiling component. On condensation the liquid formed will now be richer in the lower boiling component. This condensate will not be pure, however, and in the case of closely boiling components it may show only slight enrichment. If the condensate is volatilized a second time, the vapor in equilibrium with this liquid will now show a further enrichment in the lower boiling component. Thus, the trick to the separation of liquids that possess similar boiling points depends on repeating the vaporization–condensation cycle many times. Each cycle is termed a *theoretical plate*. A number of different column designs

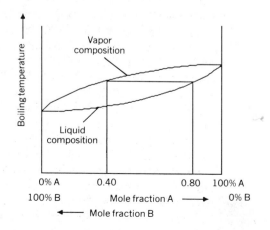

Fig. 5.5 *Liquid-vapor composition curve.*

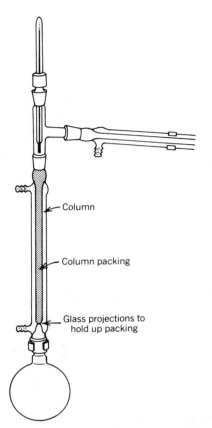

Fig. 5.6 *The fractional distillation setup. From James W. Zubrick,* The Organic Chem Lab Survival Manual *(New York: Wiley, 1984). (Reprinted by permission of John Wiley & Sons, New York, NY.)*

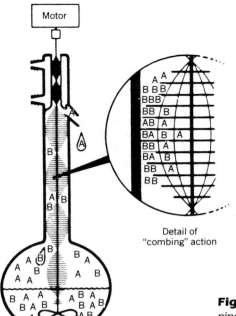

Fig. 5.7 *Schematic of a metal mesh spinning band still. (Courtesy of Perkin-Elmer Corp., Norwalk, CT.)*

are available for use at the macrolevel that achieve varying numbers of theoretical plates (see Fig. 5.6).

In most columns the design is such that increased fractionation efficiency is dependent on a very large increase in the surface area in contact with the vapor phase. This is normally accomplished by packing the fractionating column with wire gauze or glass beads. Unfortunately, a large quantity of liquid must be distributed over the column surface in equilibrium with the vapor. Furthermore, the longer the column the more efficient it becomes, but longer columns also require additional liquid phase. The column requirement of the liquid phase is termed column "hold-up." Column hold-up is defined as the amount of liquid distributed over the column packing required to maintain the system in equilibrium. This material is essentially lost from the liquid phase held in the distillation pot. The amount of column hold-up can be large compared to the total volume of material available for the distillation. With mixtures of less than 2 mL, column hold-up precludes the use of most common fractionation columns. Microfractionating columns constructed of rapidly spinning bands of metal or Teflon gauze have very low column hold-up and have a large number of plates relative to their height (Fig. 5.7). They are, however, rather expensive and normally are available only for research purposes.

An alternative to the microspinning-band distillation column is the concentric-tube column. In these columns the fractionating section is constructed of two concentric tubes in which the vapor–liquid equilibrium is established within the annular space between the two columns. The resolution of the concentric-tube system is inversely proportional to the thickness of the annular ring. Columns of this type can achieve very good separations with HETP (height equivalent theoretical plates) approaching 3/cm. In addition, column hold-up can be close to 10 μL/HETP. The major constraint in the use of these columns is the very low throughput, which can be as low as 100 μL/hour. Concentric-tube columns also have the nasty habit of flooding in inexperienced hands. These latter factors result in long resonance times at elevated temperatures for the liquid components involved. In cases where time and thermal stability are not a problem, the concentric tube column can be a powerful system for mixture separations.

Experiment 1B utilizes the Hickman still in a simple example of fractional distillation. The system is arranged with a thermometer positioned directly down the center of the still column with the bulb extending just to the bottom of the joint. As assembled, the system functions as a rough concentric tube fractionating column. It is very important to have both the still and the column positioned as close to the vertical as possible and under no circumstances should the two elements come into direct contact (see Fig. 5.4). A clean, two-theoretical-plate distillation is obtained with a two-component mixture by carrying out sequential fractional distillations with this system.

Experiment 1B
Fractional Semi-Microscale Distillation
The Separation of Hexane and Toluene

Hexane and toluene are liquid hydrocarbons that have boiling points approximately 40°C apart. The liquid–vapor composition curve in Fig. 5.5 very closely resembles the diagram expected for this system, and, therefore, it is apparent that a two-plate distillation should yield nearly pure components. The procedure outlined below consists of two parts. The first deals with the initial distillation (first plate), which separates the liquid mixture into three separate fractions. The second deals with *redistillation* of the first and third fractions (second plate). Exercising careful technique during the first distillation should provide a fraction rich in the lower boiling component, a middle fraction, and a fraction rich in the higher boiling component. Then careful redistillation of these fractions can be expected to complete the separation of the two components and to produce fractions of relatively pure hexane and toluene. The Hickman still employed in the microscale laboratory is a simple, short-path column, and, therefore, one would not expect the complete separation of the hexane and toluene in one cycle.

Physical Properties of Reactants and Products

Compound	MW	Wt/Vol	bp(°C)	Density	n_D
Hexane	86.18	1.0 mL	69	0.66	1.3751
Toluene	92.15	1.0 mL	111	0.87	1.4961

In a clean, dry, stoppered 5-mL conical vial is placed 1.0 mL of hexane and 1.0 mL of toluene using an automatic delivery pipet.
Place the vial in a small beaker to prevent tipping.
A boiling stone is added, and the Hickman still assembled with the thermometer positioned directly down the center of the column (see above discussion), and the system mounted in a sand bath (see Fig. 5.4). The temperature of the sand bath is raised to 80–90°C at a rate of 5°C/minute using a hotplate. *Do not let the temperature of the still rise too rapidly.*
Once gentle boiling begins, the heating rate should be lowered to 2°C/minute. It is absolutely crucial that the distillation rate be kept below 100 μL/3 minute in order to achieve the necessary fraction enrichment that will permit good separation during the second stage of the experiment. The distillate is collected in *three fractions* over the temperature ranges (1) 65–85°C (bath temperature 95–110°C); (2) 85–105°C (bath temperature ~140°C); (3) 105–110°C (bath temperature ~170°C) in amounts of approximately 800 μL, 400 μL, and 800 μL, respectively. Remove each fraction from the still with a bent-tip Pasteur

pipet (as in Experiment 1A). Store the liquid condensate (fractions) in clean dry 1-dram screwcap vials. *Remember to number the vials in order and use an aluminum foil cap liner.*

For each of the three fractions, record the refractive index. Fraction 1 has been enriched in one of the two components. Which one? Does the refractive index agree with that found in the literature? Fraction 3 has been enriched in the other component. Does the refractive index of that fraction support your first conclusion? If partial enrichment has been achieved, then proceed to the second part of the experiment.

REDISTILLATION OF FRACTION 1

Redistill fraction 1 in a clean Hickman still with a thermometer arranged as before (Fig. 5.4) using a 3-mL conical vial and the procedure outlined above. Collect an initial fraction over the boiling range, 68–71°C (~100–200 μL). Remove it from the collar using the Pasteur pipet, and place it in a 1-dram screwcap vial.

Determine the ultramicro boiling point and the refractive index of this lower boiling fraction. Compare the experimental values obtained with those of pure hexane reported in the literature. *If time permits, the boiling point and refractive index of pure hexane may be determined for comparison purposes.*

REDISTILLATION OF FRACTION 3

Fraction 3 is placed in a clean Hickman still using a thermometer and a 3-mL conical vial (Fig. 5.4), and redistilled using the procedure outlined above. Collect an initial fraction over the boiling range 95–108°C (~500 μL), and transfer this fraction by Pasteur pipet to a screwcap vial. Collect a final fraction at 108–110°C (~250 μL), and transfer the material to a second vial. *This second fraction is the highest boiling fraction to be collected in the three distillations and should be the richest in the high-boiling component.*

Determine the refractive index and boiling point of the second fraction, and compare your results with those found in the literature for toluene. *If time permits, determine the refractive index and boiling point of pure toluene for comparison purposes.*

Report the results obtained on the separation of hexane and toluene. Show a comparison of the experimental refractive index and boiling point data versus those values found in the literature for hexane and toluene. Give a brief discussion indicating the efficiency of the separation based on the data recorded in your laboratory notebook.

ENVIRONMENTAL DATA

Substance	Amount	TLV (mg/m^3)	Emissions (mg)	Volume (m^3)
Ethyl acetate (dist., transfer, ref. index)	1.0 mL	1400	170	0.1
Hexane (fract. dist., transfer, ref. index, bp)	1.0 mL	180	250	1.4
Toluene (fract. dist., transfer, ref. index, bp)	1.0 mL	750	200	0.3

TXDS: 1,2-Dibenzoylethylene-ipr-mus LD50:25 mg/kg

QUESTIONS

5-1. The boiling point of a liquid is affected by several factors. What effect does each of the following conditions have on the boiling point of a given liquid?

a. The pressure of the atmosphere.

b. Use of an uncalibrated thermometer.

c. Rate of heating of the liquid in a distillation flask.

5-2. Calculate the vapor pressure of a solution containing 30 mol% hexane and 70 mol% octane at 90°C assuming Raoult's law is obeyed.

Given: vapor pressure of the pure compounds at 90°C. Vapor pressure of hexane = 1390 torr; vapor pressure of octane = 253 torr.

5-3. In any distillation for maximum efficiency of the column, the distilling flask should be approximately half full of liquid. Comment on this fact in terms of (a) a flask that is too full and (b) a flask that is nearly empty.

5-4. Occasionally during a distillation a solution will foam rather than boil. A way of avoiding this problem is to add a surfactant to the solution.

a. What is the chemical constitution of a surfactant?

b. How does a surfactant reduce the foaming problem?

5-5. Explain why packed and spinning-band fractional distillation columns are more efficient at separating two liquids having close boiling points than are unpacked columns.

5-6. Why are boiling stones added to a liquid that is to be heated to the boiling point?

TECHNIQUE 2: MICROSCALE SEPARATION OF LIQUID MIXTURES BY PREPARATIVE GAS CHROMATOGRAPHY

Theory

One of the principal hurdles in dealing with experimental chemistry is the isolation of materials in their pure state. In order to fully characterize a substance, we require a pure sample of the material. In organic chemistry this is a particularly difficult demand since most organic reactions generate several products. We are generally satisfied if the desired product is the major component of the mixture obtained. As you proceed down the trail of developing experimental technique, you will note the heavy emphasis placed on separation techniques. Experiments 1A and 1B dealt directly with distillation routines focused on the separation of mixtures. In Experiment 2 we continue the development of separation techniques with an introduction to microscale preparative gas chromatography.

The methods of chromatography have revolutionized experimental organic chemistry over the past 30-odd years. It is by far the most powerful technique for separating mixtures and isolating pure substances. Chromatography is defined as the resolution of a multicomponent mixture (several hundred in some cases) by distribution between two phases, one stationary and one moving. The various types of chromatography are categorized by the phases involved: column, thin-layer, and paper (solid–liquid); partition (liquid–liquid); and vapor phase (gas–liquid) chromatography. The principal mechanism upon which these separations depend is differential solubility or adsorptivity of the mixture components with respect to the two phases involved. That is, the components exhibit different partition coefficients.

Gas chromatography is an extraordinarily powerful technique for the separation of mixtures. In this case the stationary phase is a liquid and the moving phase is a gas (the carrier gas). Fairly sophisticated instrumentation is required to carry out this type of chromatography. A diagram of a common laboratory gas chromatograph is given in Fig. 5.8.

The system starts with a heated injection port. The sample mixture is introduced by syringe through a high-temperature septum in the port, into the heated chamber through which the carrier gas (the gas phase) is flowing. The solubility of the sample in the partitioning gas depends to a large extent on the vapor pressure of the substances in the mixture. Helium and nitrogen are commonly used as carrier gases. Heating the injection system ensures vaporization of the sample. Thus, two major constraints of gas chromatography are (1) that the sample must be stable at the temperature required to cause vaporization and (2) that the sample must possess sufficient vapor pressure to be completely soluble in the gas phase at the operating temperature.

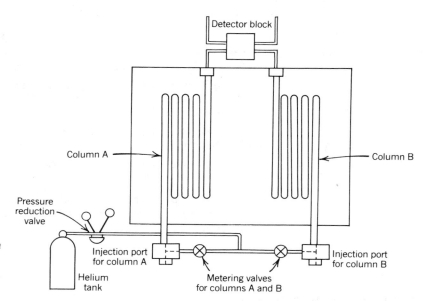

Fig. 5.8 *Block diagram of a dual-column gas chromatograph showing essential parts. (Courtesy of Gow-Mac Instrument Co., Bound Brook, NJ.)*

The vaporized mixture is swept from the injection block onto the column. This constitutes the key component in the separation process. The stationary liquid phase, in which the sample will dissolve and partition with the moving gas phase, is physically and/or chemically bound to inert packing material contained in the column. Gas chromatographic columns come in a variety of sizes and shapes. In the diagram of the Gow-Mac instrument (Fig. 5.8), two parallel coiled columns are mounted in a well-insulated oven. Considerable oven space may be saved and better temperature regulation is achieved if the columns are coiled. This is important since the column must be kept under very precise temperature control. In order to maintain reasonable vapor pressures in the gas phase during the course of the separation, most liquid mixtures will require elevated oven temperatures. Separation of the mixture occurs as the carrier gas sweeps the sample through the column. Most columns are constructed of stainless steel, glass, or fused silica. The diameter and length of the column are critical factors in determining how the internal part of the column is designed to achieve separation of the sample mixture. The surface area of the liquid phase, in contact with the sample contained in the moving gas phase, is maximized by coating the finely divided inert support with a nonvolatile liquid material (stationary phase). The coated support is carefully loaded into the column. Columns prepared in this fashion are termed "packed columns." Packed columns usually are ¼ in. or ⅛ in. in diameter and are 4 to 12 ft long. These columns are particularly attractive for use in the microscale laboratory. Simple mixtures in the 20–80 μL range usually can be resolved and the pure components collected at the exit port of the column.

Research columns are also available that have no packing but simply have the liquid phase directly applied to the walls of the column. These are referred to as wall-coat or open-tubular columns. The reduction in surface area is compensated for by making the diameter very small (0.1 mm) and the length very long (several hundred meters would not be uncommon). Termed "capillary columns," they constitute the most efficient columns employed for analytical separations. Mixtures of several hundred compounds can be completely resolved in a single pass through one of these systems. Capillary columns, however, are complicated to operate and require very small samples (0.1 μL or less). Thus, they cannot be used in a preparative routine.

Once introduced on the column, the sample will undergo partition with the liquid phase. The choice of the liquid phase is particularly important since it directly affects the relative distribution coefficients. The temperature of the column will also affect the separation in that the retention time of the lower boiling component will decrease as the temperature is increased. That is, the residence time of the component in the column will decrease. If the oven temperature is too high, equilibration with the stationary phase will not be established and the component mixture may elute together or at best undergo incomplete resolution. The flow rate of the carrier gas constitutes another important parameter. The rate must be slow enough to allow equilibration but sufficiently rapid to ensure that diffusion will not overcome resolution of the components. As noted, the length of the column also is an important factor in separation performance. We discussed earlier in the distillation experiment that distillation column efficiency is proportional to column height and determines the number of evaporation-condensation cycles. In a similar manner, increasing the length of the gas chromatographic column allows more partition cycles to occur. Difficult-to-separate mixtures, such as the xylenes (very similar boiling points) will have a better chance of being resolved on longer columns.

A successfully resolved mixture will elute as individual components, sequentially with time, at the exit port of the instrument. To monitor the exit vapors, a detector is placed in the gas stream (Fig. 5.8). After passing through the detector, the carrier gas and the separated sample components are vented through a heated exit port. Sequential collection of the separated materials can be made by attaching suitable sample-condensing tubes to the exit port (see Fig. 3.32).

A widely used detector is the nondestructive thermal conductivity sensor often referred to as a hot-wire detector. A heated element in the gas stream changes electrical resistance when a substance dilutes the carrier gas and changes its thermal conductivity. Helium possesses a higher thermal conductivity than most organic substances. When samples other than helium are present, the conductivity of the gas stream decreases and the resistance of the heated wire changes. The change in resistance is measured by differences (Wheatstone bridge), with a reference detector mounted on the second (parallel) column. This signal is plotted by a recorder. The horizontal axis is time and the vertical axis is the magnitude of the resistance difference. This plot is the chromatogram. Retention time t_R is defined as the time from sample injection to the time of maximum recorder pen deflection. The baseline width W_b of a peak is defined as the distance between two points where tangents to the points of inflection cross the baseline (Fig. 5.9).

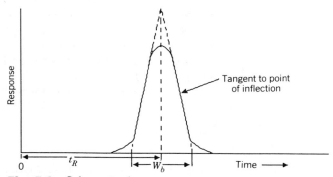

Fig. 5.9 *Schematic chromatogram.*

It is possible to estimate the number of theoretical plates (directly related to the number of distribution cycles) present in a column for a particular substance. The parameters are given in the following relationship[2]

$$n = 16[t_R/W_b]^2$$

The units of t_R and W_b are the same (minutes, seconds, centimeters). As in distillation columns, the larger the number of theoretical plates, the higher the resolution of the column.

The efficiency of a system may also be expressed as the *height equivalent to a theoretical plate,* HETP. This parameter is related to the number of theoretical plates n by

$$\text{HETP} = \frac{L}{n}$$

where L is the length of the column, usually reported in centimeters. The smaller the HETP, the more efficient the column.

The number of theoretical plates available in fractional distillation columns at the microscale level is severely limited by column hold-up (see Experiment 1A and 1B). Gas chromatographic columns, on the other hand, operate most *efficiently* at the *microscale level.* The power of this technique is demonstrated in the following experiment.

Experiment 2
The Separation of a 25-μL Mixture of Heptanal (bp = 153°C) and Cyclohexanol (bp = 160°C) by Gas Chromatography

In this experiment you will attempt to separate 25 μL of a mixture consisting of heptanal and cyclohexanol into the pure components. The volume of the mixture is approximately that of a single drop. The materials boil within 7°C of each other. This mixture would be difficult to separate even by the best microdistillation techniques available. The purity of the fractions collected will be assessed by boiling points and refractive indices.

Before starting the experiment, let us observe the sensitivity of gas chromatographic separations to experimental conditions. For example, two sets of experimental data on the heptanal-cyclohexanol mixture are included to demonstrate the results of variations in oven temperature on retention time.

In Data Set A, the oven temperature was allowed to slowly rise from 160°C to ~170°C during a series of sample collections. The retention time of the first component, heptanal, dropped from ~3 minutes to close to 2 minutes, while the retention time of the second component, cyclohexanol, was reduced from ~5.5 minutes to nearly 4 minutes. The significant decrease in resolution over this series of collections is reflected in the calculated theoretical plates, which were over 300 for heptanal and ~500 for cyclohexanol in the first trial, but which declined to below 200 for both compounds toward the last run (see Fig. 5.10 and Data Set A).

[2]Berg, E.W. "Physical and Chemical Methods of Separation"; McGraw-Hill: New York, 1963; p 111.

EXPERIMENTAL DATA: SET A

Trial No.	Heptanal				Cyclohexanol			
	Retent. Time (min)	Baseline Width (min)	No. of Theor. Pl.	Yield (mg)	Retent. Time (min)	Baseline Width (min)	No. of Theor. Pl.	Yield (mg)
1	3.1	0.7	314	8.0	5.6	1.0	502	8.0
2	2.9	0.7	275	8.0	5.3	1.0	449	8.0
3	3.0	0.7	294	7.0	5.7	1.0	520	8.0
4	2.8	0.7	256	8.0	5.1	1.1	344	8.0
5	2.5	0.6	278	8.0	4.3	1.1	244	9.0
6	2.7	0.5	467	7.0	4.6	1.0	339	10.0
7	2.5	0.6	278	10.0	4.2	1.0	282	8.0
8	2.2	0.5	310	9.0	3.5	1.0	196	8.0
9	1.8	0.5	207	8.0	3.0	1.0	144	8.0
10	2.3	0.7	173	8.0	3.9	1.0	243	8.0
Ave.	2.6 ± 0.4	0.6 ± 0.09	285 ± 78	8.1 ± 0.8	4.5 ± 0.9	1.0 ± 0.05	326 ± 129	8.3 ± 0.7

Fig. 5.10 *Data Set A: separation of heptanal-cyclohexanol.*

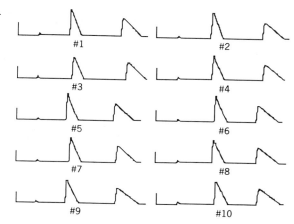

COLLECTION YIELD

Cyclohexanol

Density of cyclohexanol = 0.963 mg/μL.

In 25 μL of 1:1 cyclohexanol-heptanal, there are 12.5 μL of cyclohexanol.

Therefore, (12.5 μL) × (0.963 mg/μL) = 12 mg of cyclohexanol injected.

Percentage yield = (8.3 mg/12.0 mg)100 = 69% cyclohexanol collected.

Heptanal

Density of heptanal = 0.850 mg/μL.

Therefore, (12.5 μL) × (0.85 mg/μL) = 10.6 mg of heptanal.

Percentage yield = (8.1 mg/10.6 mg)100 = 76% heptanal.

In the B series of collections, stable oven temperatures and flow rates were maintained, and the data exhibit excellent reproducibility. Oven temperature was held at 155°C isothermal throughout the sampling process. The retention time of heptanal was observed to be slightly longer than 3 minutes with a variance of 6 seconds, while the cyclohexanol retention time was found to be slightly

EXPERIMENTAL DATA:
SET B

Trial No.	Heptanal Retent. Time (min)	Baseline Width (min)	No. of Theor. Pl.	Yield (mg)	Cyclohexanol Retent. Time (min)	Baseline Width (min)	No. of Theor. Pl.	Yield (mg)
1	3.5	0.7	400	8.0	6.6	1.1	576	8.0
2	3.2	0.7	334	9.0	6.0	1.1	476	7.0
3	3.5	0.7	400	7.0	6.6	1.2	484	10.0
4	3.2	0.7	334	9.0	6.1	1.0	595	9.0
5	3.1	0.6	427	8.0	6.0	1.1	476	8.0
6	3.2	0.7	334	9.0	6.0	1.1	476	9.0
7	3.3	0.8	272	9.0	6.1	1.1	492	8.0
8	3.1	0.7	313	8.0	6.0	1.1	476	10.0
9	3.2	0.7	334	8.0	6.1	1.1	492	8.0
10	3.2	0.7	334	8.0	6.2	1.1	508	8.0
Ave.	3.2±0.1	0.7±0.05	348±47	8.3±0.7	6.2±0.2	1.1±0.05	505±44	8.5±1.0

Fig. 5.11 *Data Set B: separation of heptanal-cyclohexanol.*

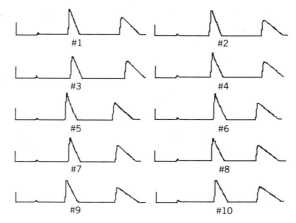

over 6 minutes with a variance of 12 seconds. The resolution remained essentially constant throughout the series, and the calculated theoretical plates were ~350 for heptanal and ~500 for cyclohexanol (see Fig. 5.11 and Data Set B).

COLLECTION YIELD

Cyclohexanol

Density of cyclohexanol = 0.963 mg/μL.

In 25 μL of 1:1 cyclohexanol–heptanal, there are 12.5 μL of cyclohexanol.

Therefore, (12.5 μL) × (0.963 mg/μL) = 12 mg of cyclohexanol injected.

Percentage yield = (8.5 mg/12.0 mg)100 = 71% cyclohexanol collected.

Heptanal

Density of heptanal = 0.850 mg/μL.

Therefore, (12.5 μL) × (0.85 mg/μL) = 10.6 mg of heptanal.

Percentage yield = (8.3 mg/10.6 mg)100 = 78% heptanal.

The results described above demonstrate that the resolution of GC peaks may be very sensitive to changes in retention time resulting from instability in oven temperatures. As the calculated theoretical plates are highly dependent on resolution values, significant degradation in column plate values can occur with variations in oven temperatures. When you consider the time and effort required to obtain a 2 plate fractional distillation on a 2-mL mixture in Experiment 1B compared to the speed and ease of obtaining a 500 plate separation on 12.5 μL of cyclohexanol in this experiment, it is hard not to be impressed with the enormous power of this technique.

Physical Properties of Reactants and Products

Compound	MW	Wt/Vol	bp(°C)	Density	n_D
Heptanal	114.19	12.5 μL	153	0.85	1.4113
Cyclohexanol	100.16	12.5 μL	160	0.96	1.4641

PROCEDURE

This procedure involves injection of 25 μL, heptanal–cyclohexanol, 1:1 (v/v), onto an $\frac{1}{4}$ in. × 8 ft stainless steel column packed with 10% carbowax 80/100 20M PAW-DMS and installed in a gas chromatograph (GC). Experimental conditions are: He flow rate, 50 mL/minute; current, 150 mA; attenuator setting, 256; chart speed, 1 cm/minute; temperature, 155°C.

The liquid effluents are collected in an uncooled 4-mm diameter collection tube (double reservoirs; overall tube length 40–50 mm). See Figs. 3.32 and 5.12.

The collection tube (oven-dried until 5 minutes before use) is attached to the heated exit port by the 5/5 ꝫ joint. Sample collection is initiated 0.5 minutes prior to detection on the recorder of the expected peak (time based on previously determined retention values—see your lab instructor) and continued until 0.5 minutes following return to baseline. After detaching the collection tube, the sample is transferred to the GC 0.1-mL conical collection vial. The transfer is facilitated by the 5/5 ꝫ joint on the conical vial. After the collection tube is joined to the vial (preweighed with stopper), the system is centrifuged. The collection tube is then removed, and the vial stoppered and reweighed.

Calculate the percentage yield. Amounts should range between 7 and 10 mg. Determine the boiling point (see p. 33) of each fraction and obtain the refractive index (see p. 37) if possible. This latter measurement will require most, if not all, of the remaining sample.

Assess the purity and efficiency of the separation from your tabulated data and the GC chromatogram.

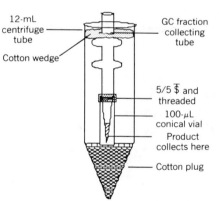

12-mL centrifuge tube

GC fraction collecting tube

Cotton wedge

5/5 ꝫ and threaded

100-μL conical vial

Product collects here

Cotton plug

Fig. 5.12 *Gas chromatographic collection tube and 0.1-mL conical vial.*

ENVIRONMENTAL DATA

Substance	Amount	TLV (mg/m³)	Emissions (mg)	Volume (m³)
Heptanal (GC, ref. index)	12.5 μL	—	12.5	
Cyclohexanol (GC, ref. index)	12.5 μL	200	11.3	0.1

QUESTIONS

5-7. Based on the data presented in the Series A chromatographic separation, can you explain why there is such a steep decline in column efficiency with temperature change?

5-8. Given the following gas chromatogram for a mixture of analytes X and Y:

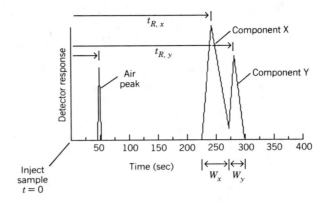

a. Calculate the number of theoretical plates for the column in reference to the peaks of each component (X and Y).

b. If the column is 12 ft long, calculate the HEPT for this column.

5-9. The number of theoretical plates a column has is important, but the crucial factor is the ability to separate two or more substances. That is, how well resolved are the peaks? The resolution of two peaks depends not only on how far apart they are (t_R) but also on the peak width (W).

Baseline resolution (R) is defined by the following equation:

$$R = \frac{2\Delta t_R}{(W_x + W_y)}$$

Because of tailing of most species on the column, a value of 1.5 is required to give baseline resolution.

a. Calculate the resolution for the peaks in question 5-8.

b. Do you think a quantitative separation of the mixture is possible based on your answer?

c. Has baseline resolution been achieved?

5-10. Discuss at least two techniques you might employ to increase the resolution of the column in question 5-9 without changing the column.

5-11. Retention times for several organic compounds separated on a gas chromatographic column are given below.

Compound	t_R (sec)
Air	75
Pentane	190
Heptane	350
2-Pentene	275

a. Calculate the relative retention of 2-pentene with respect to pentane.

b. Calculate the relative retention of heptane with respect to pentane.

TECHNIQUE 3:
SOLVENT EXTRACTION

Solvent extraction is a technique frequently used in the organic laboratory to separate or isolate a desired species from a mixture of compounds or from impurities. It is used extensively in the experiments given in this text. Solvent extraction methods are readily adapted to microscale work, since small quantities are easily manipulated in solution. This method is based on the solubility characteristics of the organic substances involved in relation to the solvents used in a particular separation procedure.

Solubility

Substances vary greatly in their solubilities in various solvents. One of the most useful approaches to aid the chemist in the prediction of the solubility of a substance in a given solvent is that *a substance tends to dissolve in a solvent that is chemically similar.* In other words, the rule is, *like dissolves like.*

For example, solubility in water requires that a species have some characteristics of water. An important class of compounds, the organic alcohols, have the hydroxyl group —O—H bonded to a hydrocarbon chain or framework. The hydroxyl group is polar, due to the difference in electronegativity of the hydrogen and oxygen atoms. In other words, the bond has *partial ionic character.*

$$\overset{\delta-}{\underset{\cdot\cdot}{\overset{\cdot\cdot}{O}}}\text{—}\overset{\delta+}{H}$$

Partial ionic character
of the hydroxyl group

This *polar* or partial ionic character leads to relatively strong hydrogen bond formation between molecules having this entity. Strong hydrogen bonding is also evident in molecules that contain groups having a hydrogen atom attached to an oxygen, nitrogen, or halogen atom. This is diagrammed below for the water system. This polar nature of a functional group is present in those compounds that have sufficient electronegativity difference between the atoms making up the group.

$$CH_3\text{—}CH_2\text{—}\overset{\delta-}{\underset{\underset{H}{\overset{|}{\diagdown}}}{O:}}\quad\quad\quad CH_3\text{—}CH_2$$

Ethanol   Hydrogen bond formation

In ethanol it is apparent that the hydroxyl end of the molecule is very similar to that of water. Therefore, when ethanol is added to water they are miscible in all proportions. This is because the attractive forces set up between the two molecules are nearly as strong as in water itself; however, the attraction is somewhat weakened by the presence of the nonpolar hydrocarbon unit, CH_3CH_2—. Hydrocarbon substrates attract each other only weakly as evidenced by their low melting and boiling points. Three examples of the contrast in boiling points between compounds of different structure, but similar molecular weights, are summarized in Table 5.1. Clearly, the nonassociative species have much lower boiling points.

If a comparison of the solubility of ethanol and a higher member of the alcohol series, octanol, in water is made, octanol is found to have less than 2% solubility

Table 5.1 Comparison of Boiling Point Data

Name	Formula	MW	bp(°C)
Ethanol	CH_3CH_2OH	46	78.3
Propane	$CH_3CH_2CH_3$	44	−42.2
Methyl acetate	CH_3COOCH_3	74	54
Diethyl ether	$(CH_3CH_2)_2O$	74	34.6
Ethylene	$CH_2{=}CH_2$	28	−102
Methyl amine	CH_3NH_2	31	−6

while ethanol dissolves completely. Why the difference? It lies in the fact that the *dominant* structural feature in octanol is the *nonpolar* hydrocarbon unit.

$$CH_3-CH_2-CH_2-CH_2-CH_2-CH_2-CH_2-CH_2-\overset{\delta-}{\overset{..}{\underset{\overset{\delta+}{\diagdown}}{O}}} \quad\quad\quad CH_3-CH_2-\overset{..}{\underset{..}{O}}-CH_2-CH_3$$
$$\quad\quad\quad\quad\quad\quad\quad\quad\quad\quad\quad\quad\quad\quad\quad\quad\quad\quad\quad\quad H$$

Octanol   Diethyl ether

The attraction of the polar hydroxyl group in the alcohol for the water is not sufficiently strong to overcome the hydrophobic character of the nonpolar hydrocarbon unit. On the other hand, since octanol has such a large nonpolar group as its dominant structural feature, we would expect enhanced solubility in nonpolar solvents. Such is the case. This is demonstrated in Experiment 13, where octanol is prepared and isolated by an extraction procedure using diethyl ether, a common, relatively nonpolar solvent. It is found that the octanol is completely miscible with diethyl ether since the nonpolar characteristics predominate in both molecules.

Experimentation has demonstrated that, in general, if a compound has both polar and nonpolar groups present in its structure, those having five or *more carbon atoms* in the hydrocarbon portion of the molecule will be more soluble in nonpolar solvents such as diethyl ether, pentane, or methylene chloride. In Fig. 5.13 the solubility of a number of straight-chain alcohols, carboxylic acids, and hydrocarbons in water are summarized. Those compounds having more than five carbon atoms have solubilities similar to the hydrocarbons.

Several general observations should be made in relation to the above discussion.

1. Branched-chain compounds have greater water solubility than their straight-chain counterparts. This fact is illustrated in Table 5.2 for a series of alcohols.
2. The presence of more than one polar unit in a compound increases solubility in water. For example, high-molecular-weight sugars, which contain multiple hydroxyl, acetal, and ketal units, are water soluble and ether insoluble. On the other hand, cholesterol, having only one hydroxyl unit, is water insoluble, ether soluble.

Maltose   Cholesterol

3. The presence of a chloro group, even though it lends some partial ionic character to the C—Cl bond, does not impart water solubility to a compound. In fact, compounds such as methylene chloride, chloroform, and carbon tetrachloride have long been used as extraction solvents. It should be noted that use of the latter two solvents has been curtailed, unless strict safety precautions are exercised, because of their carcinogenic nature.

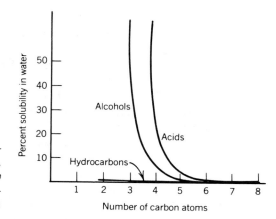

Fig. 5.13 *Solubility curve of acids, alcohols, and hydrocarbons. From Oliver Kamm, Qualitative Organic Analysis, 2nd ed. (New York: Wiley, 1932). (Reprinted by permission of John Wiley & Sons, New York, NY.)*

Table 5.2   Water Solubility of Alcohols[a]

Name	Formula	Solubility (g/100 g H_2O)
1-Pentanol	$CH_3(CH_2)_3CH_2OH$	4.0
2-Pentanol	$CH_3(CH_2)_2CH(OH)CH_3$	4.9
2-Methyl-2-butanol	$(CH_3)_2C(OH)CH_2CH_3$	12.5

[a]Data at 20°C.

4. Any functional group capable of hydrogen bond formation with water, if it constitutes the *dominant* structural feature of the system, lends water-soluble characteristics to the substance (do not forget the five-carbon rule). For example, amine derivatives of the hydrocarbons (organic relatives of ammonia) would be expected to have water solubility. This is found to be the case. In Table 5.3 the water-solubility data for a series of amines are summarized.

Table 5.3   Water Solubility of Amines[a]

Name	Formula	Solubility (g/100 g H_2O)
Ethylamine	$CH_3CH_2NH_2$	∞
Diethylamine	$(CH_3CH_2)_2NH$	∞
Trimethylamine	$(CH_3)_3N$	91
Triethylamine	$(CH_3CH_2)_3N$	14
Aniline	$C_6H_5-NH_2$	3.7
p-Phenylenediamine	$H_2N-C_6H_4-NH_2$	3.8

[a]Data at 25°C.

Partition Coefficient

A given substance, if placed in a mixture of two immiscible solvents, will distribute (*partition*) itself in a manner that is a function of its solubility between the two solvents. For example, a solute X will be distributed between two immiscible solvents according to the following equation.

$$X_{(solvent\ 1)} \rightleftharpoons X_{(solvent\ 2)}$$

$$K = \frac{[X_{(solvent\ 2)}]}{[X_{(solvent\ 1)}]}$$

The *ratio* of the concentration of the species in each solvent is a constant K for a given system at a given temperature. This equilibrium constant expression is designated the *partition coefficient*. This coefficient is similar to the partitioning of a species that occurs in chromatographic separations. The procedure for determining the partition coefficient is demonstrated in the accompanying experiment. The partition coefficient K for benzoic acid is determined for (1) a methylene chloride-water mixture and (2) a methylene chloride-5% sodium bicarbonate mixture.

The basic equation used to express the coefficient K is

$$K = \frac{[\text{g/100 mL] } organic\ layer}{[\text{g/100 mL] } water\ layer}$$

The above expression uses grams per 100 milliliter or grams per deciliter (g/dL), but grams per liter (g/L), parts per million (ppm), or molarity are also valid. The partition coefficient is dimensionless so that any concentration units may be used, provided the units are the same for both phases. If equal volumes of both solvents are used, the equation reduces to the ratio of the weights of the given species in the two solvents.

$$K = \frac{[\text{g] } organic\ layer}{[\text{g] } water\ layer}$$

Determination of the partition coefficient for a particular compound in various immiscible solvent combinations can often give valuable information to aid in the isolation and purification of the species using extraction techniques.

Extraction

The two major types of extractions utilized in the organic laboratory are the (1) solid–liquid and (2) liquid–liquid methods.

Solid–Liquid Extraction

This type of extraction is often used in natural product studies. The extraction of usnic acid from its native lichen using acetone is described in Experiment 10. This approach is useful and simple, since only one main chemical species is soluble in the solvent. The extraction of caffeine from tea (Experiment 11) is accomplished by heating the tea in an aqueous solution of sodium carbonate. These methods work well because of the solubility characteristics of the compounds involved.

It is also possible to carry out the extraction of a solid on a continuous basis at the microscale level. This approach depends upon the evaporation of the solvent, followed by its condensation and slow filtration through the solid sample. In this manner the desired species is collected in the distillation pot containing the solvent. This process is repeated over and over. The extractant solution is then concentrated to isolate the desired compound. A device for accomplishing such continuous extraction is shown in Fig. 5.14.

This apparatus, developed by Garner, consists of a small cold finger condenser inserted into a test tube. The test tube has indentations near the bottom to support a small funnel. The sample is carefully wrapped in filter paper and placed in the funnel.[3]

Liquid–Liquid Extraction

This is the more common type of extraction procedure and is used extensively in the experiments presented in this text. It is a very powerful method for the

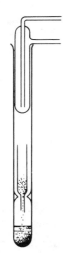

Fig. 5.14 *Apparatus for continuous extraction of a solid. From Frank L. Schneider,* Qualitative Organic Microanalysis *(vol. 2), A. A. Benedetti, ed. (Wein, Austria: Springer-Verlag). (Courtesy of Springer-Verlag, Wien, Austria.)*

[3]Schneider, F. L. "Qualitative Organic Microanalysis"; Vol. II of Monographien aus dem Gebiete der Qualitativen Mikroanalyse, Benedetti-Pichler, Ed.; Springer-Verlag: Wien, Austria, 1964; p 60.

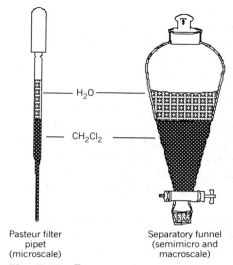

Fig. 5.15 *Extraction devices.*

Pasteur filter pipet (microscale)

Separatory funnel (semimicro and macroscale)

separation and isolation of materials encountered at the microscale level.

In the majority of extractions, a capped centrifuge tube, a 10 mm × 75 mm test tube, or a conical vial is used as the container. In any extraction technique employed, it is essential that complete mixing of the two immiscible solvents be realized.

A typical extraction procedure is described in Experiment 46. Benzanilide, formed in an aqueous-acid solution, is separated from the product mixture by extraction with three 1.0-mL portions of methylene chloride solvent.

Note. *This wording is the accepted manner of indicating that three extractions are performed, each using 1.0 mL of methylene chloride.*

At the microscale level the extraction process consists of two parts: (1) *mixing* of the two immiscible solutions and (2) *separation* of the two layers after the mixing process.

1. *Mixing.* As outlined in Experiment 46, the methylene chloride solvent (1.0 mL) is added to the 5.0 mL conical vial containing the aqueous phase and product (~1.5 mL). The procedure is outlined in the following steps.

   a. The vial is *capped*.

   b. The vial is *shaken* to thoroughly mix the two phases.

   c. The vial is carefully *vented* by loosening the cap to release any pressure that may develop.

   d. The vial is allowed to *stand* on a level surface to allow the two phases to separate. A sharp boundary should be evident.

2. *Separation.* At the microscale level the two phases are separated with a Pasteur filter pipet (in some cases a Pasteur pipet can be used), which acts as a miniature separatory funnel. The separation of the phases is shown in Fig. 5.15.

A major difference between macro and micro techniques is that at the microscale level the mixing and separation are done in two steps, whereas at the macroscale level with the separatory funnel, the mixing and separation are both done in the funnel in one step. It is important to note that the separatory funnel is an effective device for extractions at the semi-micro- or macroscale, but at the microlevel it is *not* practical to use with the small quantities involved. The recommended procedure is diagrammed in Fig. 5.16.

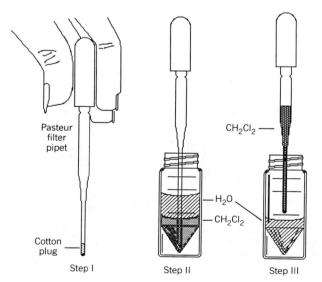

Fig. 5.16 *Pasteur filter pipet separation of two immiscible liquid phases, with denser layer containing the product.*

Pasteur filter pipet

Cotton plug

Step I

Step II

Step III

Continuing the example, it is known that benzanilide is more soluble in methylene chloride than in water. Multiple extractions are performed to ensure complete removal of the benzanilide from the aqueous phase. The methylene chloride solution is the *lower* layer since it is heavier than water.

The following steps outline the general method (refer to Fig. 5.16).

1. The pipet bulb is squeezed to force air from the pipet.
2. The pipet is then inserted into the vial until close to the bottom. Be sure to hold the pipet in a vertical position.
3. *Carefully* allow the bulb to expand, drawing only the lower methylene chloride layer into the pipet. This is done in a smooth, steady manner so as not to disturb the boundary between the layers. With practice, one can judge the amount that the bulb must be squeezed so as to just separate the layers.
4. Holding the pipet in a vertical position, place it over an empty vial and gently squeeze the bulb to transfer the methylene chloride solution into the vial. A second extraction can now be performed after addition of another portion of methylene chloride to the original vial. The identical procedure is repeated. In this manner multiple extractions can be performed with each methylene chloride extract being transferred to the same vial; that is, the extracts are combined. The reaction product has now been transferred from the aqueous to the methylene chloride layer and the phases separated.

In the case when a diethyl ether-water extraction is performed, the ether layer is less dense and thus is the top phase. An example of this type of extraction is presented in Experiment 48D. Two extractions of the aqueous phase (~2.0 mL) using 0.5-mL portions of diethyl ether are carried out. The reaction product dissolves in the ether layer and is thus separated from by-products and other impurities. The procedure followed to separate the water-ether phases is identical to that outlined above with the *exception* that it is the *top layer* that is transferred to the new container. The steps are shown in Fig. 5.17.

1. *Both* phases are drawn into the pipet as outlined above (steps 1 and 2). Try not to allow air to be sucked into the pipet since this will tend to mix

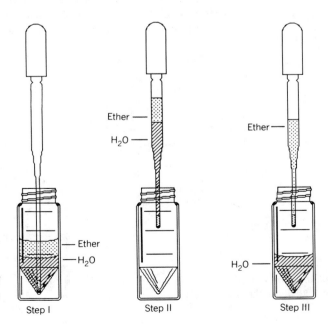

Fig. 5.17 *Pasteur filter pipet separation of two immiscible liquid phases, with less dense layer containing the product.*

Step I   Step II   Step III

the phases in the pipet. If mixing does occur, allow time for the boundary to reform.

2. The *bottom aqueous layer* is *returned* to the *original* container by gently squeezing the pipet bulb.

3. The separated ether layer is then transferred to a new vial.

It is important to realize that the organic extracts separated above (methylene chloride or diethyl ether) are wet. Before evaporation of the solvent or before further purification steps are taken, the extracts are dried to remove any residual water. This is achieved conveniently with an inorganic anhydrous salt such as sodium or magnesium sulfate. These materials readily form insoluble hydrates, thus removing the water from the wet organic phase. There are two basic requirements for an effective solid drying agent: (1) it should not react with the organic material in the system and (2) it can be easily and completely separated from the dried liquid phase. The amount of drying agent used depends on the amount of water present and on the capacity of the solid desiccant to absorb water. As a general rule, use an amount of the drying agent to just cover the bottom of the container. Additional agent can be added if necessary. Swirling the contents of the container increases the rate of drying since it aids in establishment of the equilibrium for hydration.

The drying agent can be added directly to the vial containing the organic extract (Experiment 48D) or the extract can be passed through a Pasteur filter pipet packed with the drying agent (Experiment 48B). A funnel fitted with a cotton, glasswool, or polyester plug to hold the drying agent can also be used (Experiment 47A).

The separation of organic acids or bases constitutes another important area where the extraction method has utility. An organic acid *reacts* with dilute aqueous sodium hydroxide to form a salt. The salt, having an ionic charge, dissolves and passes into the water phase. The reaction changes the solubility characteristics of a water-insoluble acid. The water phase can then be extracted with an organic solvent to remove any impurities, leaving the relatively pure acid-salt in the water phase. Neutralization of the soluble salt with hydrochloric acid causes precipitation of the insoluble acid in a relatively pure state. An example of this technique is demonstrated in Experiments 7A and 7B.

In a similar fashion, organic bases such as amines can be rendered completely water soluble by treatment with dilute hydrochloric acid to form hydrochloride salts. This is used to advantage in purifying the starting reagents in Experiments 26A and 26B.

It was emphasized in the solubility section that organic compounds with fewer than five carbon atoms and also one or more polar groups tend to be water soluble. Transfer of this type of compound into an organic phase by extraction from water is difficult. A technique often used is addition of a salt such as sodium chloride to the water phase. As a result, the attraction of the water molecules for the inorganic ions is much stronger than toward the organic species. The organic compound is then free to be extracted into an organic solvent. This approach is often referred to as *salting out*.

Experiment 3
Determination of a Partition Coefficient: Benzoic Acid

This experimental exercise illustrates the procedure that is used to determine a partition coefficient. Experience in weighing small quantities on an electronic balance, the use of automatic delivery pipets, and the transfer of solutions with the Pasteur pipet are techniques encountered in this experiment.

Physical Properties of Reactants and Products Parts A and B

Compound	MW	Wt/Vol	mmol	mp(°C)	bp(°C)	Density
Benzoic acid	122.13	100 mg	0.82	122		
Methylene chloride		2.4 mL			40	1.33
Water		600 μL			100	1.00
Sodium bicarbonate (5% solution)		600 μL				

PART A: Benzoic Acid—Methylene Chloride and Water

To a 3.0-mL conical vial fitted with a screwcap, add 50 mg (0.41 mmol) of benzoic acid. There is then added 600 μL of methylene chloride followed by 600 μL of water.

These solvents are delivered by aid of automatic delivery pipets. The methylene chloride is dispensed in the hood.

The mixture is shaken until the benzoic acid dissolves and the two layers are allowed to separate after being thoroughly mixed.

Draw the lower layer into a Pasteur filter pipet and carefully transfer the methylene chloride solution to a vial containing 100 mg of anhydrous, granular sodium sulfate and then cap the vial.

If the methylene chloride layer is not transferred totally in the first operation, a second transfer is done.

After drying for a few minutes, the methylene chloride solution is transferred to a previously tared vial using a Pasteur filter pipet. The sodium sulfate is rinsed with an additional 600 μL of methylene chloride and the rinse combined with the solution in the tared vial. The solvent is now evaporated under a gentle stream of nitrogen gas or in a sand bath in the **HOOD**.

HOOD

If a sand bath is used, a boiling stone is placed in the vial before it is tared.

Weigh the vial and determine the weight of benzoic acid in the methylene chloride layer and in the water layer.

For the methylene chloride layer, the weight of the benzoic acid and vial minus the weight of the vial equals the weight of the benzoic acid.

For the water layer, the original weight of benzoic acid minus the amount of benzoic acid in the methylene chloride layer equals the weight of the benzoic acid.

Since equal volumes of both solvents were used, the partition coefficient may be determined from the ratio of the weight of benzoic acid in the methylene chloride solvent to the weight of benzoic acid in the water layer.

Calculate the partition coefficient for benzoic acid in the solvent pair (equal volumes) used in this exercise.

PART B: Benzoic Acid—Methylene Chloride and 10% Sodium Bicarbonate Solution: An Example of Acid-Base Extraction Techniques

Benzoic acid reacts readily with sodium bicarbonate to form sodium benzoate, carbon dioxide and water. The sodium derivative has salt-like characteristics. It is very soluble in water and nearly insoluble in methylene chloride. The efficiency of the basic aqueous extraction procedure can be measured by recovering any unreacted benzoic acid from the organic layer. Obtain the melting point of any recovered residue. Sodium benzoate has a melting point above 300°C, whereas benzoic acid melts near 122°C.

QUESTIONS

5-12. Explain why diethyl ether would be expected to be a satisfactory solvent for the straight-chain hydrocarbons hexane and heptane.

5-13. The solubility of *p*-dibromobenzene, BrC_6H_4Br, in benzene is 80 $\mu g/100$ μL at 25°C. Would you predict the solubility of the dibromo compound to increase, decrease, or remain the same in 100 μL of acetone solvent at this temperature? Explain.

5-14. Each of the solvents listed below is used in the experiments of this text to extract organic compounds from aqueous solutions.

   a. Methylene chloride

   b. Pentane

   c. Toluene

   d. Diethyl ether

Will the organic phase be the upper or lower layer when each of these organic solvents is mixed with water?

5-15. A 25-mg sample of an organic compound (MW 84) is dissolved in 20 mL of water. This aqueous solution is extracted with 5.0 mL of hexane. Separation and analysis of the aqueous phase shows that it now contains only 6.3 mg of the organic compound. Calculate the partition coefficient for the compound.

5-16. A qualitative test often used to determine whether an organic compound contains oxygen is to test its solubility in concentrated sulfuric acid. Almost all oxygen-containing compounds are soluble in this acid. Explain.

5-17. Which of the two solvents, water or 10% sodium bicarbonate solution, is preferred for extracting benzoic acid from methylene chloride solution? Explain, using the data obtained in the exercise.

TECHNIQUE 4: INTRODUCTION TO INFRARED SPECTROSCOPY

The infrared spectrometer is the most complex and expensive instrument that you will encounter on a regular basis in the microscale laboratory. TREAT THE IR SPECTROMETER WITH RESPECT. You are being introduced to this sophisticated piece of equipment early in the first semester because it is particularly adapted to the characterization of microscale products. Data obtained from the IR spectrometer will be used many times throughout the year. You will become more proficient at obtaining this type of spectral information as you gain experience with the instrument. Practice in preparing the sample for instrumental analysis can significantly improve the quality of the spectra.

The late Robert B. Woodward, Nobel Laureate, one of the most outstanding synthetic organic chemists of this century, once stated: "But no single tool has had more dramatic impact upon organic chemistry than infrared measurements. The development . . . [of easily] . . . operated machines for the determination of infrared spectra has permitted a degree of immediate and continuous analytical and structural control in synthetic organic work which was literally unimaginable. . . . The power of the method grows with each day, and further progress may be expected for a long time to come. Nonetheless, its potentialities are even now greater than many realize. . . ."[4]

That statement was made some time ago, and while the dramatic impact has shifted to high-field NMR and high-resolution mass spectrometry, infrared data still play the dominant role in compound identification. The infrared spectrum is a reflection of the vibrational energy levels present in a molecule. As no two substances will have the same set of vibrational frequencies, no two substances will have identical infrared spectra. There are at present a number of excellent collections of infrared spectra that may be used for reference comparisons.[5]

The use of infrared spectra to identify the presence of particular functional groups in a molecule is one of the applications of this technique. Many of the absorption bands present in an infrared spectrum can be related to specific arrays of atoms in that material. These "group frequencies" can be extremely helpful in the interpretation of experimental results (see Experiment 4A).

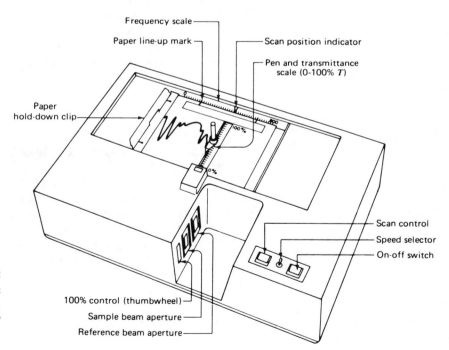

Fig. 5.18 *The Perkin-Elmer Model 710B IR. From James W. Zubrick,* The Organic Chem Lab Survival Manual *(New York: Wiley, 1984). (Reprinted by permission of John Wiley & Sons, New York, NY.)*

[4]Woodward, R.B. in "Perspectives in Organic Chemistry," Todd, A., Ed.; Interscience: New York, 1956; p 157.

[5]For example, see References, p. 71.

An introduction to the theory of this effect is given in Appendix B and the interpretation of infrared spectroscopic data is developed in some detail in Chapter 7.

Instrumentation

The workhorse "low-cost" infrared instrument used for routine characterization of materials in the undergraduate organic laboratory is the optical null, double-beam grating spectrometer (Fig. 5.18).

This instrument will rapidly (3–5 minutes) determine and record the infrared absorption spectrum of a sample from 4000 to 600 cm^{-1}. The system (Fig. 5.19) operates as follows: A source of infrared radiation is generated by a hot wire. The source output is divided into the two parallel and converging beams by the two spherical mirrors, m_1 and m_2. (Mirrors are used throughout infrared instruments instead of lenses, as glass or quartz absorb in this region of the spectrum.)

The beam reflected from m_1 is the sample beam, and the beam from m_2 is the reference beam. The two beams then pass through the sample compartment of the instrument with the beam from mirror m_1 also passing through the sample to be analyzed. Most substances to be examined are located near the convergence point of the sample beam. At this position the instrument requires the minimum quantity of sample. The transmitted beam undergoes a certain amount of instrument distortion, but generally follows the Beer-Lambert absorption law:

$$I/I_0 = e^{-acl}$$

where I = intensity of the transmitted beam
I_0 = intensity of incident beam
a = apparent absorption coefficient
c = sample concentration, mol/L
l = path length (sample thickness)

The two beams are then combined at the chopper C. The chopper is a rotating sector mirror (either half or quarter), which alternately passes the sample beam or reflects the reference beam to the entrance slit. The combined beams then follow the same path, but alternate with time. The frequency of alternation equals

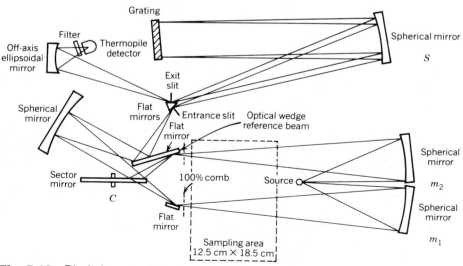

Fig. 5.19 *Block diagram of a typical IR spectrometer, showing the light source, beam chopper, diffraction grating, and attenuator wedge. Courtesy of Perkin-Elmer, Norwalk, CT.*

the frequency of chopping for the half-sector chopper. The chopping frequency generally is in the 10–15 Hz range. The beam then passes through the entrance slit (the slit forms a long and very narrow vertical image) and into that part of the instrument called the monochromator (the term literally means "making single colors"). In this section of the instrument the beam of radiation is dispersed. Thus, this type of spectrometer is often referred to as a dispersive spectrometer. That is, the different frequencies of electromagnetic radiation are separated from each other (or at least partially separated) in space. The dispersing agent in all modern instruments is a reflection grating. The beam enters the monochromator with the narrow image of the slit and is collected on a parabolic mirror or large spherical mirror S, which collimates the beam (makes all the beam rays parallel, just as a searchlight mirror converts all the light directed outward into parallel rays).

The collimated beam then strikes the reflection grating and is dispersed by diffraction. For one angle of the grating, a particular wavelength of radiation will be diffracted back onto the collimating mirror and out the exit slit. The grating is mounted so that it can be rotated. As the grating angle changes with respect to the collimating mirror, the dispersed radiation is sequentially swept by the exit slit, one wavelength (or frequency) after another. After exiting the monochromator, the dispersed radiation is tightly focused by an ellipsoidal mirror onto a small thermocouple detector. The thermocouple is evacuated and has a blackened junction that heats up as the infrared radiation strikes the black detector target.

When no sample is in either beam, the intensity of radiation reaching the detector and the temperature of the detector remain constant as the two beams alternately strike the detector. When a substance that absorbs radiation at a particular frequency is placed in the sample beam, that beam will be less intense than the reference beam at the frequency where absorption is taking place. When the radiant energy decreases, the detector will cool down and change the voltage potential across the junction. (The chopping frequency is dictated by the ability of the detector to heat up and cool down, or to "track" the chopping frequency.) When the reference beam is switched back on, the detector temperature goes up again. The detector is connected to an amplifier that will respond only to an alternating current identical in phase and frequency to the chopper rotation.

The amplifier drives a servomotor that is mechanically connected to the pen drive on the spectrum chart. The pen drive is also connected to an "optical wedge" that is simultaneously driven into the reference beam. The wedge reduces the intensity of the reference beam. The wedge is shaped so that the linear travel across the beam corresponds to the percentage transmission recorded by the pen. That is, if the optical wedge blocks 10% of the reference beam, the pen reads 90% transmission on the chart. The servomotor drives the pen as long as it detects a signal. When the optical wedge is driven into the reference beam to the point where the two beams are of equal intensity, the signal at the detector drops to zero and the pen stops. As the grating continues to turn and the absorption of the sample diminishes, a new signal develops at the detector. This signal will be of the opposite phase since the reference beam is now less intense than the sample beam. The signal is amplified and a phase-sensitive recorder motor drives the pen in the opposite direction. At the same time the wedge is pulled out of the reference beam. Thus, the instrument has recorded an absorption band in the sample over that particular set of frequencies. The servo system functions to drive the optical wedge to zero out or "null" the signal at the detector, thus the name optical null spectrometer.

The double-beam arrangement has some distinct advantages. For example, absorption signals from any substances present in both beams in equal quantities

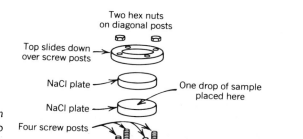

Fig. 5.20 *IR salt plates and holder. From James W. Zubrick, The Organic Chem Lab Survival Manual (New York: Wiley, 1984). (Reprinted by permission of John Wiley & Sons, New York, NY.)*

are automatically canceled. Since atmospheric water and carbon dioxide have strong absorption bands in the infrared, the double-beam instrument automatically subtracts the absorption of these substances from the recorded spectrum.

Sample-Handling in the Infrared

In order to obtain a spectrum in the infrared region, the sample must be mounted in a cell that is transparent to the radiation. Glass and quartz absorb in this region and cells constructed of these materials cannot be employed. Alkali metal halides have large regions of transmission in the infrared as do the silver halides. Of these materials, sodium chloride, potassium bromide, and silver chloride are most often used as cell windows in infrared sampling.

Liquid Samples

For materials boiling over 100°C, the procedure is very simple. Using a syringe or Pasteur pipet, place 3–5 μL of sample on a polished plate of NaCl or AgCl. Then cover with a second plate of the same material and clamp in a holder that can be mounted vertically in the instrument. *Be sure the plates are clean when you start and when you are through!* Obviously the sodium chloride cannot be cleaned with water. Silver chloride is very soft and scratches easily; it also must be kept in the dark when not in use because it will darken quickly in direct light. Spectra obtained in this fashion are referred to as capillary film spectra (see Fig. 5.20).

Solution Spectra and the Spectra of Materials Boiling Under 100°C

These samples generally require a sealed cell constructed of either NaCl or KBr. Such cells are expensive and need careful handling and maintenance. They are assembled as shown in Fig. 5.21. Only in rare cases do microlaboratory experiments require the use of sealed infrared cells.

Solid Samples

Solid powders can be mounted on horizontal NaCl plates, and the beam diverted through the sample by mirrors. This would make sample preparation very easy for solids. Unfortunately, powders tend to scatter the entering radiation very efficiently by reflection, refraction, and molecular scattering. Some of these effects become rapidly magnified at higher frequencies since they vary as the fourth power of the frequency. The result of solid-sample scattering is that large amounts of energy are removed from the sample beam. This results in very poor absorption spectra, as the instrument is forced to operate at very low energies. Remember that if there is a large reduction in energy in the sample

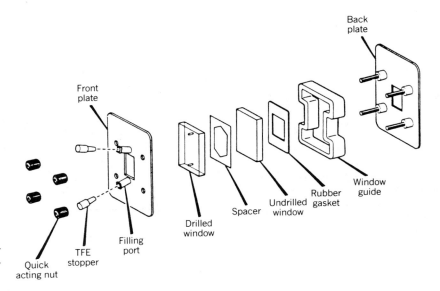

Fig. 5.21 *Sealed demountable cell or demountable cell with ports. (Courtesy of Perkin-Elmer, Norwalk, CT.)*

beam, the optical wedge is driven a large distance into the reference beam. This results in very small amounts of energy reaching the detector from either beam. The detector cannot differentiate between a drop in energy from absorption or from scattering.

The sampling routine most often employed to avoid these problems is the potassium bromide disk. The sample (2–3 mg) is finely ground in a mortar, the finer the better (lower reflection or refraction losses). Then 150 mg of previously ground and dried KBr is added to the mortar and quickly mixed by stirring, *not* grinding, with the sample. (KBr is very hygroscopic and will rapidly pick up water while being ground in an open mortar). When mixing is complete the mixture is transferred to a die and pressed into a solid disk. Potassium bromide will flow under high pressure and seal the solid sample in the alkali metal halide matrix. Potassium bromide is transparent to infrared radiation in the region of interest. Most important, however, the KBr makes a much better match of the refractive indices between the sample and its matrix then does air. Thus, reflection and refraction effects at the crystal faces of the sample are greatly suppressed. Several styles of dies are commercially available. For routine student use a die consisting of two stainless steel bolts and a barrel is the simplest to operate. The ends of the bolts are polished flat to form the die faces. The first bolt is seated to within a turn or two of the head. Then the sample mixture is added (avoid breathing over the die while adding the sample). The second bolt is firmly seated in the barrel, and then the clamped assembly is tightened by a torque wrench to 20 ft-lb. After standing for 1.5 minutes, the two bolts are removed, leaving the KBr disk mounted in the center of the barrel, which can then be mounted in the instrument. After running the spectrum of the sample, the disk can be retrieved and the sample recovered if necessary (Fig. 5.22). *Always clean the die immediately after use. KBr is highly corrosive to steel.*

For a discussion of an alternate matrix material for use with solid samples, see Question 7-22, in Chapter 7.

The standard techniques of sample preparation to obtain infrared spectra of microscale laboratory products are use of capillary films of liquids on NaCl or AgCl plates and use of KBr disks of solids.

When obtaining infrared spectra it is important to establish that the wavenumber values have been accurately recorded. Successful interpretation of the data often depends on very small shifts in these values. Calibration of the frequency scale is usually accomplished by obtaining the spectrum of a reference

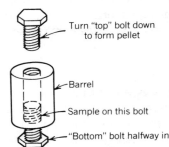

Fig. 5.22 *The KBr pellet minipress. From James W. Zubrick, The Organic Chem Lab Survival Manual (New York: Wiley, 1984). (Reprinted by permission of John Wiley & Sons, New York, NY.)*

compound such as polystyrene film. To save time, it is necessary only to record absorption peaks in the region of particular interest. As these instruments are quite stable, determination of the calibration curve once a day is usually a satisfactory procedure.

REFERENCES
1. Sadtler Library. About 80,000 spectra of single compounds; about 12,000 spectra of commercial products. Sadtler Research Labs., 3316 Spring Garden Street, Philadelphia, PA 19014.
2. D.M.S. System (Documentation of Molecular Spectra). About 15,000 spectra. IFI/Plenum Data Corp., 227 West 17th Street, New York, NY 10011.
3. Japanese collection. About 17,000 spectra. Good quality, but labeled in Japanese. Infrared Data Committee of Japan. Nankodo Co., Haruki-Cho, Tokyo.
4. Coblentz Society. 10,000 spectra. Marketed through Sadtler Research Labs., 3316 Spring Garden Street, Philadelphia, PA 19104.
5. A. P. I. collection (American Petroleum Institute). About 4500 spectra. M. C. A. collection (Manufacturing Chemists' Association). About 3000 spectra. Chemical Thermodynamics Property Center, Texas A & M College, Department of Chemistry, College Station, Texas 77843.
6. Aldrich Library of Infrared Spectra, Aldrich Chemical Co., Inc., 940 W. Saint Paul Ave., Milwaukee, WI 53233. 3rd ed., 1981. 10,000 spectra arranged by chemical type.
7. J. G. Grasselli "Atlas of Spectral Data and Physical Constants for Organic Compounds"; CRC Press: Boca Raton, FL; 1973.

Experiments 4A and 4B
Reduction of Ketones Using a Metal Hydride Reagent: Cyclohexanol; *cis*- and *trans*-4-*tert*-Butylcyclohexanol

(cyclohexanol; cyclohexanol, 4-*tert*-butyl-)

These reactions illustrate the reduction of a ketone carbonyl to the corresponding alcohol by use of sodium borohydride. The *cis*- and *trans*- isomers formed in the reduction of the substituted cyclohexanone are separated by gas chromatography.

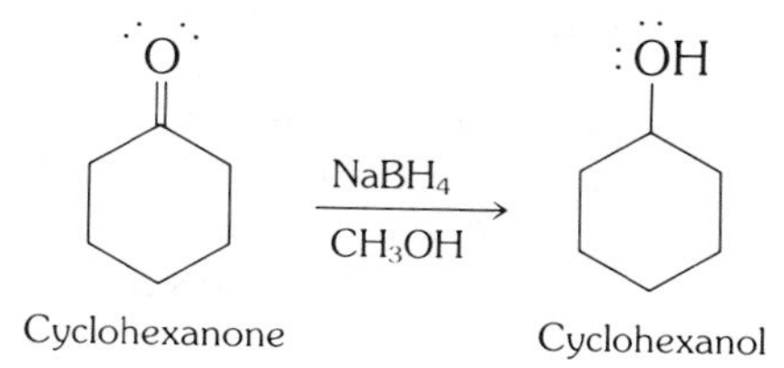

Cyclohexanone $\xrightarrow[\text{CH}_3\text{OH}]{\text{NaBH}_4}$ Cyclohexanol

DISCUSSION An important route for the synthesis of primary and secondary alcohols is the reduction of aldehydes and ketones, respectively. Reduction involves the addition of the elements of H—H across the carbonyl functional group.

There are numerous ways to accomplish the process, but the most common laboratory method involves the use of complex metal hydride reagents. The two commercially available reagents most often used are lithium aluminum hydride, $LiAlH_4$, and sodium borohydride, $NaBH_4$.

Lithium aluminum hydride is a very active reducing agent that reduces not only aldehydes and ketones but many other functional groups as well, such as esters, lactones, carboxylic acids and anhydrides, amides, alkyl halides, alkyl azides, alkyl isocyanates, and nitriles. Of great importance is the fact that it *must* be used in aprotic solvents such as diethyl ether or tetrahydrofuran. Lithium aluminum hydride reacts *violently* with water. It is not uncommon for the hydrogen gas generated in the reaction to ignite. *This reagent should not be employed unless specific instructions are available for its proper use.*

Sodium borohydride is a much more selective reducing reagent, and it is for this reason usually employed for the reduction of aldehydes and ketones. It does not react with the vast majority of common organic functional groups including C=C, C≡C, nitro, cyano, amide, and carboxylic acids and esters. Sodium borohydride reacts at an appreciable rate with water but slowly with aqueous alkaline solution or with methanol. For small-scale reactions an excess of the sodium borohydride reagent is generally used to compensate for that amount that reacts with the protic solvent (methanol). This technique is preferred to that of using a solvent in which the sodium borohydride is less soluble. On the other hand, sodium borohydride reacts rapidly with strong acid to generate hydrogen gas. This reaction can be used to advantage for the reduction of C=C bonds (see Experiment 12). The relatively high cost of the metal hydride reducing agents is offset by their low molecular weight and the fact that one mole of reducing agent reduces four moles of aldehyde or ketone.

The key step in the reduction of a carbonyl group by sodium borohydride is the transfer of a hydride ion, $:H^-$, from boron to the carbon atom of the carbonyl unit. In the reaction the hydride ion is acting as a *nucleophile*.

(The hydride may make an **e** or **a** attack depending on steric factors.)

The overall reduction process requires two hydrogen atoms, but only one comes from the borohydride reagent (attached to the carbon atom). The other hydrogen atom is derived from the protic solvent (methanol).

On reduction of an acyclic aldehyde or ketone, the attack of the hydride species occurs with equal probability from either side of the planar carbonyl unit.

In the 4-*tert*-butylcyclohexanone example, the steric environment is similar on each side of the carbonyl unit and thus the more stable equatorial alcohol (trans) is formed in larger amount. This reduction occurs by *axial* attack of the hydride species on the carbonyl group, which is possible with the relatively small sodium borohydride (or lithium aluminum hydride) reagent. Reduction of this compound with sodium borohydride yields a mixture of *cis* and *trans* isomers of the corresponding alcohol. This type of reaction wherein more than one stereoisomer product is formed, one of which predominates, is termed a *stereoselective* reaction.

Highly substituted hydride reducing reagents, such as lithium tris(sec-butyl)borohydride, must make an equatorial attack, due to steric factors, and thus produce the less stable isomer (cis) in greater amounts.

Table 5.4 Reduction of 4-*tert*-Butylcyclohexanone

Reagent	Trans %	Cis %
Sodium borohydride	80	20
Lithium aluminum hydride	92	8
Lithium tris(*sec*-butyl)-borohydride	7	93

Table 5.4 summarizes data relating to the stereochemistry of the reduction with metal hydride reagents.

PART A: Cyclohexanol

The reaction is shown above.

EXPERIMENTAL Estimated time for the experiment: 1.5 hours.

Physical Properties of Reactants and Products

Compound	MW	Wt/Vol	mmol	bp(°C)	Density	n_D
Cyclohexanone	98.15	100 μL	0.97	156	0.95	1.4507
Methanol		250 μL		65		
Sodium borohydride reducing solution		300 μL				
Cyclohexanol	100.16			161	0.96	1.4641

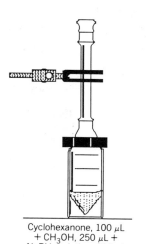

Cyclohexanone, 100 μL
+ CH$_3$OH, 250 μL +
NaBH$_4$ solution, 300 μL

Reagents and Equipment

In a 3.0-mL conical vial equipped with an air condenser, place 100 μL (95 mg, 0.97 mmol) of cyclohexanone followed by 250 μL of methanol.(■) The vial is gently shaken to obtain a homogeneous solution.

Add 300 μL of sodium borohydride reducing solution dropwise, with swirling.

HOOD [HOOD]

Note. *The reactants are dispensed using automatic delivery pipets. It is suggested that the vial be tared and the cyclohexanone be weighed after delivery so as to get an accurate weight for the percentage yield calculations.*

The stock-reducing solution should be prepared just prior to conducting the experiment.

Instructor Preparation

In a 10-mL Erlenmeyer flask is placed 50 mg of anhydrous sodium methoxide and 2.5 mL of methanol. To this solution is added 100 mg of sodium borohydride and the flask is tightly stoppered and swirled gently to dissolve the solid phase (100 μL of this solution provides approximately 2.0 mg NaOCH$_3$ and 4.0 mg NaBH$_4$).

Reaction Conditions

The resulting solution is allowed to stand at room temperature for a period of 10 minutes.

Isolation of Product

Using a calibrated Pasteur pipet, 1.0 mL of *cold* dilute hydrochloric acid is added dropwise and the aqueous mixture extracted with three 0.5-mL portions of methylene chloride. Upon each addition of methylene chloride, the vial is capped, shaken gently, and then carefully vented by loosening the cap. After separation of the layers, the bottom methylene chloride layer is removed using a Pasteur filter pipet.

Each methylene chloride extract is transferred to a Pasteur filter pipet containing 250 mg of anhydrous magnesium sulfate. The eluate is collected in a tared 5.0-mL conical vial containing a boiling stone. An additional 0.5 mL of methylene chloride is used to rinse the magnesium sulfate.(■) *Additional rinsings of the magnesium sulfate may be made if desired.* The methylene chloride is removed by careful evaporation in the hood by gentle warming in a sand bath. *Do not use a stream of nitrogen gas to hasten the evaporation. This results in loss of product.*

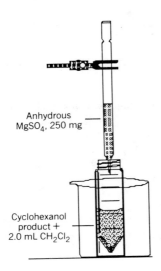

Anhydrous MgSO$_4$, 250 mg

Cyclohexanol product + 2.0 mL CH$_2$Cl$_2$

Purification and Characterization

The cyclohexanol product remaining after evaporation of the methylene chloride solvent is sufficiently pure for characterization.

Determine the weight of the liquid material and calculate the percentage yield.

Determine the refractive index and boiling point of the cyclohexanol and compare the results with the literature values.

Obtain an IR spectrum by capillary film technique, and compare it to an authentic sample. If the spectrum indicates the presence of small quantities of unreacted starting ketone (1715 cm^{-1}), purification can be achieved by preparative gas chromatography. Use the following conditions and refer to Experiment 2 for the collection technique.

Separation of Small Quantities of Cyclohexanone from Cyclohexanol

Example

9:1 (v/v) cyclohexanol/cyclohexanone.   10% carbowax 20M (stationary phase).

Injection volume: 15 μL.   Temperature: 130°C; He flow rate: 50 mL/minute.

Column: $\frac{1}{4}$ in. × 8 ft stainless steel.   Attenuator = 512; chart speed = 1 cm/minute.

	Cyclohexanol		Cyclohexanone
Run	Retention time (min)	Yield (mg)	Retention time (min)
1	15.3	6.5	11.6
2	17.3	7.3	12.5
3	17.2	8.5	12.6
4	16.0	9.6	12.0
5	14.6	7.3	11.2
6	14.5	8.7	11.2
7	15.5	8.9	11.7
8	15.5	8.9	11.8
9	16.4	8.8	12.3
10	15.4	8.4	12.7
Ave.	15.8 ± 1	8.3 ± 0.9	12.0 ± 0.6

Cyclohexanol injected = 0.9(15 μL)(0.963 mg/μL) = 13.0 mg

Percentage yield = 8.3/13.0 × 100 = 63.8%

PART B: *cis-* and *trans-4-tert-*Butylcyclohexanol

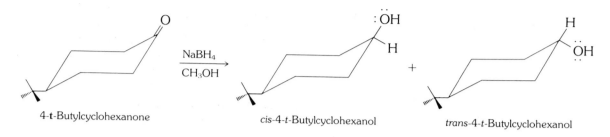

4-*t*-Butylcyclohexanone   cis-4-*t*-Butylcyclohexanol   *trans*-4-*t*-Butylcyclohexanol

EXPERIMENTAL Estimated time of the experiment: 2.0 hours. For the GC analysis, 15 minutes per student.

Physical Properties of Reactants and Products

Compound	MW	Wt/Vol	mmol	mp(°C)	bp(°C)
4-*t*-Butylcyclohexanone	154.25	50 mg	0.33	47–50	
Methanol		50 μL			
Sodium borohydride reducing solution		100 μL			65
4-*t*-Butylcyclohexanol (mixed isomers)	156.27			62–70	

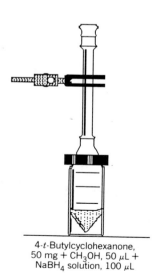

4-*t*-Butylcyclohexanone,
50 mg + CH₃OH, 50 μL +
NaBH₄ solution, 100 μL

Reagents and Equipment

A tared 3.0-mL conical vial equipped with an air condenser is charged with 50 mg (0.33 mmol) of 4-*tert*-butylcyclohexanone followed by 50 μL of methanol. (■) The vial is gently shaken to obtain a homogeneous solution.

Now add 100 μL of sodium borohydride reducing solution dropwise, with swirling.

Note. *The liquid reagents are dispensed by use of automatic delivery pipets.*

The preparation of the reducing solution is given in part A, Reagents and Equipment.

Reaction Conditions

The solution is now allowed to stand at room temperature for a period of 10 minutes.

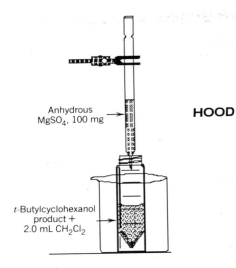

Anhydrous
MgSO₄, 100 mg

HOOD

t-Butylcyclohexanol
product +
2.0 mL CH₂Cl₂

Isolation of Product

The resulting solution is worked up using the procedure described in Part A, Isolation of Product, with the exception that 100 mg of magnesium sulfate is placed in the Pasteur filter pipet.(■)

The methylene chloride may be removed from the final solution using a gentle stream of nitrogen gas impinging on the surface of the solution while warming the vial in a sand bath in the **HOOD**.

Purification and Characterization

The product remaining after removal of the methylene chloride is sufficiently pure for characterization.

Weigh the solid product and calculate the percentage yield.

Determine the melting point of your material and compare it with the value reported in the literature.

Obtain an IR spectrum and compare it with that of an authentic sample. IR sampling is accomplished by capillary film-melt (use heat lamp) technique.

The cis-trans isomers of 4-tert-butylcyclohexanol may be separated by gas chromatography using an $\frac{1}{4} \times 8$ ft in 20% carbowax column set at 170°C.

A methylene chloride solution of the alcohol mixture having a concentration of 0.5 mg/μL is prepared and a 5.0-μL sample injected into the GC apparatus. At a flow rate of 50 mL/minute (He) the cis isomer has a retention time of 13 minutes; the trans isomer 16 minutes.

Determine the percentage of each isomer present in the sample by determining the area under the curve. The starting ketone has a retention time similar to that of the cis isomer alcohol. Therefore, if the reaction does not go to completion, the cis/trans ratio will not be accurate.

Area Under a Curve. *There are several techniques that may be used. The following method gives reproducible results of* ± 3–4%.

The peak height (mm) times the width at $\frac{1}{2}$ height (mm), measured from the baseline of the curve, yields the area under the curve.

ENVIRONMENTAL DATA

Substance	Amount	TLV (mg/m³)	Emissions (mg)	Volume (m³)
Experiment 4A				
Methylene chloride (Max., extract, evap.)	2.0 mL	350	2670	7.6
Methanol (SKIN) (Solv.)	0.55 mL	260	435	1.7
Cyclohexanone (Max. from yield, transfer,	100 μL	100	15	0.1
Experiment 4B				
Methanol (SKIN) (Max., solv.)	0.15 mL	260	120	0.5
Methylene chloride (Max., extract, evap.)	2.0 mL	350	2670	7.6
4-tert-Butylcyclohexanone (Max. from yield, transfer)	50 mg	—	3.5	

TLV: Cyclohexanol—50 ppm
    Hydrochloric acid—5 ppm

TXDS: Sodium borohydride—orl-rat LD50: 160 mg/kg
    4-tert-Butylcyclohexanone—orl-rat LD50: 5000 mg/kg
    4-tert-Butylcyclohexanol—orl-rat LD50: 4200 mg/kg

QUESTIONS

5-18. In the section of the discussion related to the mechanism of attack by the boro-hydride reagent on the carbonyl unit, it was stated that the reagent may make an *e* or *a* attack depending on steric factors. To what do the letters *e* and *a* refer? Demonstrate by use of a suitable structure.

5-19. You recall that in the discussion it was stated that NaCl and AgCl plates are used for liquid sample IR analysis. With silver chloride plates one must take precautions since they are very soft and scratch easily. Furthermore, they must be kept in the dark when not in use because they will darken quickly in direct light. Explain.

5-20. Why are there axial and equatorial hydroxyl isomers for 4-*tert*-butylcyclohexanol, but not for cyclohexanol itself?

5-21. Sodium cyanoborohydride $NaBH_3CN$ is a close relative of $NaBH_4$, but is considerably less reactive toward carbonyl groups at neutral pH. Explain.

REFERENCES

1. General references on metal hydride reduction:
  a. Walker, E. R. H. *Chem. Soc. Revs.* **1976**, *5*, 23.
  b. House, H. O. "Modern Synthetic Reactions"; Benjamin: Reading, MA, 1972.
2. Sodium borohydride as a reducing agent:
  a. Fieser, L. F.; Fieser, M. "Reagents for Organic Synthesis"; Wiley: New York, 1967; Vol. I, p 1050 and subsequent volumes.
  b. Cragg, G. M. W. "Organoboranes in Organic Synthesis"; Marcel Dekker: New York, 1973.
  c. Brown, H. C. "Boranes in Organic Chemistry"; Cornell Univ. Press: Ithaca, NY, 1972.
3. Lithium aluminum hydride as a reducing agent:
  a. Brown, W. G. *Org. React.* **1951**, *6*, 469.
  b. See reference 2a, p 581 and subsequent volumes.
4. *trans-tert*-Butylcyclohexanol has been prepared from the ketone using $LiAlH_4$ as the reducing agent: Eliel, E. L.; Martin, R. J. L.; Nasipuri, D. "Organic Syntheses"; Wiley: New York, 1973; Collect. Vol. V, p 175.
5. IR Spectral Collection: The Aldrich Library of IR Spectra; Pouchert, C. J., Ed.; 3rd ed., 1981; Aldrich Chemical Co., Milwaukee, WI.
6. IR Spectral Collection: Sadtler Research Laboratory, Inc. Spectra; Philadelphia, PA.

TECHNIQUE 5: CRYSTALLIZATION

This experiment introduces the basic strategy involved in achieving the purification of solid organic substances by crystallization. The technique of crystallizing an organic compound is one of fundamental importance and must be mastered in order to deal successfully with the purification of these materials. *It is not an easy art to acquire.* Organic solids tend not to crystallize with the ease of inorganic substances. Indeed, in earlier times an organic chemist occasionally would resist an invitation to leave a well-worn laboratory for new quarters. This concern arose from the suspicion that the older facility (in which many crystallizations had been carried out) harbored seed crystals carried by dust from the earlier work. Dust presumably aided the successful initiation of crystallization of reluctant materials. In any event, a reaction very often will be viewed a failure unless an amorphous sludge can be enticed to become a collection of beautiful white crystals. The melting point of an amorphous substance is ill-defined, and if this material is mixed with a crystalline reference compound, large melting point depressions usually result.

In a number of areas of organic chemistry, particularly those dealing with natural products, the success or failure of an investigation can depend to a large extent on the ability of the research chemist to isolate tiny quantities of crystalline substances. Often the compounds of interest must be extracted from enormous

Lipoic acid

amounts of extraneous material. In one of the more spectacular examples, Reed[6] in 1953, isolated 30 mg of the crystalline coenzyme lipoic acid from 10 tons of beef liver residue.

Procedure

The essentials of this purification technique are outlined as follows. First, dissolve the material (primarily made up of the compound of interest along with smaller quantities of contaminating substances) in a warm solvent. Second, once the solid mixture is fully dissolved, the heated solution is filtered and then brought to the point of saturation by evaporation of a portion of the solvent. Third, the warm saturated solution is cooled to cause a drop in solubility of the dissolved substance. This results in the subsequent precipitation of the solid material. Fourth, the precipitate is isolated by filtration and the last traces of solvent removed.

The technique is considered successful if the solid is recovered in high yield and is obtained in a state of higher purity than the material initially dissolved. The cycle of going from the solid state to solution and back to the solid state is termed *recrystallization* if the initial and final solid materials are crystalline.

While the technique sounds fairly simple, in reality it is demanding. The successful purification of microscale quantities of solids will require your utmost attention.

The first major problem to be faced is the choice of solvent system. In order to achieve *high recoveries*, the compound to be crystallized would ideally be very soluble in the solvent of choice at elevated temperatures, but nearly insoluble when cold. For the crystallization to result in *increased purity* of the compound, however, the impurities should be either very soluble in the solvent at all temperatures or not soluble at any temperature. In addition, the solvent should possess as low a boiling point as possible so that traces can be easily removed from the crystals following filtration.

Thus, the choice of solvent is critical to a good crystallization. Table 5.5 is a list of common solvents used in the purification of most organic solids. (The list has contracted significantly in the past few years as health concerns about these very volatile compounds has risen.)

Seldom are the solubility relationships ideal for crystallization. Most often a compromise will have to be made. If there is no suitable single solvent available, it is possible to employ a mixture of two solvents, termed a "solvent pair." In this situation, a solvent is chosen that will readily dissolve the solid. Following dissolution, the system is filtered. A second solvent miscible with the first, but in which the solute has lower solubility, is then added dropwise to the hot solution to achieve saturation. In general, polar organic molecules have higher solubilities in polar solvents, and nonpolar materials are more soluble in nonpolar solvents ("like dissolves like"). Considerable time can be spent in the laboratory working out an appropriate solvent system for a particular reaction product. In most instances with known compounds, the optimum solvent system has been established. Thus, in the large majority of cases in the microlab text, the best recrystallization solvent system will be suggested.

Because many impurities have similar solubilities to the compound of interest, most crystallizations are not very efficient. Yields of 50–70% are not uncommon. It is important that the purest possible material be isolated prior to undertaking recrystallization.

A number of microscale crystallization routines are available.

Table 5.5 Common Solvents

Solvent	bp(°C)
Water	100
Methanol	65
Ethanol, 95%	78
Ligroin	60–90
Dioxane	101
Acetone	56
Diethyl ether	35
Methylene chloride	41
Petroleum ether	30–60

[6]Reed, L. J.; Gunsalus, I. C.; Schnakenberg, G. H. F.; Soper, Q. F.; Boaz, H. E.; Kern, S. F.; Parke, T. V. *J. Am. Chem. Soc.* **1953**, *75*, 1267.

Simple Crystallization

This works well with large quantities of material (100 mg and up), and it is essentially identical to that of the macrolaboratory program.

1. Place the solid in a small Erlenmeyer flask or test tube.
2. Add a minimum amount of solvent and bring the mixture to the boiling point in a sand bath.
3. Stir and add solvent dropwise with continued heating until all of the material has dissolved.
4. Add a decolorizing agent, if necessary (powdered charcoal), to remove colored minor impurities and other resinous by-products.
5. Filter hot into a second Erlenmeyer flask (preheat the funnel with hot solvent). This operation removes the decolorizing agent and any insoluble material initially present in the sample.
6. Evaporate enough solvent to reach saturation.
7. Cool to allow crystallization (better crystal formation will occur if this step takes place slowly).
8. Collect the crystals by filtration.
9. Wash (rinse) the crystals.
10. Dry the crystals.

Filtration Techniques:

Use of the Hirsch Funnel

The standard filtration system for collecting products purified by recrystallization in the microscale laboratory is vacuum filtration with an 11-mm Hirsch funnel. In addition, many reaction products that do not require the crystallization step are collected directly by this technique. The funnel is shown in Fig. 5.23a.

The Hirsch filter funnel is composed of a ceramic cone with a circular flat bed perforated with small holes. The diameter of the bed is 11 mm, and in operation it is covered by a flat piece of filter paper of the same diameter. The funnel is sealed into a filter flask with a Neoprene adapter (see Fig. 5.23b).

The filter flask, which is heavy walled and especially designed to operate under vacuum, is constructed with a side-arm (they are often called "side-arm pressure flasks" (Fig. 5.23c)).

The side-arm is connected with heavy-walled rubber vacuum tubing to a water aspirator or water pump. When water is running through the aspirator, a partial vacuum is formed, which creates a flow of air down the vacuum tubing from the filter flask. With the rubber adapter in place, the entering air is forced through the filter paper, which is held flat by suction. The mother liquors of the crystallization are rapidly forced into the filter flask, while the crystals retained by the filter are quickly dried by the stream of air passing through them (Fig. 5.24).

When using a water pump it is very important to have a safety trap mounted in the vacuum line leading from the filter flask. Any drop in water pressure (easily created by one or two other students on the same water line turning on their aspirators at the same time) can result in water being backed up into the system as the flow through the aspirator decreases (see Fig. 5.25).

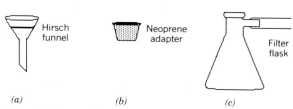

Fig. 5.23a–c *Component parts for vacuum filtration.*

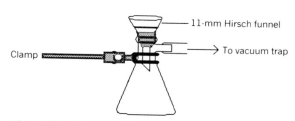

Fig. 5.24 *Vacuum filtration apparatus.*

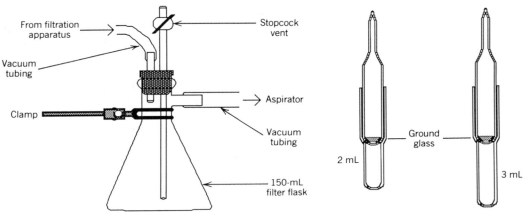

Fig. 5.25   *Vacuum trap.*   **Fig. 5.26**   *Craig tubes.*

Craig Tube Crystallizations

The Craig tube is commonly used for microscale crystallizations in the range of 10–100 mg of material (see Fig. 5.26). The following steps are involved.

1. The sample is placed in a small test tube (10 × 75 mm).
2. The solvent (0.5–2 mL) of choice is added, and the sample dissolved by heating in the sand bath. Rapid stirring with a microspatula (roll the spatula rod between your fingers) greatly aids the dissolution and protects against boilover. A modest excess of solvent is added after the sample is completely dissolved. It will be easy to remove this excess at a later stage, as the volumes involved are very small. The additional solvent ensures that the solute will stay in solution during the hot transfer.
3. The heated solution is transferred to the Craig tube by Pasteur filter pipet (the pipet is preheated with hot solvent). This transfer automatically filters the solution (if decolorizing charcoal has been added, two filtrations by the pipet may be required).
4. The hot filtered solution is then concentrated to saturation by gentle boiling in the sand bath. Constant agitation of the solution with a microspatula during this short period will avoid the use of a boiling stone and guarantee that a boilover will not occur. The ready crystallization of product on the microspatula just above the solvent surface serves as a good indication that saturation is close at hand.
5. The upper section of the Craig tube is set in place, and the system allowed to cool in a safe place. As cooling commences, seed crystals, if necessary, may be added by crushing them against the side of the Craig tube with a microspatula just above the solvent line. A good routine, if the time is available, is to place the assembly in a small Erlenmeyer, then place the Erlenmeyer in a beaker, and finally cover the first beaker with a second inverted beaker. This will ensure slow cooling, which will enhance good crystal growth (Fig. 5.27). A Dewar flask may be used when very slow cooling and large crystal growth is required (as in X-ray crystallography).
6. After the system reaches room temperature, cooling in an ice bath will further improve the yield.
7. Solvent is now removed by inverting the Craig tube assembly into a centrifuge tube and spinning the mother liquors away from the crystals (Fig. 5.28).

   This operation takes the place of the usual filtration step in simple crystallizations. It avoids another transfer of material and also avoids product contact with filter paper.

Fig. 5.27 *Apparatus for slow crystallization.*

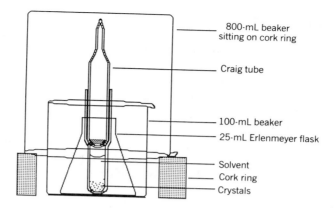

- 800-mL beaker sitting on cork ring
- Craig tube
- 100-mL beaker
- 25-mL Erlenmeyer flask
- Solvent
- Cork ring
- Crystals

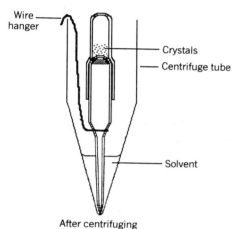

- Wire hanger
- Crystals
- Centrifuge tube
- Solvent

After centrifuging

Fig. 5.28 *Crystal collection with a Craig tube.*

8. After removal from the centrifuge, the Craig tube is disassembled and any crystalline product clinging to the upper section is scraped into the lower section. If the lower section is tared, it can be left to air dry to constant weight (wrap a piece of filter paper over the open end secured by a rubber band to avoid dust collecting on the product while drying). The yield then can be directly calculated.

The cardinal rule in carrying out the purification of small quantities of solids is: *keep the transfers to an absolute minimum!* The Craig tube is very helpful in this regard.

The above routine will maximize the crystallization yield. If time is important, the process can be shortened considerably. Shortcuts, however, invariably will lead to a corresponding drop in yield.

In the following experiment the techniques of simple crystallization, vacuum filtration, and Craig tube recrystallization are introduced.

Experiment 5
Photochemical Isomerization of an Olefin: *cis*-1,2-Dibenzoylethylene

(*cis*,-1,4-diphenyl-2-butene-1,4-dione)

The isomerization of a trans olefin to the corresponding cis isomer under photochemical conditions is demonstrated in this experiment.

trans-1,2-Dibenzoylethylene $\xrightarrow{h\nu}$ *cis*-1,2-Dibenzoylethylene

DISCUSSION   The photochemical geometric cis-trans isomerization reaction is demonstrated in this experiment by the conversion of *trans*-1,4-diphenyl-2-butene-1,4-dione to the corresponding cis isomer. The reaction proceeds by way of an excited state of the trans isomer.

The bonding between two carbon atoms of a C=C consists of a sigma bond and the pi bond created by overlap of the carbon *p* orbitals. This *p*-orbital overlap forming the pi bond imparts a certain rigidity to the C—C and, thus, free rotation of the parts of the molecule connected through the C=C is prohibited. The isomers do not interconvert unless enough energy (60–65 kcal/mol) is supplied to break the pi bond.

The nonrotation situation about the C=C gives rise to the possibility of stereoisomerism. Cis and trans geometrical isomers are not mirror images of one another and are thus not enantiomers, but diastereoisomers. The isomers differ only in the arrangement of the atoms in space and cannot exist if either carbon carries two identical groups. Diastereoisomers have different physical properties such as melting point, boiling point, dipole moment, density, and solubility. Because of the differences in physical properties, the possibility exists that the isomers may be separated, as in this experiment.

PART A: Purification of *trans*-1,2-Dibenzoylethylene

Estimated time to complete the reaction: 1.0 hour actual laboratory time. The reaction requires approximately 12 hours of irradiation.

Physical Properties of Reactants and Products

Compound	MW	Wt/Vol	mmol	mp(°C)	bp(°C)
trans-1,2-Dibenzoylethylene	236.27	150 mg	0.62	111	
Ethanol (95%)		6.0 mL			78.5
cis-1,2-Dibenzoylethylene	236.27			134	

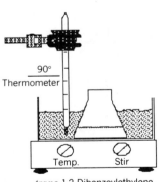

90°
Thermometer

Temp.   Stir

trans-1,2-Dibenzoylethylene,
150 mg + 95% CH₃CH₂OH, 3.0 mL

The starting olefin is purified by recrystallization.

To a 10-mL Erlenmeyer flask are added 150 mg (0.62 mmol) of *trans*-1,2-dibenzoylethylene and 6.0 mL of 95% ethanol. The mixture is heated to approximately 90°C using a sand bath.(■) Decolorizing charcoal (10 mg) is added, and the hot solution filtered by gravity (preheat funnel with hot ethanol). The filtrate is collected in a 10-mL Erlenmeyer flask and concentrated to 3 mL.(■)

The solution is allowed to cool to room temperature for a period of 15 minutes and then placed in an ice bath for an additional 10 minutes. The yellow needles are collected by vacuum filtration using a Hirsch funnel (■) and then air dried on a porous clay plate.

Trans isomer +
CH₃CH₂OH, 3 mL

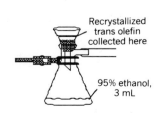

Recrystallized
trans olefin
collected here

95% ethanol,
3 mL

Weigh the olefin and calculate the percent recovery. Determine the melting point and compare your result with the literature value. Also compare the melting point with the one obtained on the distillation residue of Experiment 1A, in which a simple crystallization was performed without the aid of decolorizing charcoal.

PART B: Isomerization of the Olefin

EXPERIMENTAL

Reagents and Equipment
To a 13 × 100 mm test tube are added 50 mg (0.21 mmol) of recrystallized *trans*-1,2-dibenzoylethylene and 3.0 mL of 95% ethanol.

Reaction Conditions
GENTLY A sand bath is used to warm the mixture [GENTLY] until a homogeneous solution is obtained. The test tube is *loosely* stoppered and then placed approximately 2 in. from a 150-watt sunlamp. The solution is irradiated overnight.(■)

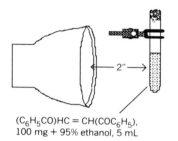

(C₆H₅CO)HC = CH(COC₆H₅),
100 mg + 95% ethanol, 5 mL

Isolation of Product
The test tube is now removed from the light source and allowed to cool to room temperature.

Note. *Be careful when removing the test tube from the light source. IT IS HOT!*

The resulting mixture is placed in an ice bath to complete crystallization of the colorless *cis*-1,2-dibenzoylethylene product. The solid is collected by vacuum filtration using a Hirsch funnel, washed with 0.5 mL of cold 95% ethanol and dried on a porous clay plate.

Purification and Characterization
A portion of the isolated product can be further purified by recrystallization from 95% ethanol using a Craig tube. Weigh the product and calculate the percentage yield. Determine the melting point and compare your result with the literature value. Obtain an IR spectrum and compare it with an authentic sample and also to the trans isomer.

ENVIRONMENTAL DATA

Substance	Amount	TLV (mg/m³)	Emission (mg)	Volume (m³)
Ethanol, 95% (Max. filter hot soln., vac. filtration.)	3.0 mL	1900	2250	1.2
Ethanol, 95% (Max., vac. filtration.)	5.5 mL	1900	4120	2.2

TXDS: 1,2-Dibenzoylethylene—ipr-mus LD50: 25 mg/kg

QUESTIONS

5-22. If the trans isomer of the compound used in this experiment is the most stable, how is it possible to make this reaction provide exclusively the cis isomer?

5-23. List the properties that an ideal solvent should have to perform the purification of an organic compound by the crystallization technique.

5-24. What is meant by the term ''solvent pair'' as it relates to crystallization of solid materials?

5-25. What is the purpose of adding a decolorizing agent such as powdered charcoal to the solvent system during a recrystallization sequence?

5-26. Why is better crystal formation observed if a solution of an organic compound is allowed to cool slowly?

5-27. List several advantages of using the Craig tube to separate crystalline products in contrast to using the Hirsch funnel filtration method.

5-28. The stereochemistry of the more highly substituted alkenes is difficult to define using the cis and trans designations. Therefore, chemists have developed a more systematic manner of indicating stereochemistry in these systems using the symbols **E** and **Z**.

    a. What do these letters signify?

    b. Draw the structures of the **E** and **Z** stereoisomers of 1,4-diphenyl-2-butene-1,4-dione used in this experiment.

REFERENCES   The above reaction was adapted from the following references:

1. Pasto, D. J.; Ducan, J. A.; Silversmith, E. F. *J. Chem. Educ.* **1974,** *51,* 277.

2. Silversmith, E. F.; Dunsun, F. C. *J. Chem. Educ.* **1973,** *50,* 568.

Reviews on photochemical isomerization reactions may be found in the following references.

1. Crombie, L. *Quart. Revs.* **1952,** *6,* 101.

2. DeMayo, P. *Adv. Org. Chem.,* **1960,** *2,* 367.

3. Fonken, G. L. In "Organic Photochemistry"; Chapman, O. L., Ed.; Marcel Dekker: New York, 1967; Vol. I, p 197.

TECHNIQUE 6:
ISOLATION, PURIFICATION, RECRYSTALLIZATION, AND MELTING POINTS (REVIEW)

Experiment 6
The Benzoin Condensation with Benzaldehyde: Benzoin

(ethanone, 2-hydroxy-1,2-diphenyl-)

The condensation of an aromatic aldehyde in the presence of a cyanide catalyst to yield an α-hydroxy ketone is known as the benzoin condensation.

Benzaldehyde   Benzoin

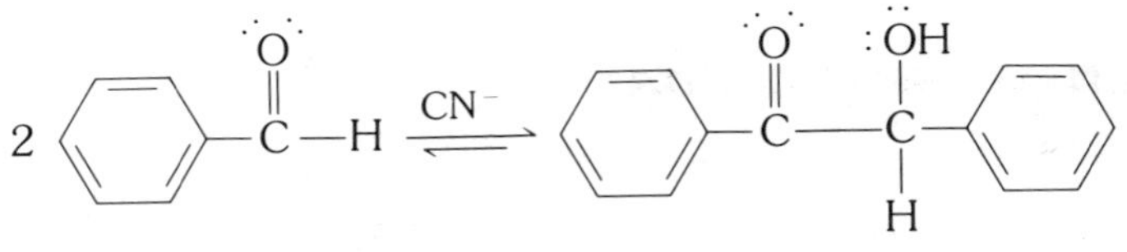

DISCUSSION   Aromatic aldehydes in the presence of cyanide ion catalyst dimerize to form the corresponding α-hydroxy ketone (acyloin). The reaction, which is reversible, is known as the benzoin condensation. Cyanide ion is a *specific* catalyst for the reaction and can function in this capacity because it is a good nucleophile, it stabilizes the anion intermediate, and it is also a good leaving group. In the mechanism outlined below, it is observed that the cyanide ion makes a nucleophilic attack on one molecule of the aromatic aldehyde to form the conjugate base of a cyanohydrin. The effect of the CN group is to increase the acidity of the aldehydic hydrogen atom, thus allowing the formation of the anion (I).

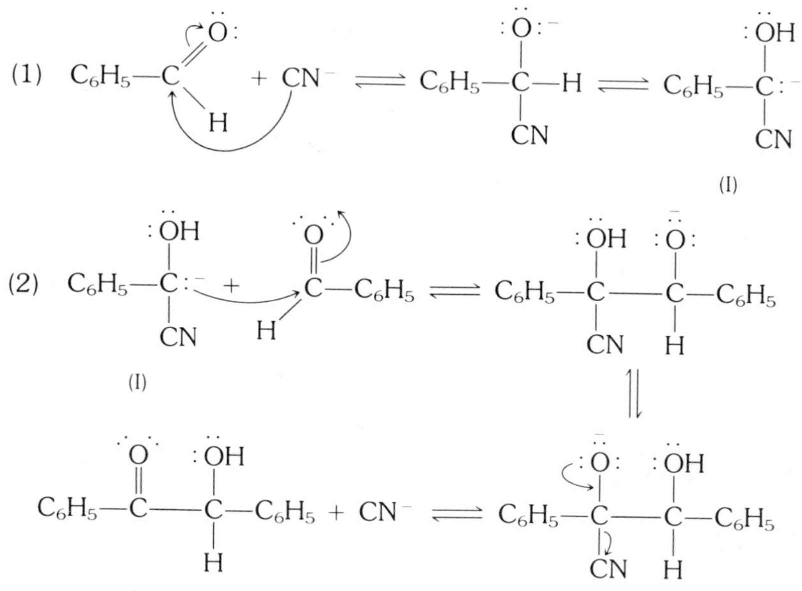

(1) reaction scheme

(2) reaction scheme

An acyloin

Once generated, the nucleophilic carbanion (I) attacks a second molecule of the aromatic aldehyde to yield a substituted cyanohydrin. This species can be stabilized by loss of cyanide ion to form the acyloin product.

The electrical effects of various substituents on the aromatic ring have been investigated. If a strongly electron donating group is in the para position of the ring, the reaction fails. As seen for the N,N-dimethylamino group, the effect is to render the carbonyl carbon atom less electrophilic.

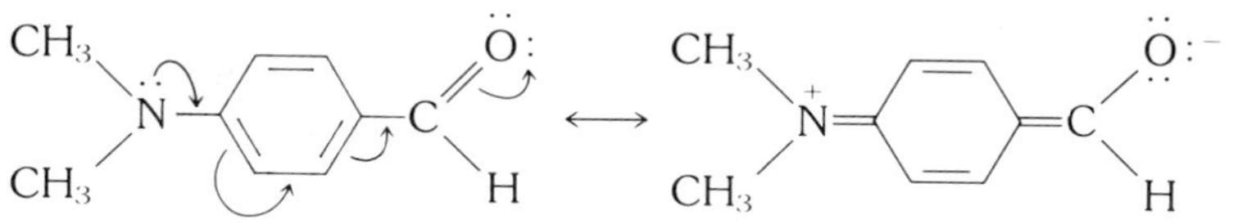

Thus, the nucleophilic attack on the carbonyl carbon by the cyanide ion is retarded. Of interest is the fact that a mixed benzoin condensation with benzaldehyde and p-N,N'-dimethylaminobenzaldehyde led to the formation of compound II, but not III. This result indicates that the cyanohydrin anion, IV, formed from p-N,N'-dimethylbenzaldehyde may add to benzaldehyde, but that the anion (I) formed from benzaldehyde will not add to p-N,N'-dimethylaminobenzaldehyde. This is due to the increase in electron density at the carbonyl carbon brought about by the presence of the p-N,N'-dimethyl substituent.

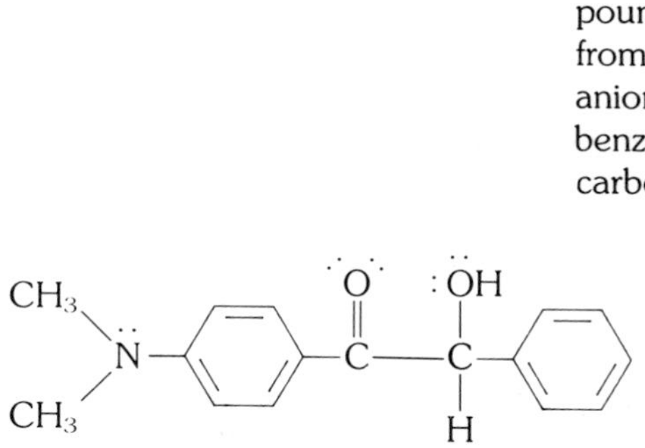

(II) 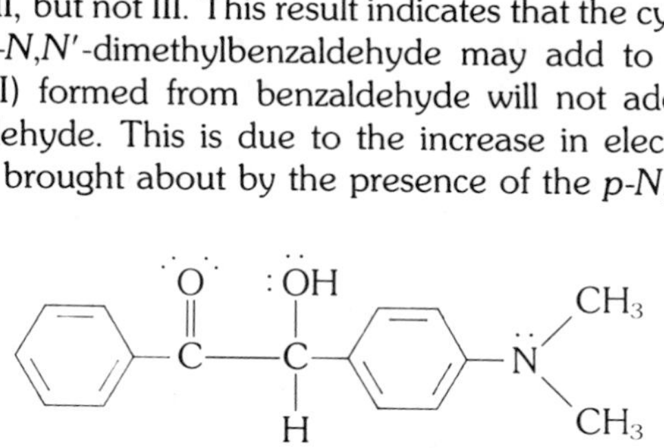   (III) 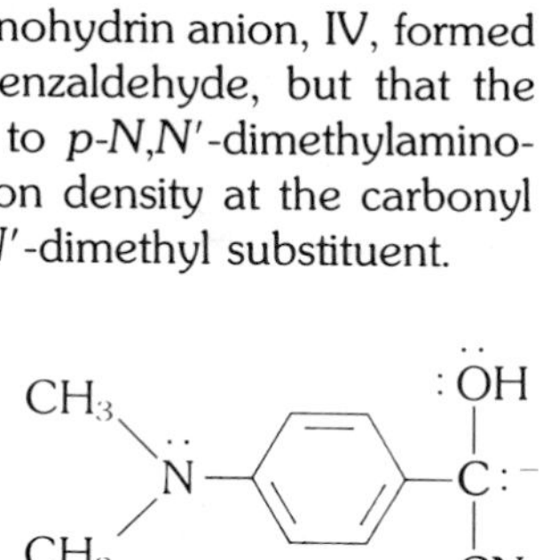   (IV)

The benzoin condensation is also hindered by strong electron withdrawing groups on the ring. The presence of a p-nitro group decreases the electron density on the carbonyl carbon atom in the cyanohydrin anion, making it less nucleophilic. This effect retards the addition of the anion to the second molecule of aldehyde.

EXPERIMENTAL Estimated time to complete the experiment: 2.0 hours.

Physical Properties of Reactants and Products

Compound	MW	Wt/Vol	mmol	mp(°C)	bp(°C)	Density	n_D
Benzaldehyde	106.13	200 μL	1.96		178	1.04	1.5463
Sodium cyanide (0.54M)		1 mL					
Ethanol (95%)		1 mL					
Benzoin	212.25			137			

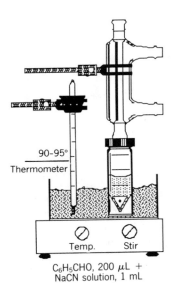

90–95°
Thermometer

Temp.   Stir

C_6H_5CHO, 200 μL +
NaCN solution, 1 mL

HOOD

Reagents and Equipment

In a tared 5.0-mL conical vial containing a magnetic spin vane and equipped with a reflux condenser are placed 200 μL (208 mg, 1.96 mml) of benzaldehyde and 1 mL of a 0.54M sodium cyanide solution.(■)

The 0.54M sodium cyanide solution should be prepared by the instructor.

Instructor Preparation

The solution is prepared by adding 0.75 g of NaCN to 10 mL of deionized water and 15 mL of 95% ethanol.

> **WARNING:** *Sodium cyanide is extremely toxic.*

Note. *The conical vial is tared so that the exact weight of benzaldehyde used can be determined. The vial is capped during this operation. It is recommended that both reagents be dispensed in the* **HOOD** *using automatic delivery pipets.*

Reaction Conditions

The mixture is now heated with stirring at a sand bath temperature of 90–95°C. It is maintained at this temperature for a period of 30 minutes. The reaction solution turns yellow in color and in approximately 5 minutes may become cloudy.

> **CAUTION:** *Do not overheat the reaction mixture. If the solution begins to darken, remove the vial from the heat source immediately.*

Benzoin
product
collected
here

H_2O, 2 mL

Isolation of Product

At the end of the reflux period, the solution is cooled to room temperature and then placed in an ice bath for an additional 10 minutes. The benzoin product is now collected by filtration under reduced pressure using a Hirsch funnel.(■) The filter cake is washed with two 0.5-mL portions of cold water and air dried under suction for 5 minutes. The crude material is further dried on a porous clay plate.

Purification and Characterization

This crude material is sufficiently pure to use in the preparation of benzoin acetate (see Experiment 8E). It may be purified by recrystallization from methanol or ethanol (95%) by use of a Craig tube.

Weigh the benzoin product and calculate the percentage yield. Determine the melting point and compare your value to that found in the literature. Obtain the IR spectrum of the compound by KBr pressed disk. Compare your spectrum to that of an authentic sample.

ENVIRONMENTAL DATA

Substance	Amount	TLV (mg/m³)	Emissions (mg)	Volume (m³)
Benzaldehyde (Max., from yield, transfer)	200 μL	—	74	
Methanol (SKIN) (Max., Craig tube)	1.0 mL	260	800	3.1
Ethanol (Max., Craig tube)	1.0 mL	1900	800	0.4

TXDS: Sodium cyanide—orl-hmn LDLo: 2857 μg/kg TWA 5 mg (CN)/m³ (SKIN)
Benzoin—orl-rat TDLo: 5460 mg/kg
Benzaldehyde—orl-rat LD50: 1300 mg/kg

QUESTIONS

5-29. The benzoin product produced in this experiment contains a chiral carbon atom, but the product itself is not optically active. Explain.

5-30. The cyanide ion is a highly specific catalyst for the benzoin condensation. Can you list three functions this ion performs in this catalytic role?

5-31. Can you suggest a reason why *p*-cyanobenzaldehyde does not undergo the benzoin condensation?

REFERENCES

1. Ide, W. S.; Buck, J. S. *Org. React.* **1948**, *4*, 269.
2. Kuebrich, I. P.; Schowen, R. L.; Wang, M.; Lupes, M. E. *J. Am. Chem. Soc.* **1971**, *93*, 1214.
3. Adams, R.; Marvel, C. S. "Organic Syntheses"; Wiley: New York, 1941; Collect. Vol. I, p 94.

TECHNIQUE 7: EXTRACTION: ACID–BASE TECHNIQUE; NUCLEAR MAGNETIC RESONANCE: SAMPLING TECHNIQUE

Extraction

For a discussion of acid–base extraction technique see Experiment 3.

Nuclear Magnetic Resonance

The NMR spectrometer is one of the most powerful experimental tools available to the modern organic chemist. The cost and complexity of operation preclude actual, "hands-on," student operation of these instruments at the sophomore level except in rare instances. Since the focus in the microscale laboratory is on student-derived data, NMR spectral data are not included in the majority of the experiments described in this text. Experiments 7A and 7B are exceptions. This experiment presents the option for students to obtain and interpret NMR data if the local opportunity exists. It is of interest to note that the alcohol products, 4-bromo- and 4-chlorobenzyl alcohol, exhibit different splitting patterns in the aromatic region.

The theory of the effect and the basic instrument design are briefly introduced in Appendix C. The interpretation of the data is presented in Chapter 7.

NMR Sampling

NMR sample preparation is fairly routine. Most samples have been measured in solution on 60-MHz instruments in thin-walled tubes approximately 5-mm diameter and 15–20 cm long. The sample size compatible with this field strength is in the range 30–50 mg dissolved in 0.3–0.5 mL of solvent.

The most practical solvents are carbon tetrachloride (CCl_4) and deuterio-chloroform ($DCCl_3$). *Handle these solvents with care* [**HOOD**] *since both are* TOXIC! A number of other deuterated solvents are commercially available. The

HOOD

universally accepted internal reference compound employed in making these measurements is tetramethylsilane (TMS). Once the sample tube has been loaded with the sample solution (transfer by Pasteur filter pipet) 1–3% TMS is added. Since the TMS is particularly volatile (bp 26.5°C), the tube should be capped immediately following addition or it will vaporize. The sample tube is spun in the instrument to average out small changes in field strength over the sample volume. The tube is positioned between the pole faces by a depth gauge and the spectrum recorded. In most cases, your laboratory instructor or teaching assistant will actually record the data for you. If you do have the opportunity to work directly with your instructor, detailed information will be provided for operating the instrument. Controls vary depending on make or model of the instrument and, therefore, we will not present this information in the text.

Experiments 7A, 7B
THE CANNIZZARO REACTION WITH
✳ ## 4-CHLOROBENZALDEHYDE OR
4-BROMOBENZALDEHYDE:
4-CHLOROBENZOIC ACID and
4-CHLOROBENZYL ALCOHOL;
4-BROMOBENZOIC ACID and
4-BROMOBENZYL ALCOHOL[1]

(benzoic acid, *p*-chloro- and benzyl alcohol, *p*-chloro- or benzoic acid, *p*-bromo- and benzyl alcohol, *p*-bromo-)

This experiment illustrates the simultaneous oxidation and reduction of an aromatic aldehyde to form the corresponding benzoic acid and benzyl alcohol.

4-Chlorobenzaldehyde   4-Chlorobenzyl alcohol   4-Chlorobenzoic acid

DISCUSSION   The carbonyl group of an aldehyde represents the intermediate stage of oxidation between an alcohol and a carboxylic acid.

Primary alcohol   Aldehyde   Carboxylic acid

[1]Portions of this experiment were previously published. Mayo, D. W.; Butcher, S. S.; Pike, R. M.; Foote, C. M.; Hotham, J. R.; Page, D. S. *J. Chem. Educ.* **1985**, *62*, 149.

It is not surprising, then, to find a reaction in which an aldehyde is oxidized and reduced in a disproportionation sequence to form the corresponding alcohol and carboxylic acid. Such a reaction is the Cannizzaro reaction. In the presence of strong base, aldehydes that lack α-hydrogen atoms undergo a self oxidation–reduction reaction. One molecule of the aldehyde oxidizes a second aldehyde molecule to the acid anion and itself is reduced to the corresponding primary alcohol. Aldehydes with α-hydrogen atoms do not give the reaction because in the presence of base they undergo the aldol reaction (Experiment 22).

The first step in the mechanistic sequence is the nucleophilic attack of the hydroxide anion on the carbonyl group of the aldehyde. This is followed by the key step in the reaction; the transfer of a hydrogen atom with its pair of electrons (a hydride species) to a carbonyl group of a second molecule of aldehyde. This sequence is diagrammed below.

$$(R = Ar,\ R_3C,\ H)$$

The strong electron-donating character of the negatively charged oxygen atom in the anion greatly facilitates the ability of the aldehydic hydrogen to transfer, with its pair of electrons, to a second molecule of aldehyde. As seen in the above mechanism, this nucleophilic addition to a carbonyl group leads to the formation of a carboxylic acid and an alkoxide anion. The final stage, involving a fast acid–base equilibrium to yield the alcohol and the carboxylic acid anion, lies far to the right. Thus, even though the Cannizzaro reaction is an equilibrium reaction, it proceeds nearly to completion. The mechanism is supported by evidence obtained by running the reaction in D_2O. It was found that the product alcohol did not contain α-deuterium substitution, indicating that the transferred hydrogen atom must come from a molecule of aldehyde and not from the solvent.

High yields of alcohol can be obtained from almost any aromatic aldehyde by running the reaction in the presence of excess formaldehyde. The formaldehyde acting as the reducing agent is in turn oxidized to formic acid.

p-Methyl-
benzaldehyde   Formaldehyde   p-Methylbenzyl alcohol   Formic
                                                                (90%)   acid

This procedure is known as a *crossed* Cannizzaro reaction.

α-Keto aldehydes undergo an internal Cannizzaro reaction to yield α-hydroxycarboxylic acids.

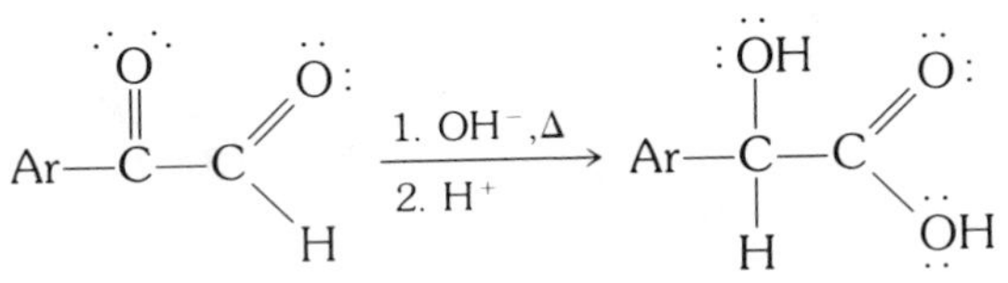

$$\text{Ar-\overset{\overset{O}{\|}}{C}-\overset{\overset{O}{\|}}{C}-H} \quad \xrightarrow[\text{2. H}^+]{\text{1. OH}^-,\Delta} \quad \text{Ar-\overset{\overset{OH}{|}}{\underset{H}{C}}-\overset{\overset{O}{\|}}{C}-OH}$$

(α-Keto aldehyde)   (α-Hydroxy acid)

PART A: Reaction with 4-Chlorobenzaldehyde

EXPERIMENTAL   Estimated time to complete the experiment: Two laboratory periods.

Physical Properties of Reactants and Products

Compound	MW	Wt/Vol	mmol	mp(°C)	bp(°C)	n_D
4-Chlorobenzaldehyde	140.57	150 mg	1.07	47.5		
Methanol	32.04	400 μL			65	1.3288
Potassium hydroxide (11*M*)	56.11	400 μL				
4-Chlorobenzoic acid	156.57			243		
4-Chlorobenzyl alcohol	142.59			75		

This reaction may be run in a centrifuge tube containing a boiling stone. The tube should be loosely fitted with a cotton plug.

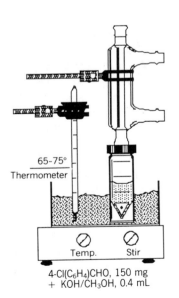

65–75°
Thermometer

Temp.   Stir

4-Cl(C₆H₄)CHO, 150 mg
+ KOH/CH₃OH, 0.4 mL

Reagents and Equipment
To a 5.0-mL conical vial containing a magnetic spin vane and equipped with a reflux condenser are added 150 mg (1.1 mmol) of 4-chlorobenzaldehyde and 0.4 mL of methanol. (■) With gentle swirling, 0.4 mL of an 11*M* aqueous solution of potassium hydroxide is then added.

Note. *It is convenient to dispense the methanol and KOH solution using automatic delivery pipets. It is recommended that the glass equipment NOT be rinsed or cleaned with acetone followed by air drying. The residual acetone undergoes the aldol reaction, which causes problems in the isolation of the desired products.*

CAUTION: *The concentrated KOH solution is very caustic. Do not allow it to come in contact with the skin or eyes.*

Reaction Conditions

The reaction mixture is heated with stirring in a sand bath at a temperature of 65–75°C for a period of one hour.

Isolation of Products

The reaction mixture is allowed to cool and 2.0 mL of chilled distilled water is added. The resulting solution is now extracted with three 0.5-mL portions of methylene chloride using a Pasteur filter pipet to transfer the extracts to a 3.0-mL conical vial. Upon each addition of the methylene chloride, the vial is capped, shaken gently, and then carefully vented by loosening the cap. After separation of the layers, the lower methylene chloride layer is removed using the Pasteur filter pipet.

Important. *Save the alkaline phase for further workup.*

Alcohol product + CH₂Cl₂, 1.5 mL

Anhydrous Na₂SO₄, 150 mg

1. *4-Chlorobenzyl alcohol.* The combined methylene chloride extracts are washed with two 0.25-mL portions of saturated sodium bicarbonate solution. The aqueous upper phase is removed (Pasteur filter pipet) and discarded. The methylene chloride layer is now dried over 150 mg of granular anhydrous sodium sulfate.(■) The solution, after drying, is transferred by use of a Pasteur filter pipet to a tared 3.0-mL conical vial. The sodium sulfate drying agent is rinsed with 0.3 mL of fresh methylene chloride and the rinse combined with the dried organic phase. The methylene chloride solvent is now evaporated using a stream of dry nitrogen gas in a warm sand bath [**HOOD**] to yield the crude 4-chlorobenzyl alcohol.

HOOD

This point in the procedure is a logical place to divide the experiment into two laboratory sessions.

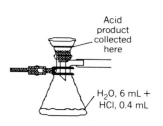

Acid product collected here

H₂O, 6 mL + HCl, 0.4 mL

2. *4-Chlorobenzoic acid.* The alkaline phase remaining from the original extraction procedure is diluted by the addition of 2.0 mL of water and acidified with the addition of 0.4 mL of concentrated hydrochloric acid. The voluminous white precipitate of the product is collected under reduced pressure by use of a Hirsch funnel and the filter cake rinsed with 2.0 mL of distilled water.(■) Air drying on a porous clay plate gives crude 4-chlorobenzoic acid.

Purification and Characterization

1. *4-Chlorobenzyl alcohol.* The crude alcohol is purified by recrystallization from a solution of 4% acetone in hexane (0.25 mL). Collection of the product under reduced pressure using a Hirsch funnel, followed by washing of the filter cake with 0.2 mL of cold hexane, gives the desired 4-chlorobenzyl alcohol. The product is air dried on a porous clay plate.

Weigh the 4-chlorobenzyl alcohol and calculate the percentage yield. Determine the melting point and compare your value to that found in the literature. Obtain the IR and NMR spectra (CDCl₃) and compare them to those of an authentic sample.

2. *4-Chlorobenzoic acid.* The crude acid may be purified by recrystallization from methanol by use of a Craig tube.

Weigh the dried material and calculate the percentage yield. Determine the melting point and compare your value to that found in the literature. Obtain the IR and NMR spectra [DMSO(D₆)] and compare them to those of an authentic sample.

PART B: Reaction with 4-Bromobenzaldehyde

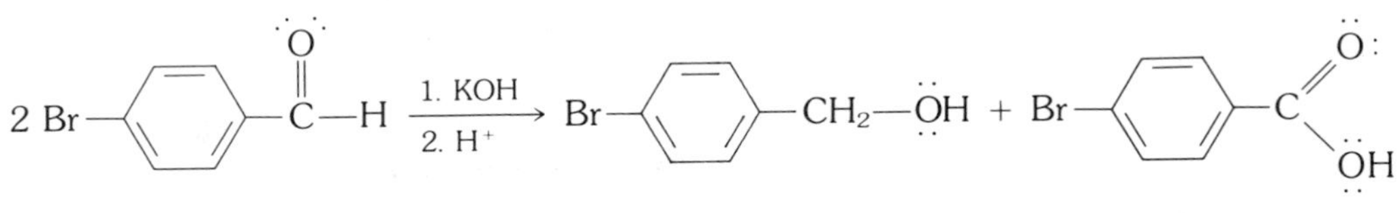

4-Bromobenzaldehyde   4-Bromobenzyl alcohol   4-Bromobenzoic acid

EXPERIMENTAL   Estimated time to complete the experiment: Two laboratory periods.

Physical Properties of Reactants and Products

Compound	MW	Wt/Vol	mmol	mp(°C)	bp(°C)
4-Bromobenzaldehyde	185.05	150 mg	0.81	67	
Methanol	32.04	400 μL			65
Potassium hydroxide (11M)	56.11	400 μL			
4-Bromobenzoic acid	201.03			254.5	
4-Bromobenzyl alcohol	187.04			77	

The amount of reagents (on a weight basis), reaction conditions, isolation technique, purification procedures, and characterization methods used in this experiment are identical to those employed above using 4-chlorobenzaldehyde.

The NMR spectra of the 4-bromo- and 4-chlorobenzyl alcohol products may be compared in order to observe the splitting pattern in the aromatic region.

ENVIRONMENTAL DATA

Substance	Amount	TLV (mg/m³)	Emissions (mg)	Volume (m³)
Experiment 7A				
4-Chlorobenzaldehyde (Max. from yield, transfer)	150 mg	—	54	
Methanol (SKIN) (Max., Craig tube)	1.4 mL	260	1120	4.3
Methylene chloride (Max., extract, evap.)	1.8 mL	350	2400	6.9
Hexane (Max., recryst.)	0.45 mL	180	300	1.7
Experiment 7B				
4-Bromobenzaldehyde (Max. from yield, transfer)	150 mg	—	54	
Methanol (SKIN) (Max., Craig tube)	1.4 mL	260	1120	4.3
Methylene chloride (Max., extract, evap.)	1.8 mL	350	2400	6.9
Hexane (Max., recryst.)	0.45 mL	180	300	1.7

TLV: Potassium hydroxide—2 mg/m³
 Hydrochloric acid—5 ppm
TXDS: Sodium bicarbonate—orl-rat LD50: 4220 mg/kg
 Sodium Sulfate—orl-mus LD50: 5989 mg/kg

EXPLAIN EACH RE EXTRACTION STEP (MECHANISMS!)

QUESTIONS **5-32.** The discussion mentions that the crossed Cannizzaro reaction can be realized when one of the components is formaldehyde. Predict the product(s) of the reaction below and give a suitable name to the reactants and products.

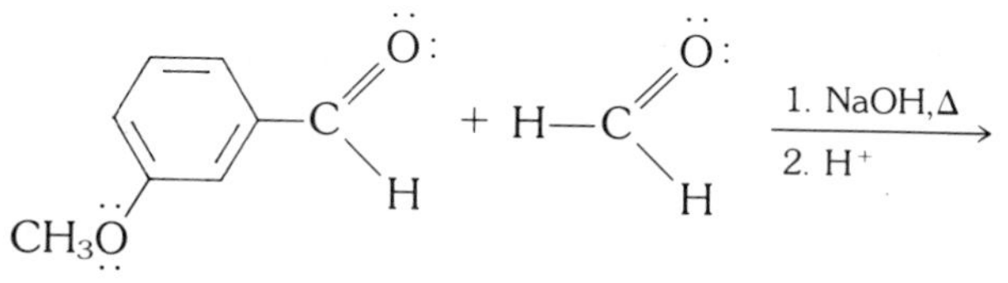

5-33 One group of investigators has suggested that a dianion

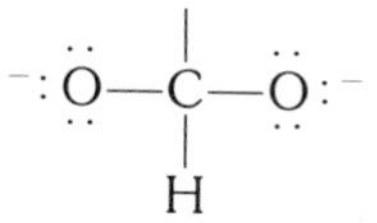

might be the source of hydride in the Cannizzaro reaction. Explain why this species would be a better source of hydride in comparison to the species depicted in the mechanism presented in the discussion section.

5-34. The Cannizzaro reaction is an oxidation–reduction sequence. In reference to question 5-32, formaldehyde is acting as what type of reagent?

5-35. Sketch the NMR spectrum you would expect to observe for the following compounds.

   a. Acetone
   b. 1,1,2-Tribromoethane
   c. Ethyl chloride
   d. 2-Iodopropane
   e. 1-Bromo-4-methoxybenzene

REFERENCES
1. Reviews on the Cannizzaro reaction:
   a. Geissman, T. A. *Org. React.* **1944,** *2,* 94.
   b. Swain, C. G.; Powell, A. L.; Sheppard, L. A.; Morgan, C. R. *J. Am. Chem. Soc.* **1979,** *101,* 3576.
2. Examples of the Cannizzaro reaction.
   a. Wilson, W. C. ''Organic Syntheses''; Wiley: New York, 1941; Collect. Vol. I, p 256.
   b. Davidson, D.; Weiss, M. ''Organic Syntheses''; Wiley: New York, 1943; Collect. Vol. II, p 590.

TECHNIQUE 8: CHROMATOGRAPHY: COLUMN AND THIN-LAYER; CONCENTRATION OF SOLUTIONS

The technique of chromatography is defined in Experiment 2 during the development of gas-phase separations. The term is derived from the Greek word for color, *chromatos*. Tswett discovered the technique (1903) during studies centered on the separation of mixtures of natural plant pigments.[7] The chromatographic zones were detected simply by observing the visual absorption bands (see Experiment 32). Thus, as originally applied the name was not an inconsistent use of terminology. Today, however, the very large majority of mixtures that are chromatographed are colorless materials. The separated zones in these cases are established by other methods. Interestingly, Tswett means ''color'' in Russian (the discoverer's nationality).

In this experiment two additional chromatographic techniques are explored.

[7]Tswett, M. *Ber. Deut. Botan. Ges.,* **1906,** *24,* 235.

Both of these procedures depend on adsorption and distribution between a stationary solid phase and a moving liquid phase. The first to be discussed is "column chromatography," a very powerful technique used extensively throughout the microscale experiments. It was one of the earliest of the modern chromatographic methods to be applied to the separation of organic mixtures by Tswett. The second procedure, "thin-layer" chromatography, was developed in the late 1950s. TLC is particularly effective in generating rapid identification and purification for use in obtaining analytical data.

COLUMN CHROMATOGRAPHY

The term *column chromatography* is derived from the use of a glass column packed with a solid stationary phase, a relationship similar to the liquid phase of gas chromatography. The moving liquid phase descends by gravity through the column.

A wide variety of substances have been employed as the stationary phase in this technique. In practice, however, two materials have become dominant in carrying out this type of separation chemistry. Finely ground (100–200 mesh) alumina (aluminum oxide, Al_2O_3) and silicic acid (silica gel, SiO_2) are by far the most useful of the known adsorbents. The liquids, which act as the moving phase and elute (wash) sample materials through the column, are many of the common organic solvents. Table 5.6 lists the better known column packings and elution solvents.

Silica gel impregnated with silver nitrate (usually 5–10% $AgNO_3$) is an attractive solid-phase adsorbent. The silver salt selectively binds to unsaturation via a silver ion pi-complex. Traces of olefinic materials are easily removed from saturated reaction products by chromatography with this system (Experiment 12). This adsorbent must be protected from light until used or the mixture will rapidly darken and become ineffective.

The procedure is usually carried out as follows:

Packing the Column

The quantity of stationary phase required is determined by the sample size. A common rule of thumb is to use 30–100 times the weight of packing to the amount of sample to be chromatographed. The size of the column is chosen to give roughly a 10:1 ratio of height to diameter for the amount of adsorbent required (Fig. 5.29).

In the microscale laboratory program two standard chromatographic columns are employed:

1. A Pasteur pipet modified by shortening the capillary tip is used for the separation of smaller mixtures (10–100 mg). Approximately 0.5–2.0 g of packing is used in the pipet column.

Table 5.6 Column Chromatography Materials

Stationary Phase		Moving Phase	
Alumina	↑ Increasing adsorption of polar materials	Water	↑ Increasing solvation of polar materials
Silicic acid		Methanol	
Magnesium sulfate		Ethanol	
Cellulose		Acetone	
Paper		Ethyl acetate	
		Diethyl ether	
		Methylene chloride	
		Cyclohexane	
		Pentane	

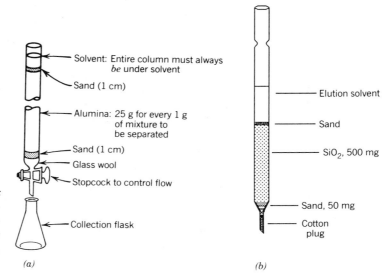

Solvent: Entire column must always *be* under solvent

Sand (1 cm)

Alumina: 25 g for every 1 g of mixture to be separated

Sand (1 cm)

Glass wool

Stopcock to control flow

Collection flask

Elution solvent

Sand

SiO$_2$, 500 mg

Sand, 50 mg

Cotton plug

(a)

(b)

Figs. 5.29a and 5.29b *Chromatographic columns. Figure 5.29a from James W. Zubrick,* The Organic Chem Lab Survival Manual *(New York: Wiley, 1984). (Reprinted by permission of John Wiley & Sons, New York, NY.)*

2. A 50-mL titration buret modified by reducing the length of the column (10 cm beyond the stopcock) is used for the larger sample mixtures (50–200 mg) and for the difficult-to-separate mixtures. Approximately 5–20 g of packing is employed in the buret column (Fig. 5.29).

Both columns are prepared by first clamping the empty column in a vertical position and then seating a small cotton or glasswool plug at the bottom. The cotton is covered with a thin layer of sand in the case of the buret. The Pasteur pipets are loaded by adding the adsorbent with gentle tapping, "dry packing." The column is then premoistened just prior to use. The burets are packed by a slurry technique. In this procedure the column is filled part way with solvent, then the stopcock is opened slightly, and as the solvent slowly drains from the column a slurry of the adsorbent-solvent is poured into the top of the column. The column should be gently tapped while adding the slurry. The solvent is then drained to the top of the adsorbent level and held at that level until used.

Sample Application

The sample is applied in a minimum amount of solvent (the least polar solvent in which the material is readily soluble) to the top of the column by Pasteur pipet. The pipet is rinsed and the rinsings added to the column just as the sample solution drains to the top of the adsorbent layer.

Elution of the Column

This is the critical step in resolving the sample mixture. Once the sample has been applied to the top of the column, the elution begins (a small layer of sand can be added to the top of the buret column following addition of the first charge of elution solvent). *It is very important not to let the column run dry.* The Pasteur pipet is free flowing (the flow rate is controlled by the size of the capillary tip), and once the sample is on the column, the chromatogram will require constant attention. The buret column flow is controlled by the stopcock. The flow rate should be set to allow time for equilibrium to be established between the two phases. The choice of solvent is dictated by a number of factors. A balance between the adsorption power of the stationary phase and solvation power of the elution solvent will govern the rate of travel of the material descending through the stationary phase. If the material travels rapidly down the column, then too few adsorption–elution cycles will occur and the materials will elute together in one fraction. If the sample travels too slowly, diffusion broadening takes over and resolution is degraded. In the latter case, samples then elute over many fractions with overlapping broad bands. The ideal solvent and elution rate strikes a balance between these two situations and maximizes the separation. The solvent or mixtures of solvents used to produce a satisfactory separation of a particular mixture can take a considerable amount of time to develop. The optimum elution parameters for the experiments in the microscale laboratory have been established in order to conserve time during the laboratory period.

Fraction Collection

As the solvent elutes from the column, it is collected in a series of "fractions" using small Erlenmeyer flasks or vials. Under ideal conditions, as the mixture of material travels down the column, it will separate into several individual bands of pure substances. By careful collection of the fractions, these bands can be separated as they sequentially elute from the column (similar to the collection of GC fractions in Experiment 2). The detection of the bands of material being eluted can be done by a number of techniques (weighing fraction residues, visible absorption bands, thin-layer chromatography, etc.). The collection protocol is usually established in the microscale experiments.

CONCENTRATION OF SOLUTIONS

The removal of solvent from the chromatographic fractions can be carried out by employing a number of different methods.

Distillation

The concentration of solvent by distillation is straightforward, and the standard routine is described in Experiment 1A. This allows for high recovery of volatile solvents and often can be carried on outside a hood. The Hickman still head and the 5-mL round-bottom flask are useful in this purpose. Distillation should be used primarily for concentration of the solution followed by transfer of the concentrate with a Pasteur filter pipet to a vial for final isolation.

Evaporation with Nitrogen Gas

A very convenient method for removal of final solvent traces is to concentrate the last 0.5 mL of solution by evaporation with a gentle stream of nitrogen gas while warming the sample in a sand bath. This is usually done at a hood station where several Pasteur pipets can be attached to a manifold leading to a tank of compressed gas. Gas flow to the individual pipets is controlled by needle valves. *Always test the gas flow with a blank vial of solvent.* This is a handheld operation. The sample vial will cool rapidly on evaporation of the solvent and gentle warming of the vial with agitation will thus aid removal of the last traces of the volatile material. This procedure avoids possible moisture condensation on the sample residue. *Do not leave the heated vial in the gas flow after the solvent is removed!* This is particularly important in the isolation of liquids. Remember to tare the vial before loading the solution to be concentrated, because achievment of constant weight is the best indication of total solvent removal.

Removal of Solvent Under Reduced Pressure

The concentration of solvent under reduced pressure is very efficient. It reduces the time of solvent removal in microscale experiments to a few seconds or at most a few minutes. In contrast, distillation or evaporation procedures require several minutes to tens of minutes for even relatively small volumes. Vacuum concentration, however, is tricky and should be practiced prior to committing hard-won reaction product to this test. The procedure is most beneficial when applied to fairly large chromatographic fractions (5–10 mL).

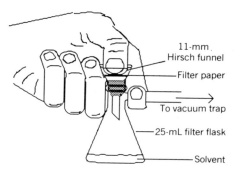

Fig. 5.30 *Removal of solvent under reduced pressure.*

The sequence of operations is described below (see also Fig. 5.30).

1. Transfer the chromatographic fraction to the 25-mL filter flask.
2. Insert the 11-mm Hirsch funnel and rubber adapter into the flask.
3. Turn on the water pump (with trap) and connect the vacuum tubing to the pressure flask side-arm while holding the flask in one hand (this is another handheld operation).
4. Place the thumb of the hand holding the filter flask over the Hirsch funnel filter bed to shut off the air flow through the system (Fig. 5.30). This will result in an immediate drop in pressure. The volatile solvent will rapidly come to a boil at room temperature. Thumb pressure adjusts air leakage through the Hirsch funnel and thereby controls the pressure in the system. Manipulate the pressure so that the liquid does not foam up into the side-arm of the filter flask.

It is essential that the filter flask be warmed by the sand bath during this operation since rapid evaporation of the solvent will quickly cool the solution. The air leak used to control the pressure results in a stream of moist laboratory air being rapidly drawn over the surface of the solution. If the evaporating liquid becomes cold, water will condense over the interior of the filter flask and eventually contaminate the isolated residue. Warming the flask while the evaporation process is being carried out will avoid this problem and help to speed solvent removal. The temperature of the flask should be checked from time to time by touching it to the palm of the free hand. The flask is kept slightly above room temperature by adjusting the heating and evaporation rates. It is best to practice this operation a few times with solvent blanks to see whether you can avoid boilovers and the accumulation of water residue in the flask.

Column chromatography is a powerful technique for the purification of organic materials. It is far more efficient than crystallization procedures. Thus, this technique is used extensively in microscale laboratory experiments. Recrystallization is avoided until the last stages of purification, where it will be most efficient, but chromatography does the real dirty work. *One major advantage of working with small amounts of product is that the chromatographic times are shortened dramatically.* Column chromatography of a few milligrams of product usually takes no more than half an hour, but to chromatograph 10 g of product might take a whole afternoon or the better part of a day. Large-scale chromatograms (50–100 g) may take several days or even several weeks to complete.

THIN-LAYER CHROMATOGRAPHY

Thin-layer chromatography is another solid–liquid partition technique of more recent development. It is a close relative to column chromatography in that the phase materials used in both techniques are essentially identical. That is, alumina and silica gel are used as stationary phases and the moving phases are the usual solvents. There are, however, some distinct operational differences between thin-layer and column chromatography. While the moving phase descends in column chromatography, in thin-layer chromatography the solvent front ascends. The column of stationary-phase material used in column chromatography is replaced, in thin-layer chromatography, by a very thin layer (100 μm) of the material spread over a flat surface. The technique has some distinct advantages at the microscale level. It is very rapid (10–20 minutes), and it employs *very* small quantities of material (2–20 μg). The chief disadvantage of this type of chromatography is that it is not very amenable to preparative scale work. Even using large surfaces and thicker layers, separations are still restricted to the 5–10 mg range.

The sequence of operations is described below.

1. A piece of window glass, a microscope slide, or a sheet of plastic can be used as a support for a thin layer of adsorbent spread over the surface. It is possible to prepare locally the glass surfaces, but plastic-supported thin-layer systems are now only commercially available. The plastic-supported layers are particularly attractive because they possess very uniform coatings and are highly reproducible in operation. Another convenient feature of the plastic-backed plates is that they can be cut with scissors into very economical 1 × 3 in. strips. Exclusive use of the latter style is made in the microscale laboratory.

2. A pencil line is drawn parallel to a short side of the plate 1.0 cm from the edge. One or two points evenly spaced are marked on the line. The sample (1 mg) to be analyzed is placed in a 100 μL conical vial and a few drops of solvent added. A micropipet (prepared by the same technique used for constructing the capillary insert used in the ultramicro boiling point determination, (see Chapter 4, p. 34) is used to apply a small fraction of the solution from the vial to the plate (Fig. 5.31).

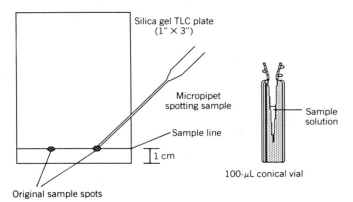

Fig. 5.31 *Sample application to a TLC plate.*

3. The chromatogram is carried out by placing the spotted thin-layer plate in a screwcapped wide-mouth jar or a beaker with a watch glass cover containing a small amount of developing solvent (Fig. 5.32). The material spot on the TLC plate initially must be positioned above the solvent line. The jar is quickly recapped or the watch glass replaced in order to maintain an atmosphere saturated with the developing solvent. The elution solvent rapidly ascends the plate by capillary action. The choice of solvent will be similar to that used in column chromatography, but need not be identical. The spotted material becomes eluted vertically up the plate. Resolution of mixtures into individual spots along the vertical axis occurs by precisely the same mechanism as in column chromatography. The development is interrupted when the solvent line nears the top of the plate.

Fig. 5.32 *Development of a TLC plate. From James W. Zubrick, The Organic Chem Lab Survival Manual (New York: Wiley, 1984). (Reprinted by permission of John Wiley & Sons, New York, NY.)*

4. Visualization of colorless separated components is achieved by placing the plate in an iodine vapor chamber for a few seconds. Iodine forms a re-

versible complex with most organic substances. Thus, dark spots will develop in those areas containing sample material. On removal from the iodine chamber, the spots are marked by pencil, because they will fade rather rapidly. The elution characteristics are reported as R_f values. The R_f value is a measure of the travel of a substance up the plate during the chromatogram relative to the solvent movement. These values are defined as the length of migration by the substance divided by the distance traversed by the solvent front (the position of the solvent front should be quickly marked on the plate when the chromatogram is terminated (Fig. 5.33).

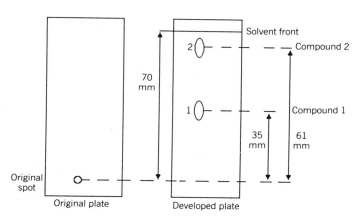

Fig. 5.33 *A sample calculation of R_f values.*

$$R_f \text{ (compound 1)} = \tfrac{35}{70} = 0.50 \qquad R_f \text{ (compound 2)} = \tfrac{61}{70} = 0.87$$

Thin-layer chromatography is used in a number of applications. The speed of the technique makes it quite useful for monitoring large-scale column chromatograms. Analysis of fractions can guide decisions on the solvent elution sequence. TLC analysis of column-derived fractions can also give an indication of how best to combine collected fractions. Following the progress of a reaction by periodically removing small aliquots for thin-layer analysis is another useful application of thin-layer chromatography. The technique is utilized in several microscale experiments. For example, see Experiments 32A and 32B.

Experiments 8A, 8B, 8C, 8D, 8E
The Esterification Reaction: Ethyl Laurate; Isopentyl Acetate; Octyl Acetate; 4-t-Butylcyclohexyl Acetate; Benzoin Acetate

(lauric acid ethyl ester; acetic acid isopentyl ester; acetic acid octyl ester; cyclohexanol, 4-t-butyl-acetate; ethanone, 2-(acetyloxy)-1,2-diphenyl-)

The experiments described below demonstrate the preparation of organic esters by treatment of a carboxylic acid or anhydride with an alcohol in the presence of an acid catalyst.

DISCUSSION   Esterification is one of the important reactions in organic chemistry and has been studied extensively.

$$R-C\overset{\displaystyle O:}{\underset{\displaystyle OH}{\Big\langle}} \; + \; R'-\overset{..}{\underset{..}{O}}H \; \underset{}{\overset{H^+}{\rightleftharpoons}} \; R-C\overset{\displaystyle O:}{\underset{\displaystyle OR'}{\Big\langle}} \; + \; H_2O$$

Carboxylic acid   Alcohol   Ester

A wide range of esters are important as chemical intermediates in organic synthesis and find extensive use in industrial product applications from fingernail polish remover and pain-relieving drugs to polymeric fiber compositions and surfactants. They are also distributed widely in nature in the form of fatty acid esters, waxes, and oils.

The majority of esters are prepared by one of four basic routes.

1. Direct esterification of a carboxylic acid with an alcohol in the presence of an acid catalyst.
2. Alcoholysis of acid chlorides, anhydrides, or nitriles.
3. Reaction of a carboxylic acid salt with an alkyl halide or sulfate.
4. The transesterification reaction.

Direct esterification, known as Fischer esterification, is the method used for the preparation of ethyl laurate in Part A of this experiment. The reaction proceeds by nucleophilic attack of the alcohol on the protonated carbonyl group of the carboxylic acid to form a tetrahedral intermediate. Regeneration of the carbonyl group produces the ester and water. The overall sequence is outlined below.

$$R-C\overset{\displaystyle \overset{..}{O}:}{\underset{\displaystyle OH}{\Big\langle}} \; + \; H^+ \; \rightleftharpoons \; R-C\overset{\displaystyle \overset{+}{O}-H}{\underset{\displaystyle OH}{\Big\langle}}$$

$$R'-\overset{..}{\underset{\displaystyle H}{O}}: \; + \; R-C\overset{\displaystyle \overset{+}{O}:-H}{\underset{\displaystyle OH}{\Big\langle}} \; \rightleftharpoons \; R'-\overset{\displaystyle \overset{+}{R}}{\underset{\displaystyle H \; :OH}{O-C-\overset{..}{O}H}} \; \rightleftharpoons \; R'-\overset{\displaystyle R}{\underset{\displaystyle \overset{..}{O}:^+ \; H \; H}{O-C-\overset{..}{O}H}}$$

$$H_3O^+ \; + \; R'-\overset{\displaystyle R}{O-C\overset{\displaystyle}{\underset{\displaystyle \overset{..}{O}:}{\Big\backslash}}} \; \rightleftharpoons \; R'-\overset{\displaystyle R}{O-C\overset{\displaystyle}{\underset{\displaystyle \overset{..}{O}-H}{\Big\backslash}}} \; + \; H_2O$$

The Fischer method of preparing esters is an equilibrium reaction and, therefore, to obtain substantial yields of the ester product, the reaction must be shifted to the side of the ester. This is accomplished in several ways. If the alcohol is relatively cheap, such as methanol or ethanol, an excess of the alcohol is used to shift the position of equilibrium. This technique is used in the preparation of ethyl laurate (Part A). An alternative is to use an excess of the carboxylic acid. A third option is removal of the ester or water formed in the reaction by distillation. The acid catalyst employed in the reaction is generally dry hydrogen chloride, concentrated sulfuric, or p-toluenesulfonic acid.

When the carboxylic acid and hydroxyl groups are present in the same molecule, a cyclic ester (called a lactone) may be formed. This is especially prevalent in the formation of five- or six-membered ring systems.

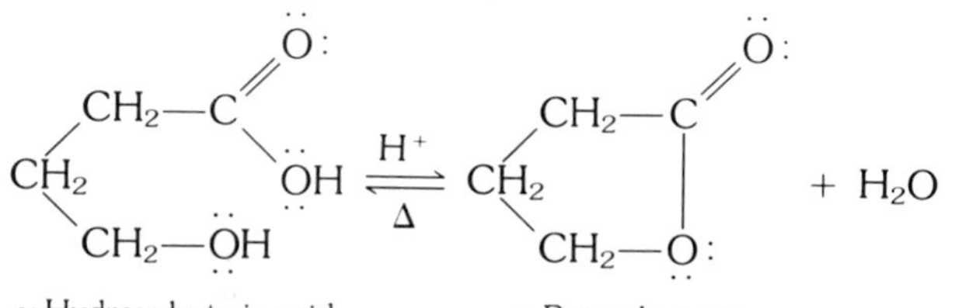

γ-Hydroxybutyric acid   γ-Butyrolactone

As noted above, the Fischer method of preparing esters is an equilibrium reaction. Heating an ester in the presence of an acid catalyst with water regenerates the corresponding carboxylic acid and alcohol. This reaction is called *acid hydrolysis* of an ester. The rate-determining step in both the esterification and hydrolysis reaction is the formation of the tetrahedral intermediate. It is therefore evident that the rate will be determined by the ease with which the nucleophile (alcohol on esterification and water on hydrolysis) approaches the carbonyl group. Steric and electrical factors have a large impact on the rate. The increase of bulky substituents on the α and β positions of the carboxylic acid or ester decreases the rate. Electron-withdrawing groups near the carbonyl group increase the rate because they increase the electrophilicity of the carbonyl carbon atom. Conversely, electron-donating groups retard the rate.

Another of the methods for the preparation of esters is also illustrated in this experiment. Isopentyl acetate, octyl acetate, 4-*t*-butyl cyclohexyl acetate, and benzoin acetate are prepared by the reaction of an alcohol with acetic anhydride, usually in the presence of an acid catalyst. This is a common technique for the preparation of acetate esters. One portion of the anhydride acylates the alcohol; the other forms the carboxylic acid.

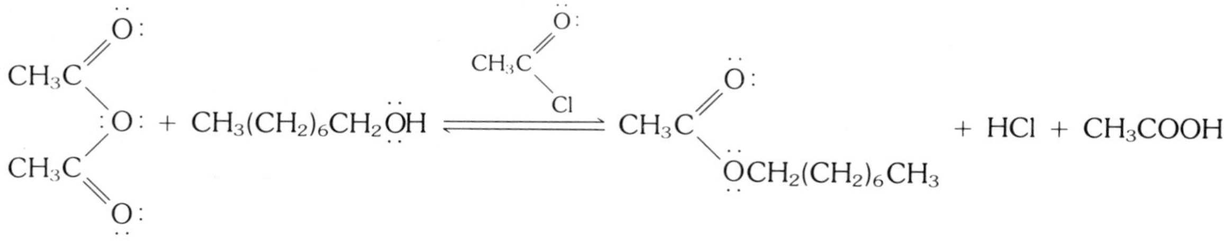

Acetic anhydride   1-Octanol   1-Octyl acetate

If a cyclic anhydride (prepared in Experiments 28A and 28B) is used as a reactant, it may react with one mole of alcohol to form a half-acid ester.

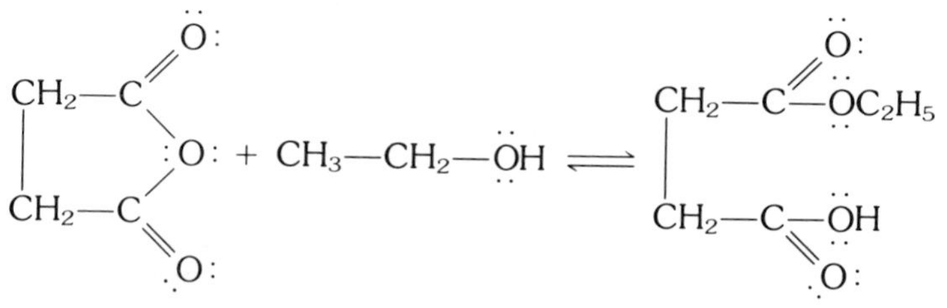

Succinic anhydride   Ethanol   Ethyl hydrogen succinate

In the preparations given in this experiment for the acetate esters listed above, three use acetyl chloride as a catalyst source. Acid chlorides are more reactive than anhydrides (why?). The acetyl chloride added to the reaction mixture reacts with a molecule of the alcohol to produce HCl *in situ*. The formation of the HCl is shown below. Notice that the other product of this reaction is a molecule of the desired ester.

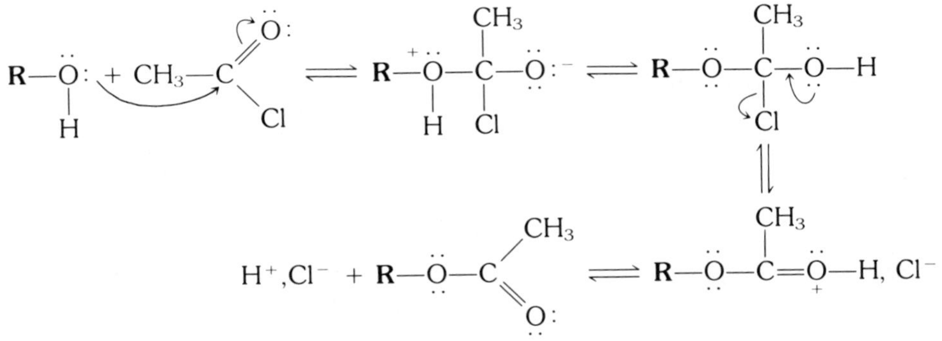

The generated hydrogen chloride acts as the catalyst (how?) for the reaction of the anhydride and alcohol to form the ester.

If the hydrolysis of an ester is carried out in the presence of base, the reaction is called *saponification*.

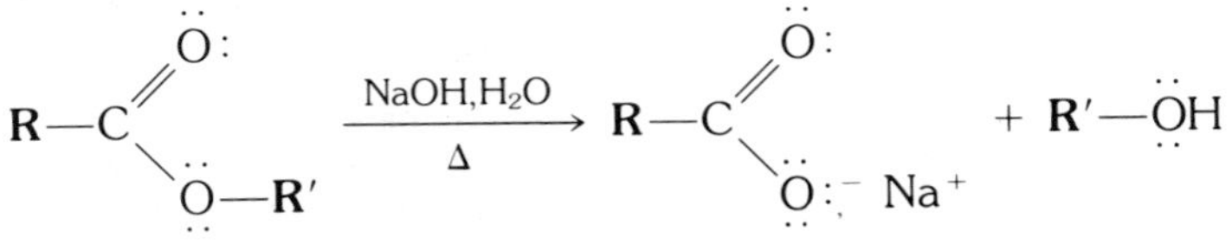

Ester   Carboxylic acid salt   Alcohol

Saponification is essentially an irreversible reaction in which one mole of base is consumed per mole of ester to generate a carboxylic acid anion (as its salt). It is possible to carry out the reaction in a quantitative manner so that a value called the *saponification equivalent* can be obtained. This value is the molecular weight of the ester divided by the number of ester groups in the species being analyzed. A weighed amount of the ester is heated with an excess (known volume) of standard alkaline hydroxide solution, and the excess base is then determined by titration with standard acid.

When the saponification reaction is carried out on a fat or oil belonging to the naturally occurring class of compounds called *lipids*, a soap is obtained.

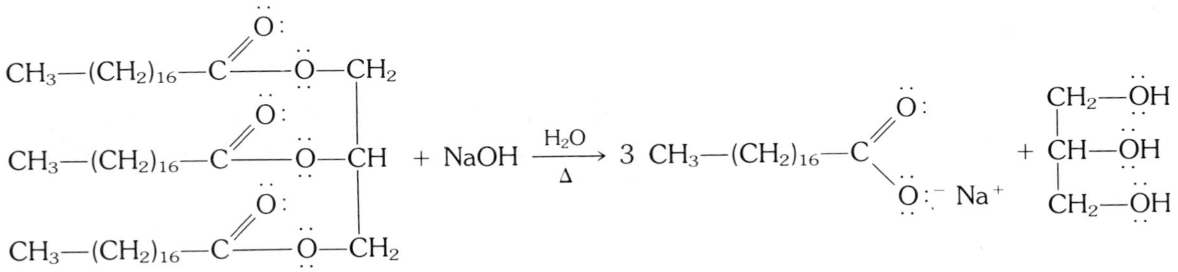

Stearic acid ester of glycerol   Sodium stearate   Glycerol
(a triglyceride)   (a soap)

Fats and oils are esters of glycerol with long-chain carboxylic acids. We should note that waxes are esters of long-chain carboxylic acids and long-chain primary alcohols.

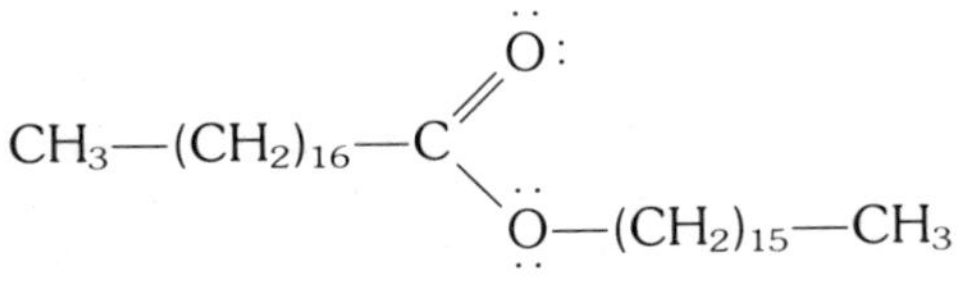

Cetyl stearate 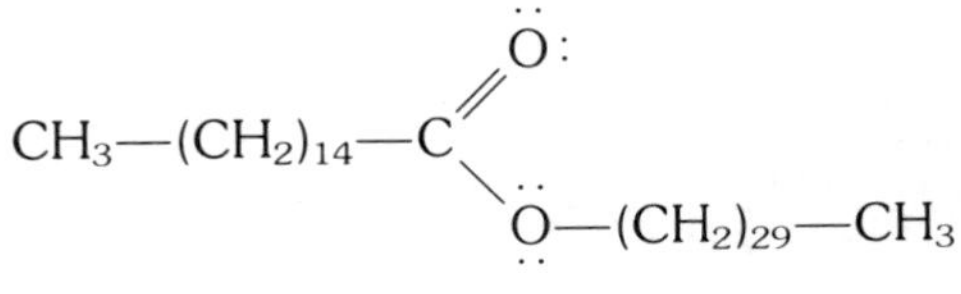   Triacontyl palmitate

The condensation of difunctional carboxylic esters with difunctional alcohols using a reaction called *transesterification* has led to the synthesis of high-molecular-weight polyesters. Many of these polymeric materials have achieved industrial importance. The reaction used to prepare the textile fiber named Dacron is outlined below.

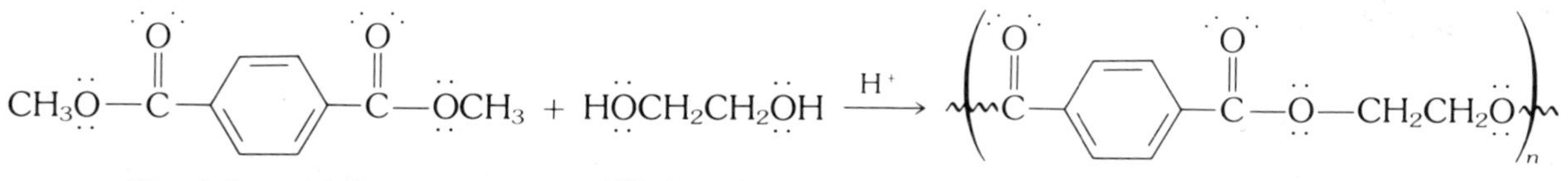

| Dimethyl terephthalate | Ethylene glycol | A polyester (Dacron) |

The transesterification reaction is acid catalyzed and is used to convert one ester into another by heating with an excess of an alcohol or carboxylic acid. In the example given below, the equilibrium is shifted to the right by removal of the ethanol as it is formed.

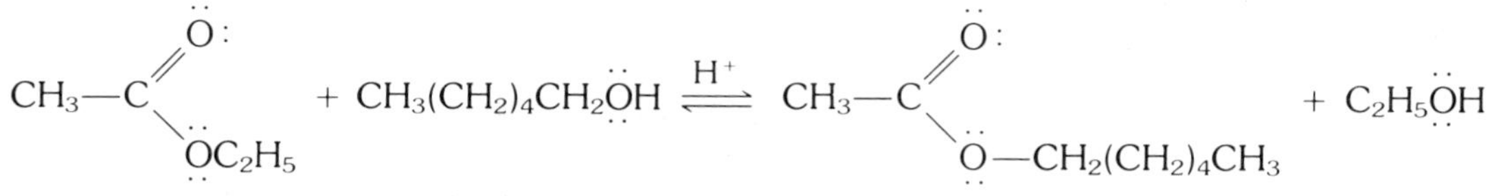

PART A: Ethyl Laurate

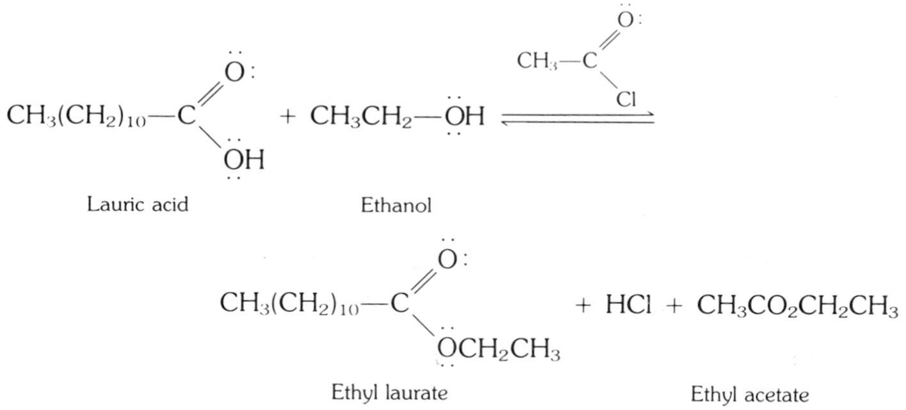

| Lauric acid | Ethanol |

| Ethyl laurate | Ethyl acetate |

DISCUSSION See above.

EXPERIMENTAL Estimated time to complete the experiment: 3.0 hours.

Physical Properties of Reactants and Products

Compound	MW	Wt/Vol	mmol	mp(°C)	bp(°C)	Density	n_D
Lauric acid	200.33	70 mg	0.35	44			
Ethanol	46.07	1.0 mL	17		78.5	0.79	1.3611
Acetyl chloride	78.50	30 μL	0.42		50.9	1.11	1.3898
Ethyl laurate	228.36				273	1.43	1.4311

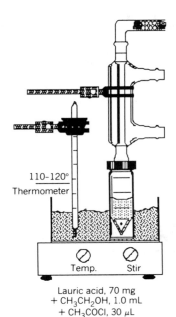

110–120°
Thermometer

Temp.   Stir

Lauric acid, 70 mg
+ CH₃CH₂OH, 1.0 mL
+ CH₃COCl, 30 μL

Reagents and Equipment

To a 3.0-mL conical vial containing a magnetic spin vane and equipped with a reflux condenser protected by a calcium chloride drying tube, add 70 mg (0.35 mmol) of lauric acid. (■) Using a graduated 1.0-mL pipet, add 1.0 mL of absolute ethanol followed by 30 μL (0.43 mmol) of acetyl chloride.

> **WARNING:** *Acetyl chloride is an irritant and should be dispensed in the* Hood *using an automatic delivery pipet. The vial should be capped immediately since this reagent is moisture sensitive.*

Reaction Conditions

The reaction mixture is now heated with stirring at reflux using a sand bath temperature of 110–120°C for a period of one hour. The resulting mixture is then cooled to room temperature and the spin vane removed with forceps.

Isolation of Product

HOOD

A boiling stone is added to the vial and the reaction solution concentrated to a volume of ~0.25 mL by warming in a sand bath [**HOOD**].

To the resulting product mixture is added 0.5 mL of diethyl ether and 0.25 mL of 5% sodium bicarbonate solution using a 1.0-mL graduated pipet. The conical vial is capped and shaken gently, and the cap carefully loosened to vent the two-phase mixture. The aqueous layer is removed with a Pasteur filter pipet and discarded. The organic phase is then extracted with three additional 0.25-mL portions of 5% sodium bicarbonate solution.

This extraction procedure also insures removal of any traces of ethyl acetate which might remain in the reaction mixture following concentration.

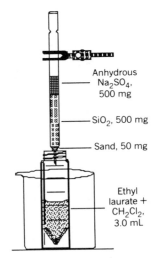

Anhydrous
Na₂SO₄,
500 mg

SiO₂, 500 mg

Sand, 50 mg

Ethyl
laurate +
CH₂Cl₂,
3.0 mL

The wet, crude ethyl laurate is dried and purified by column chromatography. In a Pasteur filter pipet in which the tip has been shortened to about 3.0 cm is placed 500 mg of activated silica gel followed by 500 mg of anhydrous sodium sulfate. (■) The column is first wet with 0.5 mL of methylene chloride and the crude ethyl laurate placed on the column using a Pasteur pipet. A tared 5-mL conical vial containing a boiling stone is used as a collection flask. The reaction vial is rinsed with two 0.5-mL portions of methylene chloride and each rinse is also transferred to the column using the same pipet. An additional 1.0 mL of methylene chloride is then added directly to the column to ensure complete elution of the product.

HOOD   The methylene chloride is removed by evaporation in the **HOOD** by using a stream of nitrogen gas and by gentle warming in a sand bath.

Purification and Characterization

Ethyl laurate, a clear, viscous, pleasant-smelling ester, is sufficiently pure as isolated for characterization. Weigh the material and calculate the percentage yield. Determine the refractive index and boiling point and compare the results to the literature values. Obtain an IR spectrum of the ester and compare it with an authentic sample.

PART B: Isopentyl Acetate

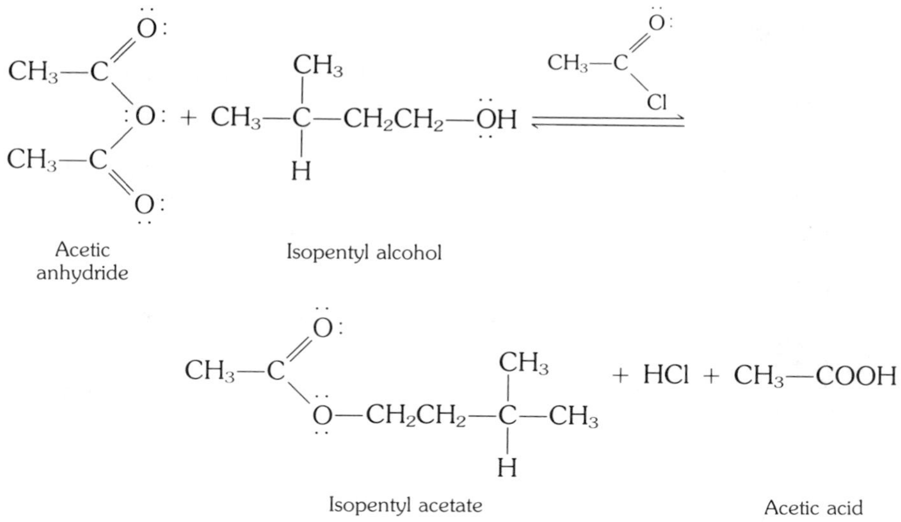

$$CH_3-C(=O)-O-C(=O)-CH_3 \;+\; CH_3-\underset{H}{\overset{CH_3}{C}}-CH_2CH_2-OH \;\;\underset{\overset{\displaystyle CH_3-C(=O)-Cl}{}}{\rightleftharpoons}$$

Acetic anhydride	Isopentyl alcohol

$$CH_3-C(=O)-O-CH_2CH_2-\underset{H}{\overset{CH_3}{C}}-CH_3 \;+\; HCl \;+\; CH_3-COOH$$

Isopentyl acetate		Acetic acid

DISCUSSION   See above.

EXPERIMENTAL   Estimated time to complete the experiment: 3.0 hours.

Physical Properties of Reactants and Products

Compound	MW	Wt/Vol	mmol	bp(°C)	Density	n_D
Isopentyl alcohol	88.15	100 μL	0.92	132	0.81	1.4053
Acetic anhydride	102.09	150 μL	1.59	139.6	1.08	1.3901
Acetyl chloride	78.50	60 μL	0.84	50.9	1.11	1.3898
Isopentyl acetate	130.19			142	0.87	1.4003

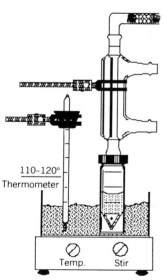

110–120°
Thermometer

Temp.   Stir

Isopentyl alcohol, 100μL
+ (CH₃CO)₂O, 150 μL
+ CH₃COCl, 60 μL

Reagents and Equipment
In a 1.0-mL conical vial containing a magnetic spin vane and equipped with a reflux condenser protected by a calcium chloride drying tube is placed 100 μL (81.0 mg, 0.92 mmol) of isopentyl alcohol, 150 μL of acetic anhydride, and 60 μL (0.84 mmol) of acetyl chloride.(■)

> **WARNING:** *The vial should be capped immediately after addition of each reagent. They should be dispensed in the* **HOOD** *using automatic delivery pipets. Acetyl chloride and acetic anhydride are moisture sensitive and are also irritants.*

Reaction Conditions
The reaction mixture is heated with stirring using a sand bath temperature of 110–120°C for a period of one hour. The resulting mixture is then cooled to room temperature and the spin vane removed with forceps.

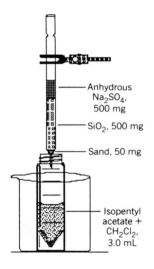

Anhydrous
Na₂SO₄,
500 mg

SiO₂, 500 mg

Sand, 50 mg

Isopentyl
acetate +
CH₂Cl₂,
3.0 mL

Isolation of Product
The identical extraction procedure is used as outlined in Part A, Isolation of Product.(■)

Purification and Characterization
The clear, viscous, pleasent-smelling ester isopentyl acetate is sufficiently pure for characterization. Weigh the product and calculate the percentage yield. Determine the refractive index and boiling point and compare your values with those found in the literature. Obtain an IR spectrum of the ester and compare it with that of an authentic sample.

PART C: Octyl Acetate

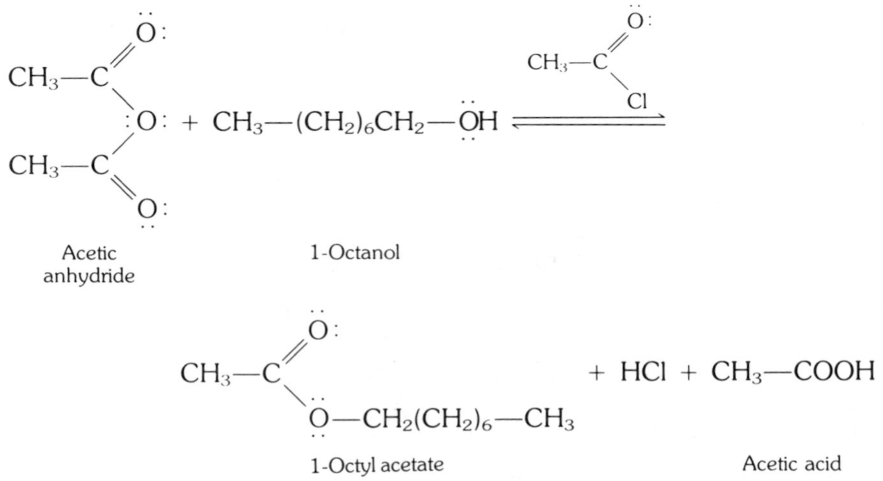

$$CH_3-C(=O)-O-C(=O)-CH_3 \ (\text{Acetic anhydride}) + CH_3-(CH_2)_6CH_2-OH \ (\text{1-Octanol}) \rightleftharpoons CH_3-C(=O)-Cl$$

$$CH_3-C(=O)-O-CH_2(CH_2)_6-CH_3 \ (\text{1-Octyl acetate}) + HCl + CH_3-COOH \ (\text{Acetic acid})$$

DISCUSSION See above.

EXPERIMENTAL Estimated time to complete the experiment: 3.0 hours.

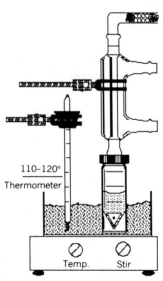

110–120°
Thermometer

Temp.   Stir

1-Octanol, 61 μL
+ (CH₃CO)₂O, 121 μL
+ CH₃COCl, 24 μL

Physical Properties of Reactants and Products

Compound	MW	Wt/Vol	mmol	bp(°C)	Density	n_D
1-Octanol	130.23	61 μL	0.38	194.4	0.83	1.4295
Acetyl chloride	78.50	24 μL	0.34	50.9	1.11	1.3898
Acetic anhydride	102.09	121 μL	1.28	139.6	1.08	1.3901
1-Octyl acetate	172.27			210	0.87	1.4150

Reagents and Equipment
In a 1.0-mL conical vial containing a magnetic spin vane and equipped with a reflux condenser protected by a calcium chloride drying tube is placed 61 μL (50 mg, 0.38 mmol) of 1-octanol followed by 121 μL (1.28 mmol) of acetic anhydride and 24 μL (0.34 mmol) of acetyl chloride.(■)

> **WARNING:** *Acetic anhydride and acetyl chloride are irritants. They are dispensed in the HOOD using automatic delivery pipets. The vial is capped after addition of each reagent, since they are moisture sensitive.*

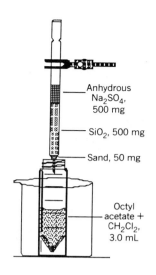

Anhydrous
Na₂SO₄,
500 mg

SiO₂, 500 mg

Sand, 50 mg

Octyl
acetate +
CH₂Cl₂,
3.0 mL

Reaction Conditions

The reaction mixture is now heated in a sand bath maintained at a temperature of 110–120°C with stirring for a period of one hour. The resulting mixture is then cooled to room temperature and the spin vane removed with forceps.

Isolation of Product

The product is isolated using the extraction procedure identical to that given in Part A.(■)

Purification and Characterization

The clear, viscous, pleasant-smelling ester octyl acetate is sufficiently pure for characterization.

Weigh the ester and calculate the percentage yield. Determine the refractive index and boiling point and compare your values to those in the literature. Obtain an IR spectrum of your octyl acetate and compare it with an authentic sample.

PART D: 4-*t*-Butylcyclohexyl Acetate

Acetic
anhydride

4-*t*-Butylcyclohexanol

4-*t*-Butylcyclohexyl acetate

+ HCl + CH₃—COOH

Acetic acid

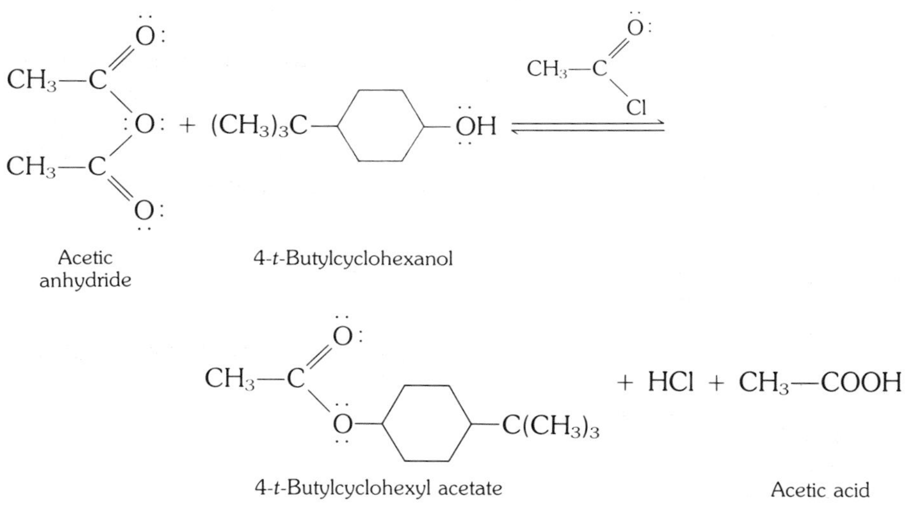

DISCUSSION   See above.

EXPERIMENTAL   Estimated time to complete the experiment: 3.0 hours.

Physical Properties of Reactants and Products

Compound	MW	Wt/Vol	mmol	mp(°C)	bp(°C)	Density	n_D
4-*t*-butylcyclohexanol (mixture of isomers)	156.27	80 mg	0.51	62–70			
Acetic anhydride	102.09	100 μL	1.06		139.6	1.08	1.3901
Acetyl chloride	78.50	60 μL	0.85		50.9	1.11	1.3898
4-*t*-Butylcyclohexyl acetate	198.30				100–105 (8 mm)	0.94	1.4507

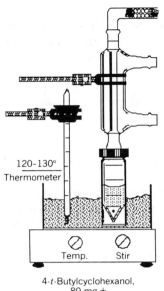

120–130°
Thermometer

Temp.   Stir

4-t-Butylcyclohexanol,
80 mg +
(CH₃CO)₂O, 100 μL +
CH₃COCl, 60 μL

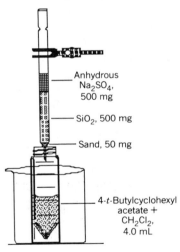

Anhydrous
Na₂SO₄,
500 mg

SiO₂, 500 mg

Sand, 50 mg

4-t-Butylcyclohexyl
acetate +
CH₂Cl₂,
4.0 mL

Reagents and Equipment

To a 3.0-mL conical vial containing a magnetic spin vane and equipped with a reflux condenser protected by a calcium chloride drying tube add 80 mg (0.51 mmol) of 4-t-butylcyclohexanol, 100 μL (1.06 mmol) of acetic anhydride, and 60 μL (0.85 mmol) of acetyl chloride.(■)

> **WARNING:** *Acetyl chloride and acetic anhydride are irritants. They are dispensed in the HOOD using automatic delivery pipets. The vial should be capped immediately after each addition since these substances are also moisture sensitive.*

Reaction Conditions

The reaction mixture is heated with stirring at a sand bath temperature of 120–130°C for a period of one hour. It is then cooled to room temperature and the spin vane removed with forceps.

Isolation of Product

The identical extraction procedure is used as outlined in Part A, Isolation of Product. The only exception is the last step in the chromatographic sequence; 2.0 mL of methylene chloride are used in the final elution step in place of 1.0 mL of this solvent.(■)

Purification and Characterization

The clear, pleasant-smelling ester 4-t-butylcyclohexyl acetate is sufficiently pure for characterization. Weigh the ester and calculate the percentage yield. Determine the refractive index and boiling point and compare your values to those reported in the literature. Obtain an IR spectrum of the ester and compare it with that shown in Fig. 5.34.

PART E: Benzoin Acetate

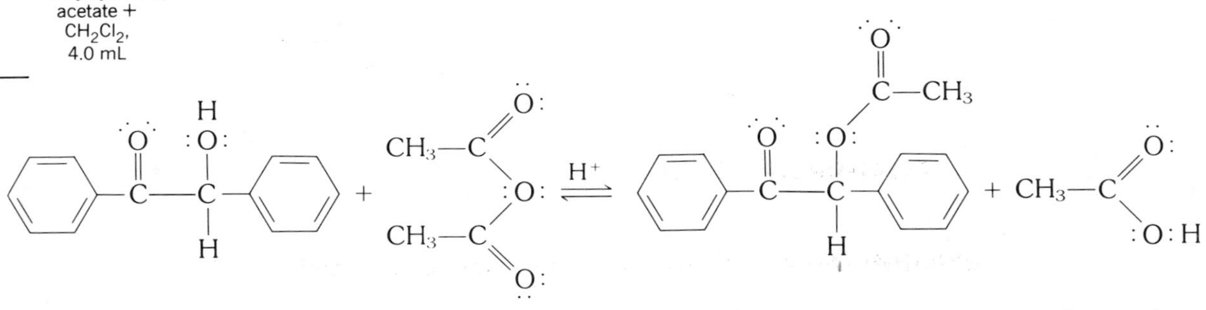

Benzoin   Acetic   Benzoin acetate   Acetic acid
                 anhydride

The formation of a solid acetate derivative of an alcohol by the esterification reaction is demonstrated in this experiment.

DISCUSSION   See above.

EXPERIMENTAL   Estimated time to complete the experiment: 0.5 hour.

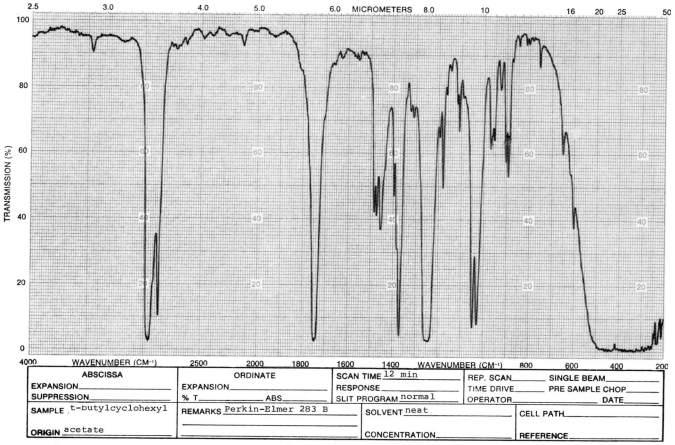

ABSCISSA	ORDINATE	SCAN TIME 12 min	REP. SCAN_____ SINGLE BEAM_____
EXPANSION_____	EXPANSION_____	RESPONSE_____	TIME DRIVE_____ PRE SAMPLE CHOP_____
SUPPRESSION_____	% T_____ ABS_____	SLIT PROGRAM normal	OPERATOR_____ DATE_____
SAMPLE t-butylcyclohexyl	REMARKS Perkin-Elmer 283 B	SOLVENT neat	CELL PATH_____
ORIGIN acetate		CONCENTRATION_____	REFERENCE_____

Fig. 5.34 *IR spectrum: 4-t-butylcyclohexyl acetate.*

Physical Properties of Reactants and Products

Compound	MW	Wt/Vol	mmol	mp(°C)	bp(°C)	Density
Benzoin	212.25	20 mg	0.09	137		
Acetic anhydride	102.09	48 μL	0.51		139.6	1.08
Glacial acetic acid	60.05	48 μL	0.84		117.9	
Conc. sulfuric acid	98.08	2 μL	0.04			
Benzoin acetate	254.29			83		

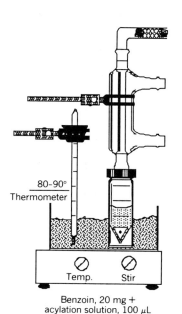

80–90°
Thermometer

Benzoin, 20 mg +
acylation solution, 100 μL

HOOD

Reagents and Equipment

In a 3.0-mL conical vial containing a magnetic spin vane and equipped with a reflux condenser protected by a calcium chloride drying tube is placed 20 mg of benzoin followed by 100 μL of acylation solution.(■)

Purification of the benzoin used in the experiment is not necessary if the melting point is above 132°C. Benzoin is prepared in Experiment 6.

Instructor Preparation

The acylation solution is prepared by mixing 5.0 mL of acetic anhydride, 5.0 mL of glacial acetic acid, and 0.5 mL of concentrated sulfuric acid. It should be dispensed in the **HOOD** by use of an automatic delivery pipet.

Reaction Conditions

The reaction mixture is warmed in a sand bath at a temperature of 80–90°C for 5 minutes.

Isolation of Product

The solution is cooled to room temperature and 0.25 mL of water added using a calibrated Pasteur pipet. The resulting mixture is cooled in an ice bath for a period of 5 minutes and then extracted with three 0.5-mL portions of methylene chloride. Upon each addition of methylene chloride, the vial is capped, shaken, and vented and the layers are allowed to separate. Using a Pasteur filter pipet, the bottom methylene chloride layer is separated and transferred to a Pasteur filter pipet containing 0.5 g of activated silica gel (100 mesh) and 0.3 g of anhydrous sodium sulfate. This chromatographic column should be previously wet with methylene chloride solvent. After the transfer of the third portion of the methylene chloride, an additional 0.5 mL of solvent is added to the column. The eluate is collected in a 10-mL Erlenmeyer flask.(■)

HOOD

The eluate is concentrated in the **HOOD** by warming in a sand bath. The resulting white, solid product is removed from the flask and dried on a porous clay plate.

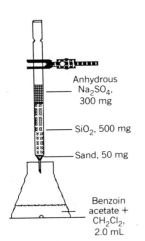

Anhydrous
Na$_2$SO$_4$,
300 mg

SiO$_2$, 500 mg

Sand, 50 mg

Benzoin
acetate +
CH$_2$Cl$_2$,
2.0 mL

Purification and Characterization

The product is generally pure enough for characterization. However, it may be recrystallized from methanol/water if desired, using the Craig tube.

Note. *In the recrystallization sequence, the material is dissolved in warm methanol. Water is then added dropwise until the solution becomes cloudy. The mixture is again warmed until a clear solution is observed and then allowed to cool. It is not necessary to wash the collected crystals before drying.*

Weigh the benzoin acetate and calculate the percentage yield. Determine the melting point and compare your value to that found in the literature. Obtain an IR spectrum of the ester and compare it with that of the starting material and also with that of an authentic sample.

ENVIRONMENTAL DATA

Substance	Amount	TLV (mg/m³)	Emissions (mg)	Volume (m³)
Experiment 8A				
Ethanol (reflux only up to Max.)	1.0 mL	1900	70–800	0.4
Methylene chloride (Max. col. chromatog.)	2.5 mL	350	3338	9.5
Acetyl chloride (Max.)	30 μL	10[a]	33	3.3
Experiment 8B				
Isopentyl alcohol (Max. from yield, transfer)	81 mg	360	52	0.14
Acetic anhydride (transfer, reflux)	150 μL	20	12	0.6
Acetyl chloride (Max.)	60 μL	10[a]	66	6.6
Methylene chloride (Max., col. chromatog., evap.)	2.5 mL	350	3338	9.8
Experiment 8C				
1-Octanol (Max. from yield, transfer)	50 mg	—	30	—
Acetic anhydride (reflux, liquid transfer)	121 μL	20	12	0.6
Acetyl chloride (Max.)	24 μL	10[a]	26	2.6
Methylene chloride (Max., col. chromatog., evap.)	2.5 mL	350	3338	9.5

ENVIRONMENTAL DATA
(continued)

Substance	Amount	TLV (mg/m³)	Emissions (mg)	Volume (m³)
Experiment 8D				
4-t-Butylcyclohexanol (Max. from yield, transfer)	80 mg	—	28	
Acetic anhydride (reflux, liquid transfer)	100 µL	20	12	0.6
Acetyl chloride (Max.)	60 µL	10^a	66	6.6
Methylene chloride (Max., col. chromatog., evap.)	2.5 mL	350	3338	9.5
Experiment 8E				
Acetic anhydride (Max., heat in open)	100 µL	20	108	5.4
Methanol (SKIN) (Max., Craig tube)	1.0 mL	260	800	3.1

^{a}This TLV value is calculated assuming the generation of equivalent amounts of acetic acid and hydrochloric acid.

TLV: Acetic acid—10 ppm
Isopentyl acetate—100 ppm
TXDS: Lauric acid—orl-rat LD50: 12 g/kg
Acetyl chloride—inl-hum TCLo: 2 ppm/1 m TFX:IRR
Sodium sulfate—orl-mus LD50: 5989 mg/kg
1-Octanol—orl-mus LD50: 1790 mg/kg
1-Octyl acetate—orl-rat LD50: 3000 mg/kg
4-t-Butylcyclohexanol—orl-rat LD50: 4200 mg/kg
4-t-Butylcyclohexyl acetate—orl-rat LD50: 5000 mg/kg
Benzoin—orl-rat LDLo: 5460 mg/kg
Sulfuric acid—orl-rat LD50: 2140 mg/kg

QUESTIONS

5-36. A series of dyes is separated by TLC. Calculate the R_f value for each dye based on the data summarized below.

Dye	Distance Moved (mm)
Solvent	66
Methyl red	56
Rhodamine B	38
Bismark brown	16
Congo red	0.5

5-37. To combat the problem of slow mass transfer and relatively slow flow rates to achieve a large number of theoretical plates, the technique of high-performance liquid chromatography (HPLC) was developed. This technique involves the use of a high pressure (~5000 psi) at the top of the column, which causes the mobile phase to be driven through the column at a faster rate. Suitable packings and pressure systems have been developed to accommodate the high pressures. The method is particularly useful for biochemical compounds that would decompose under gas chromatographic conditions.

Trace amounts of phenol in water can be detected by HPLC. If a 50-nanogram phenol standard gives a peak area of 140 units and a 20-µL water sample gives a peak area due to phenol of 26 units, calculate the phenol content of the water sample as parts per million (ppm) (microgram/milliliter).

5-38. Fatty acids are long-chain carboxylic acids, usually of more than 12 carbon atoms, isolated from saponification of fats (esters of glycerol). Draw the structure of each of the fatty acids named below and also give its common name.

hexadecanoic acid (Z)-9-octadecenoic acid
octadecanoic acid (9Z,12Z)-9,12-octadecadienoic acid

5-39. Each of the esters below undergoes the saponification reaction when treated with sodium hydroxide. Place them in order of increasing rate of reaction and offer an explanation for your choice. Also, name each ester.

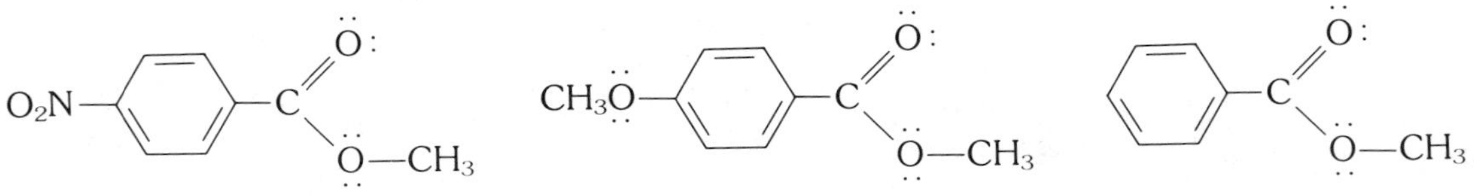

5-40. Can you list several advantages of using acetic anhydride as a reagent to prepare esters even though acetyl chloride is more reactive toward alcohols?

5-41. Discuss the difference between gas chromatography and liquid chromatography in relation to diffusion rates and the interaction of the mobile phase with the sample.

5-42. Write a mechanism for the acid-catalyzed transesterification reaction of ethyl acetate with butyl alcohol to prepare butyl acetate.

REFERENCES

These references are selected from the large number of examples given in "Organic Syntheses" of esterification.

1. Fuson, R. C.; Wojcik, B. H. "Organic Syntheses"; Wiley: New York, 1943; Collect. Vol. II, p 260.
2. Mic'ovic', V. M. *Ibid.*, p 264.
3. Bowden, E. *Ibid.*, p 414.
4. McCutcheon, J. W. "Organic Syntheses"; Wiley: New York, 1955; Collect. Vol. III, p 526.
5. Weissberger, A.; Kibler, C. J. *Ibid.*, p 610.
6. Eliel, E. L.; Fisk, M. T. "Organic Syntheses"; Wiley: New York, 1963; Collect. Vol. IV, p 169.
7. Emerson, W. S.; Longely, R. I., Jr. *Ibid.*, p 302.

TECHNIQUE 9: A CONTROLLED ATMOSPHERE EXPERIMENT

Experiment 9
The Dehydration of a Secondary Alcohol, 2-Butanol: 1-Butene; *trans*-2-Butene; *cis*-2-Butene

(1-butene; 2-butene, E-; 2-butene, Z-)

This experiment illustrates the acid-catalyzed elimination reaction of the secondary (2°) alcohol 2-butanol. The mixture of gaseous products formed in the reaction are separated and analyzed by the use of gas chromatography.

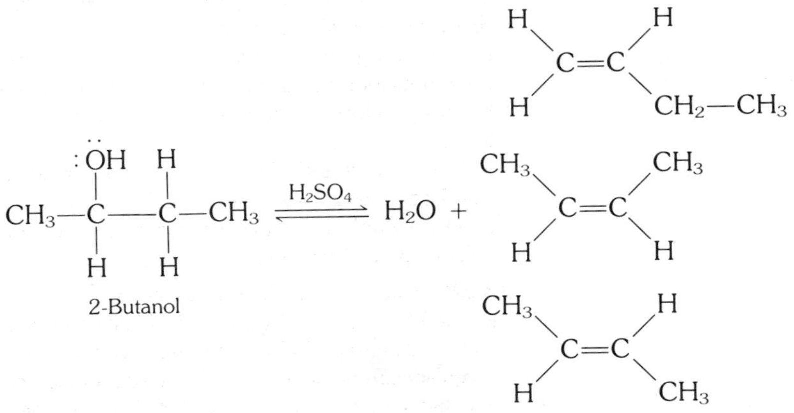

DISCUSSION   The introduction of unsaturation into an aliphatic hydrocarbon generally involves the loss of a proton and a nucleophile from adjacent carbon atoms in a chain or ring.

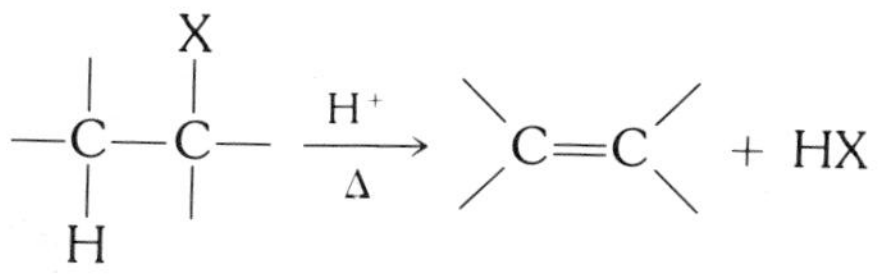

$$-\underset{H}{\underset{|}{\overset{|}{C}}}-\overset{\overset{X}{|}}{\underset{|}{C}}- \xrightarrow[\Delta]{H^+} \quad \underset{/}{\overset{\backslash}{C}}=\underset{\backslash}{\overset{/}{C}} + HX$$

One of the common methods of introducing unsaturation into an organic compound is by dehydration of the corresponding alcohol. The dehydration reaction (loss of water) belongs to the important class of organic reactions called *elimination reactions*, which are often used in synthesis and in the determination of molecular structure.

Dehydration of alcohols is an acid-catalyzed elimination reaction. Experimental evidence shows that alcohols react in the order $3° > 2° > 1°$, which relates to the stability of the carbocation intermediate formed in the reaction. Generally, sulfuric or phosphoric acid are used as catalysts in the laboratory. Lewis acids such as aluminum oxide or silica gel are frequently used at fairly high temperature as catalysts on the industrial scale.

The E_1, acid-catalyzed elimination dehydration reaction proceeds in several steps, as outlined below.

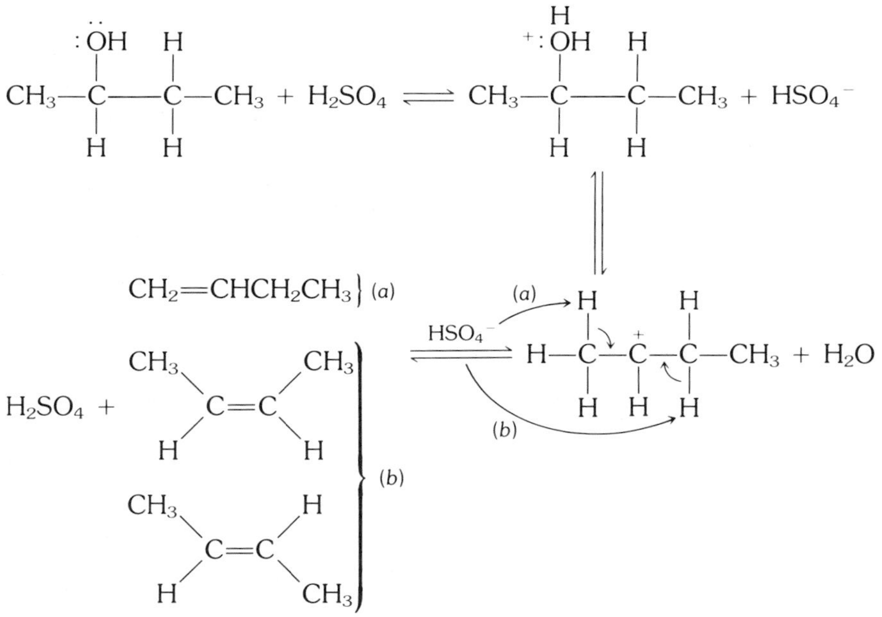

The first step involves the protonation of the oxygen atom of the alcohol to form an oxonium ion in a fast equilibrium reaction. This protonation step is important because it turns a poor leaving group, —OH, into a good leaving group, H_2O (a small neutral molecule). The second step of the reaction is the decomposition of this oxonium ion to yield a carbocation intermediate and water. This is the rate-determining step of the reaction. In the third step, the carbocation stabilizes itself by transferring a proton located on an adjacent carbon atom to a molecule of water or other base present in the system, to regenerate the catalyst and form an alkene.

E_1 elimination reactions involve equilibrium conditions and, thus, to maximize the yield, the alkene is generally removed from the reaction site. Distillation is often used since the alkene *always* has a lower boiling point than the corresponding alcohol. In the present reaction, the alkenes are gases and therefore are easily removed and collected as described.

It is important to realize that many 1° alcohols undergo the dehydration reaction by the E_2 mechanism, mainly because of the fact that the primary carbocation that would be generated in an E_1 process is relatively unstable.

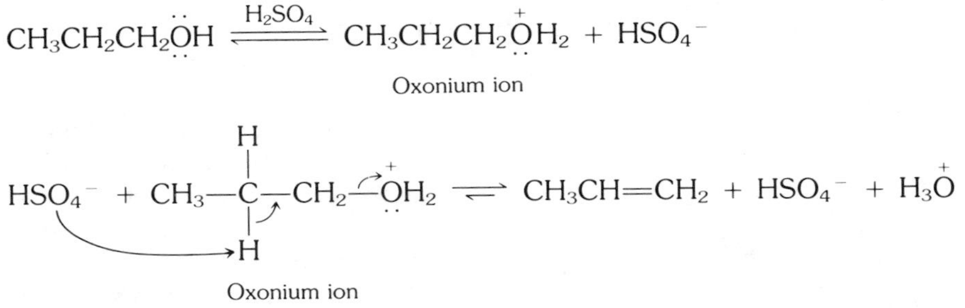

$$CH_3CH_2CH_2\overset{..}{\underset{..}{O}}H \underset{}{\overset{H_2SO_4}{\rightleftharpoons}} CH_3CH_2CH_2\overset{+}{O}H_2 + HSO_4^-$$

Oxonium ion

$$HSO_4^- + CH_3-\underset{\underset{H}{|}}{\overset{\overset{H}{|}}{C}}-CH_2-\overset{+}{\underset{..}{O}}H_2 \rightleftharpoons CH_3CH=CH_2 + HSO_4^- + H_3\overset{+}{O}$$

Oxonium ion

E_1 elimination is usually accompanied by an S_N1 substitution reaction. This competing reaction involves the same carbocation intermediate; however, since the reactions are reversible, and the alkene is easily removed from the reaction site, the substitution reaction is not too troublesome.

Certain alcohols dehydrate to yield more than one alkene. In the present reaction involving the dehydration of 2-butanol, three alkenes are formed. As seen in the mechanism scheme, this is due to the fact that the hydrogen atom removed to form the alkene can come from different adjacent carbon atoms. According to the Saytzeff rule, the alkene formed in the larger amount is that having the highest degree of substitution. Also, trans alkenes are thermodynamically more stable than their cis counterparts. Knowing these general observations, one is often able to predict the ratio of alkenes to expect from a given elimination reaction.

Rearrangement of alkyl groups and hydrogen is often observed in the dehydration of alcohols, especially in the presence of strong acid. The dehydration of 3,3-dimethyl-2-butanol with sulfuric acid catalyst results in the formation of a mixture of alkenes as shown below. Can you predict which olefin is formed in the greatest amount?

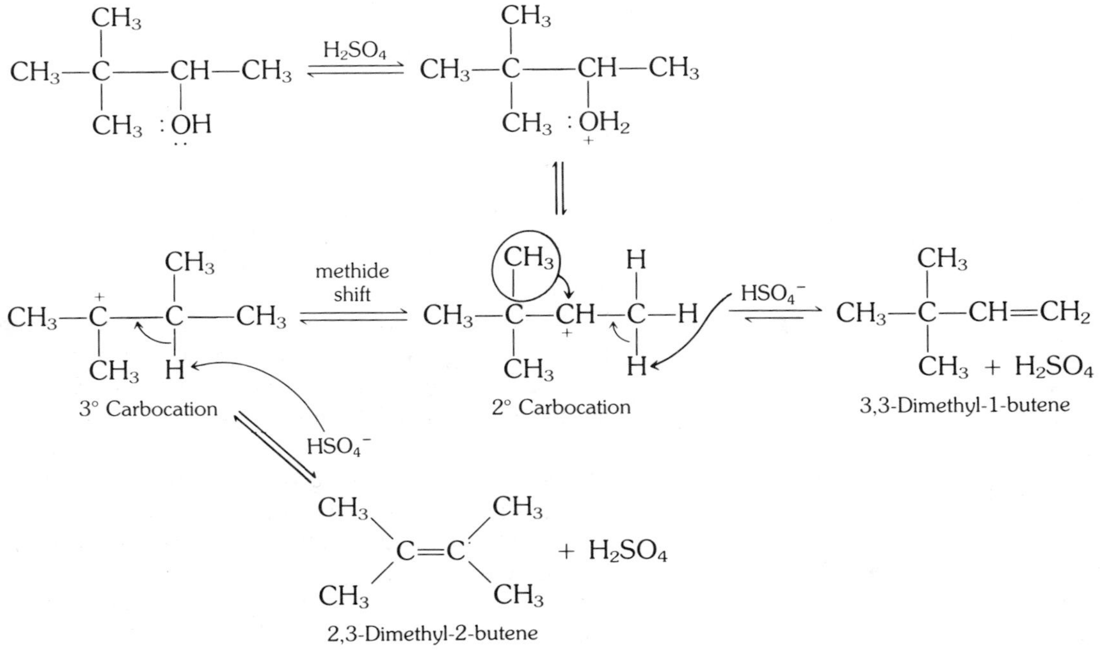

3° Carbocation   2° Carbocation   3,3-Dimethyl-1-butene

2,3-Dimethyl-2-butene

Reactions of this type are very susceptible to rearrangement because a more stable carbocation can be formed in the system. In the example above, a 2° carbocation rearranges to a more stable 3° carbocation. The rearrangement involves the movement of an alkyl group (in this case a methyl) *with its pair of electrons* to the adjacent carbon atom. This type of shift is referred to as a 1,2-shift and is very common in aliphatic systems involving carbocation intermediates. Hydride shifts are also frequently observed.

EXPERIMENTAL

Estimated time to complete the experiment: 2.5 hours.

Note. *It is suggested that the starting times of the reaction be staggered to allow access to the GC instrument by the students when the product gases are analyzed.*

Physical Properties of Reactants and Products

Compound	MW	Wt/Vol	mmol	bp(°C)	Density	n_d
2-Butanol	74.12	100 μL	1.1	99.5	0.81	1.3978
Conc. sulfuric acid	98.08	50 μL				
1-Butene	56.12			−6.3	0.60	
trans-2-Butene	56.12			0.9	0.61	
cis-2-Butene	56.12			3.7	0.62	

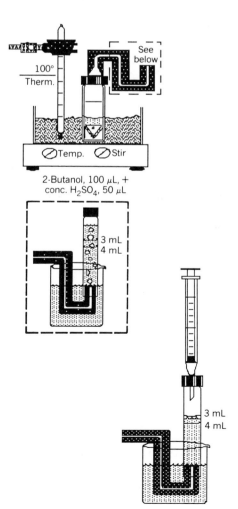

2-Butanol, 100 μL, + conc. H$_2$SO$_4$, 50 μL

Reagents and Equipment

Assemble the gas-collection apparatus shown in the figure *before* the reactants are mixed.(■)

Determine the capacity of the gas-collection reservoir by capping the tube, inverting it, and then adding 3.0 mL of water. Mark the level at 3.0 mL. Add an additional 1.0 mL and mark the 4.0 mL level.

To position the collection reservoir, fill it with water, place a finger over the open end, invert it, and place the reservoir open end down, into a beaker (100–250 mL) filled with water. When the finger is removed, the column of water should remain in the reservoir.

Place 100 μL (81 mg, 1.1 mmol) of 2-butanol and 50 μL of concentrated sulfuric acid in a clean, dry, 1.0-mL conical vial containing a magnetic spin vane. The vial is then attached to the gas-collection apparatus. Position the delivery tube under water into the open end of the collection reservoir and clamp the reservoir in place.

> **CAUTION:** *Sulfuric acid is a strong, corrosive material. Contact with the skin or eyes can cause severe burns. It is convenient to dispense the reactants using automatic delivery pipets.*

Reaction Conditions

With stirring, the reaction mixture is heated using a sand bath until the evolution of gas takes place (~100°C sand bath temperature). The mixture should be warmed slowly to prevent foaming.

Isolation of Product

Collect about 3–4 mL of gas in the collection reservoir and then, using a hypodermic syringe, withdraw a $\frac{1}{2}$-cm^3 sample through the rubber septum for GC analysis.(■)

Remove the delivery tube from the collecting reservoir and then from the water before discontinuing the heat on the reaction vial. This order prevents water from being sucked back into the reaction flask.

Purification and Characterization

The collected gas is analyzed by GC without further purification.

GC Conditions

Column: $\frac{1}{4}$ in. $\times$ 8 ft packed with 20% silicone DC-710 room temperature

20 mL/minute flow rate (He gas)

Sample size: $\frac{1}{2}$ cm^3 of collected gas

Assuming that the amount of each substance in the gas is proportional to the area of its corresponding peak, determine the percentage ratio of the three components in the gas sample.

Area Under a Curve. *There are several techniques that may be used. The following method gives reproducible results of ± 3–4%: the peak height (mm) $\times$ the width at half height (mm), measured from the baseline of the curve.*

The order of elution of the butenes is 1-butene, *trans*-2-butene, and *cis*-2-butene. *If the reaction mixture is heated strongly, rearrangement can occur and isobutene is also formed. If present, four peaks are observed, the isobutene being the first peak (lowest boiling point).*

ENVIRONMENTAL DATA

Substance	Amount	TLV (mg/m^3)	Emissions (mg)	Volume (m^3)
2-Butanol	100 μL	305	10	<0.1
Sulfuric acid	50 μL	1		
Losses unknown, but assumed small due to gas trapping.				

QUESTIONS

5-43. Gas chromatographic analysis of a mixture of organic compounds gave the following peak areas (cm^2): hexane = 2.7; heptane = 1.6; hexanol = 1.8; toluene = 0.5.

    a. Calculate the mole percent composition of the mixture. Assume that the response of the detector (area/mole) is the same for each component.

    b. Calculate the weight percent composition of the mixture under the same assumptions as in part a.

5-44. It is noted at the end of the experiment that if the mixture is heated strongly, rearrangement can occur and isobutene is also formed. Suggest a mechanism to account for the formation of this compound.

5-45. When *t*-pentyl bromide is treated with 80% ethanol, the following amounts of olefinic products are detected on analysis.

$$CH_3CH_2-\underset{\underset{CH_3}{|}}{\overset{\overset{CH_3}{|}}{C}}-Br \xrightarrow{C_2H_5OH(80\%)} CH_3CH{=}C(CH_3)_2 + CH_3CH_2\underset{\underset{CH_3}{|}}{C}{=}CH_2 + \begin{cases} t\text{-Pentyl alcohol} \\ t\text{-Pentylethyl ether} \end{cases}$$

I	II	
32%	8%	60%

Offer an explanation of why compound I is formed in far greater amount than the terminal alkene.

5-46. The —SR$_2^+$ group is easily removed in elimination reactions, but the —SR group is not. Explain.

5-47. Why is sulfuric acid used to catalyze the dehydration of alcohols rather than hydrochloric acid?

REFERENCES

1. Several dehydration reactions of secondary alcohols using sulfuric acid as the catalyst are given in "Organic Syntheses."
   a. Coleman, G. H.; Johnstone, H. F. "Organic Syntheses"; Wiley: New York, 1941; Collect. Vol. I, p 183.
   b. Norris J. F. *Ibid.*, p 430.
   c. Bruce, W. F. "Organic Syntheses"; Wiley: New York, 1943; Collect. Vol. II, p 12.
   d. Adkins, H.; Zartman, W. *Ibid.*, p 606.
   e. Wiley, R. H.; Waddey, W. E. "Organic Syntheses"; Wiley: New York, 1955; Collect. Vol. III, p 560.
   f. Grummitt, O.; Becker, E. I. "Organic Syntheses"; Wiley: New York, 1963; Collect. Vol. IV, p 771.
2. The present experiment is adapted from the method given by Helmkamp, G. K.; Johnson, H. W. Jr. "Selected Experiments in Organic Chemistry," 3rd ed.; Freeman: New York, 1983; p 99.

TECHNIQUE 10: MEASUREMENT OF SPECIFIC ROTATION

Solutions of optically active substances, when placed in the path of a beam of polarized light, may rotate the plane of the polarized light clockwise or counterclockwise. The observed optical rotation is measured using a *polarimeter*. This technique is one of the oldest instrumental procedures used to characterize chemical compounds. The results obtained from the magnitude of the observed angle of rotation α are generally expressed in terms of *specific rotation* $[\alpha]$.

Theory

Natural light behaves as if it is composed of a large number of electromagnetic waves vibrating in all possible orientations around the direction of propagation (see Fig. 5.35).

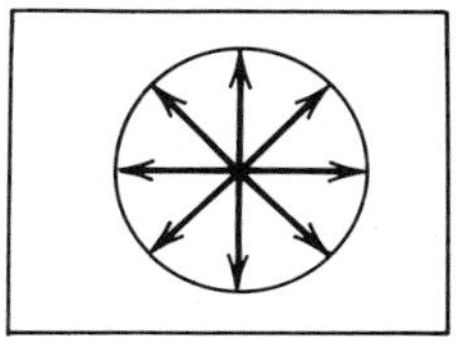

Note. *A beam of light behaves as if it were composed of two mutually perpendicular oscillating fields: electrical and magnetic. The oscillating magnetic field is not considered in the following discussion.*

The planes in which the electrical waves oscillate are perpendicular to the direction of propagation of the light beam. If one separates one particular plane of oscillation from all other planes by passing the beam of light through a polarizer, the resulting radiation is said to be *plane polarized* (see Fig. 5.36). In the interaction of light with matter, this plane-polarized radiation is represented as the vector sum of two circularly polarized waves. The electric vector of one of the waves moves in a clockwise direction while the other moves in a counterclockwise direction, both waves having the same amplitude of vibration (see Fig. 5.37). These two components add vectorially to produce plane-polarized light.

If the passage of plane-polarized light through a material results in one of the circularly polarized components being slowed down more than the other by interaction with bonding and nonbonding electrons, the transmitted beam of radiation has the plane of polarization rotated from its *original* position (Figs. 5.38 and 5.39).

The Polarimeter

The *polarimeter* measures the amount of rotation by an optically active compound of the plane-polarized light. The principal parts of the instrument are diagrammed in Fig. 5.38. Two Nicol prisms are used in the instrument. The prism by which the original light source is polarized is called the polarizer. The second prism, called the analyzer, is used to examine the polarized light after it passes through a solution of the optically active species.

Fig. 5.35 *Oscillation of the electrical field of ordinary light occurs in all possible planes perpendicular to the direction of propagation. From T. W. Graham Solomons, Organic Chemistry, 3rd ed. (New York: Wiley, 1984). (Reprinted by permission of John Wiley & Sons, New York, NY.)*

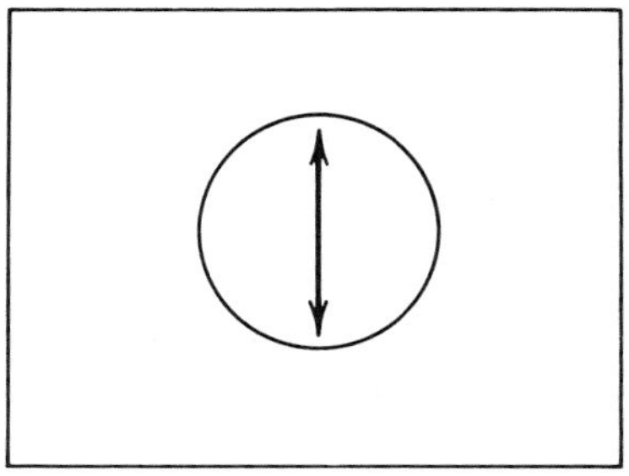

Fig. 5.36 *The plane of oscillation of the electrical field of plane polarized light. In this example the plane of polarization is vertical. From T. W. Graham Solomons, Organic Chemistry, 3rd ed. (New York: Wiley, 1984). (Reprinted by permission of John Wiley & Sons, New York, NY.)*

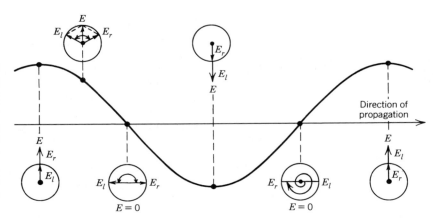

Fig. 5.37 *A beam of plane polarized light viewed from the side (sine wave) and along the direction of propagation at specific times (circles) where the resultant electric vector E and the circularly polarized components E_l and E_r are shown. From Bodie Douglas, Darl H. McDaniel, and John J. Alexander, Concepts and Models of Inorganic Chemistry, 2nd ed. (New York: Wiley, 1983). (Reprinted by permission of John Wiley & Sons, New York, NY.)*

When the axes of the analyzer and polarizer prisms are parallel and no optically active substance is present, the maximum amount of light is passed and the instrument dial is set to zero degrees. However, if the axes of the analyzer and polarizer are at right angles to each other, no transmission of light is observed and the field is dark. The introduction of a solution of an enantiomer in the wave path of the plane-polarized light causes one of the circularly polarized components, through dissymmetric interaction, to be slowed more than the other. The refractive indices are, therefore, different in the two circularly polarized

Fig. 5.38 *Operation of a polarimeter. From T. W. Graham Solomons, Organic Chemistry, 3rd ed. (New York: Wiley, 1984). (Reprinted by permission of John Wiley & Sons, New York, NY.)*

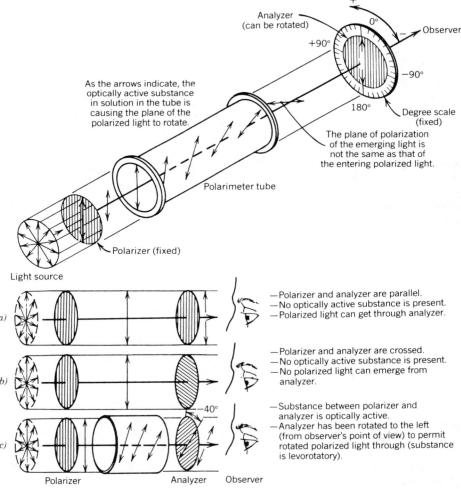

Fig. 5.39 *Plane polarized light before entering and after emerging from an optically active substance. From Bodie Douglas, Darl H. McDaniel, and John J. Alexander, Concepts and Models of Inorganic Chemistry, 2nd ed. (New York: Wiley, 1983). (Reprinted by permission of John Wiley & Sons, New York, NY.)*

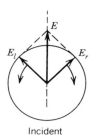

Incident

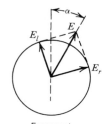

Emergent

beams. Figure 5.38 demonstrates that the left-hand component has been effected the most.

Note. *In this simplified figure the effect on only one of the circularly polarized waves is diagrammed. See Fig. 5.39 for a more accurate description (view from behind the figure).*

As seen, this results in a tilt of the plane of polarization. The analyzer prism must be rotated to the left to maximize the transmission of radiation. If rotation is counterclockwise, the angle of rotation is defined as (−) and the enantiomer that caused the effect is termed *levorotatory (l)*. Conversely, clockwise rotation is (+) and the enantiomer is *dextrorotatory (d)*. It is important to note that if a solution of equal amounts of a *d* and an *l* enantiomorphic pair is placed in the beam path of the polarimeter, no rotation is observed. Such a solution is called a *racemic* mixture if, as in this case, it is a mixture of enantiomers.

The magnitude of optical rotation depends on several factors: (1) the nature of the substance; (2) the path length through which the light passes; (3) the wavelength of light used as a source; (4) the temperature. It also depends on the concentration of the solution of the optically active material. Optical rotation is usually expressed as *specific rotation* [α], defined as:

$$[\alpha]_D^t = \frac{\alpha}{(l)\ (c)}$$

where

$[\alpha]$ = specific rotation
α = the angle of rotation
l = the path length (dm)
c = concentration (g/mL)
t = temperature (°C)
D = the D line of sodium

The observed angle of rotation α must be less than 360°. However, the *specific rotation* [α] can be very large because of the concentration effect. Indeed, this is the case with usnic acid, which has a specific rotation value near 460°.

For increased sensitivity, most polarimeters are equipped with an optical device that divides the viewed field into three adjacent parts (triple-shadow polarimeter; Fig. 5.40). A very slight rotation of the analyzer will cause one portion to become dimmer, the other lighter. The angle-of-rotation reading is taken when the sections of the fields all have the same intensity. An accuracy of ± 0.01° is obtained using this technique.

Instructions should be provided on the use of the particular model of polarimeter available in each laboratory.[8]

Fig. 5.40 *View through the eyepiece of the polarimeter. The analyzer should be set so that the intensity of all parts of the field is the same (b). When the analyzer is displaced to one side or the other, the field will appear as in (a) or (c).*

(a)   (b)   (c)

[8]For further discussion on optical activity and the use of the polarimeter see Solomons, T. W. G. "Organic Chemistry", 3rd ed.; Wiley; New York, 1984; p 313; and Willard, H. H.; Merritt, Jr., L. L.; Dean, J. A.; Settle, F. A. "Instrumental Methods of Analysis" 6th ed.; Van Nostrand: New York, 1981; p 421.

Experiment No. 10
Isolation and Characterization of an Optically Active Natural Product: Usnic Acid

[1,3 (2H,9bH)-Dibenzofurandione, 2,6-diacetyl-7,9-dihyroxy-8,9b-dimethyl-]

This experiment illustrates the extraction of a natural product from its native source, a lichen. It also introduces the method used to measure the specific rotation of an optically active compound.

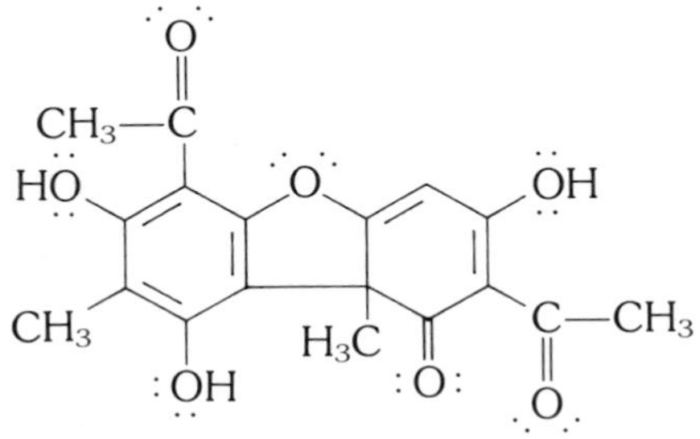

Usnic acid

DISCUSSION The common method of extracting chemical constituents from natural sources is presented in this experiment. In the present case, only one main chemical species, usnic acid, is soluble in the extraction solvent. Therefore, the purification and characterization sequence is fairly straightforward.

Usnic acid has a chiral carbon atom (see structure) and therefore exists as optically active isomers (enantiomers). Generally, in a given lichen, only one of the isomers (**R** or **S**) is present. Usnic acid has a very high specific rotation (~460°) and for this reason is an ideal candidate to measure rotation on the microscale level.

PART A: Isolation of Usnic Acid

EXPERIMENTAL Estimated time for completion of the experiment: 2.5 hours.

Physical Properties of Reactants and Products

Compound	MW	Wt/Vol	mp(°C)	bp(°C)
Lichen		1.0 g		
Acetone	58.08	7.0 mL		56.2
Usnic acid	344.31		204	

Reagents and Equipment

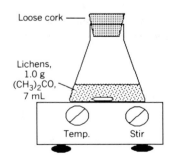

Loose cork

Lichens, 1.0 g (CH₃)₂CO, 7 mL

Temp.   Stir

In a 25-mL Erlenmeyer flask containing a magnetic stirrer and loosely capped with a cork stopper are placed 1.0 g of oven dried (40°C), crushed lichens, and 7.0 mL of acetone.(■) *The lichens used in this experiment are* Usnea sp. *(commonly called "Old Man's Beard").*

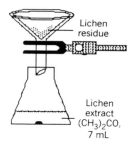

Lichen residue

Lichen extract (CH₃)₂CO, 7 mL

HOOD

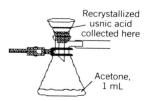

Recrystallized usnic acid collected here

Acetone, 1 mL

Reaction Conditions

The mixture is stirred for 0.5 hour at room temperature.

Isolation of Product

The resulting mixture is now filtered by gravity and the filtrate collected in a 10-mL Erlenmyer.(■) The acetone solvent is evaporated under a stream of air [HOOD], yielding the crude usnic acid.

Purification and Characterization

The crude material is recrystallized from acetone-95% ethanol (10:1). Dissolve the crystals in the minimum amount of hot acetone and add the ethanol. The yellow crystals are collected by vacuum filtration,(■) washed with *cold* acetone, and dried on a porous clay plate.

Weigh the yellow needles of usnic acid and calculate the percentage of the acid in the lichen. Determine the melting point (use the evacuated mp technique) and compare your value to that found in the literature. Obtain an IR spectrum and compare it with that of an authentic sample.

PART B: Determination of the Specific Rotation

Usnic acid is an optically active compound with a very high specific rotation. For determination of its rotation on the microscale, a low-volume, long-path-length cell must be used.

Usnic acid (80 mg) is dissolved in 4.0 mL of tetrahydrofuran (THF) solvent and transferred to the polarimetry cell using a Pasteur pipet.

This amount of usnic acid is obtained by having eight or nine students pool their crystals. The THF used is spectral grade.

The cell is placed in the polarimeter and the angle of rotation determined. The specific rotation is then calculated using the equation given in the discussion section.

ENVIRONMENTAL DATA

Substance	Amount	TLV (mg/m³)	Emissions (mg)	Volume (m³)
Acetone (Max., extract., evap.)	7.0 mL	1780	5537	3.1
Acetone (Max., recryst.)	3.0 mL	1780	2375	1.3
Tetrahydrofuran (Max.)	0.5 mL	590	438	0.7

TXDS: Usnic acid—inv-mus LD50: 25 mg/kg

QUESTIONS   5-48. Give an **R** or **S** designation to each of the following molecules.

$$CH_3$$
$$HO \cdots C \diagdown D$$
$$| \atop H$$

$$H$$
$$CH_3 \cdots C \diagdown F$$
$$| \atop Cl$$

$$CH_2CH_3$$
$$HO \cdots C \diagdown Br$$
$$| \atop Cl$$

5-49. The structure originally proposed for cordycepic acid, which has $[\alpha] = +40.3°$, is at left.

Do you agree that this is a plausible structure? If not, why not?

5-50. A sample of 150 mg of an organic compound is dissolved in 7.5 mL of water. The solution is placed in a 20-cm polarimeter tube and the rotation read in a polarimeter. The value obtained was $+2.676°$. Distilled water, in the same tube, gave a reading of $+0.016°$. Calculate the specific rotation for the compound.

5-51. Compound A is optically active and has the molecular formula $C_5H_{10}O$. On catalytic hydrogenation (addition of hydrogen) of A, compound B is obtained. B has the molecular formula $C_5H_{12}O$ and is optically inactive. Give the structures for compounds A and B.

REFERENCES

1. This experiment is adapted from that given by Todd, D. "Experimental Organic Chemistry"; Prentice-Hall: Englewood Cliffs, NJ, **1979**; p 57.
2. Usnic acid has been synthesized:
   a. Barton, D. H. R.; DeFlorin, A. M.; Edwards, O. E. *J. Chem. Soc.* **1956**, 530.
   b. Penttila, A., Fales, H. M. *Chem. Commun.* **1966**, 656.
3. A large scale method of isolation of usnic acid has been reported: Stark, J. B.; Walter, E. D.; Owens, H. S. *J. Am. Chem. Soc.* **1950**, *72*, 1819.
4. Optical, crystallographic and X-ray diffraction data has been reported for usnic acid: Jones, F. T.; Palmer, K. J. *J. Am. Chem. Soc.* **1950**, *72*, 1820.

TECHNIQUE 11: SUBLIMATION

Sublimation is a technique that is especially suitable for the purification of solid substances at the microscale level. It is particularly advantageous when the impurities present in the sample are nonvolatile under the condition employed. Sublimation is a relatively straightforward method in that the impure solid need only be heated and mechanical losses can be kept to a minimum.

Materials sublime easily only when heated below their melting points under reduced pressure. Substances that can be purified by sublimation are those that do not have strong intermolecular attractive forces. Caffeine, isolated in Experiment 11, and ferrocene, used as a reactant in Experiments 32A and 32B, meet these requirements.

Theory

The processes of *sublimation* and *distillation* are closely related. Crystals of a solid substance, in an evacuated container, will gradually change into gas molecules by the process of *evaporation*. Occasionally, one of these gas molecules will strike the crystal surface and be held by attractive forces of the cyrstalline molecules. This process is termed *condensation*.

Sublimation is the complete process of *evaporation* of crystal molecules and *condensation* of gas molecules *directly* to crystals without passing through the liquid state.

A typical single-component phase diagram is shown in Fig. 5.41 relating the solid, liquid, and vapor states of a substance with temperature and pressure. Where two of the areas (solid, liquid, or vapor) touch, there is a line, and along each line the two phases exist in *equilibrium*. BO is the sublimation vapor pressure curve of the substance in question and *only* along line BO can solid and vapor exist together in equilibrium. At temperatures and pressures along the BO curve the liquid state is thermodynamically unstable. Where the three lines representing pairs of phases intersect, all three phases exist together in equilibrium. This point is called the triple point.

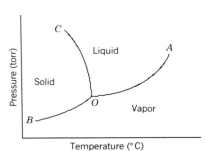

Fig. 5.41 *Single-component phase diagram.*

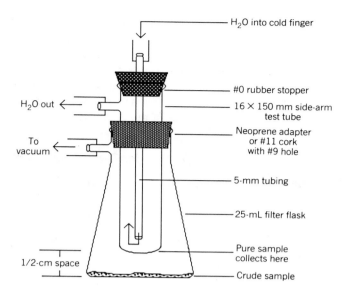

Fig. 5.42 *Sublimation apparatus.*

Fig. 5.43 *Vacuum sublimator. (Courtesy of Ace Glass Inc., Vineland, NJ.)*

Many solid substances have a sufficiently high vapor pressure near their melting point that they can be sublimed easily under reduced pressure in the laboratory. Sublimation occurs when the vapor pressure of the solid equals the applied pressure. A simple apparatus suitable for sublimation of small quantities of material is used for the purification of caffeine. This device is shown in Fig. 5.42.

The sample is placed in the 25-mL filter flask, which is then fitted with a one-hole rubber stopper through which is inserted a cold finger. The side-arm of the flask is connected to a water aspirator using a water-trap bottle between the flask and the aspirator. After evacuation of the system, cold water is run through the cold finger. Heating of the sample with a microburner or a sand bath to just below the melting point of the solid causes sublimation to occur. The gas molecules condense on the cold finger surface while any nonvolatile material remains at the bottom of the flask. Detailed instructions of the procedure to follow are outlined in the Experiment 11A. Sublimators suitable for the sublimation of small quantities are commercially available (Fig. 5.43).

Experiment 11A, 11B
Isolation and Characterization of a Natural Product: Caffeine; Caffeine 5-Nitrosalicylate

(1,3,7-trimethyl-2,6-dioxopurine)

The extraction of a natural product, caffeine, from its native source, tea, is demonstrated in this experiment. The preparation of a caffeine derivative, 5-nitrosalicylate, is also described.

Caffeine

DISCUSSION This experiment illustrates (as does Experiment 10) the common technique of isolating natural products from their common sources. In the present case, the alkaloid (meaning alkali-like and containing nitrogen) caffeine is readily soluble in hot water and is thus easily separated from the tea source.

The molecule contains the purine ring system, which plays an important role in living systems. Caffeine, the most widely used of all stimulants, is present in tea leaves at the 2–3% level and is the major stimulant in coffee.

PART A: Isolation of Caffeine

EXPERIMENTAL Estimated time to complete the experiment: 4.0 hours.

Physical Properties of Reactants and Products

Compound	MW	Wt/Vol	mmol	mp(°C)
Tea		1.0 g		
Water		10 mL		
Sodium carbonate	105.99	1.1 g	10	851
Caffeine	194.20			238

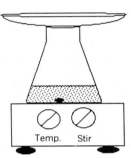

Anhydrous Na$_2$CO$_3$, 1.1 g
+ H$_2$O, 10 mL +
tea leaves, 1.0 g

Reagents and Equipment
To a 25-mL Erlenmeyer flask containing a boiling stone are added 1.1 g (0.01 mol) of anhydrous sodium carbonate and 10 mL of water.(■) The mixture is heated on a hot plate to dissolve the solid. To this solution is added 1.00 g of tea leaves.

Reaction Conditions
GENTLY A small watch glass is placed on the Erlenmeyer flask and the tea mixture then heated at boiling [GENTLY] for 30 minutes using a sand bath.

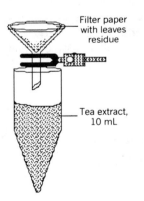

Filter paper with leaves residue

Tea extract, 10 mL

Isolation of Product
The flask is removed from the hot plate and allowed to cool to about 40–50°C (warm to the touch). The dark aqueous solution is filtered by gravity, using a fast grade filter paper, to remove the leaf residue.(■) The filtrate is collected in a 12-mL centrifuge tube. The Erlenmeyer flask is rinsed with two 1.0-mL portions of hot water, and the rinse solutions also are passed through the filter and **CAREFULLY** collected in the centrifuge tube. The liquid is then *carefully* pressed out of the tea leaves and filter paper. The leaf residue is discarded.

The filtrate is cooled to room temperature and the aqueous solution extracted with 2.0 mL of methylene chloride.

The tea solution contains some constituents that cause emulsions. Vigorous shaking will make the extraction process difficult and should be avoided.

The layers separate slowly and thus at least 15 minutes is allowed to pass before an attempt is made to remove the lower methylene chloride layer. It is then separated using a 9-in. Pasteur pipet and the extract passed through a filter containing a small plug of cotton that is covered with ~2.0 g of anhydrous sodium sulfate, previously moistened with a small amount of methylene chloride.(■) The dried filtrate is collected in a 25-mL filter flask. The aqueous solution is now extracted with four 2.0-mL portions of methylene chloride. Each extract is separated, dried as above, and collected in the same filter flask. Finally, the sodium sulfate is rinsed with an additional 2.0 mL of methylene chloride.

A boiling stone is added to the flask and the solution concentrated to dryness **HOOD** [HOOD] by warming the flask in a sand bath. The crude caffeine crystallizes as an off-white solid.

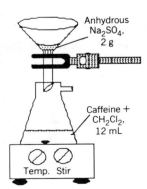

Anhydrous Na$_2$SO$_4$, 2 g

Caffeine + CH$_2$Cl$_2$, 12 mL

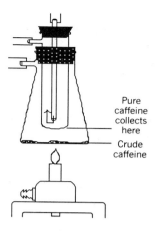

Pure caffeine collects here

Crude caffeine

CAREFULLY

Purification and Characterization

The crude caffeine is purified by the process of sublimation.

Assemble a sublimation apparatus as shown in Fig. 5.42.(■) Apply a vacuum to the system through the filter flask, using a water-trap bottle between the flask and the aspirator. After the system is evacuated, cold water is run through the cold finger. This procedure will minimize the amount of condensation that collects on the cold finger surface.

The sublimation is started by heating the flask gently with a microburner, moving the flame back and forth around the bottom and sides of the flask.

Be Careful. *Do not MELT the caffeine. If the sample does begin to melt, remove the flame for a few seconds before heating is resumed.*

When all the caffeine has completely sublimed onto the cold finger, remove the heat and allow the apparatus to cool to room temperature under reduced pressure. Carefully remove the vacuum tubing from the aspirator, returning the system to atmospheric pressure. Shut off the water to the cold finger and *carefully* remove the cold finger from the sublimation apparatus.

If this operation is done carelessly, the sublimed crystals may be dislodged from the tube and fall back into the impurities residue.

Scrape the caffeine from the cold finger onto weighing paper using a microspatula and sample brush. Weigh the caffeine product and calculate the percent by weight of caffeine in the tea leaves. Determine the melting point and compare your value to that in the literature.

The melting point must be taken in an evacuated sealed tube since caffeine sublimes and the melting point is above the sublimation temperature (see page 39).

Obtain an IR spectrum and compare it with that of an authentic sample.

PART B: Caffeine 5-Nitrosalicylate

The caffeine product may be further characterized by preparation of its 5-nitrosalicylate derivative.

Caffeine 5-nitrosalicylate

EXPERIMENTAL   Estimated time to complete the experiment: 0.5 hours.

Physical Properties of Reactants and Products

Compound	MW	Wt/Vol	mmol	mp(°C)	bp(°C)
Caffeine	194.20	11 mg	0.06	238	
5-Nitrosalicylic acid	183.12	10 mg	0.06	229–230	
Petroleum ether (60–80°)		0.5 mL			
Ethyl acetate	88.12	0.7 mL			77
Caffeine 5-nitrosalicylate	377.32			180	

Reagents and Equipment

In a 3.0-mL conical vial containing a magnetic spin vane and equipped with an air condenser are placed 11 mg (0.06 mmol) of caffeine, 10.0 mg (0.06 mmol) of 5-nitrosalicylic acid (prepared in Experiment 34D) and 0.7 mL of ethylacetate.

Reaction Conditions

GENTLY   *Gently* warm the mixture on a hot plate, with stirring, to dissolve the solids. Add 0.5 mL of petroleum ether (60–80°) to the warm ethyl acetate solution, mix, and warm for several seconds. Remove the spin vane using forceps.

Isolation of Product

The mixture is now cooled to room temperature and then placed in an ice bath for 10–15 minutes. The crystals are collected under reduced pressure using a Hirsch funnel and the filter cake washed with 0.5 mL of cold ethyl acetate. The product is dried on a porous clay plate.

Purification and Characterization

The product is sufficiently pure for characterization.

Weigh the caffeine 5-nitrosalicylate and calculate the precentage yield. Determine the melting point and compare your value with that reported above. Obtain an IR spectrum and compare it with that shown in Fig. 5.44.

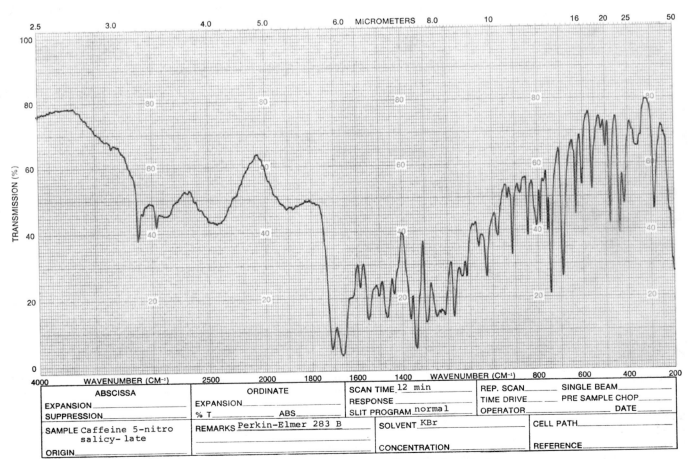

Fig. 5.44   *IR spectrum: caffeine 5-nitrosalicylate.*

ENVIRONMENTAL DATA

Substance	Amount	TLV (mg/m³)	Emissions (mg)	Volume (m³)
Experiment 11A				
Methylene chloride (Max., extract, filter, evap.)	12 mL	350	16,020	46
Experiment 11B				
Ethyl acetate	1.2 mL	1400	1080	0.8
Petroleum ether (Max., treated as hexane)	0.5 mL	180	320	1.8

TXDS: Caffeine—orl-hum LDLo: 192 mg/kg
Sodium carbonate—orl-rat LDLo: 4000 mg/kg

QUESTIONS

5-52. Compounds such as naphthalene and 1,4-dichlorobenzene find use as mothballs since they sublime at a slow rate at atmospheric pressure. Explain this behavior in terms of the structure of the molecules.

5-53. How many peaks would you expect to find in the NMR spectrum of caffeine?

5-54. The vapor pressure of 1,2-diphenylethane, *p*-dichlorobenzene, and 1,3,5-tri-chlorobenzene are 0.06 torr, 11.2 torr, and 1.4 torr, respectively, at their melting point (52–54°C). Which compound can be sublimed most rapidly at a reduced pressure of 15 torr and a temperature of 40°C?

5-55. Visualization of spots on TLC plates are often developed in an iodine chamber. Explain how the solid–vapor equilibrium operates in this instance.

REFERENCES

Several extraction procedures for isolating caffeine from tea, designed for the introductory organic laboratory, have been reported.
1. Mitchell, R. H.; Scott, W. A.; West, P. R. *J. Chem. Educ.* **1974,** *51,* 69.
2. Ault, A.; Kraig, R. *J. Chem. Educ.* **1969,** *46,* 767.
3. Laswick, J. A.; Laswick, P. H. *J. Chem. Educ.* **1972,** *49,* 708.

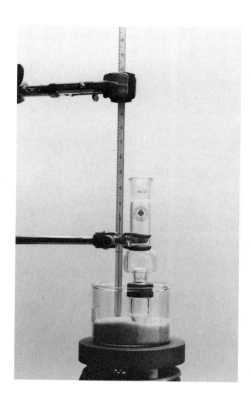

Chapter 6

Preparative Organic Chemistry

Experiment 12
Reductive Hydrogenation of an Olefin: Octane

(octane)

In this experiment the addition of hydrogen to an unsaturated hydrocarbon to produce a saturated alkane is demonstrated. This addition reaction has great utility in laboratory and industrial syntheses.

$$CH_3—(CH_2)_5—CH{=}CH_2 \underset{\substack{NaBH_4 \\ C_2H_5OH \\ HCl\ (6M)}}{\overset{H_2PtCl_6}{\rightleftharpoons}} CH_3—(CH_2)_5—CH_2—CH_3$$

1-Octene   Octane

DISCUSSION   The addition of hydrogen to an alkene is a very important reaction in organic chemistry because by use of the reaction, an alkene is converted into an alkane. The reaction is exothermic; the *heat of hydrogenation* is approximately 125 kJ/mol for most alkenes. Generally, the reaction takes place at room temperature with hydrogen gas pressures in the range of 1–4 atm., in the presence of a finely divided metal catalyst. For this reason, the reaction is referred to as *catalytic hydrogenation*. It represents a class of organic reactions known as addition reactions; one hydrogen atom of the hydrogen molecule adds to each carbon of the C=C. The metals most often used as catalysts in low-pressure hydrogenations are nickel, platinum, rhodium, ruthenium, and palladium. The addition of hydrogen to an alkene does not take place at room temperature in the absence of a metal catalyst.

In the present experiment, the metal catalyst, platinum, is generated *in situ* by the treatment of chloroplatinic acid with sodium borohydride. The hydrogen gas necessary for the reduction is generated by the addition of excess sodium borohydride in the presence of hydrochloric acid. This reduction technique is useful for easily reducible groups such as unhindered olefins, but a major limitation is that other functional groups that are normally reduced by sodium borohydride cannot be present in the molecule.

The platinum catalyst adsorbs both the molecular hydrogen and the alkene on its surface. The transfer of the hydrogen molecule to the alkene molecule on the metal surface results in reduction, producing an alkane. The addition is syn (or cis) since both hydrogen atoms add to the same side of the C=C of the alkene. The mechanistic sequence is outlined below.

H····H $C=C$ H····H → $C=C$

Metal surface   + H₂

$C-C$ H $C-C$

EXPERIMENTAL   Estimated time to complete the experiment: 2.5 hours.

Physical Properties of Reactants and Products

Compound	MW	Wt/Vol	mmol	bp(°C)	Density	n_D
1-Octene	112.22	120 μL	0.76	121.3	0.72	1.4087
Ethanol	46.07	1.0 mL		78.5		
Chloroplatinic acid (0.2M)	517.92	50 μL				
Sodium borohydride (1M)	37.83	125 μL				
Dilute HCl (6M)		100 μL				
Pentane	72.15	6.5 mL		36.1		
Octane	114.23			125.7	0.70	1.3974

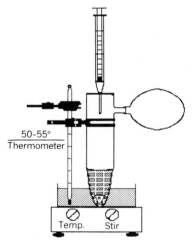

50-55°
Thermometer

Temp.   Stir

Step 1:
    0.2M H₂PtCl₆ solution, 50 μL
    + CH₃CH₂OH, 1.0 mL
    + NaBH₄ solution, 125 μL
Step 2:
    6M HCl, 100 μL
    + 1-octene, 120 μL
    + NaBH₄ solution, 1.0 mL

Reagents and Equipment

A 16 × 150 mm side-arm test tube or a 5.0 mL conical vial and Claisen head is fitted with a magnetic spin vane, a rubber balloon, and rubber septum.(■) Add 50 μL of a 0.2M solution of chloroplatinic acid, H_2PtCl_6, dispensed from an automatic delivery pipet and 1.0 mL of absolute ethanol (calibrated Pasteur pipet).

Note. *No residual acetone (perhaps from cleaning the test tube) can be present since it destroys the catalyst. If the balloon does not fit snugly, secure it with copper wire or a rubber band.*

Instructor preparation: The 0.2M H_2PtCl_6 solution is prepared by adding 41 mg (0.1 mmol) of the acid to 0.5 mL of deionized water.

A solution (125 μL) of sodium borohydride (see note below) is now added by automatic delivery pipet with vigorous stirring. The solution should turn black immediately as the catalyst, finely divided platinum, is formed.

Important. *The rubber septum is placed on the test tube IMMEDIATELY after the NaBH₄ solution is added.*

Instructor preparation: The sodium borohydride reagent is prepared by adding 0.38 g (0.01 mol) of NaBH₄ to a solution previously obtained by mixing 0.5 mL of 2.0M aqueous NaOH and 9.5 mL of absolute ethanol.

After one minute, 100 μL of 6M HCl solution is added through the septum, using a syringe. With a fresh syringe, a solution of 120 μL (86 mg, 0.76 mmol) of 1-octene (delivered from an automatic pipet to 1 mL conical vial) dissolved in 250 μL of absolute ethanol is added immediately to the acid solution.

The addition of 1-octene is followed by the dropwise addition (clean syringe) of 1.0 mL of the NaBH₄ solution over a two-minute time interval.

Important. *At this point the balloon should inflate and remain inflated for at least 30 minutes. If it does not, the procedure must be repeated.*

Reaction Conditions

The reaction mixture is vigorously stirred in a sand bath at 50°C for 45 minutes.

Isolation of Product

Water (1 mL) is added dropwise to the cooled reaction medium and the resulting mixture is extracted with three 1.0-mL portions of pentane. Each pentane extract is transferred to a stoppered 25-mL Erlenmeyer flask containing 0.5 g of anhydrous sodium sulfate.

Upon addition of each portion of pentane, the tube is capped, shaken, and vented and the layers are allowed to separate. The transfers are made using a Pasteur filter pipet.

Using a Pasteur filter pipet, the dried solution is transferred to a second 25-mL Erlenmeyer flask. The drying agent is rinsed with an additional 0.5 mL of pentane (calibrated Pasteur pipet) and the rinse also added to the second flask. The solution is concentrated using a gentle stream of nitrogen gas or by warming very gently in a sand bath **[HOOD]** to a volume of ~1.0 mL.

Purification and Characterization

The product, octane, is purified by column chromatography.

In a Pasteur filter pipet are placed 50 mg of sand, 400 mg of 10% silver nitrate on activated silica gel (200 mesh) and 50 mg of anhydrous sodium sulfate.(■)

The column is wetted with 0.5 mL of pentane (calibrated Pasteur pipet) and the crude product, obtained above, transferred to the column by Pasteur pipet. The octane is eluted using 1.5 mL of pentane and the eluate collected in a tared 5.0-mL conical vial containing a boiling stone.

The vial is fitted with an air condenser and then placed in a sand bath maintained at a temperature of 90–100°C to evaporate the pentane solvent. *The evaporation is continued until a constant weight of product is obtained.*

Record the weight of product and calculate the percentage yield. Determine the boiling point and refractive index of your material and compare the results with the literature value. Obtain an IR spectrum and compare it with an authentic sample and to the 1-octene starting material.

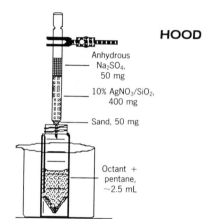

HOOD

Anhydrous Na₂SO₄, 50 mg

10% AgNO₃/SiO₂, 400 mg

Sand, 50 mg

Octant + pentane, ~2.5 mL

ENVIRONMENTAL DATA

Substance	Amount	TLV (mg/m³)	Emissions (mg)	Volume (m³)
Ethanol (solv., water soluble)	1.0 mL	1900	140	<0.1
Ethanol (NaBH₄ solution)	1.13 mL	1900	140	<0.1
Hydrochloric acid, 6M (Max., transfer)	100 μL	7	22	3.1
1-Octene (Max., reactant)	86 mg	—	85.6	
Pentane (Max., evap.)	5.5 mL	1800	5320	3.0

TLV: NaOH—2 μg/m³
Octane—300 ppm
TXDS: Sodium borohydride—orl-rat LD50: 160 mg/kg
Chloroplatinic acid—inv-rat: LD50: 49 mg/kg
Silver nitrate—unk-man LDLo: 29 mg/kg
Silica gel—inv-mus LDLo: 234 mg/kg
Anhydrous sodium sulfate—LD50: 5989 mg/kg

QUESTIONS

6-1. Squalene, first isolated from shark oil and a biological precursor of cholesterol, is a long-chain aliphatic alkene, $C_{30}H_{50}$. The compound undergoes catalytic hydrogenation to yield a compound of molecular formula $C_{30}H_{62}$. How many double bonds does a molecule of squalene have?

6-2. An optically active carboxylic acid A ($C_5H_6O_2$) reacts with one mole of hydrogen gas on catalytic hydrogenation. The product is an optically inactive carboxylic acid B ($C_5H_8O_2$). What are the structures of compounds A and B?

6-3. Two hydrocarbons, A and B, each contain six carbon atoms and one C=C. Compound A exhibits geometrical isomerism but compound B does not. However, both A and B on catalytic hydrogenation give only 3-methylpentane. Draw the structures and give a suitable name for compounds A and B.

6-4. In this experiment, the crude octane product in pentane solution was passed through a column of 10% silver nitrate on silica gel. What is the function of this column packing? [*Hint*: Ag^+ is involved.]

REFERENCES

1. Selected reviews on catalytic hydrogenation:
   (a) Adkins, H.; Shriner, R. L. in Gilman, H. "Advanced Organic Chemistry", 2nd ed.; Wiley: New York, 1943; Vol. I, p 779.
   (b) Carruthers, W. "Some Modern Methods of Organic Synthesis", 2nd ed.; Cambridge Univ. Press: London, 1978; p 437.
   (c) March, J. "Advanced Organic Chemistry", 2nd ed.; McGraw-Hill: New York, 1977; p 707.
2. Selected examples of catalytic hydrogenation of olefins in "Organic Syntheses":
   (a) Adams, R.; Kern, J. W.; Shriner, R. L. "Organic Syntheses"; Wiley: New York, 1941; Collect. Vol. I, p 101.
   (b) Bruce, W. F.; Ralls, J. O. "Organic Syntheses"; Wiley: New York, 1943; Collect. Vol. II, p 191.
   (c) Herbst, R. M.; Shemin, D. *Ibid.*, p 491.
   (d) Cope, A. C.; Herrick, E. C. "Organic Syntheses"; Wiley: New York, 1963; Collect. Vol. IV, p 304.

Experiment 13
Hydroboration-Oxidation of an Olefin: Octyl Alcohol

(1-octanol)

The hydroboration-oxidation of 1-octene to prepare the *anti*-Markownikoff addition product, 1-octanol, is performed in this experiment.

$$3\ CH_3\text{---}(CH_2)_5\text{---}CH\text{=}CH_2 \xrightarrow{\text{THF:BH}_3} [CH_3(CH_2)_7]_3B \xrightarrow[OH^-]{H_2O_2} 3\ CH_3\text{---}(CH_2)_7\text{---}OH$$

1-Octene   Trioctylborane   1-Octanol

DISCUSSION

This reaction demonstrates the addition of diborane as a BH_3-THF complex to an alkene to form an intermediate trialkylborane, which upon oxidation with alkaline hydrogen peroxide yields the corresponding alcohol.

The addition reaction is known as "hydroboration" and is a general reaction for all classes of acyclic and cyclic alkenes as well as alkynes. As depicted in the mechanism below, the boron hydride reagent adds successively to three molecules of the alkene to form a trialkylborane.

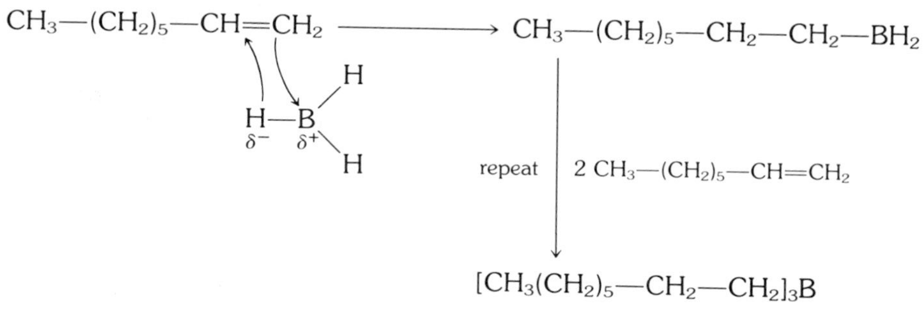

$$CH_3-(CH_2)_5-CH=CH_2 \longrightarrow CH_3-(CH_2)_5-CH_2-CH_2-BH_2$$

H—B
$\delta^- \quad \delta^+$
H

repeat | 2 $CH_3-(CH_2)_5-CH=CH_2$

$$[CH_3(CH_2)_5-CH_2-CH_2]_3B$$

It is important to note that with the 1-octene component, the boron atom becomes attached to the less substituted carbon atom of the double bond. As the alkene donates the π electron pair to the boron atom which possesses a vacant p-orbital, a partial positive charge is developed on the more highly substituted carbon. This is because of the electron-releasing effect of the alkyl groups, which enable the carbon to better accommodate the positive charge. The boron atom in turn develops a partial negative charge, which aids cleavage of a hydrogen-boron bond. The hydrogen atom is thus transferred from the boron atom to the carbon atom of the alkene, which has developed the partial positive charge. The reaction has been shown is to have a four-centered transition state wherein the four atoms involved undergo changes in bonding at the same time.

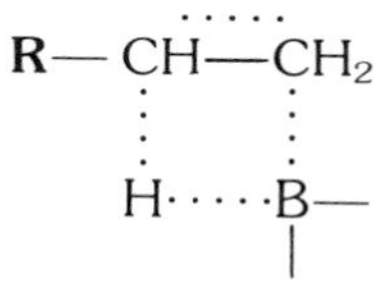

$$\mathbf{R}-\overset{\cdots\cdots}{CH}-CH_2$$
$$H\cdots\cdots B-$$

When internal alkenes with different degrees of substitution undergo hydroboration, the boron atom also attaches itself to the less substituted carbon atom. This direction of addition appears to result in part from steric factors. That is, a bulky boron-containing group can approach the least substituted carbon atom more easily. Thus we see that due to electronic and steric factors, the addition of the B—H elements to the C=C is highly selective with respect to orientation (regioselectivity). Accumulated evidence demonstrates that the reaction occurs by syn addition, a consequence of the four-centered transition state. Therefore, as a result of the mechanism operating in this reaction, the new C—B and C—H bonds are necessarily formed on the same face of the C=C bond as shown in the following example.

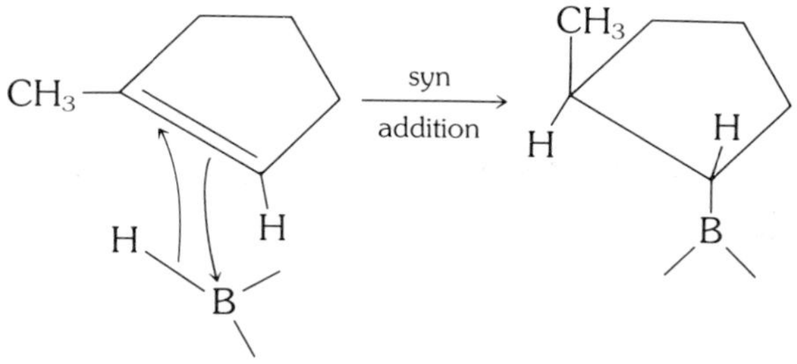

The organoboranes are important in organic synthesis because of the reactions they undergo. Procedures have been developed by which the boron atom may be replaced by a wide variety of groups such as —H, —OH, —NH$_2$, —Br, and

—I. The present experiment demonstrates the conversion of an organoborane to an alcohol by oxidation with alkaline hydrogen peroxide solution. As observed in the experimental section, it is not necessary to isolate the organoborane. This is a great advantage because most alkylboranes are pyrophoric (flammable).

There is conclusive evidence that oxidation of the C—B bond proceeds with retention of configuration of the carbon atom. That is, the hydroxyl group that replaces the boron atom has the identical orientation in the molecule as that of the boron atom.

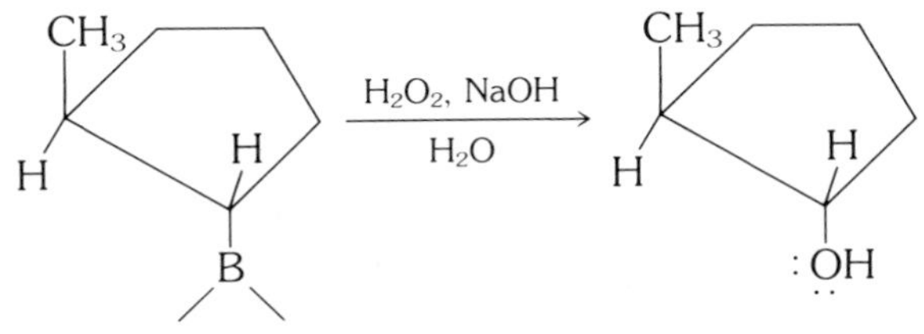

As a result of the hydroboration-oxidation sequence of reactions, the elements of H—OH are added to the original C=C in an *anti*-Markownikoff manner when the alkene is unsymmetrical.

In the oxidation reaction a hydroperoxide anion is generated in the alkaline medium. This species makes a nucleophilic attack on the boron atom. Migration of an alkyl group from boron to oxygen then occurs, forming a boron ester (a borate). Hydrolysis of the final boron ester, generated by rearrangement of the three alkyl groups, produces the desired alcohol. The mechanism of the oxidation sequence is given below.

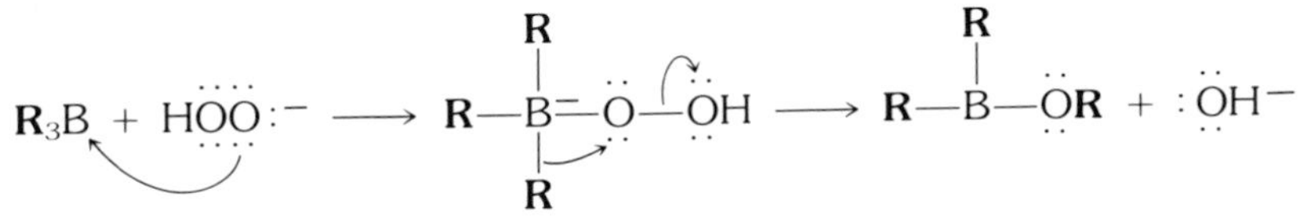

where: $R = CH_3(CH_2)_5CH_2—CH_2—$

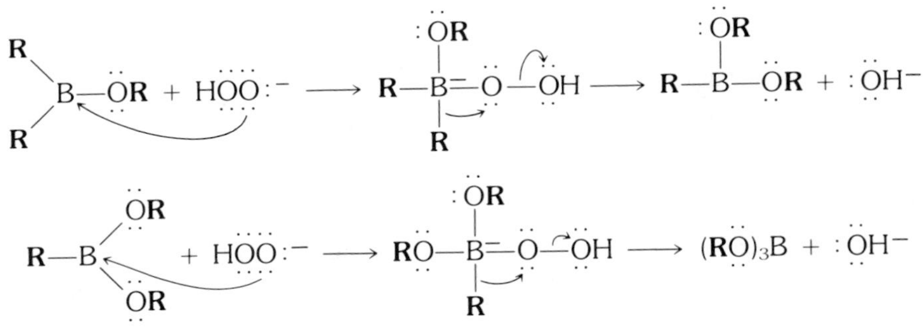

$(R\ddot{O})_3B + H_2O \longrightarrow 3\ R\ddot{O}H + B(\ddot{O}H)_3$

EXPERIMENTAL Estimated time to complete the experiment: 4.0 hours.

Physical Properties of Reactants and Products

Compound	MW	Wt/Vol	mmol	bp(°C)	Density	n_D
1-Octene	112.22	210 μL	1.34	121	0.72	1.4087
Borane-THF (1*M*)		500 μL				
Sodium hydroxide (3*M*)	40.00	300 μL				
Hydrogen peroxide (30%)	34.01	300 μL				
Octanol	130.23			194	0.83	1.4295

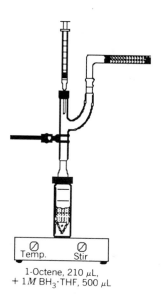

1-Octene, 210 μL,
+ 1M BH₃·THF, 500 μL

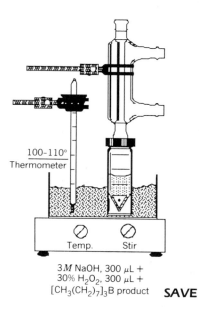

3M NaOH, 300 μL +
30% H₂O₂, 300 μL +
[CH₃(CH₂)₇]₃B product **SAVE**

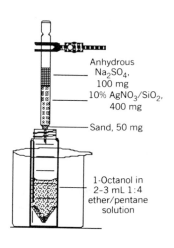

Anhydrous
Na₂SO₄,
100 mg

10% AgNO₃/SiO₂,
400 mg

Sand, 50 mg

1-Octanol in
2-3 mL 1:4
ether/pentane
solution

Reagents and Equipment

A 5.0-mL conical vial containing a spin vane and equipped with a Claisen head equipped with a rubber septum and calcium chloride drying tube.(■) Through the rubber septum is added 210 μL (150 mg, 1.34 mmol) of 1-octene (in one portion) with a 1.0-cm³ syringe.

Important. *The glassware and syringe are dried in a 100°C oven for at least 0.5 hour before use.*

The reaction vessel is cooled in an ice bath and, using the same syringe, 500 μL (0.5 mmol) of the borane-THF solution are added through the septum over a five-minute period.

> **CAUTION:** *The BH₃-THF reagent reacts violently with water.*

Reaction Conditions

The reactants are allowed to warm to room temperature and then stirred for a period of 45 minutes. Using a Pasteur pipet, two drops of water are now added to hydrolyze any unreacted borane complex.

At this stage of the procedure the vial may be removed from the Claisen head, capped, and allowed to stand until the next laboratory period.

The reaction vial is removed from the Claisen head and using a graduated 1.0-mL pipet, 300 μL of 3M sodium hydroxide solution is added, followed by the dropwise addition of 300 μL of 30% hydrogen peroxide solution over a 10-minute period. The hydrogen peroxide solution is also added with a graduated 1.0-mL pipet. The reaction vial is stirred gently after each addition.

> **CAUTION:** *Hydrogen peroxide blisters the skin. Concentrated solutions can explode!*

The vial is fitted with a reflux condenser and the reaction mixture heated at reflux with stirring for a period of one hour at a sand bath temperature of 100–110°C.(■)

The resulting two-phase mixture is cooled to room temperature and the spin vane removed by the use of forceps.

Isolation of Product

Using a Pasteur filter pipet, the bottom layer is separated and transferred to a 3.0-mL reaction vial. *Save the organic phase.*

This separated water layer is extracted with two 1.0-mL portions of diethyl ether and the ether extracts combined with the original organic phase. Upon the addition of each portion of ether, the vial is capped, shaken, and vented, and the layers are allowed to separate. The top ether layer is then separated using a Pasteur filter pipet. *If a solid forms during the extraction, a few drops of 0.1N HCl are added.*

The combined organic phase and ether extracts are now extracted with 750 μL of 0.1N hydrochloric acid solution followed by extraction with several 0.5-mL portions of water until the aqueous extract is neutral to pH paper.

A boiling stone is added to the vial and the organic phase concentrated by warming in a sand bath (60–65°C) in the hood.

Purification and Characterization

A Pasteur filter pipet is packed with 400 mg of 10% silver nitrate–treated activated silica gel followed by 100 mg anhydrous sodium sulfate.

The organic residue isolated above is dissolved in 500 μL of pentane (spectral or HPLC grade) and transferred by Pasteur pipet to the column.(■) The material is eluted from the column with 3 mL of a 1:4 diethyl ether/pentane solution. The eluate is collected in a tared 5.0-mL conical vial containing a boiling stone.

The eluate is concentrated to a constant weight by warming on a sand bath (60–65°C) in the hood. Weigh the 1-octanol and calculate the percentage yield. Determine the boiling point and compare your result with the literature value. Obtain an IR spectrum of the material and compare it with that of an authentic sample.

ENVIRONMENTAL DATA

Substance	Amount	TLV (mg/m³)	Emissions (mg)	Volume (m³)
1-Octene (Max. from yield, transfer)	150 mg	—	48	
Tetrahydrofuran (Max., solv.)	500 μL	590	444	0.8
Hydrogen peroxide (30%; Max.)	300 μL	1.5	90	65
Ether (Max., extract. to column Chromatog.)	2.6 mL	1200	1860	1.5
Pentane (Max., column chromatog.)	2.9 mL	1800	1815	1.0
Diborane (Max.)	7 mg	0.1	7	70

TLV: Hydrochloric acid—5.0 ppm
      Sodium hydroxide—2 μg/m³
TXDS: Silica gel—inv-mus LDLo: 234 mg/kg
      Sodium sulfate—orl-mus LD50: 5989 mg/kg
      Silver nitrate—unk-man LDLo: 29 mg/kg
      1-Octanol—orl-mus LD50: 1790 mg/kg

QUESTIONS

6-5. Using the hydroboration reaction, outline a reaction sequence for each of the following conversions.

    a. 1-Pentene to 1-pentanol.
    b. 1-Methylcyclopentene to *trans*-2-methylcyclopentanol.
    c. 2-Phenylpropene to 2-phenyl-1-propanol.

6-6. When diborane, B_2H_6, dissociates in ether solvents such as tetrahydrofuran (THF), a complex between BH_3 (borane) and the ether is formed. For example.

$$B_2H_6 \;+\; 2\; \overset{..}{\underset{..}{O}} \longrightarrow 2\; \overset{+}{O}{-}\overset{-}{B}H_3$$

    a. In the Lewis sense, what is the function of BH_3 as it forms the complex? Explain.
    b. Write the Lewis structure for BH_3. Diagram its expected structure indicating the bond angles in the molecule.

6-7. In reference to question 6-6a, explain why borane (BH_3) reacts readily with the π electron system of an olefin.

6-8. In an unsymmetrical alkene, the boron atom adds predominantly to the least highly substituted carbon atom. For example, 2-methyl-2-butene gives the products indicated below.

(Digylme: $CH_3\overset{..}{O}{-}CH_2CH_2{-}\overset{..}{O}{-}CH_2CH_2{-}\overset{..}{O}CH_3$
Diethyleneglycol dimethyl ether)

Offer a reasonable explanation to account for the ratio obtained.

6-9. An advantage of the hydroboration reaction is that rearrangement of the carbon skeleton does not occur. This contrasts with results obtained upon the addition of hydrogen chloride to the double bond.

For example,

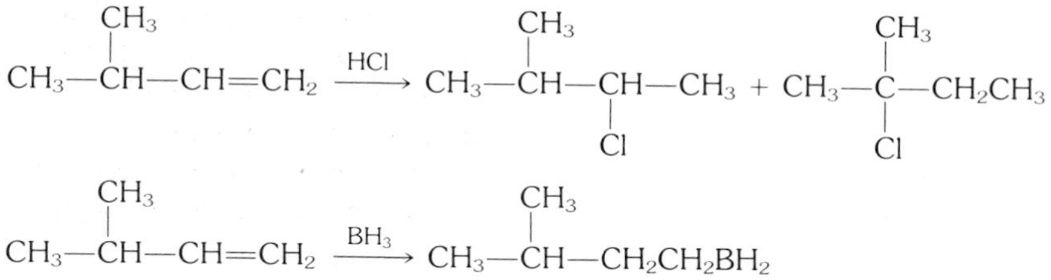

Offer an explanation for the difference in these results.

REFERENCES

1. For references relating to diborane as a hydroboration agent, see Experiment 16.
2. This experiment was based on that reported by Kabalka, G. W. *J. Chem. Educ.* **1975**, *52*, 745.

Experiment 14
Diels-Alder Reaction: 4-Cyclohexene-*cis*-1,2-dicarboxylic Acid Anhydride

(*cis*-4-cyclohexene-1,2-dicarboxylic anhydride)

This experiment illustrates the Diels-Alder reaction of 1,3-butadiene (generated *in situ*) with maleic anhydride to form a carbocyclic six-membered ring.

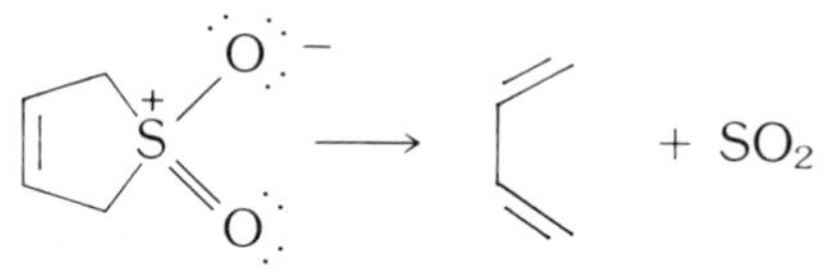

3-Sulfolene	*s-cis*-1,3-Butadiene

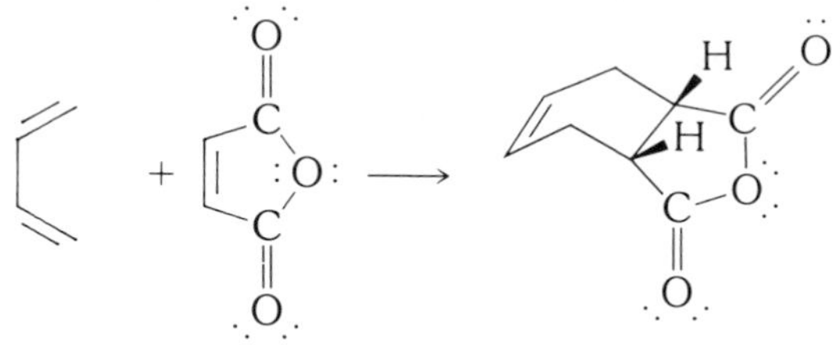

s-cis-1,3-Butadiene	Maleic anhydride	Diels-Alder adduct 4-cyclohexene-*cis*-1,2-dicarboxylic anhydride

DISCUSSION   The Diels-Alder reaction is one of the most useful synthetic conversions in organic chemistry. It is an example of a [4 + 2] cycloaddition reaction between a conjugated diene and a dienophile (an -ene or -yne), which leads to the formation of six-membered cyclic rings.

The reaction proceeds well if the dienophile bears electron-attracting groups and the diene electron-donating groups. Thus, α,β-unsaturated esters, ketones, nitriles, and so on make excellent dienophiles and are often used in the Diels-Alder reaction. By varying the nature of the diene and dienophile, a very large number of compounds can be prepared. Unsubstituted alkenes, such as ethylene, are poor dienophiles and react with 1,3-butadiene only at elevated temperatures and pressures. Since the Diels-Alder reaction is reversible, the lowest possible temperature is generally used.

The reaction is highly stereospecific; the diene component must be in the *s-cis* configuration in order to yield a cyclic product having a cis C=C. For this reason, cyclic dienes react more readily than acyclic species. The configuration of the groups on the diene and dienophile are retained in the adduct, and thus the mode of addition must be in the suprafacial–suprafacial manner.

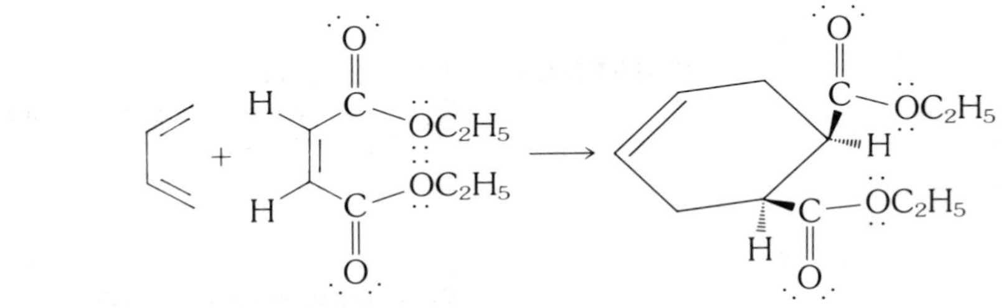

Diethyl maleate   Diels-Alder adduct
(carboethoxy groups cis)   (carboethoxy groups cis)

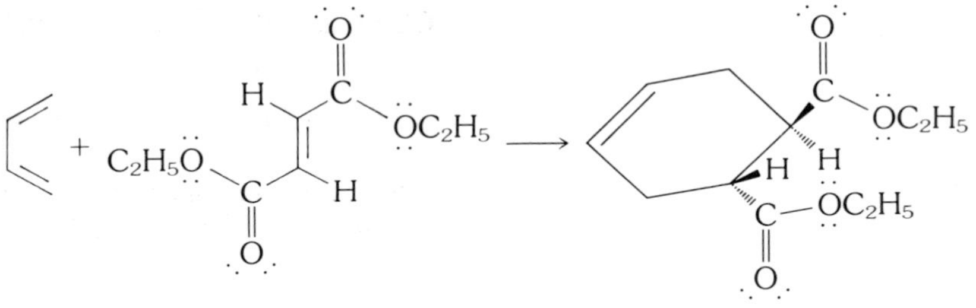

Diethyl maleate   Diels-Alder adduct
(carboethoxy groups trans)   (carboethoxy groups trans)

The reaction of cyclopentadiene with maleic anhydride demonstrates a further consequence of the mode of addition. In this situation, there are two possible ways in which the reactants may bond. This leads to the formation of two products designated the *endo* and *exo* forms.

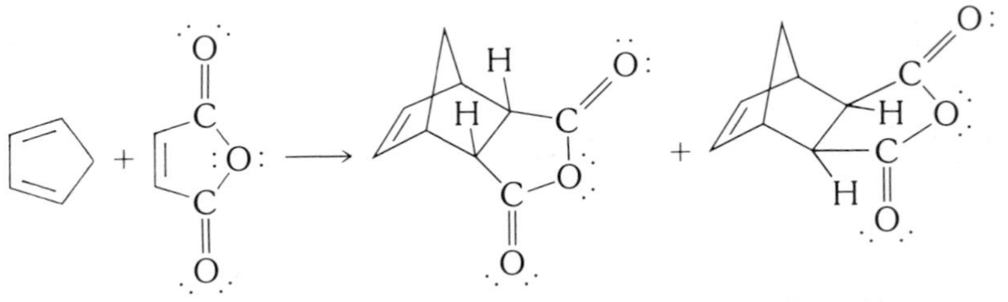

Endo adduct   Exo adduct

Generally, the *endo* form of the molecule predominates, but *endo/exo* ratios may vary depending upon reaction conditions.

The reaction is a thermal cycloaddition sequence, and the mechanism, based on orbital symmetry considerations, shows no formation of an intermediate. A simplified electron flow sequence is given below.

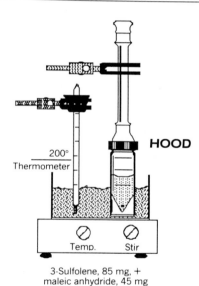

In the present reaction the 3-sulfolene decomposes at a moderate temperature to yield sulfur dioxide and 1,3-butadiene. This is an example of a retro-Diels-Alder reaction. By using this reagent, the diene component is generated *in situ*.

EXPERIMENTAL   Estimated time to complete the experiment: 1.5 hours.

Physical Properties of Reactants and Products

Compound	MW	Wt/Vol	mmol	mp(°C)	bp(°C)
3-Sulfolene	118.15	85 mg	0.72	66	
Maleic anhydride	98.06	45 mg	0.46	60	
Xylene		40 μL			137–140
cis-4-Cyclohexene-1,2-dicarboxylic acid anhydride	152.15			104	

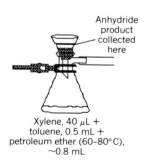

200°
Thermometer

HOOD

Temp.   Stir

3-Sulfolene, 85 mg, +
maleic anhydride, 45 mg
+ xylene, 40 μL

Reagents and Equipment

In a 1.0-mL conical vial equipped with an air condenser and containing a boiling stone are placed 85 mg (0.72 mmol) of 3-sulfolene, 45 mg (0.46 mmol) of maleic anhydride, and 40 μL of xylene.(■)

Note. *The maleic anhydride should be finely ground and protected from moisture to prevent hydrolysis to the corresponding acid. A mixture of xylenes in the bp range 137–140°C will suffice. The solvent should be freshly distilled or dried over molecular sieves before use. The xylene is dispensed in the HOOD using an automatic delivery pipet.*

Reaction Conditions

The reaction mixture is now heated at reflux using a sand bath at a temperature of 200°C for a period of 20 minutes.

> **CAUTION:** *The reaction is exothermic. Overheating should be avoided. Sulfur dioxide is evolved in the process; adequate ventilation should be provided.*

The conical vial is removed and the contents allowed to cool to room temperature.

Isolation of Product

To the resulting solution is added 0.5 mL of toluene followed by the dropwise addition of petroleum ether until a slight cloudiness persists.

The amount of petroleum ether required is in the range of 0.25–0.35 mL. During the recrystallization step, the sides of the vial may have to be scratched with a glass rod to induce crystallization.

The solution is reheated until it becomes clear and is then allowed to cool in an ice bath. The crystalline product is collected by vacuum filtration and washed with 0.5 mL of *cold* petroleum ether.(■)

Anhydride
product
collected
here

Xylene, 40 μL +
toluene, 0.5 mL +
petroleum ether (60–80°C),
~0.8 mL

Be careful. *Do not use an excess of the petroleum ether wash. Loss of product results if an excess is used.*

Purification and Characterization

The product is of sufficient purity for characterization. Weigh the product and calculate the percentage yield. Determine the melting point and compare your result with the literature value. Obtain an IR spectrum of the material and compare it with that of an authentic sample.

ENVIRONMENTAL DATA

Substance	Amount	TLV (mg/m^3)	Emissions (mg)	Volume (m^3)
3-Sulfolene (treated as SO_2)	85 mg			
Sulfur dioxide (Max.)	46 mg	5	46	9.2
Xylene (SKIN; Max., solvent)	40 μL	435	35	<0.1
Toluene (SKIN)	0.5 mL	375	434	1.2
(Max., vac. filter)				
Petroleum ether (cryst., wash)	0.85 mL	180	544	3.0
(treated as hexane)				

TXDS: 3-Sulfolene—ins-rbt LD50: 25 mg/kg
cis-4-Cyclohexene-1,2-dicarboxylic anhydride—orl-rat LDL: 4590 mg/kg

QUESTIONS   **6-10.** Predict the product in each of the following Diels-Alder reactions.

a. [structure] + $CH_3O_2C-C\equiv C-CO_2CH_3$ $\longrightarrow$

b. [structure] + [structure] $\longrightarrow$

c. [structure] + $CH_2=CH-CN$ $\longrightarrow$

6-11. Cyclopentadiene reacts as a dienophile in the Diels-Alder reaction a great deal more readily than does 1,3-butadiene. Explain.

6-12. Predict the diene and dieneophile that would lead to each of the following products.

a. [structure]   b. [structure]   c. [structure]

6-13. The dissociation of 3-sulfolene to form 1,3-butadiene generates one mole of sulfur dioxide gas per mole of 1,3-butadiene. Substantial quantities of SO_2 would be generated if this decomposition were carried out on a large scale. Suggest a method for trapping the gas to prevent its escape into the environment.

REFERENCES   **1.** The preparation of 4-cyclohexene-*cis*-1,2-dicarboxylic acid anhydride by the reaction of 1,3-butadiene with maleic anhydride is recorded:
Cope, A. C.; Herrick, E. C. "Organic Syntheses"; Wiley: New York, 1963; Collect. Vol. IV, p. 890.

2. The conditions of this reaction were adapted from those reported by Sample, T. E., Jr.; Hatch, L. F. *J. Chem. Educ.* **1968**, *45*, 55.

Experiment 15
Diels-Alder Reaction: 9,10-Dihydroanthracene-9,10-α,β-succinic Acid Anhydride

(9,10-ethanoanthracene-11,12-dicarboxylic anhydride, 9,10-dihydro-)

In this experiment the Diels-Alder reaction is demonstrated in which an aromatic ring system constitutes the diene component.

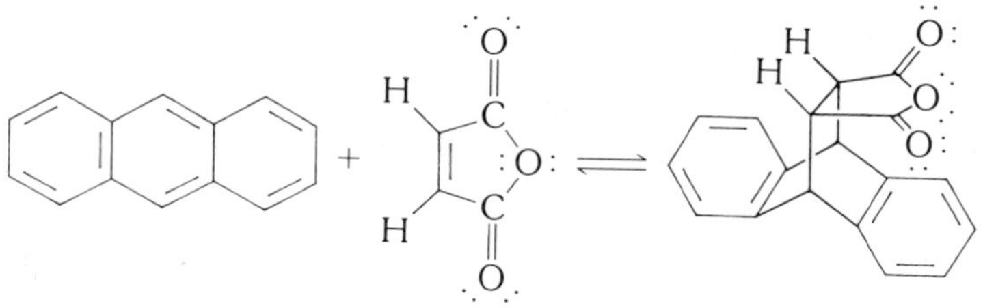

Anthracene	Maleic anhydride	9,10-Dihydroanthracene-9,10-α,β-succinic acid anhydride

DISCUSSION   This experiment is a further example of the Diels-Alder reaction. For a discussion of the Diels-Alder reaction, see Experiment 14. The central ring of anthracene has a characteristic diene system and thus reacts to form stable adducts with many dienophiles at the 9,10 position. The reaction of anthracene and maleic anhydride has been studied extensively and is a reversible reaction.

Higher hydrocarbons containing this anthracene nucleus also react with maleic anhydride, but differ widely in the rates at which they react. 1,2,5,6-Dibenzanthracene (*A*), 2,3,6,7-dibenzanthracene (*B*; pentacene), and 9,10-diphenylanthracene (*C*), are examples of aromatic hydrocarbons that undergo the Diels-Alder reaction.

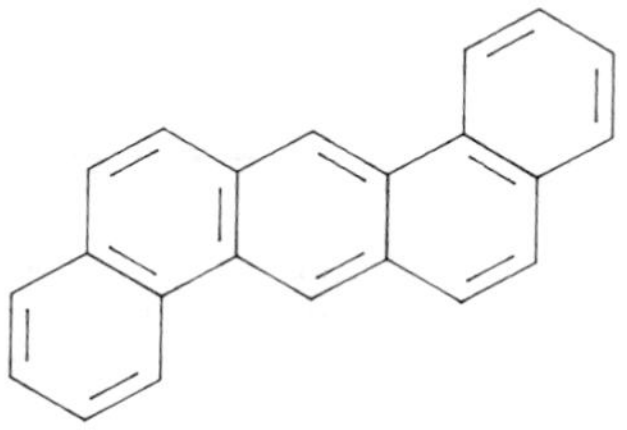

(*A*)	(*B*)	(*C*)

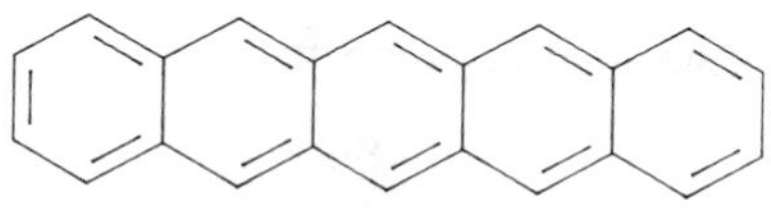

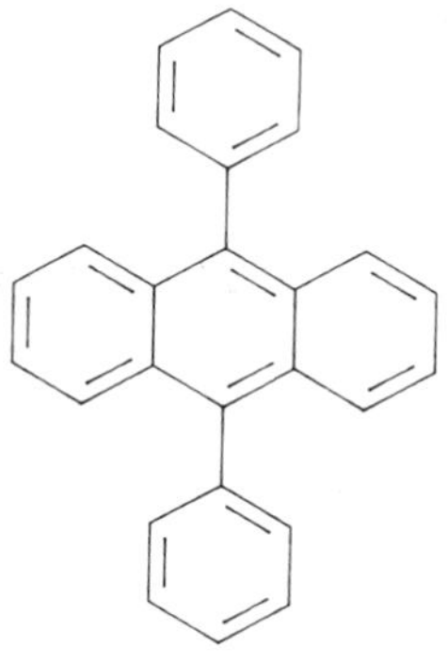

EXPERIMENTAL Estimated time to complete the experiment: 1.5 hours.

Physical Properties of Reactants and Products

Compound	MW	Wt/Vol	mmol	mp(°C)	bp(°C)
Anthracene	178.24	40 mg	0.22	216	
Maleic anhydride	98.06	20 mg	0.20	60	
Xylene		0.5 mL			137–140
9,10-Dihydroanthracene-9,10-α,β-succinic acid anhydride	276			261–262	

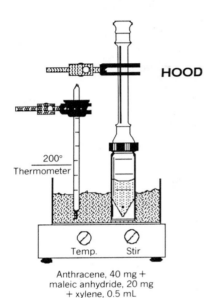

HOOD

200°
Thermometer

Temp.   Stir

Anthracene, 40 mg +
maleic anhydride, 20 mg
+ xylene, 0.5 mL

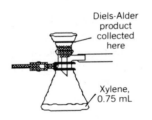

Diels-Alder
product
collected
here

Xylene,
0.75 mL

Reagents and Equipment

In a 1.0-mL conical vial containing a boiling stone and equipped with an air condenser are placed 40 mg (0.22 mmol) of anthracene and 20 mg (0.20 mmol) of maleic anhydride followed by 0.5 mL of xylene [**HOOD**] using a graduated 1.0-mL pipet.(■)

Note. *A high-purity grade of anthracene and maleic anhydride is recommended. Anthracene may be recrystallized from 95% ethanol. A mixture of xylenes with a bp range of 137–140°C is sufficient and should be dried over molecular sieves before use.*

Reaction Conditions

The reaction mixture is now heated under fairly vigorous reflux using a sand bath temperature of ~200°C for a period of 30 minutes. During this time the initial yellow color of the mixture gradually disappears. The resulting solution is allowed to cool to room temperature and then is placed in an ice bath for 10 minutes to complete crystallization of the product.

Isolation of Product

The crystals are collected by vacuum filtration using a Hirsch funnel and the filter cake washed with 0.25 mL of ice-cold xylene.(■) The material is now placed on a porous clay plate under an inverted watch glass overnight along with paraffin shavings or film to remove the xylene. *The high boiling point of the xylenes makes it difficult to remove last traces of solvent simply by air drying. Xylene, as well as other hydrocarbons used for recrystallizations, is absorbed by paraffin. Keep the crystals and paraffin separated.*

Purification and Characterization

The product is of sufficient purity for characterization. Weigh the dried Diels-Alder product and calculate the percentage yield. Determine the melting point and compare your result with the literature value. Obtain an IR spectrum of the material and compare it with that of an authentic sample.

ENVIRONMENTAL DATA

Substance	Amount	TLV (mg/m³)	Emissions (mg)	Volume (m³)
Xylene-o,m, or p (SKIN) (Max., recryst.)	0.75 mL	435	652	1.5

TLV: Maleic anhydride—0.25 ppm
TXDS: Anthracene—orl-rat TDLo: 20 g/kg/79W-1 TFX:ETA
Anthracene, 11,12-dicarboxylic acid anhydride, 9,10-dihydro—ipr-mus LDLo: 200 mg/kg

QUESTIONS   **6-14.** Given the data tabulated below for the rate of reaction of maleic anhydride with a series of substituted 1,3-butadienes, offer a reasonable explanation to account for the trend in the rates.

R	k(rel) at 25°C
—H	1
—CH$_3$	4.2
—C(CH$_3$)$_3$	<0.05

6-15. Predict the structure of the product formed in the reactions given below.

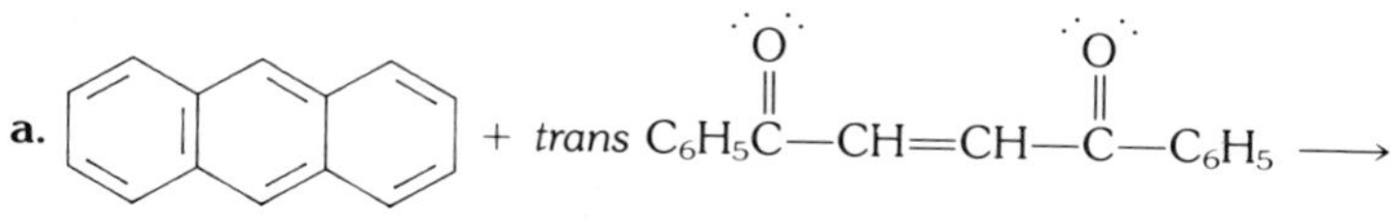

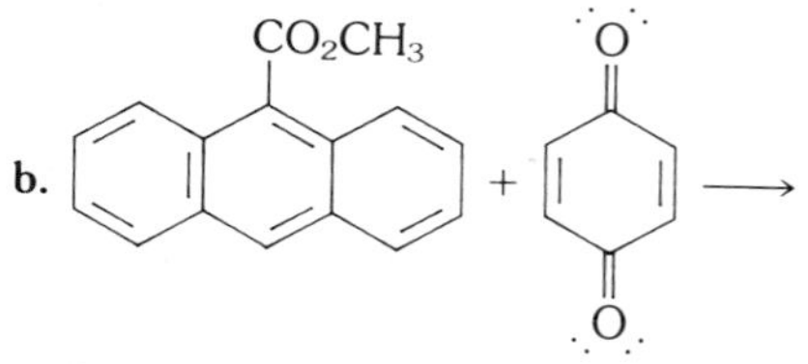

a.

$$\text{+ } trans \text{ } C_6H_5C\!-\!CH\!=\!CH\!-\!C\!-\!C_6H_5 \longrightarrow$$

b.

$$CO_2CH_3$$

6-16. Offer an explanation of why anthracene preferentially forms a Diels-Alder adduct at the 9,10 position.

REFERENCES   For references on the Diels-Alder reaction, refer to Experiment 14.

Experiments 16A and 16B
Diborane Reductions: Thioxanthene or Xanthene

(thioxanthene or xanthene)

The reduction of an aldehyde or ketone with diborane normally yields the corresponding alcohol. In the selected reactions described below, the carbonyl group is reduced to a methylene unit.

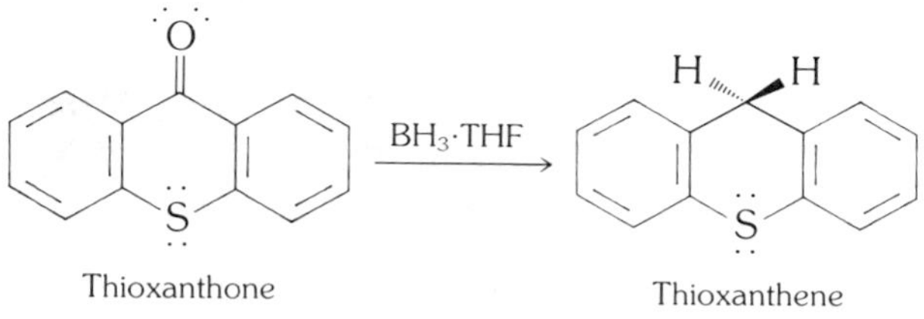

$$\xrightarrow{\text{BH}_3\cdot\text{THF}}$$

Thioxanthone   Thioxanthene

DISCUSSION   Diborane is a useful selective reducing agent and is prepared by reaction of boron trifluoride etherate with sodium borohydride. The diborane

$$3\text{NaBH}_4 + 4\text{BF}_3 \rightarrow 3\text{NaBF}_4 + 2\text{B}_2\text{H}_6$$

produced may be distilled into tetrahydrofuran to form the BH$_3$-THF complex, which is commercially available as a 1.0M solution. Diborane reacts rapidly with water, and for this reason procedures using the BH$_3$-THF complex must be run under anhydrous conditions.

Diborane is a Lewis acid that attacks electron-rich centers. Thus, when aldehydes or ketones are treated with the BH_3-THF complex, the borate ester is formed, which, upon hydrolysis, gives the corresponding alcohol. The reduction of the carbonyl group is believed to take place by addition of the electron-deficient boron atom to the oxygen atom, followed by irreversible transfer of hydride ion from boron to the carbon atom.

In the case of the xanthone ring system, the corresponding xanthydrol is not formed. It was found that the intermediate borate ester undergoes an elimination reaction, forming a borate ion and a resonance-stabilized xanthonium ion. The second stage of the reaction involves the addition of the BH_3 to the onium ion, which is a special example of the well-known hydroboration reaction. In the addition process the hydride group of the borane adds to the C-9 carbon atom forming the methylene, $-CH_2-$, group. The overall result is the reduction of a carbonyl to a methylene unit.

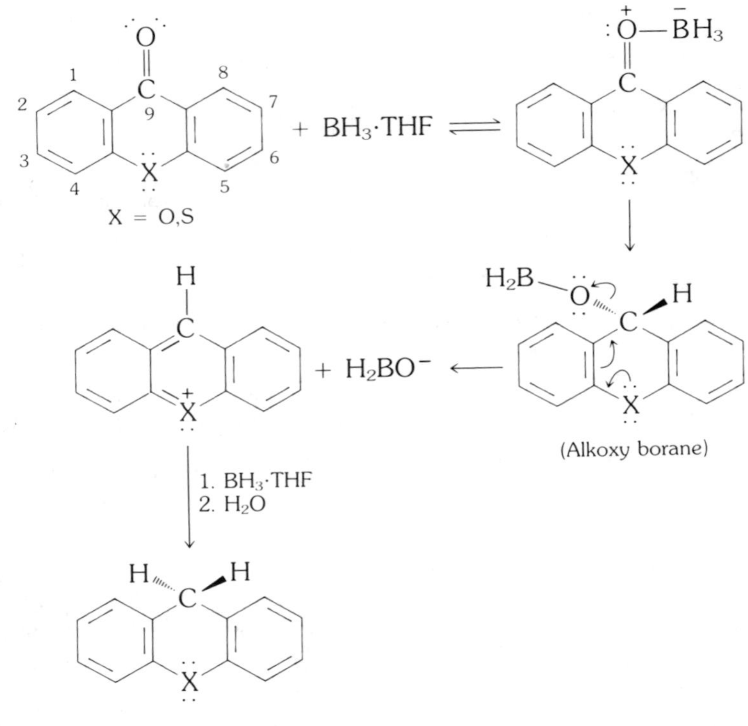

PART A: Thioxanthene

The reaction is shown above.

EXPERIMENTAL Estimated time to complete the experiment: 1.5 hours.

Physical Properties of Reactants and Products

Compound	MW	Wt/Vol	mmol	mp(°C)	bp(°C)
Thioxanthone	212.28	50 mg	0.24	209	
Tetrahydrofuran	72.12	1.7 mL			67
Borane-THF, 1M		1.0 mL			
Thioxanthene	198.29			128–129	

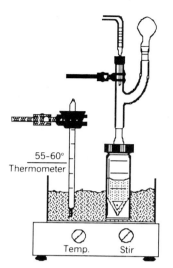

55–60°
Thermometer

Temp. Stir

Thioxanthone, 50 mg +
THF, 1.7 mL +
1*M* BH$_3$-THF, 1.0 mL

Reagents and Equipment

In a 5.0-mL conical vial containing a magnetic spin vane and equipped with a Claisen head, which in turn is fitted with a nitrogen inlet tube (prepared from a syringe) and a rubber septum, are added 50 mg (0.24 mmol) of thioxanthone and 1.7 mL of dry tetrahydrofuran. (■)

Note. *The tetrahydrofuran must be absolutely dry. It is recommended that HPLC grade reagent be used and that it be distilled once from calcium hydride and stored over molecular sieves. It may than be used for up to a week without adversely affecting the yield of product.*

The reaction vial is flushed with a gentle stream of nitrogen gas for several minutes, and then a small balloon is placed over the Claisen head outlet so as to maintain a dry atmosphere in the system.

The mixture is heated with stirring at 55–60°C until the thioxanthone dissolves, yielding a yellow solution. Then with stirring, 1.0 mL of a 1.0*M* solution of BH$_3$-THF is added in one portion through the rubber septum with a 1.0-mL syringe.

> **CAUTION: The BH$_3$-THF solution reacts violently with water.**

Reaction Conditions

The solution is heated at 55–60°C in a sand bath with stirring for a period of 5 minutes. The solution becomes colorless during this time.

Isolation of Product

CAREFUL

Thumb controls pressure; continuous shaking

THF, 2.5 mL, +
CH$_3$CH$_2$OH, 0.5 mL
+ H$_2$O, 2 mL +
thioxanthene

The reaction is quenched by the *careful*, dropwise addition of approximately 10 drops of 95% ethanol (Pasteur pipet), with stirring, until the observed foaming subsides. *Alcohol is added to decompose any unreacted BH$_3$-THF reagent.*

After the solution has cooled to room temperature, the mixture is transferred by Pasteur pipet to a 25-mL filter flask containing a boiling stone. Using a calibrated Pasteur pipet, two 1.0-mL portions of water are added to the reaction mixture. The tetrahydrofuran solvent is carefully removed under reduced pressure with frequent shaking of the flask.(■) As the tetrahydrofuran evaporates, white crystals of thioxanthene appear. The product crystals are collected under reduced pressure by use of a Hirsch funnel, and the filter cake is washed with two 1.0-mL portions of water. The crystals are then air dried on a porous clay plate.

Purification and Characterization

The thioxanthene product is essentially pure and requires no further treatment. Weigh the product and calculate the percentage yield. Determine the melting point of the material and compare it with the value reported in the literature. Obtain IR spectra of thioxanthone and thioxanthene and compare them to each other as well as to those of an authentic sample.

PART B: Xanthene

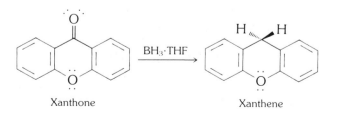

Xanthone   Xanthene

EXPERIMENTAL Estimated time to complete the experiment: 3.0 hours.

Physical Properties of Reactants and Products

Compound	MW	Wt/Vol	mmol	mp(°C)	bp(°C)
Xanthone	196.22	50 mg	0.26	174	
Tetrahydrofuran	72.12	0.7 mL			67
Borane-THF, 1M		0.75 mL			
Xanthene	182.23			100.5	

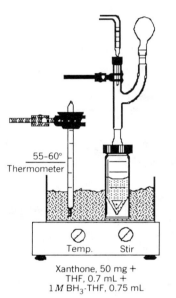

55–60°
Thermometer

Temp. Stir

Xanthone, 50 mg +
THF, 0.7 mL +
1M BH₃-THF, 0.75 mL

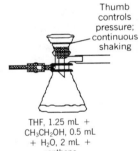

Thumb
controls
pressure;
continuous
shaking

THF, 1.25 mL +
CH₃CH₂OH, 0.5 mL
+ H₂O, 2 mL +
xanthene

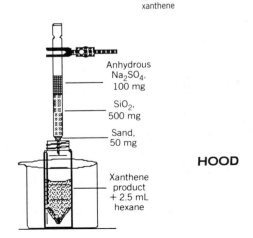

Anhydrous
Na₂SO₄,
100 mg

SiO₂,
500 mg

Sand,
50 mg

Xanthene
product
+ 2.5 mL
hexane

HOOD

Reagents and Equipment

Using the experimental apparatus described in Part A, 50 mg (0.26 mmol) of xanthone followed by 0.7 mL of dry tetrahydrofuran (see Part A, Reagents and Equipment) are placed in the reaction flask.(■) A dry nitrogen atmosphere is maintained in the system and the reaction mixture is heated with stirring to 55–60°C using a sand bath. After the xanthone dissolves, 0.75 mL of 1.0M solution of BH₃-THF is added through the rubber septum on the screwcap of the Claisen head in one portion using a 1.0-mL syringe.

> **CAUTION:** *The BH₃-THF solution reacts* violently *with water.*

Reaction Conditions

The reaction solution is stirred at 55–60°C in a sand bath for a period of one hour.

Isolation of Product

The reaction solution is now quenched by the careful addition (Pasteur pipet) with stirring of approximately 10 drops of 95% ethanol to the warm solution or until the observed foaming subsides. *Alcohol is added to decompose any un-reacted BH₃-THF reagent.*

The solution is transferred by Pasteur pipet to a 25-mL filter flask containing a boiling stone. Two 1.0-mL portions of water are now added (calibrated Pasteur pipet) to the solution. The tetrahydrofuran solvent is carefully removed under reduced pressure with frequent shaking of the flask.(■) As the solution becomes more concentrated, white crystals of xanthene appear. The crystals are collected under reduced pressure using a Hirsch funnel.

Purification and Characterization

The crude xanthene is purified by column chromatography. Place 0.5 g of activated silica gel followed by 0.1 g of anhydrous sodium sulfate in a Pasteur filter pipet.(■) The column is first wetted with a small amount of hexane and then a solution of the crude xanthene in 0.25 mL of methylene chloride is placed on the column using a Pasteur pipet. Xanthene is eluted by adding additional hexane (approximately 2.5 mL). The eluate is collected in a tared 5-mL conical vial containing a boiling stone. The hexane solvent is removed by evaporation [HOOD] by warming the solution on a sand bath to yield pure xanthene.

Weigh the solid and calculate the percentage yield of xanthene. Determine the melting point and compare it with the value found in the literature. Obtain an IR spectrum of xanthone and xanthene and compare them to each other as well as to authentic samples.

ENVIRONMENTAL DATA

Substance	Amount	TLV (mg/m^3)	Emissions (mg)	Volume (m^3)
Tetrahydrofuran (Max., evap.)	2 mL	590	1776	3.0
Ethanol (Max., reactant, evap.)	0.5 mL	1900	357	0.2
Diborane (Max.)	13.8 mg	0.1	13.8	138
Hexane (Max., column chromatog.)	3 mL	180	1980	11
Methylene chloride (Max.)	0.1 mL	350	134	0.4

TXDS: Xanthone—inv-mus LD50: 180 mg/kg
Xanthene—suc-mus LD50: 690 mg/kg
Sodium sulfate—orl-mus LD50: 5989 mg/kg
Silica gel—inv-mus LDLo: 234 mg/kg

QUESTIONS

6-17. In the reaction performed in this experiment, assume that the first stage of the reaction is the rate-controlling step. If so, would you predict that the relative rate of reduction of the carbonyl group to the methylene unit in compound A would be faster or slower than that of xanthone under the conditions of this experiment? Explain.

Compound A:

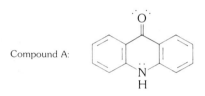

6-18. The reduction of aldehydes or ketones to the methylene group occurs with hydride reagents only when some special feature of the substrate promotes cleavage of the C—OH linkage. Suggest a suitable mechanism by which the reduction given below might occur.

$$(CH_3)_2\ddot{N} \text{—} \bigcirc \text{—} \underset{\underset{\displaystyle C}{\overset{\displaystyle H}{|}}}{C}=\ddot{O} \xrightarrow{\;BH_3\;} (CH_3)_2\ddot{N} \text{—} \bigcirc \text{—} CH_3$$

6-19. Diborane, on dissociation to borane, also forms complexes with sulfides and amines. Draw a suitable structure to represent the complex formed between BH$_3$ and dimethyl sulfide and also between BH$_3$ and triethylamine.

REFERENCES

The reactions carried out in the experiments outlined in Parts A and B were adapted from the work of Wechter, W. J. *J. Org. Chem.* **1963,** *28,* 2935.

1. Diborane as a reducing agent:
   a. Carruthers, W. "Some Modern Methods of Organic Synthesis", 2nd ed.; Cambridge Univ. Press: New York, 1978; Chapter 5.
   b. Fieser, L. F.; Fieser, M. "Reagents for Organic Synthesis"; Wiley:, New York, 1967; Vol. I, p. 199. Subsequent volumes of this series have further examples of diborane as a reducing agent.
   c. Pelter, A. *Chem. Ind.* (London) **1976,** 888.
   d. Lane, C. F. *Aldrichimica Acta* **1973,** *6,* 36.
2. Diborane as a hydroborating agent:
   a. Brown, H. C. "Boranes in Organic Chemistry"; Cornell Univ. Press: New York, 1972.
   b. Brown, H. C. "Organic Synthesis via Boranes"; Wiley: New York, 1975.
   c. Brown, H. C. "Hydroboration"; Benjamin: New York, 1962.
   d. Zweifel, G.; Brown, H. C. *Org. React.* **1963,** *13,* 1.

Experiment 17
Grignard Reaction With a Ketone: Triphenylmethanol

(methanol, triphenyl-)

This is the first of a number of Grignard reactions presented in the text. The present experiment involves the addition of a Grignard reagent to a ketone. It is a classic method for the synthesis of *tert*-alcohols.

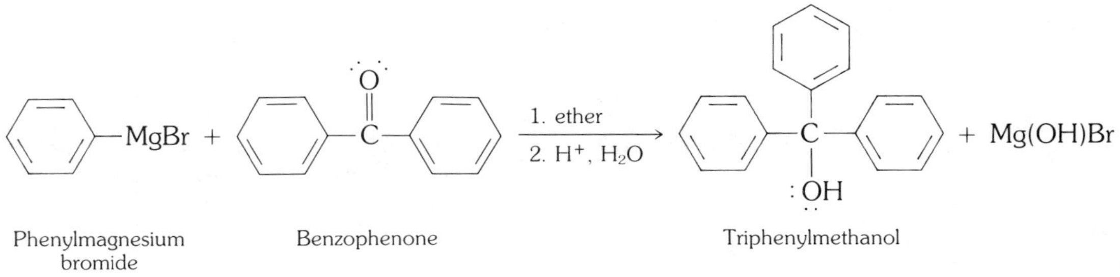

Phenylmagnesium bromide   Benzophenone   Triphenylmethanol

DISCUSSION Addition of nucleophilic Grignard reagents to electrophilic carbonyl carbon centers is an important method for the formation of C—C bonds. This experiment involves the addition of phenylmagnesium bromide to benzophenone to yield the corresponding tertiary (3°) alcohol. Because it is possible to vary the structure of the Grignard reagent and also that of the carbonyl species, a wide variety of 3° alcohols are prepared by this method.

Reaction of Grignard reagents with other carbonyl derivatives yields a number of important classes of compounds. For example, reaction with formaldehyde yields 1° alcohols; with higher aldehydes, 2° alcohols; with esters, 3° alcohols or ketones; with carbon dioxide, carboxylic acids.

The mechanism involves a nucleophilic attack of the Grignard reagent at the carbon atom center of the carbonyl group. The alkoxide ion intermediate is then hydrolyzed with dilute acid to yield the desired alcohol. The sequence is outlined below.

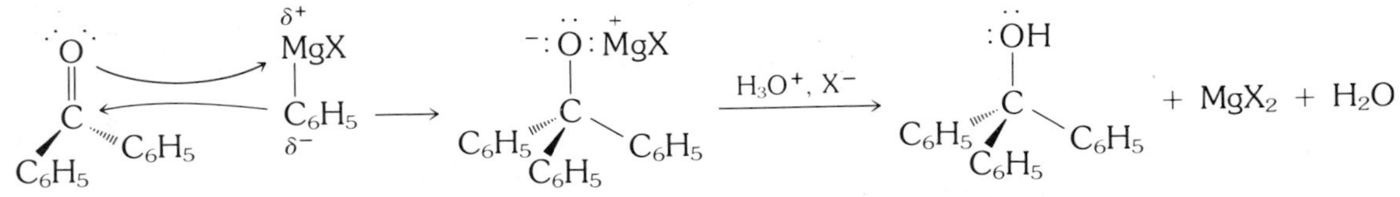

EXPERIMENTAL Estimated time to complete the experiment: Two laboratory periods.

> **CAUTION:** *Ether is a flammable liquid and also a narcotic. All flames should be extinguished during the time of this experiment.*

Physical Properties of Reactants and Products

Compound	MW	Wt/Vol	mmol	mp(°C)	bp(°C)	Density	n_D
Bromobenzene	157.02	76 μL	0.72		156	1.50	1.5597
Diethyl ether	74.12	1.3 mL			34.5		
Magnesium	24.3	18 mg	0.73				
Iodine	253.81	1 crystal					
Benzophenone	182.21	105 mg	0.58	48			
Triphenylmethanol	260.34			164			

Reagents and Equipment

Preparation of Phenylmagnesium Bromide

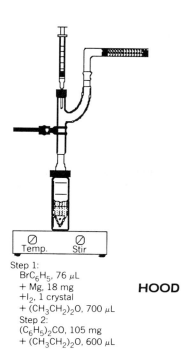

Step 1:
BrC$_6$H$_5$, 76 μL
+ Mg, 18 mg
+I$_2$, 1 crystal
+ (CH$_3$CH$_2$)$_2$O, 700 μL
Step 2:
(C$_6$H$_5$)$_2$CO, 105 mg
+ (CH$_3$CH$_2$)$_2$O, 600 μL

Note. *All the glassware used in the experiment should be cleaned, dried in an oven at 110°C for at least 30 minutes, and then cooled in a desiccator before use.*

In a 3.0-mL conical vial containing a magnetic spin vane and equipped with a Claisen head fitted with a calcium chloride drying tube and a rubber septum is placed 18 mg (0.73 mmol) of magnesium, a small crystal of iodine, and 100 μL of anhydrous diethyl ether.(■)

Note. *A 2–3 in. piece of magnesium ribbon is scraped clean of oxide coating and cut into 1-mm length sections. This freshly cut material is handled only with forceps.*

A solution of 76 μL (113 mg, 0.72 mmol) of bromobenzene in 300 μL of anhydrous diethyl ether is prepared in a dry, screwcapped vial. *Automatic delivery pipets are used to deliver these reagents in the* **HOOD**.

The bromobenzene-ether solution is drawn into a 1.0-mL syringe, and the syringe is then inserted through the rubber septum on the Claisen head.

An additional 300 μL of diethyl ether is placed in the empty vial and the vial is capped and set aside for later use.

Reaction Conditions

While stirring, 6-8 drops of the bromobenzene solution is added to initiate the formation of the Grignard reagent. Evidence of reaction is the evolution of tiny bubbles from the surface of the magnesium.

SLOWLY When the reaction has started, the remainder of the bromobenzene is added dropwise *slowly* over a 3–5 minute period and the reactants warmed slightly.

Upon completion of this addition, the rinse in the capped vial is drawn into the syringe and added through the septum in a single portion.

The resulting solution is now heated gently for a period of 15 minutes.

> **CAUTION:** *DO NOT OVERHEAT! This will cause loss of ether solvent and promote formation of by-products. Small fragments of magnesium may remain at the end of this time.*

The gray-brown mixture is cooled to room temperature.

The Benzophenone Reagent

A solution of 105 mg (0.58 mmol) of benzophenone in 300 μL of anhydrous diethyl ether is prepared in a dry shell vial. *The ether is measured using an* **HOOD** *automatic delivery pipet and is dispensed in the* **HOOD**.

The solution is immediately drawn into a 1.0-mL syringe, and the syringe then inserted into the rubber septum on the Claisen head.

An additional 300 μL of the anhydrous diethyl ether is placed in the empty vial and the vial is capped and set aside for later use.

Reaction Conditions

CAREFULLY The benzophenone solution is added *carefully*, with stirring, to the Grignard reagent over a period of 30 seconds at such a rate as to maintain the ether solvent at a steady reflux.

Upon completion of this addition, the rinse in the capped vial is added, in like manner, in one portion.

After stirring for 2–3 minutes, the reaction mixture is allowed to cool to room temperature and the reaction vial removed from the Claisen head and capped. During this cooling period the reaction mixture generally solidifies. *It is recommended that the vial be placed in a beaker to prevent loss of product by tipping.*

Note. *If the laboratory is done in two periods, one may stop at this point or after the hydrolysis sequence in the next step.*

Isolation of Product

The magnesium alkoxide salt is hydrolyzed by the *careful*, dropwise addition of 3*M* HCl from a Pasteur pipet using a small stirring rod to break up the solid. The addition is continued until the aqueous phase is acidic to litmus paper. A two-layer reaction solution forms (ether-water) as the solid dissolves.

> **CAUTION:** *The addition of the acid is accompanied by the evolution of heat. An ice bath should be handy to cool the solution if necessary. Additional ether may be added if required to maintain the volume.*

The magnetic spin vane is now removed with forceps and set aside to be rinsed with an ether wash. The vial is capped tightly, shaken, and vented, and the layers are allowed to separate.

Using a Pasteur filter pipet, the lower aqueous layer is transferred to a clean 5.0-mL conical vial.

Important. *The ether layer is saved since it contains the crude reaction product.*

The aqueous layer, after transfer to the 5.0-mL vial, is now washed with three 0.5-mL portions of diethyl ether using a calibrated Pasteur pipet. Rinse the spin vane with the first portion of ether as it is added to the vial. The vial is capped, shaken, and vented, and the layers are allowed to separate. After each extraction, the ether layer is combined with the ether solution retained above. The bottom aqueous layer is finally discarded.

The combined ether layers are now extracted with 0.5 mL of cold water to remove any acidic residue. The aqueous rinse is discarded. The ether solution is dried over 250–300 mg of anhydrous sodium sulfate for approximately 10 minutes. The drying agent is stirred intermittently with a glass rod. A larger amount of the drying agent may be used if necessary.

HOOD

Using a dry Pasteur filter pipet, the ether solution is transferred to a previously tared Craig tube containing a boiling stone. The ether solution is transferred in 0.5-mL portions, evaporating the ether by warming in a sand bath [**HOOD**] to concentrate the solution between transfers. The vial and drying agent are rinsed with an additional 0.5 mL of ether, the rinse added to the crude product in the Craig tube, and the solution concentrated to dryness.

Determine the weight of the crude triphenylmethanol product.

Purification and Characterization

The major impurity in the crude triphenylmethanol is biphenyl, formed by a coupling reaction.

$$2RX + Mg \xrightarrow{\text{ether}} R\text{—}R + MgX_2$$

These two materials can be separated by taking advantage of their differing solubility in ligroin. Ligroin, a nonpolar solvent, easily dissolves the nonpolar biphenyl molecule, but much less readily dissolves the more polar triphenylmethanol.

To the crude product in the Craig tube is added 0.5 mL of cold ligroin and the solid material scraped into suspension with a small stirring rod. Swirl the solid product with the solvent for several minutes. *Ligroin, bp 30–60°C, is used. Petroleum ether may be substituted if desired.*

Isolate the solid triphenylmethanol using the Craig tube in the usual manner. *Save the ligroin solution containing any biphenyl* by transferring it, using a Pasteur pipet, to a tared 10-mL Erlenmeyer flask.

HOOD   Repeat the above extraction with a second 0.5-mL portion of ligroin, again stirring the solid suspension and combining the recovered ligroin solution with that saved above. The open Erlenmeyer flask is now placed in the **HOOD** overnight, or warmed in a sand bath, to allow the ligroin to evaporate. Determine the amount of biphenyl (and other impurities) produced in the reaction.

The Craig tube containing the solid triphenylmethanol is heated in a 100°C oven for a period of 5 minutes and then the crystals placed on a clay plate to complete the drying process.

Weigh the product and calculate the percentage yield. Determine the melting point of the triphenylmethanol and compare your result to the literature value. If desired, the product may be purified further by recrystallization from isopropyl alcohol using the Craig tube.

Obtain an IR spectrum of the triphenylmethanol and compare it with that of an authentic sample. It may be of interest to obtain the IR spectrum of the material containing the biphenyl by-product to establish that biphenyl is formed in the reaction.

ENVIRONMENTAL DATA

Substance	Amount	TLV (mg/m^3)	Emissions (mg)	Volume (m^3)
Benzophenone	105 mg			
Bromobenzene (Max. from yield, transfer)	133 mg	—	45	
Diethyl ether (Max., evap.)	2.3 mL	1200	1642	1.4
Ligroin (recrystal., treated as hexane)	1.5 mL	180	1290	7.2
(Preparation of Grignard also included here.)				

TLV: Iodine—0.1 ppm
     Hydrochloric acid—5 ppm
TXDS: Benzophenone—orl-mus LD50: 2895 mg/kg
     Magnesium—orl-dog LDLo: 230 mg/kg
     Sodium sulfate—orl-mus LD50: 5989 mg/kg

QUESTIONS   **6-20.** Predict the product formed in each of the reactions below and give each reactant and product a suitable name.

a. $CH_3CH_2MgBr + CH_2O \xrightarrow[\text{2. H}^+]{\text{1. ether}}$

b. $p\text{-}CH_3C_6H_4MgBr + CH_3CH_2CHO \xrightarrow[\text{2. H}^+]{\text{1. ether}}$

c. $C_6H_5MgBr + D_2O \xrightarrow{\text{ether}}$

d. $CH_3\!-\!\overset{\displaystyle CH_3}{\underset{\displaystyle CH_3}{\bigcirc}}\!-\!MgBr + CO_2 \xrightarrow[\text{2. H}^+]{\text{1. ether}}$

6-21. Using the Grignard reaction, carry out the following transformations. Any necessary organic or inorganic reagents may be used. Name all reactants and products.

a. [structure: cyclohexanone → 1-phenylcyclohexanol with :ÖH]

b. [structure: bromocyclohexane → cyclohexyl-CH$_2$CH$_2$OH]

c. CH$_3$—[benzene]—Br ⟶ CH$_3$—[benzene]

6-22. Outline a balanced stepwise reaction scheme for the preparation of triphenylmethanol by the reaction of phenylmagnesium bromide with (a) methyl benzoate, and (b) diethyl carbonate.

6-23. In the experiment, ligroin is used as a solvent for the separation of the product from biphenyl.
  a. What is the composition of ligroin?
  b. Can you suggest an alternative solvent that might be used in this step?

REFERENCES
1. General references on Grignard reagents:
  a. Grignard, V. *Compt. rend.* **1900**, *130*, 1322.
  b. Jones, R. G.; Gliman, H. *Chem. Revs.* **1954**, *54*, 835.
  c. Wakefield, B. J. *Organometal. Chem. Revs.* **1966**, *1*, 131.
  d. Ashby, E. C. *Quart. Revs.* **1967**, *21*, 259.
  e. Coates, G. E.; Green, M. L. H.; Wade, K. "Organometallic Compounds", 3rd ed.; Methuen: London; Vol. II, 1968.
  f. Yoffe, S. T.; Nesmeyanov, A. N. "Handbook of Magnesium-Organic Compounds"; Pergamon: London; Vols. I–III, 1957.
  g. Kharasch, M. S.; Reinmuth, O. "Grignard Reactions of Non-metallic Substances"; Prentice-Hall: New York; 1954.
2. A synthesis of triphenylmethanol (triphenylcarbinol) is reported in "Organic Syntheses": Bachmann, W. E.; Hetzner, H. P. "Organic Syntheses"; Wiley: New York, 1955; Collect. Vol. III, p 839.
3. The preparation of a series of tertiary alcohols by the addition of Grignard reagents to diethyl carbonate is reported in "Organic Syntheses": Moyer, W. W.; Marvel, C. S. "Organic Syntheses"; Wiley: New York, 1943; Collect. Vol. II, p 602–603.

Experiment 18
Grignard Reaction With an Aldehyde: 4-Methyl-3-heptanol

(3-heptanol, 4-methyl-)

The addition of a Grignard reagent to an aldehyde is described in this experiment and constitutes a classic method for the synthesis of *secondary* alcohols.

[reaction scheme]

$$CH_3CH_2CH_2-\overset{\underset{|}{H}}{\overset{CH_3}{\underset{|}{C}}}-MgBr \ + \ CH_3CH_2-\overset{\overset{O}{\|}}{C}-H \ \xrightarrow[\text{2. H}^+,\text{ H}_2\text{O}]{\text{1. ether}} \ CH_3CH_2CH_2-\overset{\underset{|}{H}}{\overset{CH_3}{\underset{|}{C}}}-\overset{\underset{|}{\ddot{O}H}}{\overset{H}{\underset{|}{C}}}-CH_2CH_3$$

1-Methylbutylmagnesium bromide   Propanal   4-Methyl-3-heptanol

DISCUSSION In this experiment the addition of a nucleophilic Grignard reagent, 1-methylbutylmagnesium bromide, to the electrophilic carbonyl carbon atom of an aldehyde (propanal) is described. The product is a 2° alcohol, 4-methyl-3-heptanol. Because it is possible to vary the structure of both the Grignard reagent and the aldehyde, a wide variety of 2° alcohols are prepared by this route. The reaction of ethyl formate with Grignard reagents is also used as a method of preparing 2° alcohols. Primary alcohols are obtained when formaldehyde is used as the aldehyde component. The mechanistic sequence is outlined below.

$$R-MgBr + CH_3CH_2-\overset{O}{\underset{}{C}}-H \xrightarrow{ether} CH_3CH_2-\overset{+MgBr}{\underset{R}{\overset{-:\ddot{O}:}{\underset{|}{C}}}}-H \xrightarrow[H_2O]{H^+} CH_3CH_2-\overset{:\ddot{O}H}{\underset{R}{\overset{|}{C}}}-H$$

$$R = -\overset{CH_3}{\underset{H}{\overset{|}{\underset{|}{C}}}}-CH_2CH_2CH_3$$

EXPERIMENTAL Estimated time to complete the experiment: 3.0 hours. The chromatographic separation requires approximately an additional 15 minutes per student.

> **CAUTION:** *Ether is a flammable liquid and also a narcotic. All flames should be extinguished during the time of this experiment.*

Physical Properties of Reactants and Products

Compound	MW	Wt/Vol	mmol	bp(°C)	Density	n_D
2-Bromopentane	151.05	125 μL	1.0	117	1.21	1.4413
Diethyl ether	74.12	600 μL		34.5		
Magnesium	24.31	36 mg	1.48			
Iodine	253.81	1 crystal				
Propanal	58.08	50 μL	0.69	49	0.81	1.3636
4-Methyl-3-heptanol	130.23			160–161		1.4310

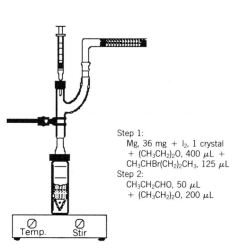

Step 1:
Mg, 36 mg + I₂, 1 crystal
+ (CH₃CH₂)₂O, 400 μL +
CH₃CHBr(CH₂)₂CH₃, 125 μL
Step 2:
CH₃CH₂CHO, 50 μL
+ (CH₃CH₂)₂O, 200 μL

Temp.   Stir

Reagents and Equipment

Preparation of 1-Methylbutylmagnesium Bromide

Note. *All the glassware used in the experiment should be cleaned, dried in an oven at 110°C for at least 30 minutes, and then cooled in a desiccator before use.*

In a 3.0-mL conical vial containing a magnetic spin vane and equipped with a Claisen head fitted with a calcium chloride drying tube and a rubber septum is placed 36 mg (1.5 mmol) of magnesium, a small crystal of iodine, and 100 μL of anhydrous diethyl ether.(■)

Note. *A 2–3 in. piece of magnesium ribbon is scraped clean of oxide coating and cut into sections 1 mm in length. This freshly cut material is handled only with forceps.*

A solution of 125 μL (153 mg, 1.0 mmol) of 2-bromopentane in 200 μL of anhydrous diethyl ether is prepared in a dry, screwcapped vial. *An automatic delivery pipet is used to deliver these reagents.*

The 2-bromopentane solution is drawn into a 1.0-mL syringe and the syringe then inserted through the rubber septum on the Claisen head. An additional 100 μL of diethyl ether is placed in the empty vial, and the vial is capped and set aside for later use.

Reaction Conditions

While stirring, 6–8 drops of the 2-bromopentane-ether solution are added to initiate the formation of the Grignard reagent. Evidence of reaction is the evolution of tiny bubbles from the surface of the magnesium.

SLOWLY   When the reaction has started, the remainder of the 2-bromopentane-ether solution is added dropwise [**SLOWLY**] over a 3–5 minute period and the reactants warmed slightly. Upon completion of this addition, the rinse in the capped vial is drawn into the syringe and added through the septum in a single portion. The resulting solution is now heated gently for a period of 15 minutes.

> **CAUTION:** *DO NOT OVERHEAT. This will cause loss of ether solvent. Small fragments of magnesium may remain at the end of this time.*

The gray-colored mixture is cooled to room temperature.

The Propanal Reagent

A solution of the aldehyde is prepared by weighing 50 μL (40 mg, 0.7 mmol) of propanal into a tared, oven-dried, capped vial followed by the addition of 100 μL of anhydrous diethyl ether. The propanal is the limiting reagent and therefore an accurate weight should be recorded for the percentage yield calculations. *The aldehyde and diethyl ether are measured by automatic delivery*
HOOD   *pipets and dispensed in the* **HOOD.**

The aldehyde solution is immediately drawn into a 1.0-mL syringe and the syringe then inserted into the rubber septum on the Claisen head.

An additional 100 μL of the anhydrous diethyl ether is placed in the empty vial, and the vial is capped and set aside for later use.

Reaction Conditions

CAREFULLY   The propanal solution is now added *carefully* with stirring to the Grignard reagent over a period of 30 seconds at such a rate as to keep the ether solvent at a steady reflux.

Following this addition, the rinse in the capped vial is added, in one portion in a similar manner.

The reaction mixture is stirred for a period of 5 minutes and allowed to cool to room temperature, and the conical vial removed and capped. *It is recommended that the vial be placed in a beaker to prevent loss of product by tipping.*

Note. *If the laboratory is done in two periods, one may stop at this point or after the hydrolysis sequence in the next step.*

Isolation of Product

The alkoxide magnesium salt is hydrolyzed by the *careful*, dropwise addition of 2–3 drops of water from a Pasteur pipet and the resulting solution stirred for a period of 5 minutes.

> **CAUTION:** *The addition of water causes the evolution of heat. An ice bath should be handy to cool the solution if necessary.*

Two to three drops of 3M HCl are now added and the capped vial is allowed to stand at room temperature for a period of 5 minutes. A two-phase reaction mixture develops (ether-water) as the magnesium salt is hydrolyzed.

The magnetic spin vane is removed with forceps and set aside to be rinsed with an ether wash. The vial is capped tightly, shaken, and vented, and the layers are allowed to separate.

SAVE Using a Pasteur filter pipet, the bottom aqueous layer is transferred to a clean 5.0-mL conical vial. *Save* the ether layer since it contains the crude reaction product.

The aqueous layer previously transferred to the 5.0-mL vial is now washed with three 0.5-mL portions of diethyl ether, and the magnetic spin vane is rinsed with the first portion as it is added to the vial. Upon addition of each portion of ether (calibrated Pasteur pipet), the vial is capped, shaken, and vented, and the layers are allowed to separate. With the aid of a Pasteur filter pipet, each ether layer is removed and combined with the ether solution retained above. After the final extraction the lower aqueous layer is discarded. The combined ether fractions are now extracted with 0.5 mL of cold water to remove any acidic material. The aqueous rinse is discarded.

HOOD The ether solution is dried by transferring it by aid of a Pasteur filter pipet to a Pasteur filter pipet containing 500 mg of anhydrous sodium sulfate. The eluate is collected in a tared 10 × 75 mm test tube. The ether solvent is removed from the eluate [**HOOD**] by warming in a sand bath to concentrate the solution to a weight less than 90 mg.

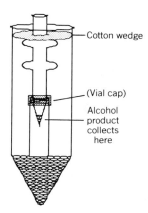

Cotton wedge

(Vial cap)

Alcohol product collects here

Purification and Characterization

The product, 4-methyl-3-heptanol, is isolated and purified using gas chromatography.

The entire sample of crude material obtained above is injected onto the GC column using a 100-μL syringe, and the components of interest are collected as they elute, employing the technique described in Experiment 2.(■)

GC conditions: Gow Mac Series 150, Thermal Conductivity Detector

20% Carbowax 20M column, $\frac{1}{4}$ in. × 8 ft

Temperature, 145°C

Flow rate, 50 mL/minute (He gas)

The retention time for 4-methyl-3-heptanol under these conditions is 7–8 minutes (2 minutes after any other peak).

Determine the weight of 4-methyl-3-heptanol collected and calculate the percentage yield. Determine the boiling point and refractive index of the alcohol and compare them with the values reported in the literature. Obtain an IR spectrum of the alcohol and compare it with that of an authentic sample. If enough of the hydrocarbon by-product is collected, obtain an IR spectrum to verify its identity.

ENVIRONMENTAL DATA

Substance	Amount	TLV (mg/m^3)	Emissions (mg)	Volume (m^3)
2-Bromopentane (Max. from yield, transfer)	125 μL	—	53	
Propanal (Max. from yield, transfer)	40 mg	—	20	
Diethyl ether (Max., evap.) (Preparation of Grignard also included here.)	2.6 mL	1200	1860	1.6

TLV: Iodine—0.1 ppm
      Hydrochloric acid—5 ppm
TXDS: Propanal—orl-rat LDLo: 800 mg/kg
       2-Bromopentane—ipr-mus LD50: 150 mg/kg
       4-Methyl-3-heptanol—ivn-mus LD50: 180 mg/kg
       Magnesium—orl-dog LDLo: 230 mg/kg
       Sodium sulfate—orl-mus LD50: 5989 mg/kg

QUESTIONS **6-24.** Carry out each of the following transformations using the Grignard reaction. Any necessary organic or inorganic reagents may be used. Name each reactant and product.

a. $CH_3(CH_2)_2CH_2Br \longrightarrow (CH_3CH_2CH_2CH_2)_2CHOH$

b.

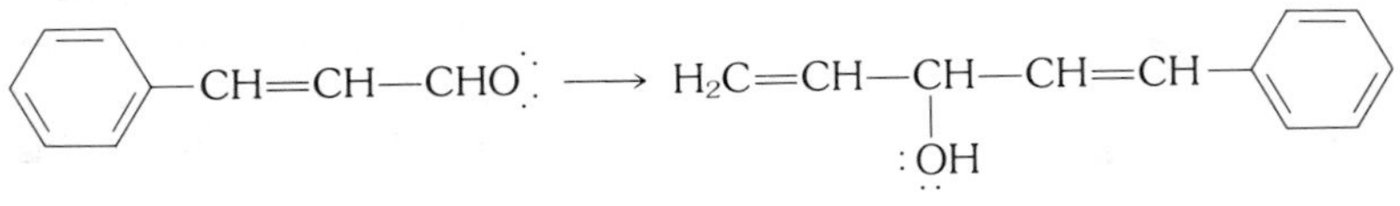

c.

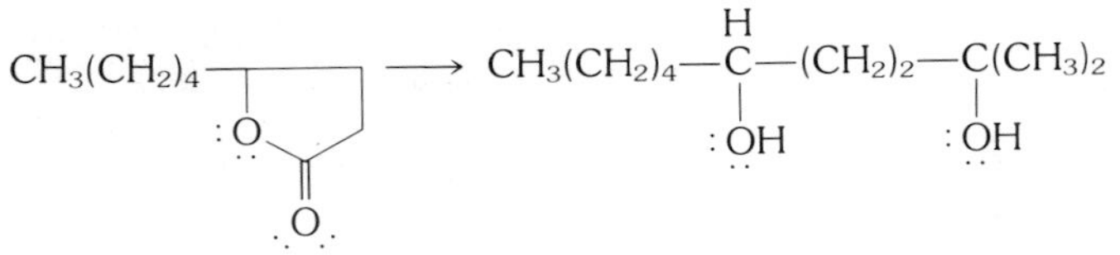

d.

6-25. Explain why Grignard reagents cannot be prepared from an organic halide that contains an hydroxyl (—OH), acid carbonyl (—COOH), thiol (—SH), or amino (—NH₂) group.

6-26. What would be the final product of the reaction between methyl benzoate and two equivalents of ethylmagnesium bromide?

6-27. Consider the same reaction as in question 6-26 except carry out the reaction with ethyl benzoate. What product would be expected in this case?

REFERENCES 1. For references relating to the preparation of Grignard reagents, see Experiment 17.
2. A list of secondary alcohol preparations presented in "Organic Syntheses" is summarized below.
    a. Coleman, G. H.; Craig, D. "Organic Syntheses"; Wiley: New York, 1943; Collect. Vol. II, p 179.
    b. Drake, N. L., *Ibid.*, p 406.
    c. Overberger, C. G.; Saunders, J. H.; Allen, R. E.; Gander, R. "Organic Syntheses"; Wiley: New York, 1955; Collect. Vol. III, p 200.
    d. Coburn, E. R. *Ibid.*, p 696.
    e. Skattebol, L.; Jones, E. R. H.; Whiting, M. C. "Organic Syntheses"; Wiley: New York, 1963; Collect. Vol. IV, p 792.

Experiment 19
The Perkin Reaction: Condensation of Rhodanine with an Aromatic Aldehyde: o-Chlorobenzylidene Rhodanine

(rhodanine, 5-(o-chlorobenzylidene)-)

In this reaction the base-catalyzed condensation of an aromatic aldehyde with an active methylene component is described. This condensation constitutes an example of the well known Perkin reaction.

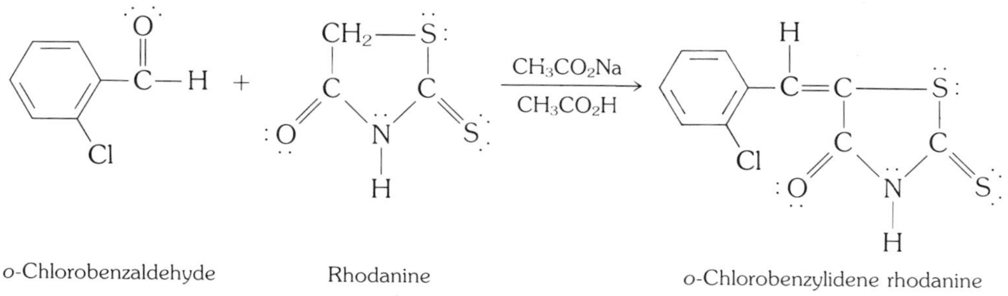

o-Chlorobenzaldehyde   Rhodanine   o-Chlorobenzylidene rhodanine

DISCUSSION The original Perkin reaction, the base-catalyzed condensation of an aromatic aldehyde with a carboxylic acid anhydride, was used extensively to prepare α,β-unsaturated carboxylic acids. The sodium or potassium salt of the carboxylic acid related to the anhydride was employed as the base. In the process, the carbanion generated by removal of a proton from the α-carbon of the anhydride makes a nucleophilic attack on the aldehydic carbonyl carbon atom. In general, the condensation is an aldol-type reaction.

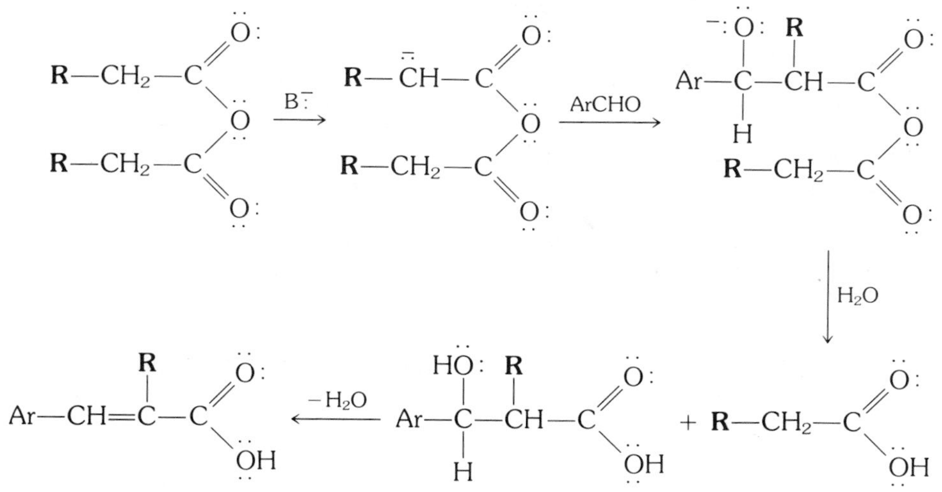

One variation of the Perkin reaction is the condensation of rhodanine with aromatic aldehydes. The resulting compounds have been shown to exhibit antibacterial, antitubercular, antimalarial, antifungal, and antiparasitic activity. Rhodanine has an active methylene group that can be deprotonated with a relatively mild base to generate a nucleophilic agent.

The reaction of rhodanine with the aldehyde group is an example of a nucleophilic attack on a trigonal carbon atom. Under the present conditions, elimination follows the initial nucleophilic addition to yield the benzylidene derivative. The mechanism is shown below.

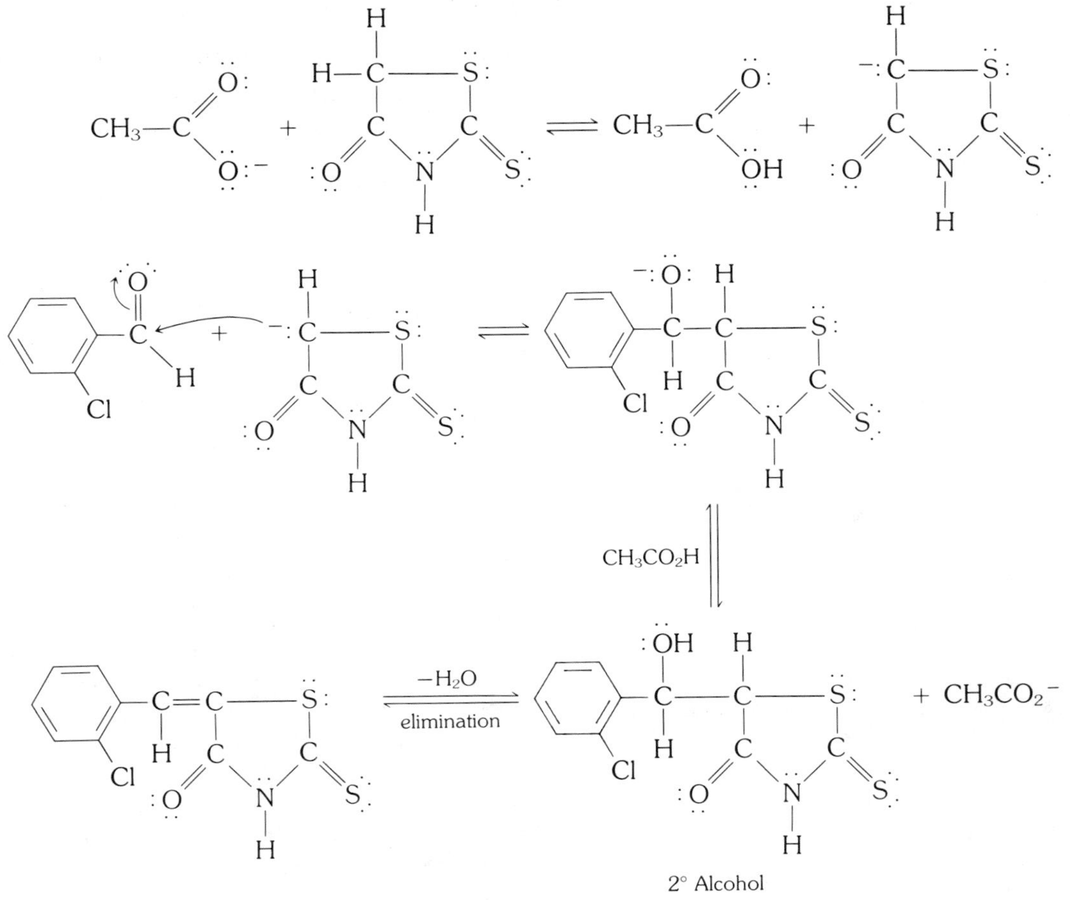

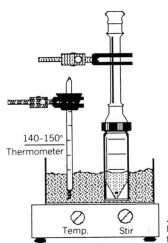

2° Alcohol

EXPERIMENTAL   Estimated time of the experiment: 1.5 hours.

Physical Properties of Reactants and Products

Compound	MW	Wt/Vol	mmol	mp(°C)	bp(°C)	Density	n_D
Rhodanine	133.19	30 mg	0.23	170			
Sodium acetate	82.03	52 mg	0.63	324			
Glacial acetic acid	60.05	1.0 mL			118		
o-Chlorobenzaldehyde	140.57	58 mg	0.41		212	1.25	1.5662
o-Chlorobenzylidene rhodanine	259.76			191			

Reagents and Equipment

HOOD   In a 3.0-mL conical vial containing a boiling stone and equipped with an air condenser are placed 30 mg (0.23 mmol) of rhodanine, 52 mg (0.63 mmol) of anhydrous sodium acetate, and 1.0 mL of glacial acetic acid **[HOOD]** dispensed from a graduated pipet.(■) To this mixture is added 58 mg (0.41 mmol) of o-chlorobenzaldehyde delivered from a Pasteur pipet.

Note. *The sodium acetate is dried in the oven for one hour before use. The aldehyde must be free from the corresponding acid or lower yields of product*

140–150°
Thermometer

Rhodanine, 30 mg +
NaOAc, 52 mg +
CH₃COOH, 1.0 mL +
o-ClC₆H₄CHO, 58 mg

Temp.   Stir

Note: It is important that the vial be immersed in sand to the level of the reaction mixture.

will result. It is recommended that the purity be checked by IR analysis. The reaction vial may be weighed before and after the addition to obtain an accurate weight of aldehyde.

Reaction Conditions

The reaction mixture is now heated at a sand bath temperature of 140–150°C for a period of 30 minutes.

Important. *The vial is immersed in the sand to the level of the reaction mixture.*

During this time, the reaction mixture becomes homogeneous and turns yellow in color. The resulting solution is allowed to cool to room temperature and then is placed in an ice bath to complete crystallization of the product.

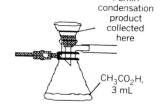

Perkin condensation product collected here

CH₃CO₂H, 3 mL

Isolation of Product

The yellow crystals are collected by vacuum filtration using a Hirsch funnel and the reaction vial and filter cake are washed with two 1.0-mL portions of cold glacial acetic acid delivered from a calibrated Pasteur pipet.(■) The crystals are dried on a porous clay plate.

Purification and Characterization

Fine, bright yellow needles are obtained on recrystallization (Craig tube) of 5 mg of the crude o-chlorobenzylidene rhodanine from 0.5 mL of glacial acetic acid.

Weigh the dried product and calculate the percentage yield. Determine the melting point and compare your results with the literature value. Obtain an IR spectrum of the product and compare it with that of an authentic sample.

ENVIRONMENTAL DATA

Substance	Amount	TLV (mg/m³)	Emissions (mg)	Volume (m³)
o-Chlorobenzaldehyde (Max. from yield.)	58 mg	—	7	
Acetic acid (Max., vap., filter)	1.0 mL	25	1049	42
Acetic acid (Max., cryst., wash)	2.0 mL	25	2098	84
Acetic acid (Recryst., Craig)	0.5 mL	25	524	21
(Note that the acetic acid is handled in the open except for the Craig tube step.)				

TXDS: Rhodanine—ipr-mus LD50: 200 mg/kg
2-Chlorobenzaldehyde—ipr-mus LDLo: 10 mg/kg

QUESTIONS

6-28. In which of the two species represented below is the underlined hydrogen atom more acidic? Explain.

6-29. A number of other active methylene compounds similar to rhodanine have been used in the Perkin condensation reaction. Two are shown in the margin.

Draw the structure of the product that would be formed if each underwent the Perkin condensation with *p*-chlorobenzaldehyde.

6-30. *p*-Nitrobenzaldehyde reacts at a faster rate than benzaldehyde in the Perkin reaction while *p*-N,N-dimethylaminobenzaldehyde is much less active toward the same substrate. Explain.

(An azlactone)   Hydantoin

6-31. Which of the two compounds below would you expect to be more reactive in the Perkin reaction toward rhodanine?

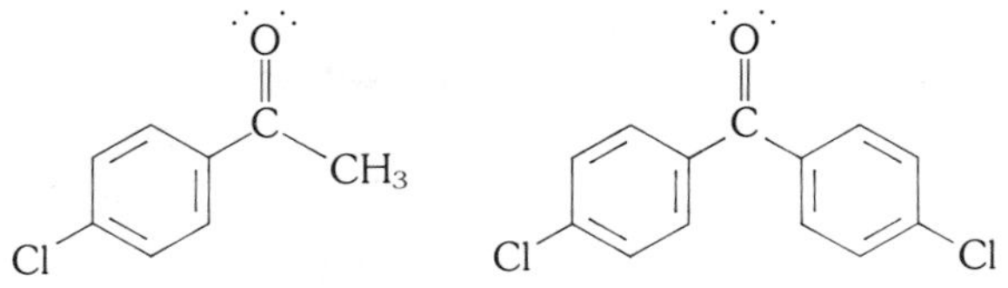

REFERENCES

1. For a review on the general Perkin reaction see Johnson, J. R. *Org. React.* **1942**, *1*, 210.
2. For examples of the general Perkin reaction see:
   (a) Thayer, F. K. "Organic Syntheses"; Wiley: New York, 1941; Collect. Vol., I, p 398.
   (b) Herbst, R. M.; Shemin, D. "Organic Syntheses"; Wiley: New York, 1943; Collect. Vol. II, p 1.
   (c) Weiss, R. *Ibid.*, p 61.
   (d) Corson, B. B. *Ibid.*, p 229.
   (e) Johnson, J. R. "Organic Syntheses"; Wiley: New York, 1955; Collect. Vol. III, p 426.
3. For references using rhodanine in the Perkin reaction see:
   (a) Brown, F. C. *Chem. Revs.* **1961**, *61*, 463.
   (b) Andreasch, R. *Monatsh. Chem.* **1928**, *49*, 122.
   (c) Campbell, N.; McKail, J. E., *J. Chem. Soc.* **1948**, 1215.
   (d) Julian, P. L.; Sturgis, B. M. *J. Am. Chem. Soc.* **1935**, *57*, 1126.
   (e) Foye, W. O.; Tovivich, P. *J. Pharm. Sci.* **1977**, *66*, 1607.

Experiment 20
Synthesis of a Triphenylmethane Dye: Crystal Violet

({4-(bis[*p*-dimethylaminophenyl]methylene)-2,5-cyclohexadiene-1-ylidine}-dimethylammonium chloride)

This reaction illustrates one of the routes for the preparation of the triphenylmethane class of dyestuffs.

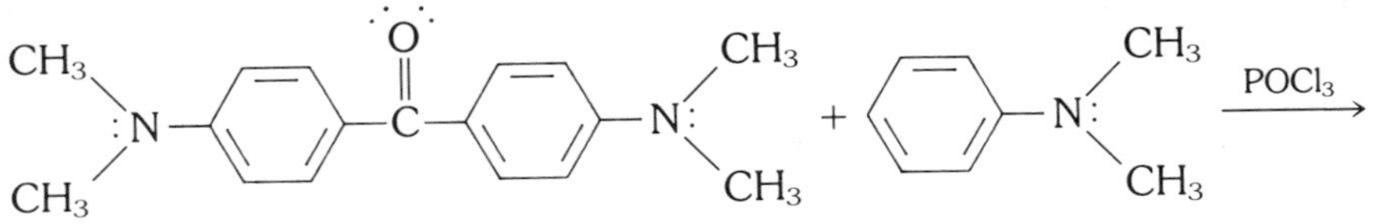

4,4′-Bis(dimethylamino)-benzophenone   *N,N*-Dimethylaniline

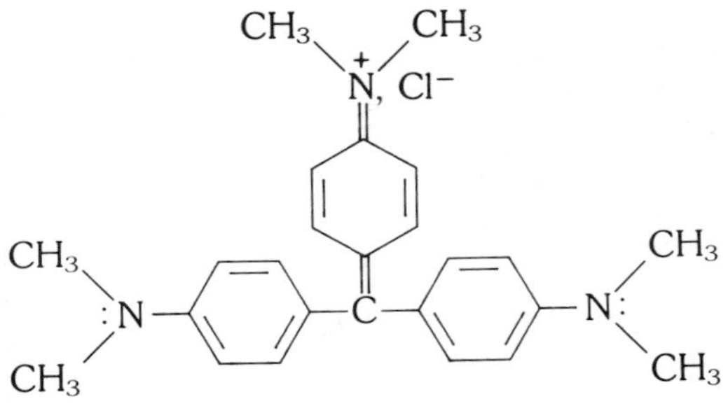

Crystal violet

DISCUSSION Triphenylmethane dyes were among the earliest synthetic coloring matters. Crystal violet belongs to the rosaniline series of triphenylmethane dyestuffs, all of which contain an amino substituent on each of the three phenyl groups. That is, crystal violet is a hexamethylpararosaniline dye (Color Index No. 681). This group of dyes is known for their intensity and brilliance; however, they have very poor light-fastness characteristics. The probable mechanistic sequence is shown below.

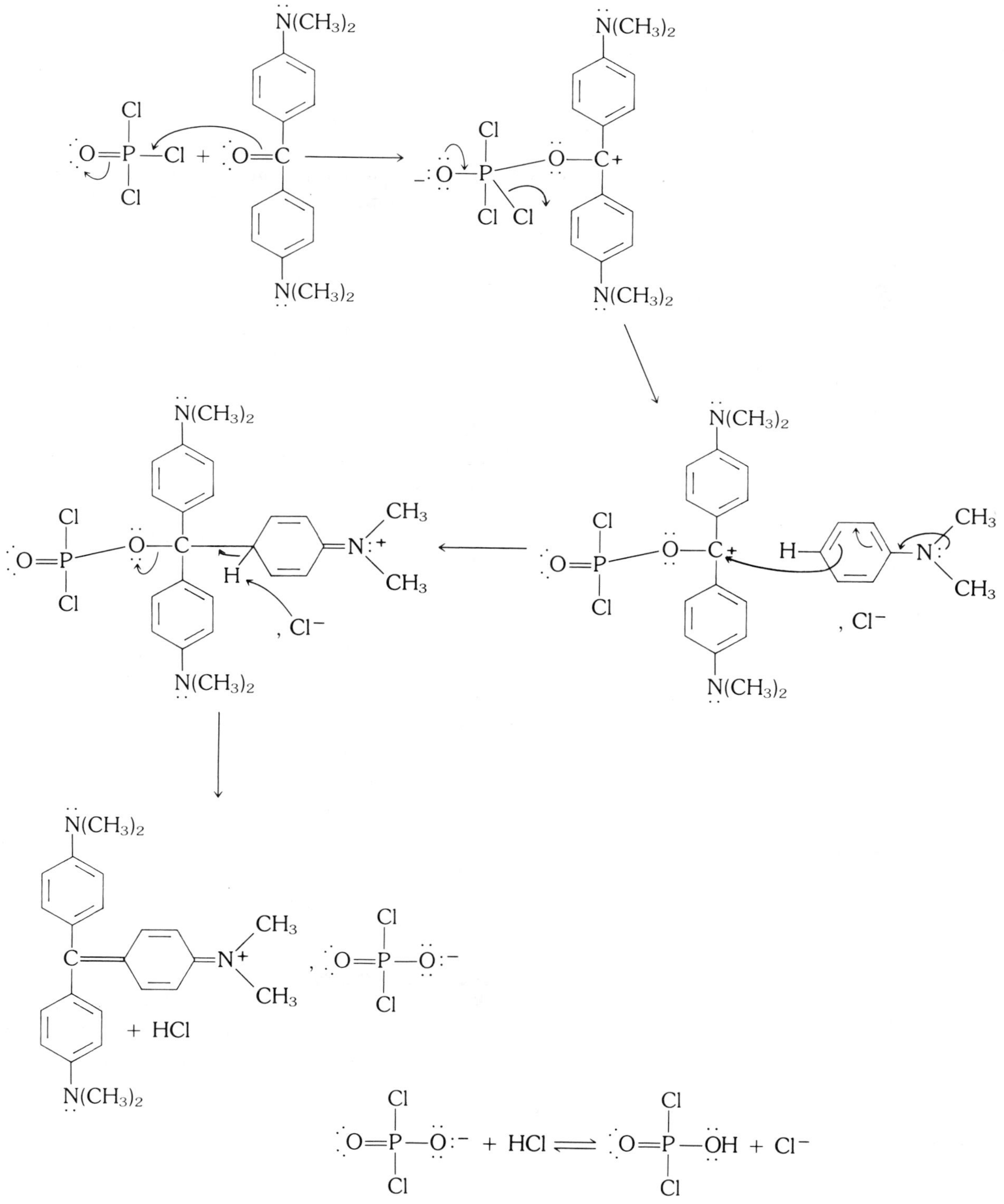

EXPERIMENTAL   Estimated time to complete the reaction: 1.5 hours.
(but overnight crystallization required)

Physical Properties of Reactants and Products

Compound	MW	Wt/Vol	mmol	mp(°C)	bp(°C)	Density	n_D
4,4′-Bis-(dimethylamino)benzophenone	268.36	50 mg	0.19	179			
N,N-Dimethylaniline	121.18	95 μL	0.75		194	0.96	1.5582
Phosphorous oxychloride	153.33	32 μL	0.35		105.3	1.68	
Crystal violet	407.99			~215			

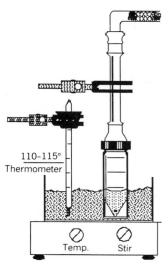

110–115°
Thermometer

Temp.   Stir

Step 1:
(CH₃)₂NC₆H₄COC₆H₄N(CH₃)₂,
50 mg +
(CH₃)₂NC₆H₅, 95 μL +
POCl₃, 32 μL
Step 2:
2.5 mL H₂O added

HOOD

Reagents and Equipment

> **WARNING:** *The glass equipment should be dried in an oven (110°C) half an hour before use as the POCl₃ reagent reacts vigorously with water.*

In a 5.0-mL conical vial containing a boiling stone and equipped with an air condenser protected by a calcium chloride drying tube, are placed 50 mg (0.19 mmol) of 4,4′-bis(dimethylamino)benzophenone, 95 μL (91 mg, 0.75 mmol) of N,N-dimethylaniline, and 32 μL (53 mg, 0.35 mmol) of phosphorous oxychloride.(■)

> **CAUTION:** *4,4′-Bis(dimethylamino)benzophenone (Michler's ketone) is a cancer suspect agent.* N,N-*dimethylaniline is highly toxic. The POCl₃ reagent is corrosive.*

The aniline and phosphorous oxychloride reagents are dispensed in the **HOOD** *using automatic delivery pipets.*

Reaction Conditions

The reactants are now heated at a sand bath temperature of 110–115°C for a period of 0.5 hour. A heavy green oil forms during this period. Water (2.5 mL) is added and the mixture heated for an additional 0.5 hour or until the thick, oily material dissolves. The solution is then cooled to room temperature and allowed to stand overnight.

Isolation of Product

The resulting mixture is placed in an ice bath for 20 minutes. The crystals are then collected by vacuum filtration using a Hirsch funnel and the filter cake is washed with three 1.0-mL portions of cold water.(■) The green crystals are then allowed to dry thoroughly on filter paper placed in a desiccator.

Dye
collected
here

H₂O,
5.5 mL +
by-products

Purification and Characterization

The dye is sufficiently pure for characterization. Weigh the product and calculate the percentage yield. Obtain an IR spectrum and compare it with that of an authentic sample. *Since the product is a salt, the melting point is variable and difficult to determine accurately. The material is therefore identified by its IR spectrum.*

ENVIRONMENTAL DATA

Substance	Amount	TLV (mg/m³)	Emissions (mg)	Volume (m³)
N,N-Dimethylaniline (SKIN) (Max., pipet, filter)	91 mg	25	91	3.6
Phosphorous oxychloride (Max., pipet, heat, filter, water will decompose)	52.6 mg	0.6	52.6	88

TXDS: 4,4'-Bis(dimethylamino)benzophenone—orl-rat LDLo: 15/kg/78 W-C TFX:CAR
Crystal violet—orl-rat LD50: 420 mg/kg

QUESTIONS

6-32. Draw the resonance forms of the crystal violet cation showing how the charge is shared by the three nitrogen atoms in the molecule.

6-33. When an alkali metal hydroxide is added to a solution of crystal violet, the color of the solution gradually fades, the electrical conductance is diminished, the alkalinity of the solution decreases, and the colorless covalent alcohol that forms may be precipitated. This species is called a pseudo-base. Suggest a structure for the pseudo-base formed by the treatment of a solution of crystal violet with sodium hydroxide.

6-34. In relation to question 6-33 other sources of reactive ions interact with the crystal violet carbocation to form similar covalent products. Predict the structure of the compounds formed by treatment of the dye carbocation with methylmagnesium bromide and with sodium cyanide.

REFERENCES

1. For a detailed overview of the triphenylmethane class of dyes see Venkataraman, K. "The Chemistry of Synthetic Dyes"; Academic Press: New York, 1952; Vol. II, p 718.

2. Crystal Violet is formed as a by-product in the preparation of 4,4'-bis-(dimethylamino)benzophenone.
Tuzun, C.; Ogliaruso; M.; Becker, E. I. "Organic Syntheses"; Wiley: New York, 1973; Collect. Vol. V, p 111. Also see Georgievies, G. *Ber.* **1905**, *38*, 884.

Experiment 21A, 21B, 21C, 21D[1]
Olefin Preparation by the Wittig Reaction: **E**-Stilbene; 1,2-Methylenedioxy-4-(2-methylpropenyl)-benzene; 1-Methylene-4-*t*-butylcyclohexane; 17β-Hydroxy-17α-methyl-3-methylene-5α-androstane

(**E**-benzene, 1,1'(1,2-ethenediyl)bis-; benzene, [1,2-methylenedioxy-4-(2-methylpropenyl)-]; cyclohexane, 1-methylene-4-(*t*-butyl)-; androstane, 17β-hydroxy-17α-methyl-3-methylene-, (5α))

These preparations illustrate the Wittig reaction, which is used extensively in organic synthesis to prepare olefins. The reaction involves the condensation of a *phosphorous ylid* with a selected carbonyl-containing compound.

[1]Portions of this experimental section have been accepted for publication. Pike, R.M.; Mayo, D.W.; Butcher, S.S.; Butcher, D.J.; Hinkle, R.J. *J. Chem. Ed.* **1986** (in press).

$$C_6H_5CHO + [(C_6H_5)_3PCH_2C_6H_5{}^+, \ Br^- + NaNH_2] \xrightarrow{\text{THF}} C_6H_5CH{=}CHC_6H_5 + \quad (C_6H_5)_3PO$$

Benzaldehyde   "Instant ylid"   E-Stilbene   Triphenylphosphine
                 benzyltriphenylphosphonium   oxide
                 bromide-sodium amide

DISCUSSION   The Wittig reaction constitutes a valuable method for the preparation of alkenes. The major advantage of this approach is that there is no ambiguity in the location of the C=C generated by the reaction.

The reaction involves the formation of a phosphorous ylid obtained by treatment of a phosphonium salt with a strong base. Phosphorous ylids are stable, highly reactive species for which resonance structures may be written; the phosphorous atom accommodates the electron donation into its 3*d*-orbitals.

$$[R_3\overset{+}{P}{-}\overset{-}{C}H_2 : \quad \longleftrightarrow \quad R_3P{=}CH_2]$$

The phosphonium salts are available through a nucleophilic displacement reaction with various alkyl halides.

$$(C_6H_5)_3P + RCH_2X \longrightarrow (C_6H_5)_3\overset{+}{P}CH_2RX^-$$

$$(X = I, \ Br, \ Cl)$$

Treatment of the phosphonium salt with a strong base such as butyl lithium in THF or sodium hydride in DMSO produces the ylid.

$$(C_6H_5)_3\overset{+}{P}CH_2R, \ \overset{-}{X} + C_4\overset{-}{H_9}{:}Li^+ \xrightarrow{\text{THF}} (C_6H_5)_3\overset{+}{P}{-}\overset{..}{\overset{-}{C}}H{-}R + C_4H_{10} + LiX$$

The ylids are generally not isolated, but treated directly with the carbonyl compound.

The "instant ylids" are solid-phase mixtures of a phosphonium salt with sodium amide and are now commercially available. Thus, the mixture need only be suspended in a suitable solvent to generate the desired ylid. This is a marked advantage over the usual methods employed to obtain these species. The ylid is then reacted with a suitable carbonyl compound to form an intermediate *betaine*, which on alkaline hydrolysis gives the olefin product. The mechanistic sequence is outlined below.

(a) Generation of the ylid

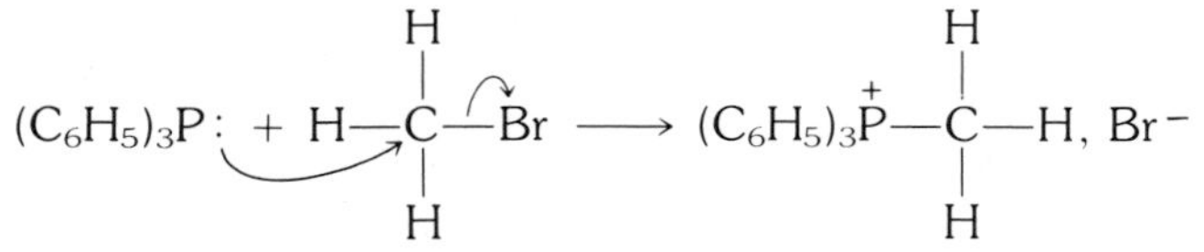

$$(C_6H_5)_3P{:} + H{-}\underset{H}{\overset{H}{\underset{|}{\overset{|}{C}}}}{-}Br \longrightarrow (C_6H_5)_3\overset{+}{P}{-}\underset{H}{\overset{H}{\underset{|}{\overset{|}{C}}}}{-}H, \ Br^-$$

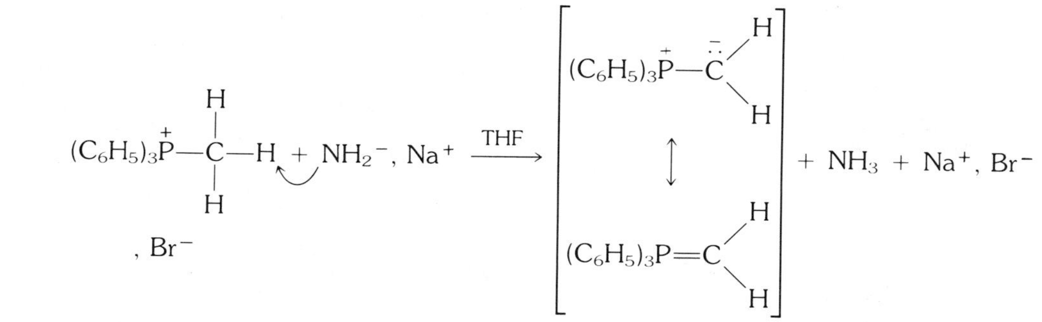

(b) Addition

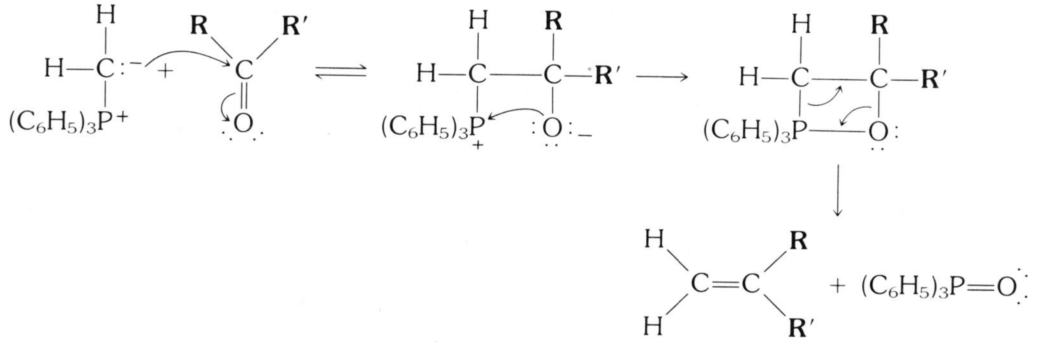

The Wittig reaction is very general. The carbonyl entity may be aliphatic or aromatic and contain C=C bonds and functional groups such as OH, O**R**, NO$_2$, N**R**$_2$, X, and so on. It is the only general method for the preparation of exocyclic carbon–carbon double bonds.

> **CAUTION:** *Tetrahydrofuran is a flammable liquid and all flames should be extinguished during the time of these experiments.*

PART A: E-Stilbene

The reaction for Part A is shown above.

EXPERIMENTAL   Estimated time to complete the reaction: 3.5 hours.

Physical Properties of Reactants and Products

Compound	MW	Wt/Vol	mmol	mp(°C)	bp(°C)	Density	n_D
Benzaldehyde	106.13	100 µL	0.95		178	1.04	1.5463
"Instant ylid"							
benzyltriphenylphosphonium							
bromide-sodium amide		600 mg	~1.2				
Tetrahydrofuran	72.12	1.0 mL			67		
E-Stilbene	180.25			124–125			

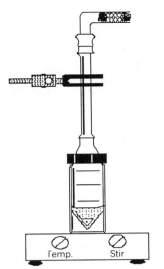

Step 1:
$(C_6H_5)_3 \cdot P^+ \cdot CH_2 \cdot C_6H_5, Br^- + NaNH_2$
600 mg + THF, 1.0 mL
Step 2:
C_6H_5 CHO, 100 μL

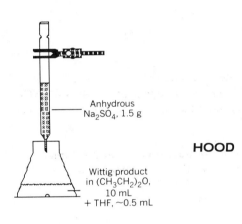

Anhydrous
Na_2SO_4, 1.5 g

HOOD

Wittig product
in $(CH_3CH_2)_2O$,
10 mL
+ THF, ~0.5 mL

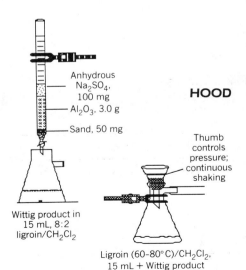

Anhydrous
Na_2SO_4,
100 mg

Al_2O_3, 3.0 g

Sand, 50 mg

HOOD

Thumb
controls
pressure;
continuous
shaking

Wittig product in
15 mL, 8:2
ligroin/CH_2Cl_2

Ligroin (60–80°C)/CH_2Cl_2,
15 mL + Wittig product

Reagents and Equipment

In a dry 5.0-mL conical vial containing a magnetic spin vane and equipped with an air condenser protected by a calcium chloride drying tube is placed 600 mg (~1.2 mmol) of benzyltriphenylphosphonium bromide-sodium amide ("instant ylid") mixture.(■) *Freshly distilled* tetrahydrofuran (1.0 mL) is now added using a calibrated Pasteur pipet and the mixture stirred for 15 minutes at room temperature. During this time the mixture turns orange in color.

Following generation of the ylid, 100 μL (0.95 mmol) of benzaldehyde (*freshly distilled*) is added to the reaction flask using an automatic delivery pipet.

Reaction Conditions

The resulting heterogeneous mixture is now stirred at room temperature for an additional 15 minutes. The system develops a light brown color during this time.

Isolation of Product

The reaction is quenched by adding 1.0 mL of a 25% aqueous NaOH solution (calibrated Pasteur pipet), and the resulting mixture is transferred to a 12-mL centrifuge tube using a Pasteur pipet. The reaction vial is rinsed with three 2.0-mL portions of diethyl ether, each of which is also transferred to the centrifuge tube (Pasteur filter pipet). The resulting two-phase mixture is now partially neutralized by careful addition of 3.0 mL of 0.1*N* HCl.

The ether layer (top) is separated using a Pasteur filter pipet and placed in a 25-mL Erlenmeyer flask. The remaining aqueous layer is extracted with two additional 2.0-mL portions of ether, and these extracts are separated as before and combined with the original ether layer.

The combined ether solution is transferred by Pasteur filter pipet to a short microcolumn prepared from a Pasteur filter pipet containing 1.5 g of anhydrous sodium sulfate.(■) The dried eluate (~10 mL) is collected in a 25-mL Erlenmeyer flask containing a boiling stone. This solution is now concentrated to dryness using a gentle stream of nitrogen or by warming in a sand bath [**HOOD**], to give a white powder.

Purification and Characterization

The crude product isolated above is purified by chromatography on an alumina column.

A short (1 cm × 10 cm) buret column is packed with 3.0 g of activated basic alumina and 100 mg of anhydrous sodium sulfate. The column is premoistened with ligroin.(■) The crude material is added directly to the column and is then eluted in a single fraction using 15 mL of 8:2 ligroin (60–80°C)/methylene chloride solution. The eluate is collected in a tared 25-mL filter flask containing a boiling stone and the solvent evaporated [**HOOD**] under vacuum in a warm sand bath with swirling.(■)

Note. *A 25-mL side-arm filter flask equipped with a Hirsch funnel and filter paper disks to control the pressure is a convenient system for the removal of a small volume (5–20 mL) of solvent.*

The product residue is sufficiently pure for characterization but may be recrystallized from a minimum amount of 95% ethanol if desired. Weigh the **E**-stilbene and calculate the percentage yield. Determine the melting point and compare your value with that reported in the literature. Obtain an IR spectrum and compare it with that of an authentic sample.

PART B: 1,2-Methylenedioxy-4-(2-methylpropenyl)-benzene

$$\text{Piperonal} + [(C_6H_5)_3PCH(CH_3)_2{}^+, \ Br^- \ + \ NaNH_2] \xrightarrow{\text{THF}} \text{1,2-Methylenedioxy-4-(2-methylpropenyl)-benzene} + (C_6H_5)_3PO$$

Piperonal

"Instant ylid"
isopropyltriphenylphosphonium
bromide-sodium amide

1,2-Methylenedioxy-
4-(2-methylpropenyl)-
benzene

Triphenylphosphine
oxide

EXPERIMENTAL Estimated time to complete the reaction: 4.5 hours.

Physical Properties of Reactants and Products

Compound	MW	Wt/Vol	mmol	mp(°C)	bp(°C)	n_D
Piperonal	150.14	150 mg	1.0	37		
"Instant ylid" isopropyltriphenylphosphonium bromide-sodium amide		750 mg	~1.7			
Tetrahydrofuran	72.12	1.0 mL			67	
1,2-Methylene-dioxy-4-(2-methylpropenyl) benzene	176.2				255	1.5603

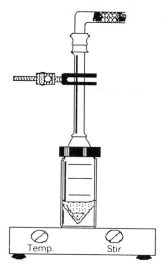

Step 1:
$(C_6H_5)_3 \cdot P^+ \cdot CH(CH_3)_2$, $Br^- + NaNH_2$,
750 mg + THF, 1.0 mL
Step 2:
Piperonal, 150 mg

Reagents and Equipment

In a dry 5.0-mL conical vial containing a magnetic spin vane and equipped with an air condenser protected by a calcium chloride drying tube is placed 750 mg (~1.7 mmol) of isopropyltriphenylphosphonium bromide-sodium amide ("instant ylid") mixture.(■) *Freshly distilled* tetrahydrofuran (1.0 mL) is now added and the mixture stirred for 30 minutes at room temperature. During this time the mixture turns an orange-brown color.

Following generation of the ylid, 150 mg (1.0 mmol) of piperonal is added to the reaction flask.

Important. *As stated above, the THF must be distilled just prior to use. It is recommended that the aldehyde be checked by IR to determine whether any acid is present. If so, the aldehyde must be purified.*

Reaction Conditions

The resulting heterogeneous reaction mixture is stirred at room temperature for an additional 15 minutes. During this time the system becomes lighter in color.

Isolation of Product

The reaction is quenched by adding 1.0 mL of 25% aqueous NaOH solution (calibrated Pasteur pipet) and the resulting mixture is transferred to a 12-mL centrifuge tube by means of a Pasteur pipet. The reaction vial is rinsed with two 1.0-mL portions of ether, which are also transferred to the centrifuge tube.

The ether layer is separated using a Pasteur filter pipet and placed in a 25-mL Erlenmeyer flask. The remaining aqueous layer is extracted with three 5-mL portions of ether, and these extracts are combined with the original ether layer.

The combined ether solution is transferred by Pasteur filter pipet to a short

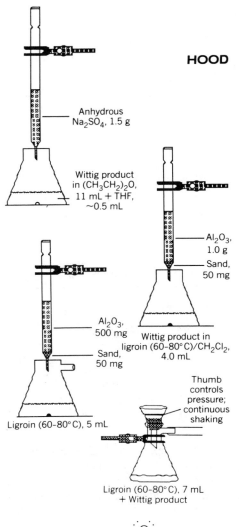

Anhydrous
Na$_2$SO$_4$, 1.5 g

Wittig product
in (CH$_3$CH$_2$)$_2$O,
11 mL + THF,
~0.5 mL

Al$_2$O$_3$,
1.0 g

Sand,
50 mg

Al$_2$O$_3$,
500 mg

Sand,
50 mg

Wittig product in
ligroin (60–80°C)/CH$_2$Cl$_2$,
4.0 mL

Ligroin (60–80°C), 5 mL

Thumb
controls
pressure;
continuous
shaking

Ligroin (60–80°C), 7 mL
+ Wittig product

HOOD

microcolumn prepared from a Pasteur filter pipet containing 1.5 g of anhydrous sodium sulfate.(■) The dried eluate is collected in a 25-mL Erlenmeyer flask containing a boiling stone. This solution is now concentrated to dryness, using a gentle stream of nitrogen or by warming in a sand bath [**HOOD**], yielding a yellow oil.

Purification and Characterization

The crude product is purified by two sequential chromatographic procedures on basic alumina.

A Pasteur filter pipet is packed with 1.0 g of activated basic alumina premoistened with 2.0 mL of ligroin (60–80°C).(■) The crude yellow oil is dissolved in an equal volume of ligroin and the resulting solution transferred to the column using a Pasteur pipet. The material is eluted directly with 4.0 mL of 8:2 ligroin (60–80°C)/methylene chloride solution. The eluate, collected in a 10-mL Erlenmeyer flask, is concentrated [**HOOD**] under a gentle stream of nitrogen.

This partially purified product is dissolved in 1.0 mL of ligroin and transferred by Pasteur pipet to a second microcolumn prepared from a Pasteur filter pipet containing 0.5 g of basic alumina premoistened with 1.0 mL of ligroin.(■) The flask is rinsed with an additional 1.0 mL of ligroin and this rinse also added to the column. The product is eluted using 5.0 mL of ligroin (60–80°C).

The eluate is collected in a tared 25-mL filter flask containing a boiling stone and the solvent removed under reduced pressure as described in Part A, Purification and Characterization.(■)

The residual oil, 1, 2-methylenedioxy-4-(2-methylpropenyl)-benzene, is sufficiently pure for characterization.

Weigh the product and calculate the percentage yield. Determine the boiling point and compare it with the literature value. Obtain an IR spectrum and assign frequencies to the major bands. Compare the results with the spectrum of the compound shown in Fig. 6.1.

PART C: Methylene-4-*t*-butylcyclohexane

4-*t*-Butylcyclohexanone

+ [(C$_6$H$_5$)$_3$PCH$_3$$^+$, Br$^-$ + NaNH$_2$] $\xrightarrow{\text{THF}}$

"Instant ylid"
methyltriphenylphosphonium
bromide-sodium amide

1-Methylene-
4-*t*-butylcyclohexane

+ (C$_6$H$_5$)$_3$PO

Triphenylphosphine
oxide

EXPERIMENTAL   Estimated time to complete the experiment: 4.0 hours.

Physical Properties of Reactants and Products

Compound	MW	Wt/Vol	mmol	mp(°C)	bp(°C)	n_D
4-*t*-Butylcyclohexanone	154.26	100 mg	0.64	50		
"Instant ylid" methyltriphenylphosphonium bromide-sodium amide		320 mg	~0.72			
Tetrahydrofuran		1.0 mL			67	
Methylene-4-*t*-butylcyclohexane	151.27				185	1.4630

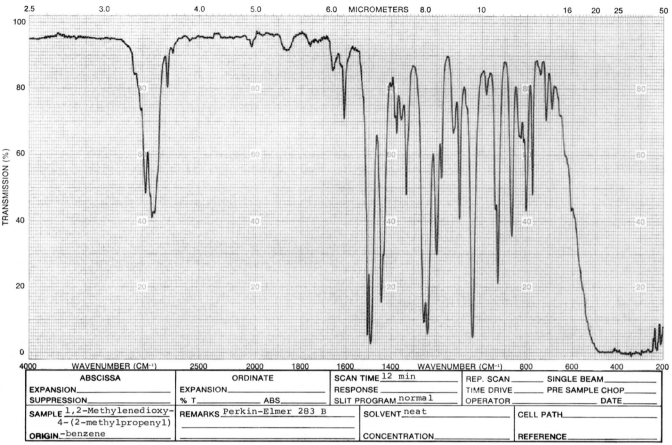

Fig. 6.1 IR spectrum: *1,2-methylenedioxy-4-(2-methylpropenyl)-benzene.*

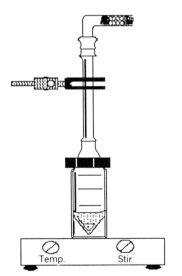

Step 1:
$(C_6H_5)_3$-P^+-CH_3, Br^- + $NaNH_2$,
320 mg + THF, 1.0 mL
Step 2:
4-*t*-Butylcyclohexanone,
100 mg

Reagents and Equipment

In a dry 5.0-mL conical vial containing a magnetic spin vane and equipped with an air condenser protected by a calcium chloride drying tube is placed 320 mg (0.72 mmol) of methyltriphenylphosphonium bromide-sodium amide ("instant ylid") mixture.(■) Freshly distilled tetrahydrofuran (1.0 mL) is now added using a calibrated Pasteur pipet and the mixture stirred for 15 minutes at room temperature. During this period it turns a bright yellow color.

Following generation of the ylid, 100 mg (0.64 mmol) of 4-*t*-butylcyclohexanone is added to the reaction flask. *The THF must be freshly distilled before use.*

Reaction Conditions

The resulting heterogeneous mixture is stirred at room temperature for an additional 90 minutes. The system develops a light tan color over this period of time.

Isolation of Product

The reaction is quenched by adding 1.0 mL of a 25% aqueous NaOH solution (calibrated Pasteur pipet) and the resulting mixture transferred to a 12-mL centrifuge tube using a Pasteur filter pipet. The reaction flask is rinsed with three 2.0-mL portions of diethyl ether, which are also transferred to the centrifuge tube. The resulting two-phase system is partially neutralized by the careful addition of 2.0 mL of 0.1N HCl.

The ether layer is separated by Pasteur filter pipet and placed in a 25-mL Erlenmyer flask. The remaining aqueous layer is extracted with three additional

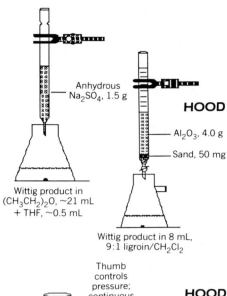

Anhydrous
Na$_2$SO$_4$, 1.5 g

HOOD

Wittig product in
(CH$_3$CH$_2$)$_2$O, ~21 mL
+ THF, ~0.5 mL

Al$_2$O$_3$, 4.0 g

Sand, 50 mg

Wittig product in 8 mL,
9:1 ligroin/CH$_2$Cl$_2$

Thumb
controls
pressure;
continuous
shaking

HOOD

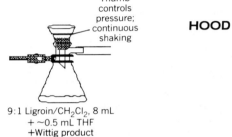

9:1 Ligroin/CH$_2$Cl$_2$, 8 mL
+ ~0.5 mL THF
+Wittig product

5-mL portions of diethyl ether, and these extracts are separated as before and combined with the original ether layer.

The combined ether fractions are transferred by Pasteur filter pipet to a short microcolumn prepared from a Pasteur filter pipet containing 1.5 g of anhydrous sodium sulfate.(■) The dried eluate is collected in a 25-mL Erlenmeyer flask containing a boiling stone. This solution is now concentrated to dryness, using a gentle stream of nitrogen or by warming on a sand bath [**HOOD**], yielding a colorless liquid residue.

Purification and Characterization

The crude product isolated above is purified by chromatography on an alumina column.

A short (1 cm × 10 cm) buret column is packed with 4.0 g of activated basic alumina and premoistened with ligroin.(■) The crude material is dissolved in 0.5 mL of 9:1 ligroin (60–80°C)/methylene chloride solvent and the solution transferred to the column using a Pasteur filter pipet. The product is eluted in a single fraction using 8.0 mL of 9:1 ligroin (60–80°C)/methylene chloride. The eluate is collected in a tared 25-mL filter flask containing a boiling stone. The solvent is evaporated [**HOOD**] under vacuum, with swirling, in a warm sand bath leaving a liquid residue. Product loss may occur if the concentrated residue is heated to excess.(■)

The product is of sufficient purity for characterization. Weigh the methylene-4-*t*-butylcyclohexane and calculate the percentage yield. Determine the boiling point and compare your result with that of the literature value. Obtain an IR spectrum and assign frequencies to the major bands. Compare results with the spectrum of the compound shown in Fig. 6.2.

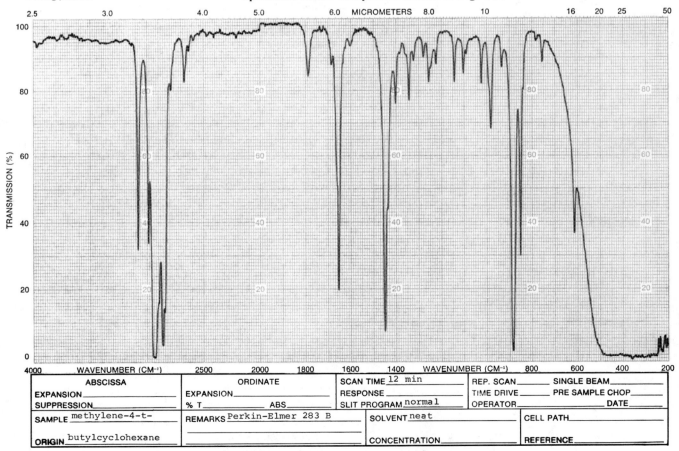

ABSCISSA		ORDINATE			SCAN TIME 12 min		REP. SCAN_____	SINGLE BEAM_____
EXPANSION_____		EXPANSION_____			RESPONSE_____		TIME DRIVE_____	PRE SAMPLE CHOP_____
SUPPRESSION_____		% T_____	ABS_____		SLIT PROGRAM normal		OPERATOR_____	DATE_____
SAMPLE methylene-4-t-		REMARKS Perkin-Elmer 283 B			SOLVENT neat		CELL PATH_____	
ORIGIN butylcyclohexane					CONCENTRATION_____		REFERENCE_____	

Fig. 6.2 IR spectrum: *1-methylene-4-t-butylcyclohexane.*

PART D: 17β-Hydroxy-17α-methyl-3-methylene-5α-androstane

3-Keto-17β-hydroxy-17α-methyl-5α-androstane

$+ [(C_6H_5)_3PCH_3^+, Br^- + NaNH_2] \xrightarrow{\text{THF}}$

"Instant ylid"
methyltriphenylphosphonium
bromide-sodium amide

17β-Hydroxy-17α-methyl-3-methylene-5α-androstane

$+ \quad (C_6H_5)_3PO$

Triphenylphosphonium
oxide

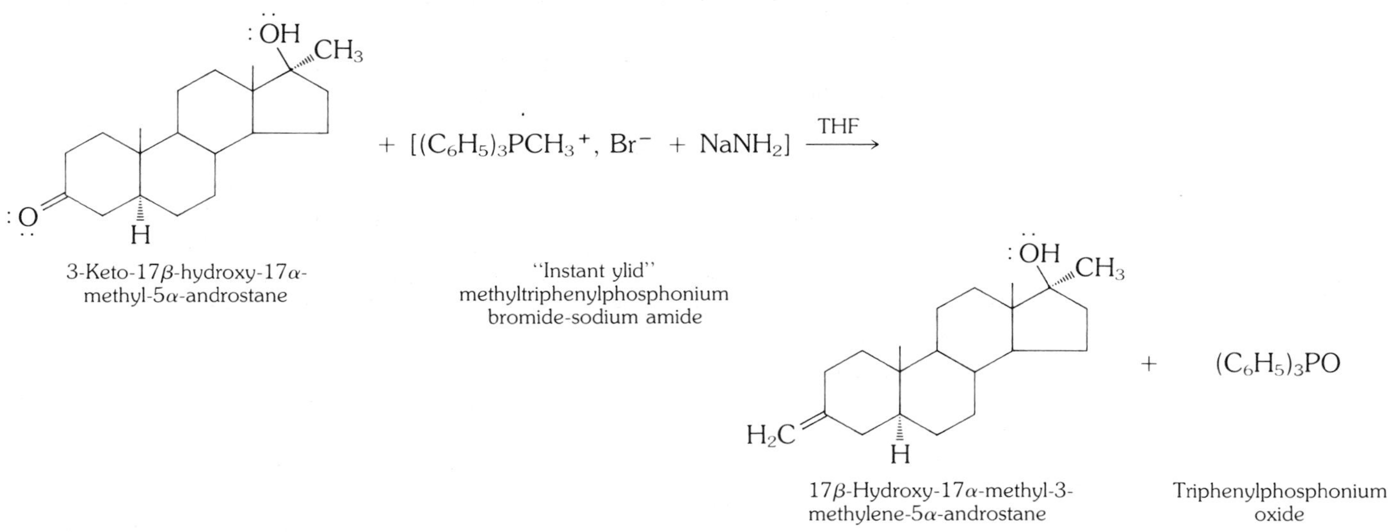

EXPERIMENTAL   Estimated time to complete the reaction: 3.5 hours.

Physical Properties of Reactants and Products

Compound	MW	Wt/Vol	mmol	mp(°C)	bp(°C)
3-Keto-17β-hydroxy-17α-methyl-5α-androstane	280.55	150 mg	0.53	189–192	
"Instant ylid" methyltriphenyl-phosphonium bromide-sodium amide		320 mg	~0.72		
Tetrahydrofuran	72.12	1.0 mL			67
17β-Hydroxy-17α-methyl-3-methylene-5α-androstane				179–181	

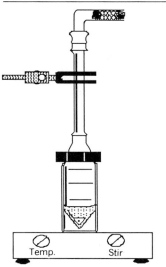

Step 1:
(C₆H₅)₃-P⁺-CH₃, Br⁻ + NaNH₂,
320 mg + THF, 1.0 mL
Step 2:
3-keto-17β-hydroxy-17α-
methyl-5α-androstane,
150 mg

Reagents and Equipment

In a dry 3.0-mL conical vial containing a magnetic spin vane and equipped with an air condenser protected by a calcium chloride drying tube is placed 320 mg (~0.72 mmol) of methyltriphenylphosphonium bromide-sodium amide ("instant ylid") mixture.(■) *Freshly distilled* tetrahydrofuran (1.0 mL) is now added from a calibrated Pasteur pipet and the mixture stirred at room temperature for 15 minutes. A bright yellow salt results.

Following generation of the ylid, 150 mg (0.53 mmol) of 3-keto-17β-hydroxy-17α-methyl-5α-androstane is added to the reaction flask.

Reaction Conditions

The resulting heterogeneous mixture is stirred at room temperature for an additional 90 minutes. The system develops a tan color over this period of time.

Isolation of Product

The reaction is quenched by adding 1.0 mL of a 25% aqueous NaOH solution (calibrated Pasteur pipet) and the resulting mixture transferred to a 12-mL cen-

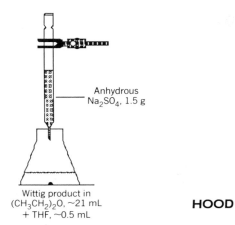

Anhydrous
Na₂SO₄, 1.5 g

Wittig product in
(CH₃CH₂)₂O, ~21 mL
+ THF, ~0.5 mL

trifuge tube using a Pasteur filter pipet. The reaction flask is rinsed with three 2.0-mL portions of diethyl ether, which are also transferred to the centrifuge tube. The resulting two-phase system is partially neutralized by the careful addition of 3.0 mL of 0.2N HCl.

The ether layer is separated by Pasteur filter pipet and placed in a 25-mL Erlenmeyer flask. The remaining aqueous layer is extracted with three additional 5-mL portions of diethyl ether, and these extracts are separated as before and combined with the original ether layer.

HOOD

The combined ether solution is transferred by Pasteur filter pipet to a short microcolumn prepared from a Pasteur filter pipet containing 1.5 g of anhydrous sodium sulfate.(■) The dried eluate is collected in a 50-mL Erlenmeyer flask containing a boiling stone. This solution is now concentrated to dryness, using a gentle stream of nitrogen in a warm sand bath [**HOOD**], yielding a white solid.

Purification and Characterization

The crude product is recrystallized from 95% ethanol using a Craig tube and dried on a porous clay plate.

Weigh the product and calculate the percentage yield. Determine the melting point and compare it with the literature value. Obtain an IR spectrum and assign frequencies to the major bands. Compare results with the spectrum of the compound shown in Fig. 6.3.

Fig. 6.3 IR spectrum: *17β-hydroxy-17α-methyl-3-methylene-5α-androstane.*

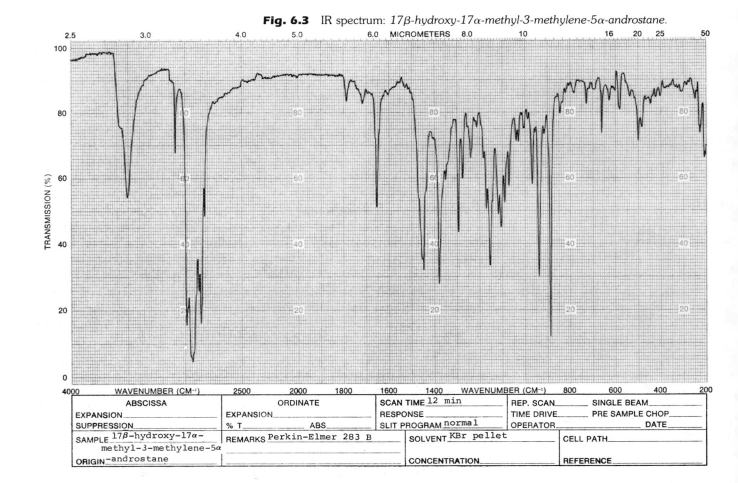

ENVIRONMENTAL DATA

Substance	Amount	TLV (mg/m³)	Emissions (mg)	Volume (m³)
Experiment 21A				
"Instant ylid"	600 mg			
Benzaldehyde	100 µL			
Tetrahydrofuran (Max., water soluble)	1.0 mL	590	890	1.5
Diethyl ether (Max., extractant, dried, evap.)	10 mL	1,200	7,140	6.0
Ligroin (Max., col. chromatog., treated as hexane)	12 mL	180	10,320	57
Methylene chloride (Max., col. chromatog., evap.)	3 mL	350	4,000	11.4
Ethanol 95% (Max., recryst.)	1.0 mL	1,900	750	0.4
Experiment 21B				
"Instant ylid"	750 mg			
Piperonal (react.)	150 mg			
Tetrahydrofuran (Max., water soluble)	1.0 mL	590	890	1.5
Diethyl ether (Max., extractant, evap.)	17 mL	1,200	12,140	10.1
Methylene chloride (Max., col. chromatog. I, evap.)	0.8 mL	350	1,070	3.1
Ligroin (Max., col. chromatog. I, evap., treated as hexane)	5.2 mL	180	4,470	25
Ligroin (Max., col. chromatog. II, evap.)	8.0 mL	180	6,880	38
Experiment 21C				
"Instant ylid"	320 mg			
4-t-Butylcyclohexanone (reactant)	100 mg			
Tetrahydrofuran (Max., water soluble)	1.0 mL	590	890	1.5
Diethyl ether (Max., rinse, extract, dry, evap.)	21 mL	1,200	15,000	12.5
Methylene chloride (Max., col. chromatog., evap.)	0.85 mL	350	1,135	3.2
Ligroin (Max., col. chromatog. evap., treated as hexane)	8.65 mL	180	7,440	41
Experiment 21D				
"Instant ylid"	320 mg			
Tetrahydrofuran (Max., water soluble)	1.0 mL	590	890	1.5
Diethyl ether (Max., rinse, extract, dry, evap.)	21 mL	1,200	15,000	12.5
Ethanol, 95% (Max., recryst.)	1.0 mL	1,900	750	0.4

TLV: Hydrochloric acid—5 ppm
　　　Sodium hydroxide—2 µg/m³
TXDS: Benzaldehyde—orl-rat LD50: 1300 mg/kg
　　　Sodium sulfate—orl-mus LD50: 5989 mg/kg
　　　Stilbene—ipr-mus LD50: 1150 mg/kg
　　　Piperonal—orl-rat LD50: 2700 mg/kg
　　　t-Butylcyclohexanone—orl-rat LD50: 5000 mg/kg

QUESTIONS

6-35. Complete each of the following reactions by giving a suitable structure for the species represented by the letters. Give a suitable name for compound B in each reaction.

a. $(C_6H_5)_3\overset{+}{P}$—$CH_2C_6H_5$, Cl^- $\xrightarrow[\text{C}_2\text{H}_5\text{OH}]{\text{NaOC}_2\text{H}_5}$ A

A + C_6H_5CH=CH—$\overset{H}{\underset{}{C}}$=$O$ $\longrightarrow$ B + $(C_6H_5)_3PO$

b. $(C_6H_5)_3\overset{+}{P}$—CH_3, Br^- $\xrightarrow[\text{dimethylsulfoxide}]{\text{NaH}}$ A

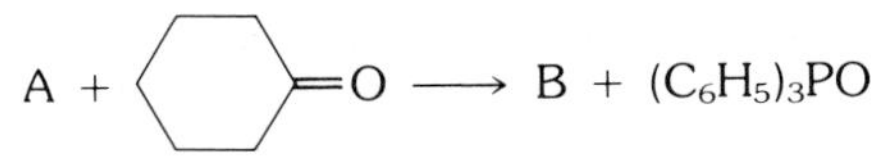

A + (hexagon)=O $\longrightarrow$ B + $(C_6H_5)_3PO$

6-36. Why is it important that any aldehyde used in the Wittig reaction be free of carboxylic acid impurity?

6-37. Reaction of triphenylphosphine with benzyl bromide produces the corresponding phosphonium salt. Suggest a suitable mechanism for this reaction.

6-38. Heteroatoms other than P are also capable of stabilizing the negative charge on C to yield ylids. For example, nitrogen is capable of forming such a system.

$$\left[\begin{array}{c}(C_6H_5)_3\overset{+}{P}-\overset{\cdot\cdot}{\overset{-}{C}}H_2 \\ \updownarrow \\ (C_6H_5)_3P=CH_2\end{array}\right] \qquad (CH_3)_3\overset{+}{N}-\overset{\cdot\cdot}{\overset{-}{C}}H_2$$

A phosphorous ylid   A nitrogen ylid

Why aren't resonance structures drawn for the nitrogen ylid as in the phosphorous system?

6-39. Would you expect that the sulfonium salt, $(C_6H_5)_2S^+$—CH_3, Br^-, is capable of forming an ylid when reacted with a strong base? If so, would its structure be represented in a manner resembling the P or N ylid? Explain.

REFERENCES

1. Of the many reviews on the Wittig reaction three are listed.
   a. Maerker, A. *Org. React.* **1965,** *14,* 270.
   b. Trippett, S. *Quart. Revs.* **1963,** *17,* 406.
   c. Schoellkopf, U. In "Newer Methods of Preparative Organic Chemistry"; Foerst, W., Ed.; Academic Press: New York, 1964; Vol. 3, p 111.
2. Below are listed selected examples where the Wittig reaction is applied in "Organic Syntheses."
   a. McDonald, R. N.; Campbell, T. W. "Organic Syntheses"; Wiley: New York, 1973; Collect. Vol. V, p 499.
   b. Jorgenson, M. J.; Thacher, A. F. *Ibid.*, p 509.
   c. Wadsworth, W. G., Jr; Emmons, W. D. *Ibid.*, p 547.
   d. Wittig, G.; Schoellkopf, U. *Ibid.*, p 751.
   e. Campbell, T. W.; McDonald, R. N. *Ibid.*, p 985.

Experiment 22
Aldol Condensations: Dibenzalacetone

(1,4-Pentadien-3-one, 1,5-diphenyl-)

The aldol condensation is used extensively in organic synthesis to form carbon—carbon bonds. The example presented here illustrates the utility of the method.

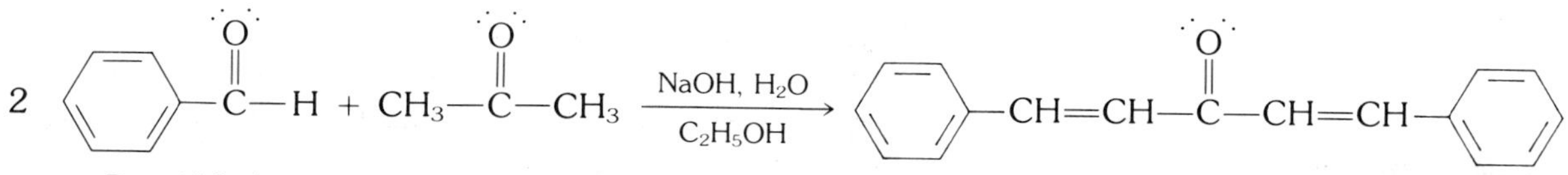

2 Benzaldehyde + Acetone $\xrightarrow[\text{C}_2\text{H}_5\text{OH}]{\text{NaOH, H}_2\text{O}}$ Dibenzalacetone

DISCUSSION The aldol condensation is one of the fundamental reactions of organic chemistry. The reaction is wide in scope and may be used to condense various combinations of aldehydes and ketones leading to formation of new carbon—carbon bonds. The mixed condensation of an aldehyde having no α-hydrogen atom with a ketone is known as the Claisen-Schmidt reaction. This variation of the aldol condensation is illustrated by the synthesis of dibenzalacetone.

The reaction conditions of this experiment favor the formation of dibenzal-acetone. This product is insoluble in the aqueous-alcohol solvent and precipitates from the reaction medium as it is formed, while the starting materials and the intermediate, benzalacetone, are soluble in aqueous alcohol. These experimental conditions assist in driving the reaction to completion.

The aldol reaction involves the addition of a deprotonated alpha-carbon atom of an aldehyde or ketone to a carbonyl carbon of an aldehyde or ketone. The reaction is generally base catalyzed and involves several mechanistic steps.

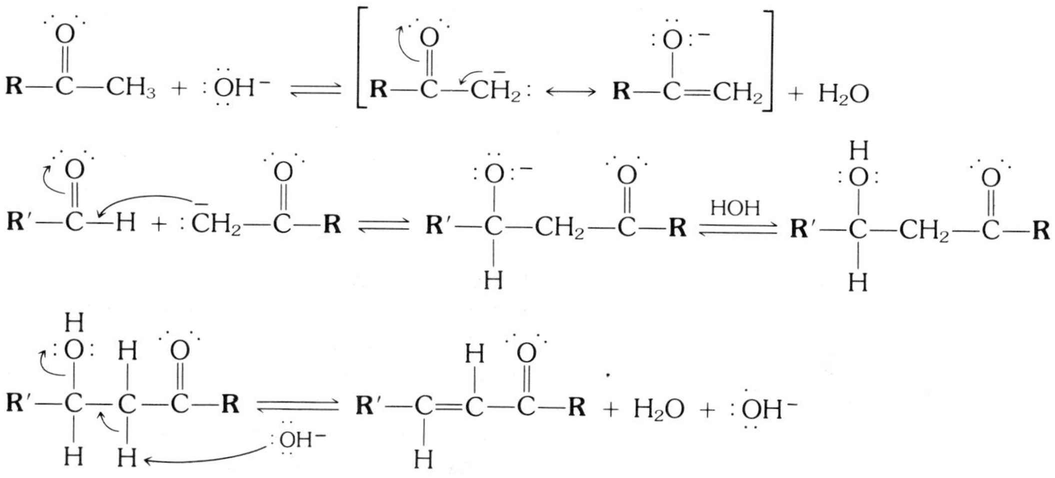

Note. *The reaction may be carried out in a 10 × 75 mm test tube. However, the reagents must be stirred efficiently with a glass rod at frequent intervals. If a larger test tube is used, a small magnetic stirring bar or vane is more efficient as an agitator.*

EXPERIMENTAL Estimated time to complete the experiment: 2 hours.

Physical Properties of Reactants and Products

Compound	MW	Wt/Vol	mmol	mp(°C)	bp(°C)	Density	n_D
Benzaldehyde	106.13	80 μL	0.79		178	1.04	1.5463
Acetone	58.08	29 μL	0.40		56	0.79	1.3588
NaOH catalyst solution		1 mL					
Dibenzalacetone	234.30			111			

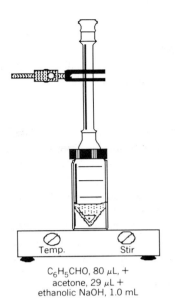

C₆H₅CHO, 80 μL, +
acetone, 29 μL +
ethanolic NaOH, 1.0 mL

Reagents and Equipment

In a 3.0-mL conical vial containing a magnetic spin vane and equipped with an air condenser is placed 80 μL (84 mg, 0.79 mmol) of benzaldehyde and 29.0 μL (23 mg, 0.40 mmol) of acetone.(■)

Important. *It is recommended that the purity of the benzaldehyde be checked by IR. The presence of benzoic acid in the benzaldehyde can substantially lower the yield of product. It may be purified by distillation under reduced pressure (bp 178–179°C; 57–59°C/8 torr).*

These reagents may be added to the vial by weight, or by volume using an automatic delivery pipet. In either case, the vial should be capped between additions, especially after the addition of the acetone to prevent loss by evaporation.

The stoichiometric quantities of reagents are used. An excess of benzaldehyde results in a more intractable product; excess acetone favors the formation of benzalacetone.

To this reaction mixture is added 1.0 mL of aqueous-ethanolic sodium hydroxide catalyst solution delivered from a calibrated Pasteur pipet.

Instructor Preparation

CAUSTIC

The catalyst solution is prepared by dissolving 0.4 g of sodium hydroxide (CAUSTIC) in 4.0 mL of water. To this solution is added 3.0 mL of 95% ethanol.

Reaction Conditions

The reaction mixture is stirred at room temperature for a period of 30 minutes. During this time a yellow solid product precipitates from solution.

Isolation of Product

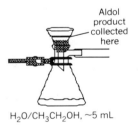

Aldol
product
collected
here

H₂O/CH₃CH₂OH, ~5 mL

The crude, yellow dibenzalacetone is collected by vacuum filtration using a Hirsch funnel.(■) The magnetic spin vane is now removed from the reaction vial with forceps. *Some of the product adheres to the magnetic spin vane. This material should be removed by carefully scraping the vane with a microspatula. The material is added to the product collected by filtration.*

The collected filter cake is returned to the reaction vial, stirred with 1.0 mL of water, and recollected as before. This procedure is repeated at least twice until the filtrate is near the neutral point as indicated by pH test paper.

Important. *It is essential to remove the NaOH completely. If it is not removed, the recrystallization step proves difficult.*

The product is air dried by maintaining the suction on the Hirsch funnel for approximately 10 minutes. During this operation a piece of filter paper may be placed over the mouth of the Hirsch funnel to prevent dust contamination.

Purification and Characterization

The crude dibenzalacetone may be purified by recrystallization from 95% ethanol using the Craig tube.

Weigh the product and calculate the percentage yield. Determine the melting point and compare your result with the literature value. Obtain an IR spectrum of the material and compare it with that of an authentic sample.

ENVIRONMENTAL DATA

Substance	Amount	TLV (mg/m³)	Emissions (mg)	Volume (m³)
Benzaldehyde (Max. from yield, transfer)	84 mg	—	37	
Acetone (Max. from yield, transfer)	23 mg	1780	20	<0.1
Ethanol (Max., solv., recryst., Craig tube)	1.43 mL	1900	1140	0.6

TLV: Sodium hydroxide—2.0 mg/m³
TXDS: Benzaldehyde—orl-mus LD50: 28 mg/kg

QUESTIONS

6-40. A key step in the total synthesis of the hydrocarbon azulene is given below. Outline a suitable mechanism to account for the reaction.

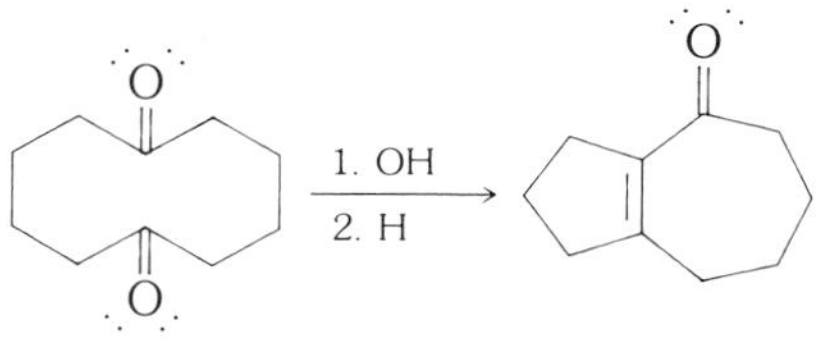

6-41. The aldol reaction has been utilized extensively for the generation of five- and six-membered rings. Suggest a suitable mechanism for the cyclization reactions shown below.

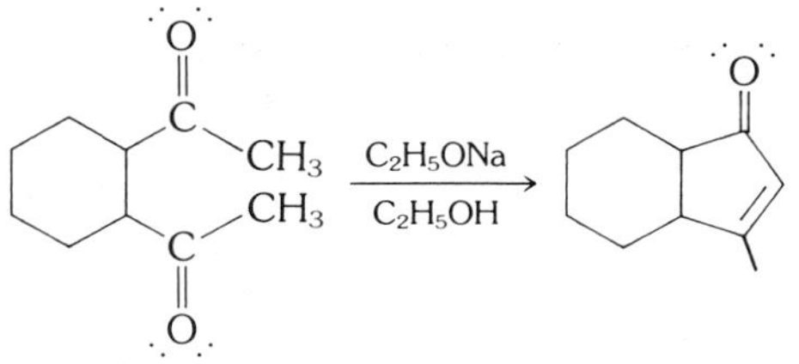

6-42. Predict the major organic product formed in each of the following reactions.

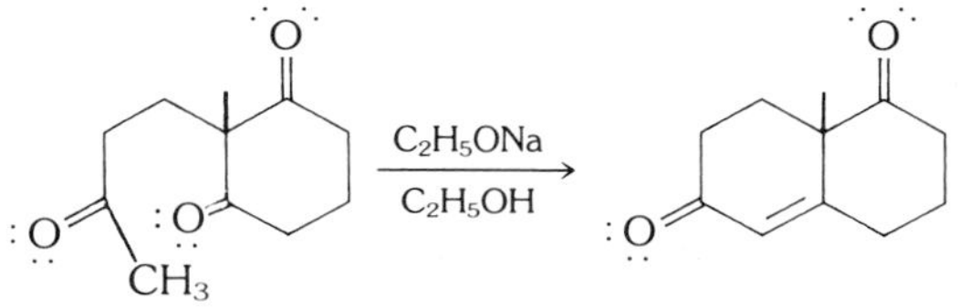

a. $CH_3CH_2NO_2 + CH_2O \xrightarrow{NaOH}$

b. $C_6H_5-CH=CH-CHO + CH_3-\overset{\overset{O}{\|}}{C}-C_6H_5 \xrightarrow[C_2H_5OH]{C_2H_5ONa}$

c. $C_6H_5CHO + C_6H_5CH_2CN \xrightarrow[C_2H_5OH]{C_2H_5ONa}$

6-43. "Crossed" or "mixed" aldol condensations are practical for synthesis, if one of the aldehydes (or ketones) has no α-hydrogen atoms. Explain.

6-44. Give several examples of aldehydes or ketones that could be used in the "crossed" aldol condensation with propanal. Assign structures and names to the products that could be formed and point out any side reactions that might occur.

6-45. In the aldol condensation using the conditions of this experiment, why is it essential that the aldehyde component contain none of the corresponding acid?

REFERENCES   **1.** Review articles:
    a. Neilsen A. T.; Houlihan, W. J. *Org. React.* **1968,** *16,* 1.
    b. Mukaiyama T. *Ibid.* **1982,** *28,* 203.
  2. Examples of the aldol and Claisen-Schmidt reactions include:
    a. Hill, G. A.; Bramann, G. M. "Organic Syntheses"; Wiley: New York, 1941; Collect. Vol. I, p 81.
    b. Leuck, G.J.; Cejka, L. *Ibid.* p 283.
    c. Kohler, E. P.; Chadwell, H. M. *Ibid,* p 78.
    d. Conrad, C. R.; Dolliver, M. A. "Organic Syntheses"; Wiley: New York, 1943; Collect. Vol. II, p 167.
    e. Russel A.; Kenyon R. L. "Organic Syntheses"; Wiley: New York, 1955; Collect. Vol. III, p 747.

Experiment 23

Quantitative Analysis of a Grignard Reagent: 1-Methylbutylmagnesium Bromide

(magnesium, bromo(1-methylbutyl-))

This experiment demonstrates the technique by which a Grignard reagent may be analyzed. Also see Experiment 35.

$$CH_3CH_2CH_2\!-\!\underset{\underset{H}{|}}{\overset{\overset{CH_3}{|}}{C}}\!-\!Br \;+\; Mg \;\xrightarrow{\text{ether}}\; CH_3CH_2CH_2\!-\!\underset{\underset{H}{|}}{\overset{\overset{CH_3}{|}}{C}}\!-\!MgBr$$

    2-Bromopentane   Magnesium   1-Methylbutylmagnesium bromide

DISCUSSION   This experiment demonstrates the formation of a Grignard reagent and a titration method by which the amount of the reagent prepared can be analyzed.

The discovery by Victor Grignard in 1900 that organic halides react with magnesium metal to give organomagnesium compounds was a landmark in organic chemistry. It is one of the most useful and versatile reagents in organic synthesis.

The reaction of the Grignard reagent with water is the basis of the analytical method used in this experiment.

$$\mathbf{RMgX + HOH \longrightarrow RH + Mg(OH)X}$$

In the Grignard reagent, the carbon atom bound to the electropositive magnesium atom has a high charge density, which is responsible for the strong nucleophilic and basic character exhibited by this organometallic reagent. The carbon, acting as a base, can abstract a proton from protic reagents such as water, carboxylic acids, alcohols, etc. In this process, the corresponding hydrocarbon (the conjugate acid of the **R** group anion) and the basic magnesium halide species are produced. This reaction sequence can be used in the laboratory as a synthetic method to convert organohalides to hydrocarbons.

$$CH_3CH_2CH_2\!-\!\underset{\underset{H}{|}}{\overset{\overset{CH_3}{|}}{\underset{\delta^-}{C}}}\!\!:\underset{\delta^+}{MgBr} + H\overset{..}{\underset{..}{O}}H \longrightarrow CH_3CH_2CH_2CH_2CH_3 + Mg(OH)Br$$

                                                          Pentane

    1-Methylbutylmagnesium bromide

Titration of the Mg(OH)X species with a standardized acid solution allows one to determine the equivalents of Grignard reagent originally present in the solution.

$$2 \text{ Mg(OH)X} + H_2SO_4 \longrightarrow 2 \text{ HOH} + MgSO_4 + MgX_2$$

An excess of the sulfuric acid is generally added to ensure that the Mg(OH)X is completely reacted. The excess acid is then neutralized with the standard sodium hydroxide solution. The difference between the total amount of sulfuric acid used and the amount of sodium hydroxide required corresponds to the number of equivalents of the acid actually used to neutralize the Mg(OH)X species. This value is then directly related to the equivalents of Grignard reagent by the equations given above.

PART A: 1-Methylbutylmagnesium Bromide

The reaction is shown above.

> **CAUTION:** *Ether is a flammable liquid and also a narcotic. All flames should be extinguished during the time of this experiment.*

EXPERIMENTAL   Estimated time for the experiment: 1.0 hour.

Physical Properties of Reactants and Products

Compound	MW	Wt/Vol	mmol	bp(°C)	Density	n_D
2-Bromopentane	151.05	125 µL	1.0	117	1.21	1.4413
Magnesium	24.31	36 mg	1.5			
Iodine	253.81	1 crystal				
Diethyl ether	74.12	400 µL		34.5		
Pentane	72.15			36	0.63	1.3575

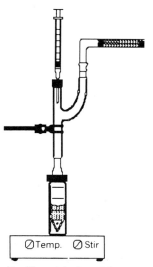

Mg, 36 mg + I$_2$, 1 crystal
+ (CH$_3$CH$_2$)$_2$O, 400 µL +
CH$_3$CHBr(CH$_2$)$_2$CH$_3$, 125 µL

Reagents and Equipment

Note. *All the glassware used in the experiment should be cleaned, dried in an oven at 110°C for at least 30 minutes, and then cooled in a desiccator before use.*

The reagent is prepared exactly as described in Experiment 18. The reagents, amounts of reagents, order of addition, workup manipulations, and precautions are the same. The equipment is also identical, with the exception that a 1-mL vial is used. (■)

The gray-colored Grignard reagent mixture is cooled to room temperature and then analyzed by the titration method outlined below in Part B.

If not used for analysis, the solution of Grignard reagent may be treated with various reagents to prepare a wide variety of compounds. For example, see Experiments 17, 18, 36, and 48B.

PART B: Analysis of the Grignard Reagent

EXPERIMENTAL   Estimated time for the experiment: 0.5 hour.

Reagents and Equipment
In a 50-mL Erlenmeyer flask is placed 10 mL of freshly boiled distilled water and one drop of phenolphthalein indicator.

Using a syringe, the cool Grignard reagent solution is transferred to the Erlenmeyer flask. The reaction vial is rinsed with 0.5 mL of diethyl ether and the rinse also added to the Erlenmeyer flask.

Note. *The addition of water to the Grignard reagent results in the hydrolysis of this reagent to form the corresponding hydrocarbon and a basic magnesium halide, Mg(OH)X. The water is initially boiled to remove any carbon dioxide present, which would react with the Grignard reagent.*

This resulting mixture is now analyzed by the titration method given below.

Analysis by Titration
From a 10-mL buret is now added 5 or 6 mL of standard 0.2N sulfuric acid solution. *The resulting solution should be acidic and colorless. If not, add an additional portion of the acid.*

A boiling stone is added to the flask and the mixture heated at a sand bath temperature of 90–95°C for a period of 5 minutes.

While the solution is still warm, a drop of phenolphthalein is added and the excess acid neutralized by back titration with 0.1N sodium hydroxide solution. Back titration produces a very light colored, pink end point. *It may be necessary to add an additional drop of acid and then more base to get the best possible end point.*

Data and Calculations
The difference between the initial and final buret readings is the volume of standard acid and base used in the titration of the Grignard reagent.

From the data, calculate the equivalents of Grignard reagent present. Also, as a percentage, determine the amount of Grignard reagent analyzed compared to the theoretical amount.

ENVIRONMENTAL DATA   The Grignard preparation emissions are listed in Experiment 18.

TLV: Iodine—0.1 ppm
       Pentane—1800 mg/m^3
       Sodium hydroxide—2 mg/m^3

TXDS: 2-Bromopentane—ipr-mus LD50: 150 mg/kg
        Magnesium—orl-dog LDLo: 230 mg/kg
        Sulfuric acid—unk-man LDLo: 135 mg/kg

QUESTIONS   **6-46.** Technical grade ether often contains ethanol. Would you recommend this material as a suitable solvent for the preparation of Grignard reagents? If not, why not?

6-47. In the synthesis of alcohols by the Grignard reaction, the final stage of the sequence to obtain the alcohol product involves hydrolysis of the magnesium alkoxide salt with dilute acid. For the preparation of 2,3,4-trimethyl-3-pentanol by the reaction of methylmagnesium iodide with diisopropyl ketone in ether solvent, why is a saturated solution of aqueous ammonium chloride preferred in the final hydrolysis step instead of the usual dilute mineral acid solution?

6-48. In the titration procedure outlined above for the analysis of the Grignard reagent, freshly boiled distilled water is used. As stated, this is to remove dissolved carbon dioxide. What reaction could occur during the titration if the CO_2 were not removed?

6-49. One equivalent of o-acetylbenzoic acid was treated with two equivalents of methylmagnesium iodide in ether solvent. The product was isolated by careful neutralization of the product formed in the reaction.
   a. What was the structure of the product formed in the reaction?
   b. Why were two equivalents of Grignard reagent required?

REFERENCES
   1. For references relating to the preparation of Grignard reagents, see Experiment 17.
   2. For the many references related to the preparation and use of the Grignard reagent cited in "Organic Syntheses," see the Reaction Index in Collect. Vols. I–V under Grignard Reactions.

Experiment 24
Preparation of an Enol Acetate: Cholesta-3,5-dien-3-ol Acetate

(cholesta-3,5-dien-3-ol acetate)

The preparation of the enol acetate of an α,β-unsaturated ketone is demonstrated in this experiment.

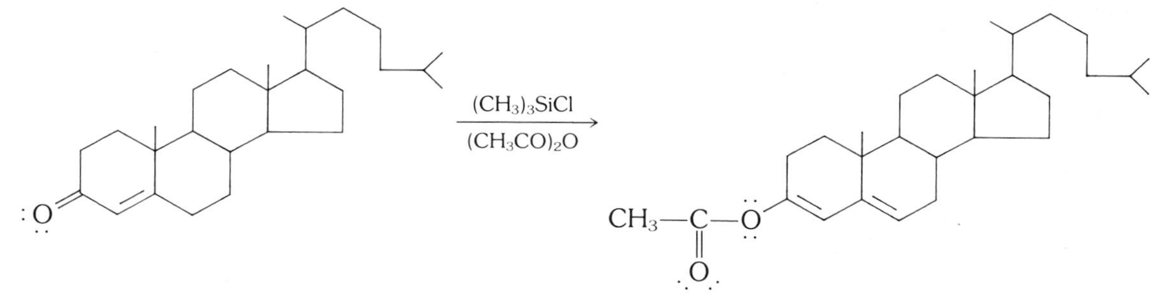

Cholesta-4-en-3-one → [$(CH_3)_3SiCl$ / $(CH_3CO)_2O$] → Cholesta-3,5-dien-3-ol acetate

DISCUSSION Enolate anions play a major role in synthetic organic chemistry since they are good nucleophilic reagents. They undergo nucleophilic substitution reactions with relatively electrophilic carbon atom centers leading to the formation of C—C bonds.

This experiment illustrates one method of stabilizing such enolates as acetate esters. An enolate is generally formed by removal of a proton from the α-carbon atom of a carbonyl compound. In the present case, the reactant is an α,β-unsaturated carbonyl species. The proton is thus removed from the γ-carbon atom to form the anion, which is stabilized by delocalization of the charge.

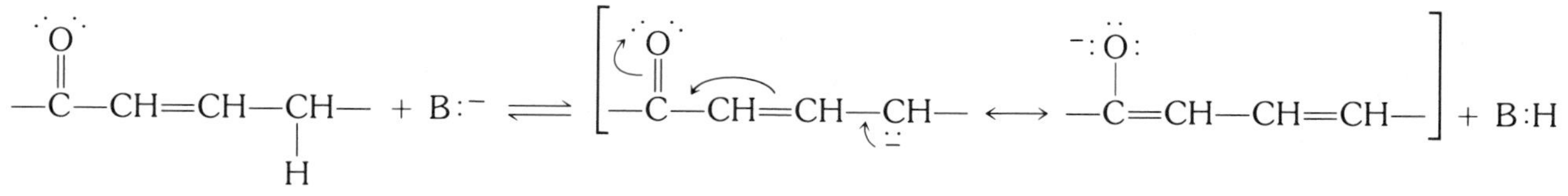

The enol acetate is then formed in the presence of the acylium ion generated from the acetic anhydride. The mechanism for the sequence is outlined below.

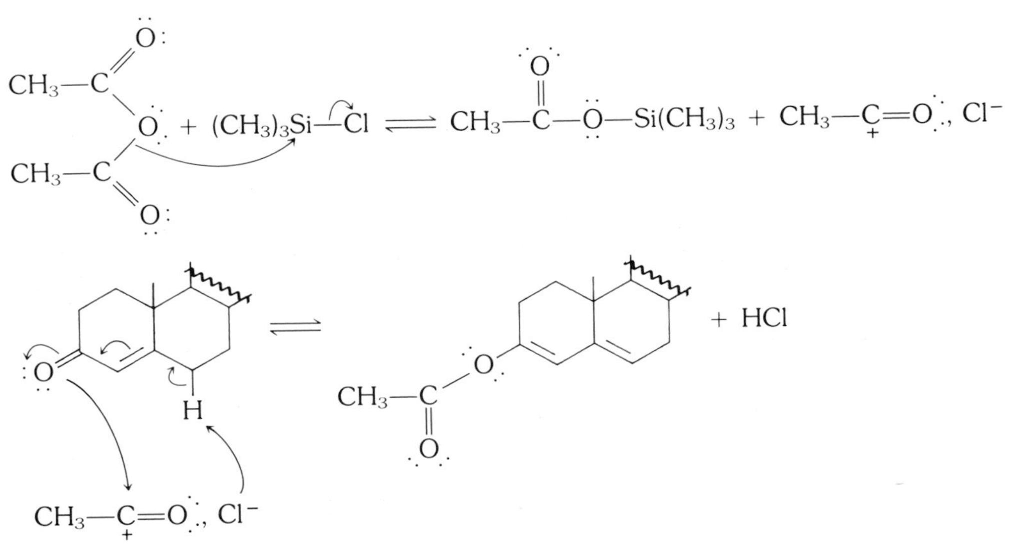

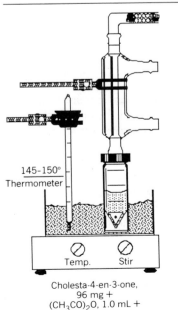

EXPERIMENTAL   Estimated time to complete the experiment: 4.0 hours.

Physical Properties of Reactants and Products

Compound	MW	Wt/Vol	mmol	mp(°C)	bp(°C)	Density	n_D
Cholesta-4-en-3-one	384.65	96 mg	0.25	81–82			
Acetic anhydride	102.09	1.0 mL	10.6		140	1.08	1.3901
Chlorotri-methylsilane	108.64	200 μL	1.58		58	0.86	
Cholesta-3,5-dien-3-ol acetate	425.68			79–80			

Reagents and Equipment

Note. *The glass equipment should be dried in an oven (110°C) before use.*

HOOD   In a dry 5.0-mL conical vial containing a magnetic spin vane and equipped with a reflux condenser protected by a calcium chloride drying tube is placed 96 mg (0.25 mmol) of cholesta-4-en-3-one, 1.0 mL of acetic anhydride and 200 μL of chlorotrimethylsilane [**HOOD**].(■)

HOOD   **Note.** *The quality of the acetic anhydride has a significant influence on the reaction. For best results the anhydride should be distilled and stored over molecular sieves before use. It is convenient to dispense [**HOOD**] the anhydride and the chlorotrimethylsilane (hydrolyzes rapidly in moist air) using automatic delivery pipets.*

Reaction Conditions

The reaction mixture is heated with stirring at reflux for a period of 1–2 hours by aid of a sand bath (145–150°C). The course of the reaction is followed using TLC.

145-150°
Thermometer

Temp.   Stir

Cholesta-4-en-3-one,
96 mg +
(CH₃CO)₂O, 1.0 mL +
(CH₃)₃SiCl, 200 μL

Directions. *Activated silica gel plates are used with 1:1 methylene chloride–hexane as the developing solvent. Visualization of the separated components is achieved by placing the plate in a closed jar containing a few crystals of iodine. See Chapter 3 for further details.*

Isolation of Product

The reaction mixture is allowed to cool to room temperature and then placed in an ice bath for 15–30 minutes. A solid product forms during this time. This solid is collected by vacuum filtration and the filter cake washed with 15 mL of 5% aqueous sodium bicarbonate followed by 5 mL of cold water.(■)

Purification and Characterization

The crude product is purified by column chromatography. In a 1 × 10 cm buret is placed 2.0 g of activated silica gel (100 mesh) packed wet (slurry) with methylene chloride.(■) The product is dissolved in about 1.0 mL of hexane and transferred by Pasteur pipet to the column. The material is eluted from the column using approximately 10 mL of 1:1 methylene chloride-hexane solvent. The eluate is collected in a tared 25-mL filter flask.

The solvent is removed by warming in a sand bath under reduced pressure to give the solid product, cholesta-3,5-dien-3-ol acetate.(■)

This material is recrystallized from methanol using a Craig tube, and the resulting crystals are dried on a clay plate. Weigh the product and calculate the percentage yield. Determine the melting point and compare it with the literature value. Obtain an IR of the material and compare it with an authentic sample as well as with the starting material.

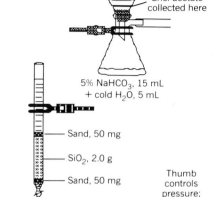

Enol acetate collected here

5% NaHCO₃, 15 mL + cold H₂O, 5 mL

Sand, 50 mg

SiO₂, 2.0 g

Sand, 50 mg

Thumb controls pressure; continuous shaking

1:1 Hexane/CH₂Cl₂, 10 mL

1:1 Hexane/CH₂Cl₂, 10 mL, enol acetate

ENVIRONMENTAL DATA

Substance	Amount	TLV (mg/m³)	Emissions (mg)	Volume (m³)
Chlorotrimethylsilane	200 μL			
Acetic anhydride (Max., vacuum filtration)	1.0 mL	20	1080	54
Methylene chloride (Max., col. chromatog., evap.)	5.0 mL	350	6675	19
Hexane (Max., col. chromatog., evap.)	6.0 mL	180	3954	22
Methanol (SKIN) (Max., Craig tube)	1.0 mL	260	800	3.1

TXDS: Chlorotrimethylsilane—ipr-mus LDLo: 750 mg/kg
Silica gel—inv-mus LDLo: 234 mg/kg

QUESTIONS

6-50. Suggest a suitable mechanism to account for the formation of the products in the reactions given below.

a. $CH_3CH_2—CH—\overset{\displaystyle O}{\overset{\|}{C}}—C_6H_5 \xrightarrow[D_2O]{NaOD} CH_3CH_2—\overset{\displaystyle D}{\underset{\displaystyle CH_3}{C}}—\overset{\displaystyle O}{\overset{\|}{C}}—C_6H_5$

with CH₃ below the first carbon

b. $(+)\ C_6H_5—CH—\overset{\displaystyle O}{\overset{\|}{C}}—C_6H_5 \xrightarrow[H_2O]{NaOH} (\pm)\ C_6H_5—CH—\overset{\displaystyle O}{\overset{\|}{C}}—C_6H_5$

with CH₃ below each first carbon

(Racemate)

6-51. Write the structure for all possible enolate anions that could form by reaction of the ketones listed below in alkaline solution.

a. $C_6H_5—CH_2CH_2—\overset{\overset{\displaystyle O}{\|}}{C}—CH(CH_3)_2$

b. $C_6H_5—\overset{\overset{\displaystyle O}{\|}}{C}—CH_2—\overset{\overset{\displaystyle O}{\|}}{C}—CH_3$

c. 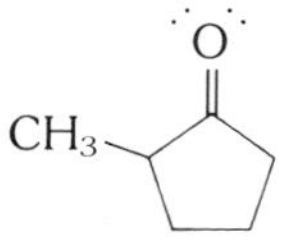 CH_3 — cyclopentanone

6-52. Reaction of $(CH_3)_3SiCl$ with alcohols produces a trimethylsilyl ether. For example,

$$CH_3CH_2CH_2\ddot{O}H + (CH_3)_3SiCl \xrightarrow[\text{ether}]{(C_2H_5)_3N:} CH_3CH_2CH_2\ddot{O}Si(CH_3)_3 + (C_2H_5)_3\overset{+}{N}H, Cl^-$$

The trimethylsilyl ether is more volatile than the corresponding alcohol. Explain.

6-53. Trimethylchlorosilane reacts with enolate anions to form stable silyl enol ethers. For example,

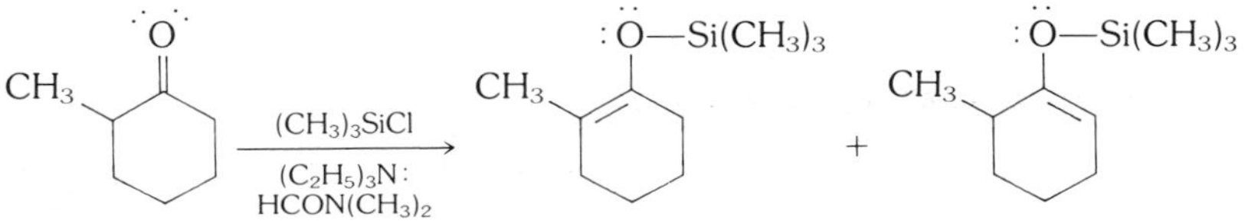

$$\xrightarrow[\substack{(C_2H_5)_3N: \\ HCON(CH_3)_2}]{(CH_3)_3SiCl}$$

Realizing that the reaction is run under conditions that allow equilibrium to be established in the system, predict which of the above silyl enol ethers is formed in the largest amount and why.

REFERENCES

1. The present experiment is adapted from the work reported by Chowdhury, P. K.; Sharma, R. P.; Barua, J. N. *Tetrahedron Letts.* **1983,** *24,* 3383.
2. For information on the formation and reactivity of enolate anions, see House, H. O. "Modern Synthetic Reactions," 2nd ed.; Benjamin: Reading, MA, 1972; Chapter 9, p 492.

Experiments 25A, 25B, 25C
Williamson Synthesis of Ethers: Propyl *p*-Tolyl Ether; Methyl *p*-Tolyl Ether; Methyl *p*-Ethylphenyl Ether

(ether, propyl *p*-tolyl; anisole, *p*-methyl; anisole, *p*-ethyl)

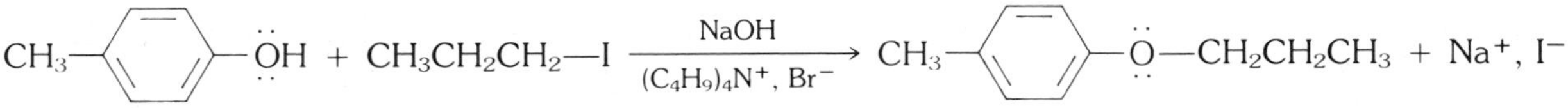

$$CH_3—\langle\ \rangle—\ddot{O}H + CH_3CH_2CH_2—I \xrightarrow[(C_4H_9)_4N^+,\ Br^-]{NaOH} CH_3—\langle\ \rangle—\ddot{O}—CH_2CH_2CH_3 + Na^+, I^-$$

p-Cresol   Propyl iodide   Propyl *p*-tolyl ether

The condensation of alkyl halides with sodium phenoxides to produce ethers is an example of the Williamson synthesis. The use of phase transfer catalysis is demonstrated.

DISCUSSION The three compounds, whose preparations are outlined below, belong to the class of ethers referred to as alkyl aryl ethers. The general method of preparation is the Williamson reaction. This reaction is often used for the synthesis of symmetrical and unsymmetrical ethers where the alkyl halide is primary. Elimination is generally observed if secondary or tertiary halides are used.

The conditions under which these reactions are conducted lend themselves to the use of phase transfer catalysis. The system involves two phases, the aqueous phase and the organic phase. In the present case, the alkyl halide acts both as the solvent and as a reactant. The phase transfer catalyst plays a very important role. In effect, it carries the phenolic ion (in the aqueous phase) as an ion-pair, across the boundary layer into the organic phase where the reaction occurs. The ether product and the corresponding halide salt of the catalyst are produced in these conversions. The halide salt then migrates back into the aqueous phase and the process repeats itself. The catalyst can play this role since the large organic groups (the four butyl groups) increase the solubility of the ion-pair in the organic phase, while the charged ionic center of the salt renders this species soluble in the aqueous phase.

In the reactions described below, the mechanism involves a nucleophilic backside attack of the phenolic ion on the alkyl halide.

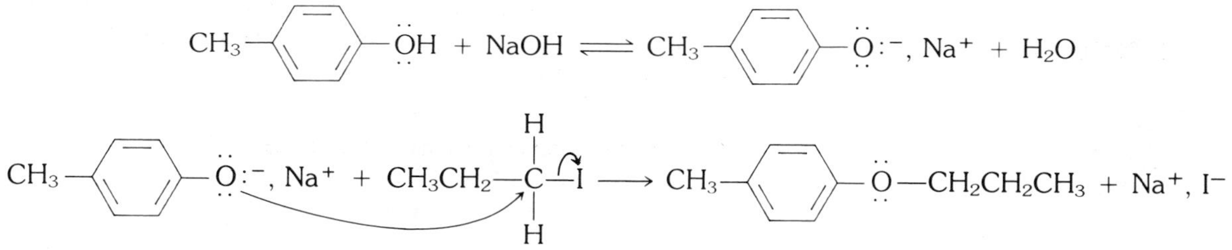

PART A: Propyl *p*-Tolyl Ether

The reaction for Part A is shown above.

EXPERIMENTAL Estimated time of the experiment: 2.5 hours.

Physical Properties of Reactants and Products

Compound	MW	Wt/Vol	mmol	mp(°C)	bp(°C)	Density	n_D
p-Cresol	108.15	80 μL	0.78		202	1.02	1.5312
25% NaOH solution		130 μL					
Tetrabutyl-ammonium bromide	322.38	9 mg	0.03	103–104			
Propyl iodide	169.99	75 μL	0.77		102	1.75	1.5058
Propyl *p*-tolyl ether	150.22				210.4	0.9497	

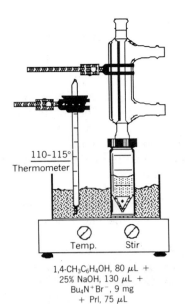

110–115°
Thermometer

Temp.   Stir

1,4-CH₃C₆H₄OH, 80 μL +
25% NaOH, 130 μL +
Bu₄N⁺Br⁻, 9 mg
+ PrI, 75 μL

Reagents and Equipment

In a 1.0-mL conical vial containing a magnetic spin vane and equipped with a reflux condenser is placed 80 μL (84 mg, 0.78 mmol) of p-cresol and 130 μL of 25% aqueous sodium hydroxide. The resulting solution is thoroughly mixed.(■) Tetrabutylammonium bromide catalyst (9 mg) is added, followed by 75 μL (131 mg, 0.77 mmol) of propyl iodide.

HOOD

Note. *The cresol reagent should be warmed in a hot water bath to obtain the liquid phase. This reagent and the propyl iodide are dispensed in the* **[HOOD]** *using an automatic delivery pipet.*

> **CAUTION:** *Propyl iodide is a cancer suspect agent.*

Reaction Conditions

The reaction vessel is placed in a sand bath at 110–115°C with vigorous stirring for a period of 45–60 minutes.

Isolation of Product

The resulting two-phase mixture is cooled to room temperature, and the spin vane removed with forceps. The aqueous layer is separated using a Pasteur filter pipet, and transferred to a 3-mL conical vial. The spin vane is rinsed with 0.5-mL diethyl ether and the rinse is then added to the aqueous fraction. The vial is capped, agitated, and vented and the water layer removed. The diethyl ether extract is transferred to the 1-mL conical vial containing the product fraction. The resulting solution is extracted with a 200 μL portion of 5% aqueous sodium hydroxide solution and following separation of the aqueous phase, washed with 100 μL of water. Removal of the aqueous phase yields the crude, wet ether product.

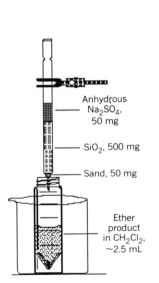

Anhydrous
Na₂SO₄,
50 mg

SiO₂, 500 mg

Sand, 50 mg

Ether
product
in CH₂Cl₂,
~2.5 mL

Purification and Characterization

The crude product is purified by chromatography on silica gel. A microchromatographic column is prepared by placing 500 mg of activated silica gel in a Pasteur filter pipet followed by 50 mg of anhydrous sodium sulfate.(■) The crude product is dissolved in 250 μL of methylene chloride and transferred to the dry column by use of a Pasteur pipet. The material is eluted with 2.0 mL of methylene chloride and the eluate collected in a tared 3.0-mL conical vial containing a boiling stone. The vial is then fitted with an air condenser and the solvent evaporated by placing the vial in a sand bath maintained at a temperature of 60–65°C.

Weigh the pure propyl p-tolyl ether and calculate the percentage yield. Determine the boiling point and compare it with the literature value. Obtain an IR and NMR spectrum of the compound and compare them to an authentic sample.

PART B: Methyl p-Tolyl Ether

$$CH_3-\!\!\!\!\diagup\!\!\!\diagdown\!\!\!\!-\ddot{O}H + CH_3-I \xrightarrow[\text{(C}_4H_9)_4N^+,\ Br^-]{\text{NaOH}} CH_3-\!\!\!\!\diagup\!\!\!\diagdown\!\!\!\!-\ddot{O}-CH_3 + Na^+,\ I^-$$

p-Cresol   Methyl   Methyl p-tolyl ether
                  iodide

EXPERIMENTAL Estimated time to complete the experiment: 2.5 hours.

Physical Properties of Reactants and Products

Compound	MW	Wt/Vol	mmol	mp(°C)	bp(°C)	Density	n_D
p-Cresol	108.15	160 μL	1.55		202	1.02	1.5312
25% NaOH solution		260 μL					
Tetrabutyl-ammonium bromide	322.38	15 mg	0.05	103–104			
Methyl iodide	141.94	228 mg	1.6		41–43	2.28	1.5304
Methyl p-tolyl ether	122.17				176.5	0.97	1.512

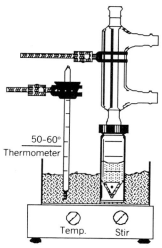

50–60°
Thermometer

Temp.   Stir

1,4-CH$_3$C$_6$H$_4$OH, 160 μL
+ 25% NaOH, 260 μL +
Bu$_4$N$^+$Br$^-$, 15 mg
+ CH$_3$I, 228 mg

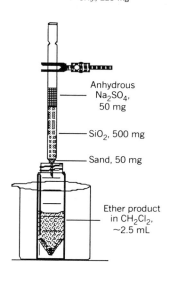

Anhydrous
Na$_2$SO$_4$,
50 mg

SiO$_2$, 500 mg

Sand, 50 mg

Ether product
in CH$_2$Cl$_2$,
~2.5 mL

Reagents and Equipment
Using the apparatus described in Part A, place 160 μL (167 mg, 1.5 mmol) of p-cresol and 260 μL of 25% aqueous sodium hydroxide solution in the reaction vial. (■) After mixing, 15 mg (0.05 mmol) of the phase transfer catalyst, tetrabutylammonium bromide, and 228 mg (1.6 mmol) of methyl iodide are added.

Note. *See Part A, Reagents and Equipment, for manner of dispensing the reagents (except for the volatile halide which is delivered dropwise by Pasteur pipet).*

Reaction Conditions
The reaction vial is now placed in a sand bath maintained at 50–60°C and the mixture stirred for a period of 45 minutes.

Important. *Maintain the sand bath at the suggested temperature range to obtain the maximum yield.*

Isolation of Product
The resulting product mixture is worked up as in Part A.

Purification and Characterization
The crude product is purified by chromatography on silica gel as described in Part A, Purification and Characterization. (■)
   Weigh the pure p-methyl anisole and calculate the percentage yield. Determine the boiling point and compare it to the literature value. Obtain IR and NMR spectra of the compound and compare them to an authentic sample.

PART C: Methyl *p*-Ethylphenyl Ether

$$CH_3CH_2-\!\!\underset{\text{4-Ethylphenol}}{\bigcirc}\!\!-\ddot{O}H \; + \; \underset{\substack{\text{Methyl}\\\text{iodide}}}{CH_3-I} \; \xrightarrow[\text{(C}_4\text{H}_9)_4\text{N}^+,\text{Br}^-]{\text{NaOH}} \; CH_3CH_2-\!\!\underset{\text{Methyl }p\text{-ethylphenyl ether}}{\bigcirc}\!\!-\ddot{O}-CH_3 \; + \; Na^+, I^-$$

EXPERIMENTAL   Estimated time to complete the experiment: 2.5 hours.

Physical Properties of Reactants and Products

Compound	MW	Wt/Vol	mmol	mp(°C)	bp(°C)	Density	n_D
4-Ethylphenol	122.17	150 mg	1.2	42–45			
25% NaOH solution		250 μL					
Tetrabutyl-ammonium bromide	322.38	15 mg	0.05	103–104			
Methyl iodide	141.94	205 mg	1.45		41–43	2.28	1.5304
Methyl *p*-ethyl-phenyl ether	136.20				195–196	0.96	1.5120

Note. *This product may be used as starting material for the preparation of p-methoxyacetophenone (see Experiment 52A).*

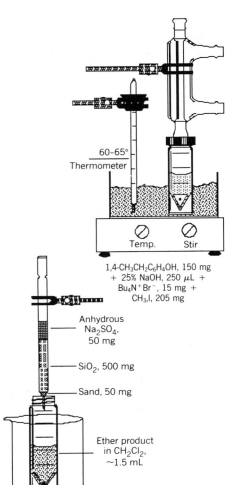

60-65°
Thermometer

Temp.   Stir

1,4-CH₃CH₂C₆H₄OH, 150 mg
+ 25% NaOH, 250 μL +
Bu₄N⁺Br⁻, 15 mg +
CH₃,I, 205 mg

Anhydrous
Na₂SO₄,
50 mg

SiO₂, 500 mg

Sand, 50 mg

Ether product
in CH₂Cl₂,
~1.5 mL

Reagents and Equipment

The apparatus in Part A is also used in this conversion. The reaction vial is charged with 150 mg (1.2 mmol) of 4-ethylphenol and 250 μL of 25% aqueous sodium hydroxide solution.(■) The mixture is stirred at room temperature until dissolution occurs. The phase transfer catalyst, tetrabutylammonium bromide (15 mg, 0.05 mmol), is now added, followed by 205 mg (1.5 mmol) of methyl iodide.

The alkaline solution is dispensed using an automatic delivery pipet, and the methyl iodide reactant by Pasteur pipet.

Reaction Conditions

The reaction vial is placed in a sand bath maintained at 60–65°C and the mixture stirred for a period of 1.0 hour.

Isolation of Product

The resulting product mixture is worked up as described in Part A.

Purification and Characterization

The crude product is purified by chromatography on silica gel as described in Part A, Purification and Characterization.(■)

Weigh the pure *p*-ethylanisole and calculate the percentage yield. Determine the boiling point and compare your result with the literature value. Obtain IR and NMR spectra of the compound and compare them with the spectra shown in Figs. 6.4 and 6.5.

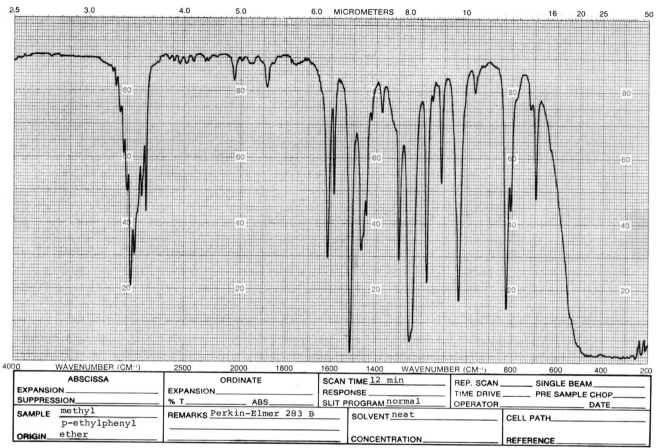

ABSCISSA	ORDINATE	SCAN TIME 12 min	REP. SCAN _____ SINGLE BEAM _____
EXPANSION_____	EXPANSION_____	RESPONSE _____	TIME DRIVE _____ PRE SAMPLE CHOP_____
SUPPRESSION_____	% T_____ ABS_____	SLIT PROGRAM normal	OPERATOR _____ DATE_____
SAMPLE methyl	REMARKS Perkin-Elmer 283 B	SOLVENT neat	CELL PATH_____
p-ethylphenyl			
ORIGIN ether		CONCENTRATION_____	REFERENCE_____

Fig. 6.4 IR spectrum: *methyl p-ethylphenyl ether.*

Fig. 6.5 NMR spectrum: *methyl p-ethylphenyl ether.*

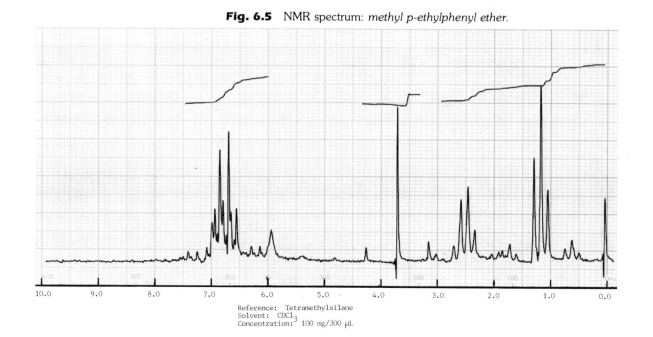

Reference: Tetramethylsilane
Solvent: CDCl$_3$
Concentration: 100 mg/300 μL

ENVIRONMENTAL DATA

Substance	Amount	TLV (mg/m³)	Emissions (mg)	Volume (m³)
Experiment 25A				
Propyl iodide (Max. from yield)	75 μL			
p-Cresol (Max. from yield, transfer)	84 mg	22	34	1.5
Methylene chloride (Max., col. chromatog., evap.)	2.25 mL	350	3000	8.6
Experiment 25B				
p-Cresol (SKIN) (Max. from yield, transfer)	167 mg	22	100	4.5
Methyl iodide (SKIN) (Max. from yield, transfer)	228 mg	10	157	16
Methylene chloride (Max. col. chromatog., evap.)	2.25 mL	350	3000	8.6
Experiment 25C				
p-Ethylphenol (Max. from yield)	150 mg	—	78	
Methyl iodide (SKIN) (Max., from yield, transfer)	205 mg	10	127	13
Methylene chloride (Max., col. chromatog., evap.)	2.25 mL	350	3000	8.6

TLV: Sodium hydroxide—2 μg/m³
TXDS: p-Methylanisole—orl-rat LD50: 1920 mg/kg
    Propyl iodide—ihl-rat LC50: 73,000 mg/m³/30M
    Silica gel—ivn-mus LDLo: 234 mg/kg
    Sodium sulfate—orl-mus LD50: 5989 mg/kg
MTDS: Tetrabutylammonium bromide—dnd-mam:lyn 50 mmol/L

QUESTIONS   **6-54.** Thiol ethers are often prepared using a Williamson synthesis. For example.

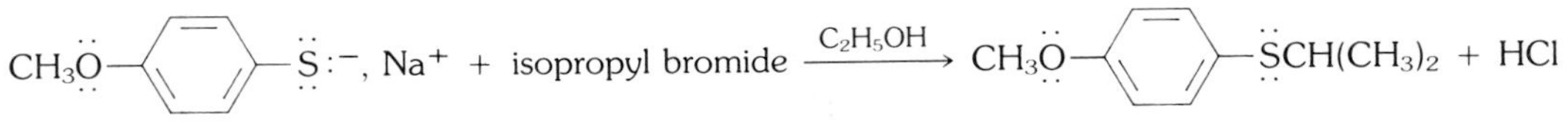

$$CH_3\ddot{O}-\langle\rangle-\ddot{S}:^-, Na^+ + \text{isopropyl bromide} \xrightarrow{C_2H_5OH} CH_3\ddot{O}-\langle\rangle-\ddot{S}CH(CH_3)_2 + HCl$$

$$CH_3\ddot{O}-\langle\rangle-\ddot{S}:^-, Na^+ + \text{2-bromo-1-nitropropane} \xrightarrow{C_2H_5OH} CH_3\ddot{O}-\langle\rangle-\ddot{S}CH(CH_3)CH_2NO_2$$

The reaction with isopropyl bromide is 16 times faster than the reaction with 2-bromo-1-nitropropane. Explain.

6-55. If 3-bromo-1-propanol is treated with NaOH, a compound of molecular formula C_3H_6O is formed. Suggest a structure for this product.

6-56. Arrange the substituted phenols given below in order of increasing reactivity toward ethyl iodide in the Williamson reaction. Explain your order.

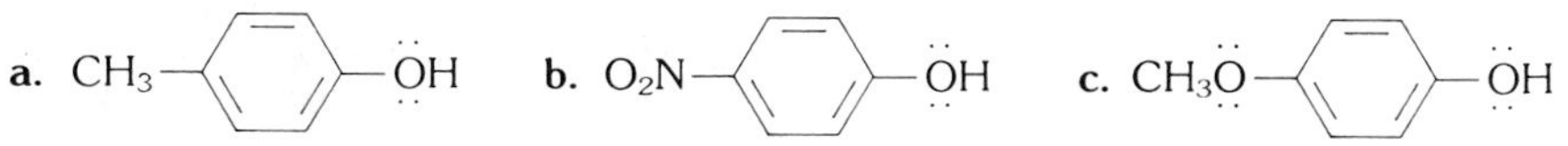

a. CH_3—⟨benzene ring⟩—$\ddot{O}H$ **b.** O_2N—⟨benzene ring⟩—$\ddot{O}H$ **c.** $CH_3\ddot{O}$—⟨benzene ring⟩—$\ddot{O}H$

6-57. *trans*-2-Chlorocyclohexanol reacts readily with NaOH to form cyclohexene oxide, but the cis isomer does not undergo this reaction. Explain.

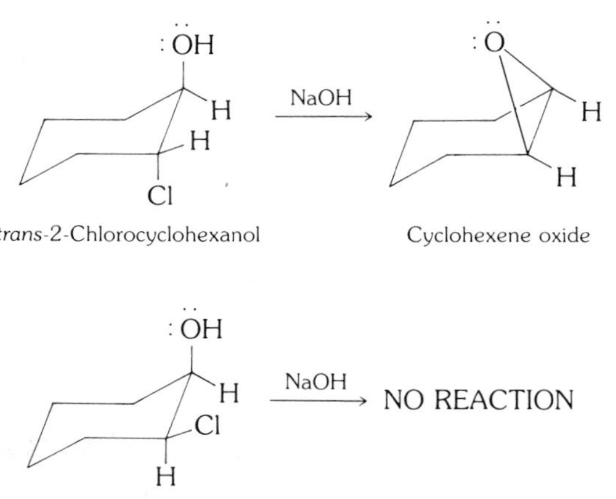

trans-2-Chlorocyclohexanol   Cyclohexene oxide

cis-2-Chlorocyclohexanol

6-58. *t*-Butyl ethyl ether might be prepared two ways using different starting materials.

$(CH_3)_3C\ddot{O}{:}^-,\ K^+\ +\ CH_3CH_2Cl$

$CH_3CH_2\ddot{O}{:}^-,\ K^+\ +\ (CH_3)_3CCl$

$(CH_3)_3C—\ddot{O}—CH_2CH_3$

Which route would you choose to prepare the above ether and why?

REFERENCES   **1.** References selected from the large number of examples of the Williamson reaction in "Organic Syntheses" include:

a. Wheeler, T. S.; Willson, F. G. "Organic Syntheses"; Wiley: New York, 1941; Collect. Vol. I, p 296.

b. Marvel, C. S.; Tanenbaum, A. L. *Ibid.*, p 435.

c. Fuson, R. C.; Wojcik, B. H. "Organic Syntheses"; Wiley: New York, 1943; Collect. Vol. II, p 260.

d. Allen, C. F. H.; Gates, J. W., Jr. "Organic Syntheses"; Wiley: New York, 1955; Collect. Vol. III, p 140.

e. *Ibid.*, p 418.

f. Boehme, W. R. "Organic Syntheses"; Wiley: New York, 1963; Collect. Vol. IV, p 590.

g. Vyas, G. N.; Shah, N. M. *Ibid.*, p 836.

h. Gassman, P. G.; Marshall, J. L. "Organic Syntheses"; Wiley: New York, 1973; Collect. Vol. V, p 424.

i. Kuryla, W. C.; Hyve, J. E. *Ibid.*, p 684.

j. Pedersen, C. J. *Org. Synth.* **1972**, *52*, 67.

2. Review articles on phase transfer catalysis:

a. Gokel, G. W.; Weber, W. P. *J. Chem. Educ.* **1978**, *55*, 350.

b. *Ibid.*, 429.

c. Varughese, P. *J. Chem. Educ.* **1977**, *54*, 666.

d. Jones, R. A. *Aldrichimica Acta* **1976**, *9*, 35.

3. The procedures used in these experiments for the preparation of the ethers were adapted from the work of:

a. Rowe, J. E. *J. Chem. Educ.* **1980**, *57*, 162.

b. McKillop, A.; Fiaud, J. C.; Hug, R. P. *Tetrahedron* **1974**, *30*, 1379.

Experiment 26A, 26B
Amide Synthesis: Acetanilide or
N,N'-Diacetyl-1,4-phenylenediamine

(acetamide, *N*-phenyl-; acetamide, *N,N'*-1,4-phenylenebis-)

One of the major routes used for the preparation of amides is the reaction of ammonia, a primary amine, or a secondary amine with an active acylating reagent. The reactions presented in Parts A and B demonstrate the use of acetic anhydride as the acylating agent.

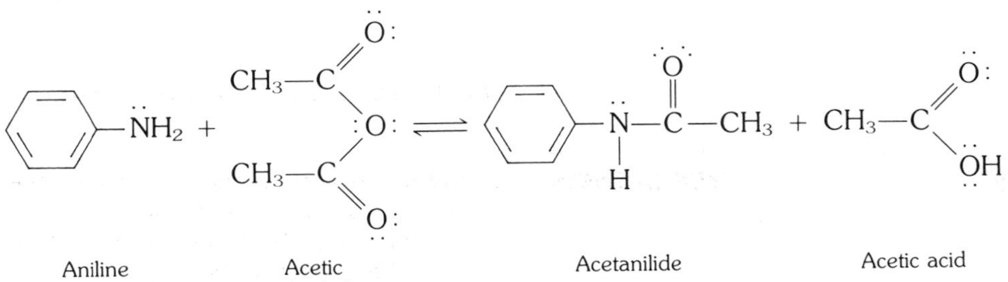

| Aniline | Acetic anhydride | Acetanilide | Acetic acid |

DISCUSSION   A number of important chemical and biochemical synthetic sequences are initiated by the addition of a nitrogen nucleophile to a carbonyl carbon atom. The experiments outlined in Parts A and B illustrate the attack of a primary amine on the acetyl group of acetic anhydride. Ammonia or secondary amines also react readily with active acylating reagents.

The mechanism is an example of the attack of a nucleophilic reagent on the trigonal carbon atom of the carbonyl unit.

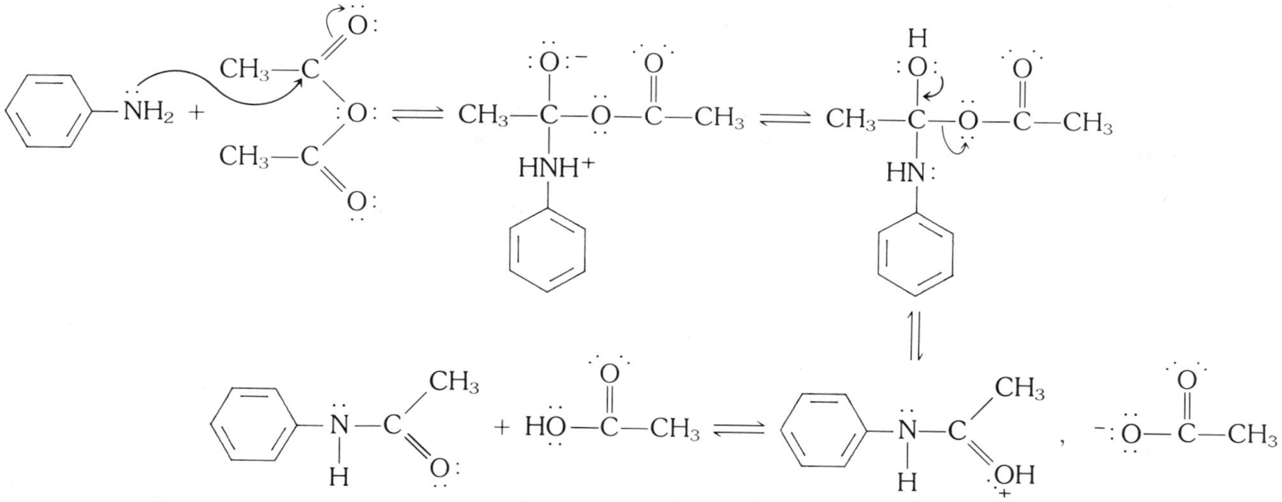

PART A: Acetanilide

The reaction is shown above.

EXPERIMENTAL Estimated time to complete the experiment: 1.5 hours.

Physical Properties of Reactants and Products

Compound	MW	Wt/Vol	mmol	mp(°C)	bp(°C)	Density	n_D
Aniline	93.13	100 μL	1.09		184	1.02	1.5863
Conc. HCl		3 drops					
Sodium acetate							
trihydrate	136.08	150 mg	1.10	58			
Acetic anhydride	102.09	150 μL	1.59		140	1.08	1.3901
Acetanilide	135.17			114			

Reagents and Equipment

HOOD In a tared 10 × 75 mm test tube fitted with a cork stopper is placed 100 μL of aniline [**HOOD**].

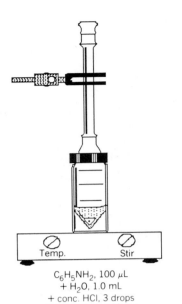

C₆H₅NH₂, 100 µL
+ H₂O, 1.0 mL
+ conc. HCl, 3 drops
NaOAc·3H₂O, 150 mg
+ (CH₃CO)₂O, 150 µL

> **WARNING:** *Aniline is a toxic material and is a cancer suspect agent.*

The aniline is dispensed using an automatic delivery pipet. The test tube is again weighed to determine the exact amount of aniline delivered.

HOOD A 1.0-mL graduated pipet is used to add, with swirling, 0.5 mL of water followed by 3 drops of concentrated hydrochloric acid [**HOOD**] (Pasteur pipet). To the resulting solution is added 10 mg of powdered decolorizing charcoal (Norit).

The well mixed suspension is transferred (Pasteur pipet) to a 25-mm funnel fitted with fast-grade filter paper to remove the charcoal by gravity filtration.

Wet the filter paper in advance with distilled water and blot the excess water from the stem of the funnel.

The filtrate is collected in a 3.0-mL conical vial containing a magnetic spin vane and equipped with an air condenser.(■) An additional 0.5-mL portion of water is used to rinse the test tube and the collected charcoal. The rinse is combined with the original filtrate.

Tap all of the filtrate from the funnel stem into the collecting vial. As a result of this purification step a clear, colorless solution of aniline hydrochloride should be obtained.

A solution of sodium acetate trihydrate is prepared by placing 150 mg (1.10 mmol) of this material in a 10 × 75 mm test tube followed by 0.5 mL of distilled water. The tube is capped and set aside for use in the next step.

HOOD Using an automatic delivery pipet, add [**HOOD**] 150 µL of acetic anhydride to the aniline hydrochloride solution, with stirring, followed quickly by addition (Pasteur pipet) of the previously prepared sodium acetate solution.

Reaction Conditions

The reaction is very rapid and the acetanilide product begins to precipitate immediately upon mixing of the reagents. After allowing the reaction mixture to stand at room temperature for approximately 5 minutes, it is placed in an ice bath for an additional 5–10 minute period to complete the crystallization process.

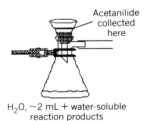

Acetanilide collected here

H₂O, ~2 mL + water-soluble reaction products

Isolation of Product

The acetanilide product is collected by filtration under reduced pressure using a Hirsch funnel.(■) The conical vial is rinsed with two 0.5-mL portions of water (calibrated Pasteur pipet) and the rinse used to wash the collected filter cake. The snow-white crystals are dried on a porous clay plate or on filter paper in a desiccator.

Purification and Characterization

Further purification of the product is generally not required. However, the acetanilide may be recrystallized from hot water or from ethanol-water.

Acetanilide (150 mg) can be recrystallized from approximately 3 mL of water or 2 mL of ethanol-water (1 : 10 v/v) with better than 80% recovery.

Weigh the dried crystals and calculate the percentage yield. Determine the melting point of the material and compare your result to the literature value. Obtain an IR spectrum of the product and compare it with that of an authentic sample.

PART B: *N,N'*-Diacetyl-1,4-phenylenediamine

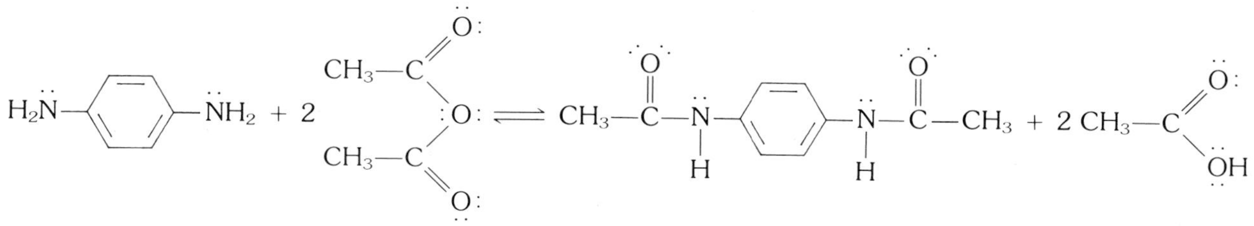

| 1,4-Phenylenediamine | Acetic anhydride | | *N,N'*-Diacetyl-1,4-phenylenediamine | Acetic acid |

EXPERIMENTAL   Estimated time to complete the experiment: 1.5 hours.

Physical Properties of Reactants and Products

Compound	MW	Wt/Vol	mmol	mp(°C)	bp(°C)	Density	n_D
1,4-Phenylenediamine	108.14	117 mg	1.08	138			
Conc. HCl		6 drops					
Sodium acetate trihydrate	136.08	300 mg	2.20	58			
Acetic anhydride	102.09	350 μL	3.71		140	1.08	1.3901
N,N'-Diacetyl-1,4-phenylenediamine	192.24			312–315			

Reagents and Equipment
In a 10 × 75 mm test tube is placed 117 mg (1.08 mmol) of 1,4-phenylene-diamine.

> **CAUTION:** *This reagent is toxic and is a cancer suspect agent.*

With gentle swirling, add 1.0 mL of distilled water and 6 drops (Pasteur pipet) of concentrated hydrochloric acid. After dissolution, add 30 mg of powdered

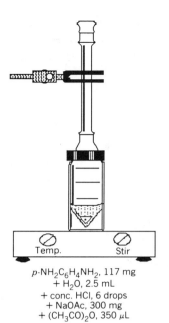

p-NH$_2$C$_6$H$_4$NH$_2$, 117 mg
+ H$_2$O, 2.5 mL
+ conc. HCl, 6 drops
+ NaOAc, 300 mg
+ (CH$_3$CO)$_2$O, 350 μL

decolorizing charcoal. The well mixed suspension is transferred, by use of a Pasteur pipet, to a 25-mm funnel fitted with a fast-grade filter paper previously wet with water. The charcoal is removed by gravity filtration. The filtrate, which is clear to slightly yellow, is collected in a 5-mL conical vial containing a magnetic spin vane and equipped with an air condenser.(■) Three 0.5-mL portions of water (calibrated Pasteur pipet) are used to rinse the test tube, and in turn are used to wash the collected charcoal. The rinse is combined with the original filtrate.

A solution of sodium acetate trihydrate is prepared by placing 300 mg (2.20 mmol) of this material in a 10 × 75 mm test tube followed by the addition of 0.5 mL of distilled water. The mixture is stirred with a spatula to aid the dissolution process. The tube is then capped and set aside for use in the next step.

Reaction Conditions
To the 1,4-phenylenediamine hydrochloride solution is added 350 μL of acetic anhydride and the mixture stirred briefly using a magnetic stirrer.

HOOD **Note.** *The acetic anhydride is dispensed in the* **HOOD** *using an automatic delivery pipet. A slight amount of white precipitate may be observed at this stage.*

The previously prepared sodium acetate solution is added by Pasteur pipet to the reaction mixture with stirring.

Isolation of Product
The reaction is very rapid and the desired product begins to precipitate almost immediately. After stirring briefly, the mixture is allowed to stand at room temperature for a few minutes and then is placed in an ice bath for an additional 5–10 minutes.

Diamide product collected here

H$_2$O, ~3.5 mL + water-soluble reaction products

Purification and Characterization
The crude N,N'-diacetyl-1,4-phenylenediamine is collected by vacuum filtration using a Hirsch funnel.(■) The reaction vial is rinsed with two 0.5-mL portions of water and the rinse used to wash the filter cake. The collected material is placed on a clay plate to dry.

If this material is utilized as the substrate in Experiment 34C, recrystallization from methanol is suggested.

Weigh the dried crystals and calculate the percentage yield. Determine the melting point and compare it with the literature value. Obtain an IR spectrum of the product and compare it with that shown in Fig. 6.6.

ENVIRONMENTAL DATA

Substance	Amount	TLV (mg/m³)	Emissions (mg)	Volume (m³)
1,4-Phenylenediamine (Emissions should be small.)	117 mg			
Aniline (SKIN) (transfer and reactant)	100 μL	19	10	0.5
Acetic anhydride (Transfer, open container)	150 μL	20	6	0.3

TLV: Hydrochloric acid—5 ppm
TXDS: Acetanilide—orl-rat LD50: 800 mg/kg
Sodium acetate—orl-rat LD50: 3530 mg/kg

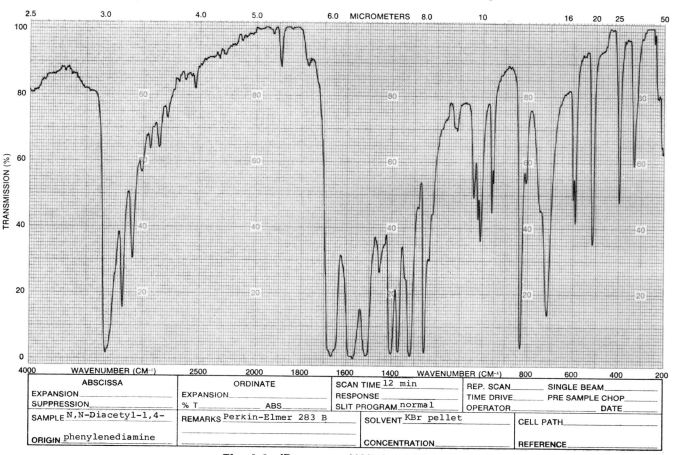

ABSCISSA
EXPANSION_____
SUPPRESSION_____
SAMPLE N,N-Diacetyl-1,4-
ORIGIN phenylenediamine

ORDINATE
EXPANSION_____
% T_____ ABS_____
REMARKS Perkin-Elmer 283 B

SCAN TIME 12 min_____
RESPONSE _____
SLIT PROGRAM normal_____
SOLVENT KBr pellet_____
CONCENTRATION_____

REP. SCAN_____ SINGLE BEAM_____
TIME DRIVE_____ PRE SAMPLE CHOP_____
OPERATOR_____ DATE_____
CELL PATH_____
REFERENCE_____

Fig. 6.6 IR spectrum: *N,N'-diacetyl-1,4-phenylenediamine.*

QUESTIONS

6-59. What is the function of the sodium acetate in the reactions outlined in this experiment?

6-60. In the first stage of the experiment, aniline (Part A) and *p*-phenylenediamine (Part B) are treated with an aqueous HCl solution. What species is formed in each of these acid solutions?

6-61. Arrange the following substituted aniline compounds in increasing order of reactivity toward acetic anhydride.

a.

$\overset{..}{N}H_2$

:OCH$_3$

b.

$\overset{..}{N}H_2$

CN:

c.

$\overset{..}{N}H_2$

N(CH$_3$)$_2$

6-62. Suggest a mechanism for the preparation of ethanoic anhydride from ethanoic acid and ethanoyl chloride in the presence of pyridine.

6-63. Anhydrides generally react more slowly than acid chlorides with an amine to prepare amides. Explain this observation.

6-64.

$$CH_3CH_2-C\overset{\overset{..}{O}:}{\underset{\overset{..}{N}H_2}{}} \quad \text{is less basic than} \quad CH_3CH_2-\overset{..}{N}H_2$$

Explain.

REFERENCES   1. Review articles:
   a. Satchell, D. P. N. *Quart. Revs.* **1963**, *17*, 160.
   b. Beckwith, A.L.J. in "The Chemistry of the Amides"; Zabicky, J. Ed.; Wiley: New York, 1966; p 86.
2. Selected acylation reactions in "Organic Syntheses" between anhydrides and amines:
   a. Noyes, W. A.; Porter, P. K. "Organic Syntheses"; Wiley: New York, 1941; Collect. Vol. I, p 457.
   b. Herbst, R. M.; Shemin, D. "Organic Syntheses"; Wiley: New York, 1943; Collect. Vol. II, p 11.
   c. Jacobs, T. L.; Winstein, S.; Linden, G. B.; Robson, J. H.; Levy, E. F.; Seymour, D. "Organic Syntheses"; Wiley: New York, 1955; Collect. Vol. III, p 456.
   d. Fanta, P. E.; Tarbell, D. S. *Ibid.*, p 661.
   e. Wiley, R. H.; Borum, O. H. "Organic Syntheses"; Wiley: New York, 1963; Collect. Vol. IV, p 5.
   f. Cava, M. P.; Deana, A. A.; Muth, K.; Mitchell, M. J. "Organic Syntheses"; Wiley: New York, 1973; Collect. Vol. V, p 944.

Experiment 27
Imide Synthesis: *N*-Phenylmaleimide

(maleimide, *N*-phenyl)

The condensation of an anhydride with aniline to form an imide is described in this experiment. The initial reaction to give the acid amide is followed by an intramolecular condensation to produce the desired imide derivative.

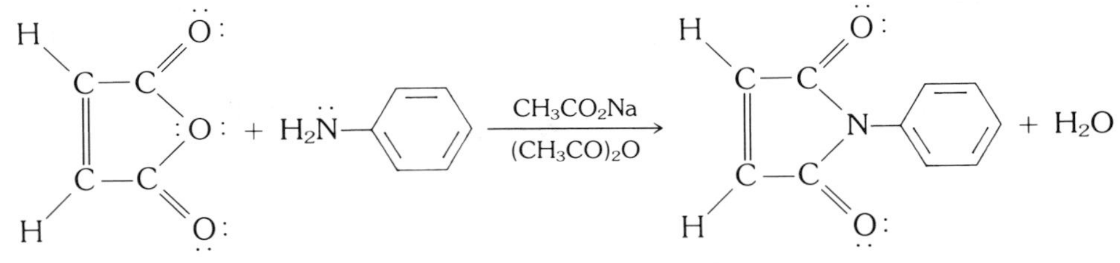

Maleic anhydride   Aniline   *N*-Phenylmaleimide

DISCUSSION   Imides are diacyl derivatives of ammonia or primary amines. The reaction is similar in its scope and mechanism to the acetylation of aniline or 1,4-phenylenediamine presented in Experiments 26A and 26B. As illustrated in the present experiment, cyclic anhydrides produce cyclic imides. Derivatives of imides have been suggested for use in the treatment of arthritis, tuberculosis, and epilepsy. Several also have been found to be growth stimulants.

The second step in the reaction is much slower than the first. That is, attack of an amide nitrogen on the carbonyl carbon of the acid is slower than the attack of the amine nucleophile on the anhydride carbonyl carbon. The mechanistic sequence is given below.

(1)

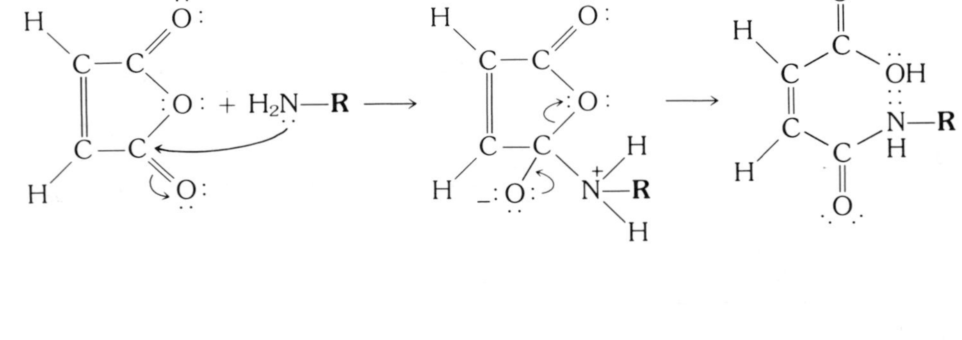

(2)

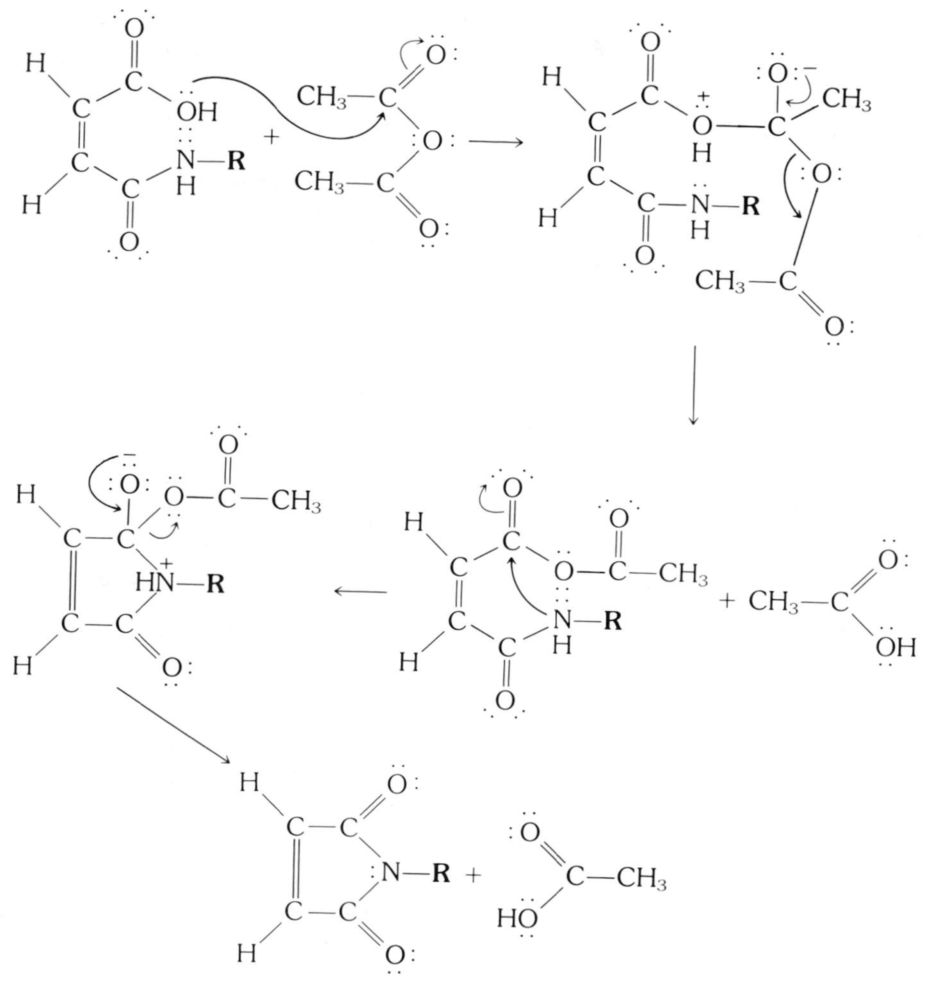

EXPERIMENTAL   Estimated time of the experiment: 2.0 hours.

PART A: Preparation of Maleanilic Acid

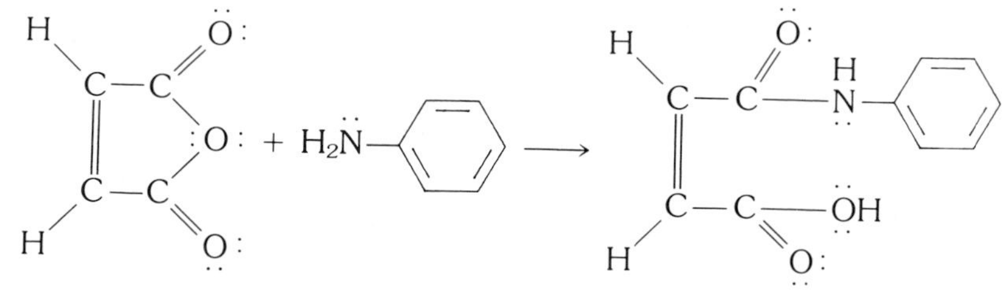

Maleic anhydride   Aniline   Maleanilic acid

Physical Properties of Reactants and Products

Compound	MW	Wt/Vol	mmol	mp(°C)	bp(°C)	Density	n_D
Maleic anhydride	98.06	60 mg	0.61	60			
Diethyl ether	74.12	1.2 mL			34.5		
Aniline	93.13	56 µL	0.62		184	1.02	1.5863
Maleanilic acid	191.18			201–202			

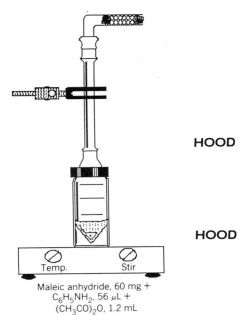

Maleic anhydride, 60 mg +
$C_6H_5NH_2$, 56 μL +
$(CH_3CO)_2O$, 1.2 mL

HOOD

HOOD

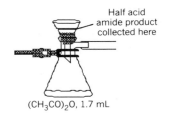

Half acid
amide product
collected here

$(CH_3CO)_2O$, 1.7 mL

Reagents and Equipment

In a 3.0-mL conical vial containing a magnetic spin vane and equipped with an air condenser protected with a drying tube, place 60 mg (0.61 mmol) of maleic anhydride and 1.0 mL of anhydrous diethyl ether.(■) The mixture is stirred at room temperature until all the maleic anhydride has dissolved.

> **WARNING.** *Diethyl ether is highly flammable. All flames in the laboratory should be extinguished. Dispense this reagent in the* HOOD *using a calibrated Pasteur pipet.*

In a separate, dry vial prepare a solution of 56 μL (57 mg, 0.62 mmol) of aniline in 100 μL of anhydrous diethyl ether.

> **WARNING:** *Aniline is highly toxic and is a cancer suspect agent. It should be dispensed in the* HOOD *using an automatic delivery pipet.*

Using a Pasteur pipet, add the aniline-ether solution in one portion to the stirred maleic anhydride-ether solution. The vial is rinsed with 100 μL of anhydrous ether, and this rinse is also transferred to the reaction solution.

Reaction Conditions

The reaction mixture is stirred at room temperature for a period of 15 minutes and then cooled in an ice bath for 5–10 minutes.

Isolation of Product

The deposit of fine, cream-colored powder is collected by vacuum filtration using a Hirsch funnel.(■) The maleanilic acid crystals are then washed with 0.5 mL of cold diethyl ether (calibrated Pasteur pipet) and air dried in the funnel for 5 minutes.

Purification and Characterization

Weigh the maleanilic acid and calculate the percentage yield. Determine the melting point and compare your value to that in the literature. Obtain an IR spectrum and compare it with an authentic sample. The air-dried product is suitable for use in the next step without further purification.

PART B: Preparation of *N*-Phenylmaleimide

Maleanilic acid $\xrightarrow[\text{(CH}_3\text{CO)}_2\text{O}]{\text{CH}_3\text{CO}_2\text{Na}}$ *N*-Phenylmaleimide + H_2O

Physical Properties of Reactants and Products

Compound	MW	Wt/Vol	mmol	mp(°C)	bp(°C)	Density	n_D
Maleanilic acid	191.18	100 mg	0.52	201–202			
Sodium acetate	82.03	25 mg	0.30	324			
Acetic anhydride	102.09	200 μL	2.12		140	1.08	1.3901
N-Phenylmaleimide	173.17			90–91			

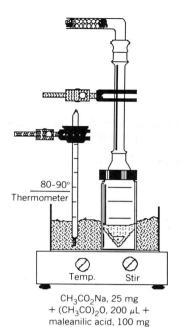

80–90°
Thermometer

CH$_3$CO$_2$Na, 25 mg
+ (CH$_3$CO)$_2$O, 200 μL +
maleanilic acid, 100 mg

Imide
product
collected
here

H$_2$O, ~2.5 mL +
water-soluble
reaction products

Reagents and Equipment

In a 3.0-mL conical vial containing a magnetic spin vane and equipped with an air condenser protected by a drying tube are placed 25 mg (0.3 mmol) of anhydrous sodium acetate and 200 μL (216 mg, 2.12 mmol) of acetic anhydride. (■)

> **CAUTION:** *Acetic anhydride is corrosive and a lachrymator. It should be dispensed in the* HOOD *by use of an automatic delivery pipet.*

Maleanilic acid, prepared in Part A (100 mg, 0.52 mmol) is now added to the reaction vial.

Reaction Conditions

The reaction mixture is heated, with stirring, at a sand bath temperature of 80–90°C for a period of 30 minutes. It is then cooled to room temperature and 1.0 mL of cold water added (calibrated Pasteur pipet). The resulting mixture is stirred for a few minutes and then placed in an ice bath for 5–10 minutes.

Isolation of Product

The solid product is collected by vacuum filtration using a Hirsch funnel and the filter cake washed with three 0.5-mL portions of cold water (calibrated Pasteur pipet). (■) It is air dried in the Hirsch funnel for a period of 5–10 minutes.

Purification and Characterization

The crude *N*-phenylmaleimide is recrystallized from 95% ethanol to yield canary-yellow needles. *Use the smallest amount of ethanol possible. An excess will lead to a poor recovery.* After drying the product on a porous clay plate, weigh the crystals and calculate the percentage yield. Determine the melting point and compare your result with the literature value. Obtain an IR spectrum and compare it with that of an authentic sample.

ENVIRONMENTAL DATA

Substance	Amount	TLV (mg/m³)	Emissions (mg)	Volume (m³)
Aniline (SKIN) (Max. from yield, transfer)	57 mg	19	5	0.3
Acetic anhydride (Max. from transfer)	216 mg	20	216	10.8
Ether (Max., solv., vac., filter)	1.2 mL	1200	857	0.7
Ether (Max., evap.)	0.5 mL	1200	357	0.3
Ethanol (Max.)	0.5 mL	1900	400	0.2

TLV: Maleic anhydride—0.25 ppm
TXDS: *N*-Phenylmaleimide—orl-rat LDLo: 100 mg/kg
Sodium acetate—orl-rat LD50: 3530 mg/kg

QUESTIONS   **6-65.** As stated in the discussion section of this experiment, the second step in the reaction to form the imide is much slower than that of the first stage (formation of the acid-amide). Explain.

6-66. Phthalimide

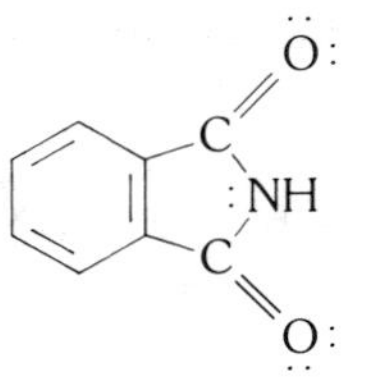

has a $K_a = 5 \times 10^{-9}$. Write an equation for the reaction of phthalimide with potassium amide (a strong base) in dimethyl formamide solvent. Name the product.

6-67. Predict which of the species below is the most acidic. Explain.

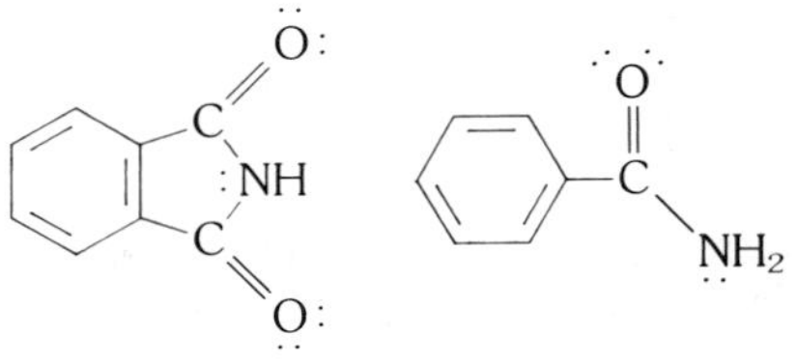

Phthalimide   Benzamide

6-68. The phthalimide anion is a strong nucleophile. It can react easily with primary alkyl halides to form substituted phthalimides.

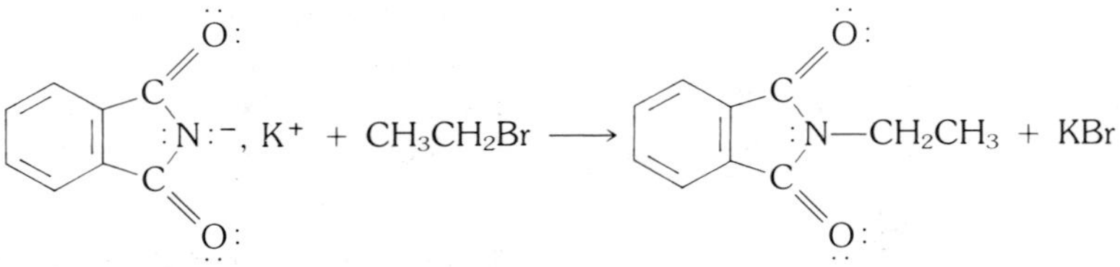

Suggest a suitable mechanism for this reaction.

REFERENCES   1. Review article on cyclic imides:
Hargreaves, M. K.; Pritchard, J. G.; Dave, H. R. *Chem. Rev.* **1970,** *70,* 439.
2. Selected imide preparations in "Organic Syntheses" include:
a. Noyes, W. A.; Porter, P. K. "Organic Syntheses"; Wiley: New York, 1941; Collect. Vol. I, p 457.
b. Smith, L. I.; Emerson, O. H. "Organic Syntheses"; Wiley: New York, 1955; Collect, Vol. III, p 151.
c. Soine, T. O.; Buchdahl, M. R. "Organic Syntheses"; Wiley: New York, 1963; Collect. Vol. IV, p 106.
d. Cava, M. P.; Deana, A. A.; Muth, K.; Mitchell, M. J. "Organic Syntheses"; Wiley: New York, 1973; Collect. Vol. V, p 944.

Experiments 28A and 28B
Synthesis of Cyclic Carboxylic Acid Anhydrides: Succinic Anhydride; Phthalic Anhydride

(2,5-furandione, dihydro-; 1,3-isobenzofurandione)

One of the important methods for preparing cyclic carboxylic acid anhydrides is presented in this experiment. The reaction demonstrates the use of acetic anhydride, an important industrial material, as a dehydrating agent.

DISCUSSION Five- and six-membered cyclic anhydrides can be easily formed when the corresponding dicarboxylic acid is heated in the presence of a dehydrating agent. Among the dehydrating agents most often used is acetic anhydride; anhydrides that have boiling points higher than acetic acid can be prepared by this method.

The formation of the anhydride from its corresponding acid in the presence of acetic anhydride is referred to as *anhydride exchange*.

It is possible to prepare five and six membered cyclic anhydrides in the absence of acetic anhydride by direct dehydration at elevated temperatures. Maleic anhydride, for example, is easily obtained by this method.

The mechanistic sequence for anhydride exchange is given below.

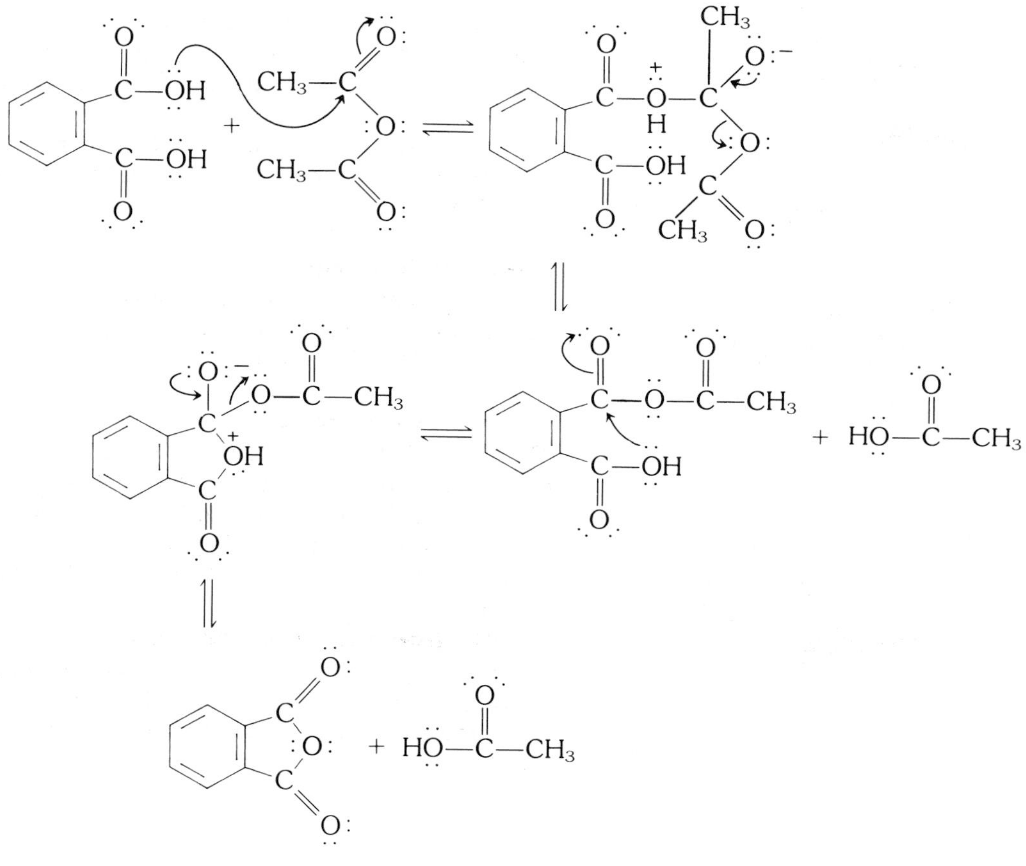

PART A: Succinic Anhydride

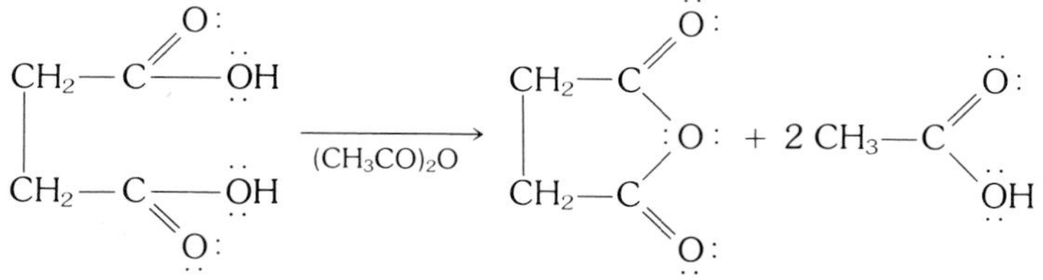

Succinic acid   Succinic anhydride

EXPERIMENTAL Estimated time to complete the experiment: 1.5 hours.

Physical Properties of Reactants and Products

Compound	MW	Wt/Vol	mmol	mp(°C)	bp(°C)	Density	n_D
Succinic acid	118.09	150 mg	1.27	188			
Acetic anhydride	102.09	200 μL	2.12		140	1.08	1.3901
Succinic anhydride	100.08			120			

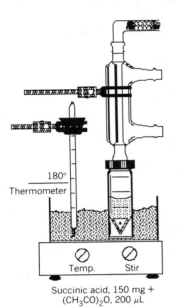

180°
Thermometer

Temp.   Stir

Succinic acid, 150 mg +
(CH₃CO)₂0, 200 μL

Succinic
anhydride
collected
here

(CH₃CH₂)₂O, 1.5 mL +
CH₃CO₂H, ~100 μL

Reagents and Equipment

Note. *All equipment must be dried in an oven (110°C) half an hour before use.*

In a 1.0-mL conical vial containing a magnetic spin vane and equipped with a reflux condenser protected by a calcium chloride drying tube, place 150 mg (1.3 mmol) of succinic acid and 200 μL of acetic anhydride.(■)

> **CAUTION:** *Acetic anhydride is a moisture-sensitive irritant. It should be dispensed in the* HOOD *using an automatic delivery pipet.*

Reaction Conditions

The reaction mixture is heated with stirring in a sand bath at a temperature of 180°C for a period of 45 minutes. The time is measured from the point at which the succinic acid has completely dissolved.

Isolation of Product

The mixture is cooled to room temperature, whereupon a voluminous precipitate of succinic anhydride deposits. The vial is cooled further in an ice bath for 5 minutes and the solid collected by vacuum filtration using a Hirsch funnel.(■) The white needles are washed with three 0.5-mL portions of diethyl ether (calibrated Pasteur filter pipet) and then placed on a porous clay plate to dry.

Purification and Characterization

The succinic anhydride crystals are sufficiently pure for characterization.

Weigh the product and calculate the percentage yield. Determine the melting point and compare your results to the literature value. Obtain the IR spectrum and compare it with an authentic sample.

PART B: Phthalic Anhydride

Phthalic acid $\xrightarrow{(CH_3CO)_2O}$ Phthalic anhydride $+ 2\ CH_3C(=O)OH$

Phthalic acid

Phthalic anhydride

EXPERIMENTAL   Estimated time to complete the experiment: 1.5 hours.

Physical Properties of Reactants and Products

Compound	MW	Wt/Vol	mmol	mp(°C)	bp(°C)	Density	n_D
Phthalic acid	166.14	100 mg	0.60	210			
Acetic anhydride	102.09	200 μL	2.12		140	1.08	1.3901
Phthalic anhydride	148.12			131			

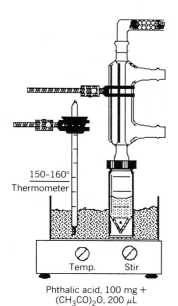

150–160°
Thermometer

Temp.   Stir

Phthalic acid, 100 mg +
(CH₃CO)₂O, 200 μL

Phthalic
anhydride
collected
here

Hexane, 0.5 mL +
CH₃CO₂H, ~100 μL

Reagents and Equipment

To a 1.0-mL conical vial containing a magnetic spin vane and equipped with a reflux condenser protected by a calcium chloride drying tube, add 100 mg (0.60 mmol) of phthalic acid and 200 μL of acetic anhydride.(■)

> **CAUTION:** *Acetic anhydride is a moisture-sensitive irritant. It should be dispensed in the* HOOD *using an automatic delivery pipet.*

Reaction Conditions

The reaction solution is heated with stirring at a sand bath temperature of 150–160°C for a period of 30 minutes.

Note. *To obtain the reaction temperature, the vial is positioned firmly on the bottom of the sand bath container.*

Isolation of Product

The mixture is cooled to room temperature, whereupon the product crystallizes from solution. The vial and contents are cooled in an ice bath for a period of 10 minutes, and the solid is collected by vacuum filtration using a Hirsch funnel.(■) The filter cake is rinsed carefully by dropwise addition of 0.5 mL of cold hexane (Pasteur pipet) and then air dried in the Hirsch funnel. Final drying is achieved by placing the crystals on a porous clay plate.

Purification and Characterization

The phthalic anhydride is sufficiently pure for characterization.

Weigh the product and calculate the percentage yield. Determine the melting point and compare it with the literature value. Obtain an IR spectrum and compare it with that of an authentic sample.

ENVIRONMENTAL DATA

Substance	Amount	TLV (mg/m³)	Emissions (mg)	Volume (m³)
Acetic anhydride (transfer in hood, reflux.)	216 mg	20	10	0.5
Diethyl ether (Max., evap.)	1.5 mL	1200	1071	0.9
Hexane (Max., evap.)	0.5 ml	180	330	1.8

TXDS: Phthalic anhydride—orl-rat LD50: 7900 mg/kg
Succinic anhydride—scu-rat TD: 2600 mg/kg/65W-1

QUESTIONS

6-69. As stated in the discussion, direct dehydration can be used as a method for the preparation of five and six membered cyclic anhydrides. Propose a suitable mechanism for the reaction below.

$$\text{maleic acid} \xrightarrow{\Delta} \text{maleic anhydride} + H_2O$$

6-70. Propose a suitable mechanism for the formation of the mixed anhydride obtained in the following reaction.

$$C_6H_5CH_2CO_2H + (CF_3CO)_2O \longrightarrow C_6H_5CH_2\overset{\overset{\displaystyle O}{\|}}{C}-O-\overset{\overset{\displaystyle O}{\|}}{C}-CF_3 + CF_3CO_2H$$

6-71. There are two 1,3-cyclobutane dicarboxylic acids. One can form an anhydride, the other cannot. Draw the structures of these compounds and indicate which one can be converted to an anhydride. Explain.

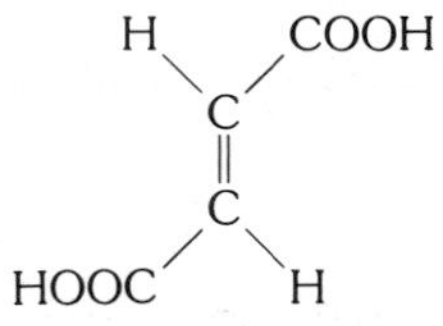

6-72. When maleic acid is heated to about 100°C, it forms maleic anhydride. However, fumaric acid requires a much higher temperature (250–300°C) before it dehydrates. In addition it only forms maleic anhydride. Explain.

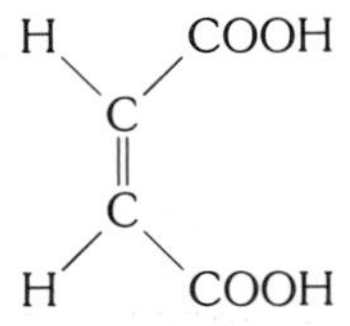

Maleic acid   Fumaric acid

REFERENCES

1. Selected references from "Organic Syntheses" in which anhydrides are prepared, using acetic anhydride as the dehydrating agent, include:
   a. Clarke, H. T.; Rahrs, E. J. "Organic Syntheses"; Wiley: New York, 1944; Collect. Vol. I, p 91.
   b. Nicolet, B. H.; Bender, J. A. *Ibid.*, p 410.
   c. Grummitt, O.; Egan, R.; Buck, A. "Organic Syntheses"; Wiley: New York, 1955; Collect. Vol. III, p 449.
   d. Shriner, R. L.; Furrow, C. L. Jr. "Organic Syntheses"; Wiley: New York, 1963; Collect. Vol. IV, p 242.
   e. Cason, J. *Ibid.*, p 630.
   f. Horning, E. C.; Finelli, A. F. *Ibid.*, p 790.
2. The synthesis of succinic anhydride is given in Fieser, L. F.; Martin, E. L. "Organic Syntheses"; Wiley: New York, 1943; Collect. Vol. II, p 560.

Experiment 29
Heterocyclic Ring Synthesis: 1-**H**-Benzimidazole

(1,3-benzodiazole)

The condensation of a 1,2-diaminobenzene with formic acid is one of the methods used to synthesize the highly aromatic benzimidazole ring system. Benzimidazole itself is prepared in this experiment.

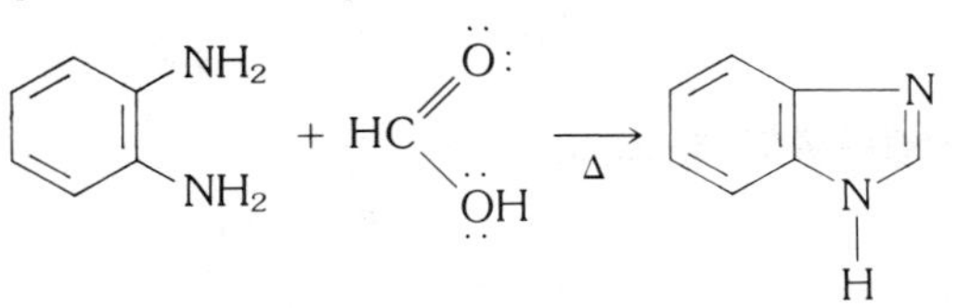

o-Phenylenediamine   Formic acid   1-**H**-Benzimidazole

DISCUSSION

This experiment illustrates the classic method of forming the benzimidazole ring system. This heterocyclic series is generally prepared from 1,2-diaminobenzenes by reaction with carboxylic acids or their derivatives, such as nitriles, under acidic conditions. The ring system is highly aromatic; it is difficult to oxidize or reduce and is stable to acids and bases. It is an important heterocyclic ring system since it occurs in vitamin B_{12} and in many biologically active compounds. Benzimidazole itself inhibits the growth of certain yeasts and bacteria.

The reaction actually proceeds in two stages. The first involves the *in situ* formation of a substituted formamide; the second, an intramolecular addition and elimination reaction to form the heterocyclic ring. The sequence is outlined below.

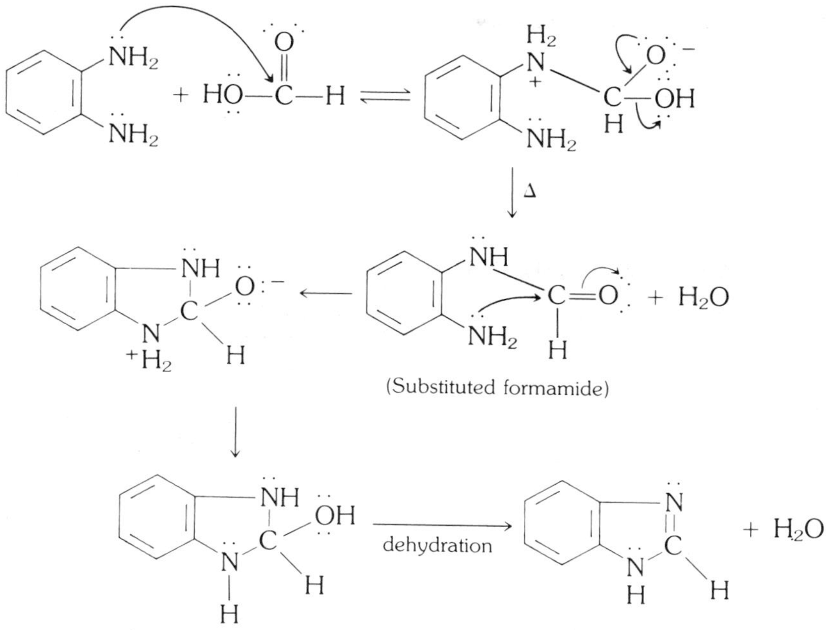

The reaction scheme shows the synthesis of benzimidazole from o-phenylenediamine and formic acid, proceeding through a substituted formamide and cyclic intermediate with dehydration to give 1-H-benzimidazole + H₂O.

(Substituted formamide)

dehydration + H₂O

EXPERIMENTAL   Estimated time to complete the experiment: 2.0 hours.

Physical Properties of Reactants and Products

Compound	MW	Wt/Vol	mmol	mp(°C)	bp(°C)	Density	n_D
o-Phenylene-diamine	108.14	108 mg	1.0	102			
90% Formic acid	46.03	64 μL	1.7		101	1.22	1.3714
1-**H**-Benzimidazole	118.14			174			

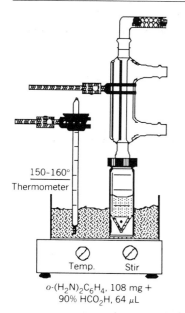

150–160°
Thermometer

Temp.   Stir

o-(H₂N)₂C₆H₄, 108 mg +
90% HCO₂H, 64 μL

Benzimidazole
product
collected
here

H₂O, ~2.0 mL

Reagents and Equipment

To a 1.0-mL conical vial containing a magnetic spin vane and equipped with a reflux condenser protected by a calcium chloride drying tube, add 108 mg (1.0 mmol) of o-phenylenediamine and 64 μL (69 mg, 1.7 mmol) of 90% formic acid.(■)

> **WARNING:** *o-Phenylenediamine is toxic and is a cancer suspect agent. Formic acid is very corrosive to the skin. It should be dispensed in the HOOD using an automatic delivery pipet.*

Reaction Conditions

The reaction mixture is heated, with stirring, at a sand bath temperature of 150–160°C for a period of 1 hour.

Isolation of Product

Crude benzimidazole is precipitated on addition of 630 μL of 10% aqueous sodium hydroxide solution from an automatic delivery pipet. The product is collected by vacuum filtration using a Hirsch funnel and the filter cake washed with three 0.5-mL portions of cold water (calibrated Pasteur pipet).(■)

Purification and Characterization

The crude material is recrystallized from water using a Craig tube and dried on a porous clay plate.

Weigh the crystals and calculate the percentage yield. Determine the melting point and compare your result with the literature value. Obtain an IR spectrum and compare it with that of an authentic sample.

ENVIRONMENTAL DATA

Substance	Amount	TLV (mg/m³)	Emissions (mg)	Volume (m³)
Formic acid, 90% (Max.)	64 μL	9	62	6.9
o-Phenylenediamine (relatively involatile)	108 mg		0	

TLV: Sodium hydroxide—2 μg/m³
TXDS: o-Phenylenediamine—orl-rat LD50: 1070 mg/kg
         Benzimidazole—orl-rat LDLo: 500 mg/kg

QUESTIONS   **6-73.** The parent compound of the imidazole series, imidazole (I) itself, was first prepared in 1858.

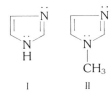

Can you account for the fact that it has a very high boiling point (256°C) whereas 1-methyl imidazole (II) has a somewhat lower boiling point of 199°C?

6-74. The imidazole ring system has a great deal of aromatic character. Can you formulate two resonance hybrid structures that account for this characteristic?

6-75. Imidazole is a weak acid and reacts with strong bases to form the corresponding anion. Show this reaction and draw resonance structures to account for the stability of the generated anion.

6-76. Suggest a mechanism for the dehydration involved in the last step in the synthesis of benzimidazole.

REFERENCES   The conditions of this reaction were adapted from those reported by Wagner, E. C.; Millett, W. H. "Organic Syntheses"; Wiley: New York, 1943; Collect. Vol. II, p 65.

Experiment 30
Heterocyclic Ring Synthesis: 4-Hydroxycoumarin; Dicoumarol

(2H-1-benzopyran-2-one, 4-hydroxy; 3,3'-methylene bis(4-hydroxycoumarin)

This reaction illustrates the condensation of a carbon nucleophile with a carbonyl carbon to form a C—C bond. In the reaction a β-ketoester is prepared, which on cyclization forms a lactone. Further condensation with formaldehyde yields dicoumarol, a material that is the prototype for the oral anticoagulants widely used in the medical field to lower blood coagulability.

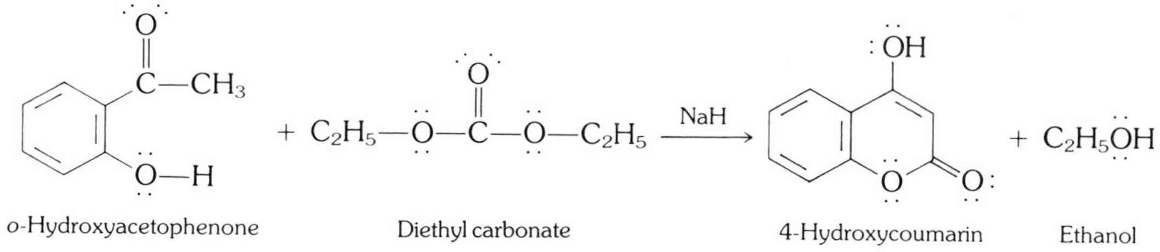

o-Hydroxyacetophenone   Diethyl carbonate   4-Hydroxycoumarin   Ethanol

DISCUSSION This reaction sequence illustrates the well-known Claisen condensation, which is widely used to form C—C bonds. The condensation is brought about by the nucleophilic attack of a carbanion on the carbonyl carbon of an ester. The carbanion is generated by removal of a slightly acidic hydrogen from the α-carbon atom of a ketone, nitrile, or ester using a relatively strong base. In the present reaction, the methyl ketone is deprotonated by the base, NaH. The resulting *enol anion* then attacks the ester, diethyl carbonate. The β-ketoester product thus formed is esterified *intramolecularly* to form the lactone product.

The condensation of 4-hydroxycoumarin with formaldehyde also involves the nucleophilic attack of an anion on a carbonyl carbon atom to produce dicoumarol. This substance is present in mouldy sweet clover. It is a blood anticoagulant and leads to the haemorrhagic sweet clover disease that kills cattle.

The mechanistic sequence for the reaction is given below.

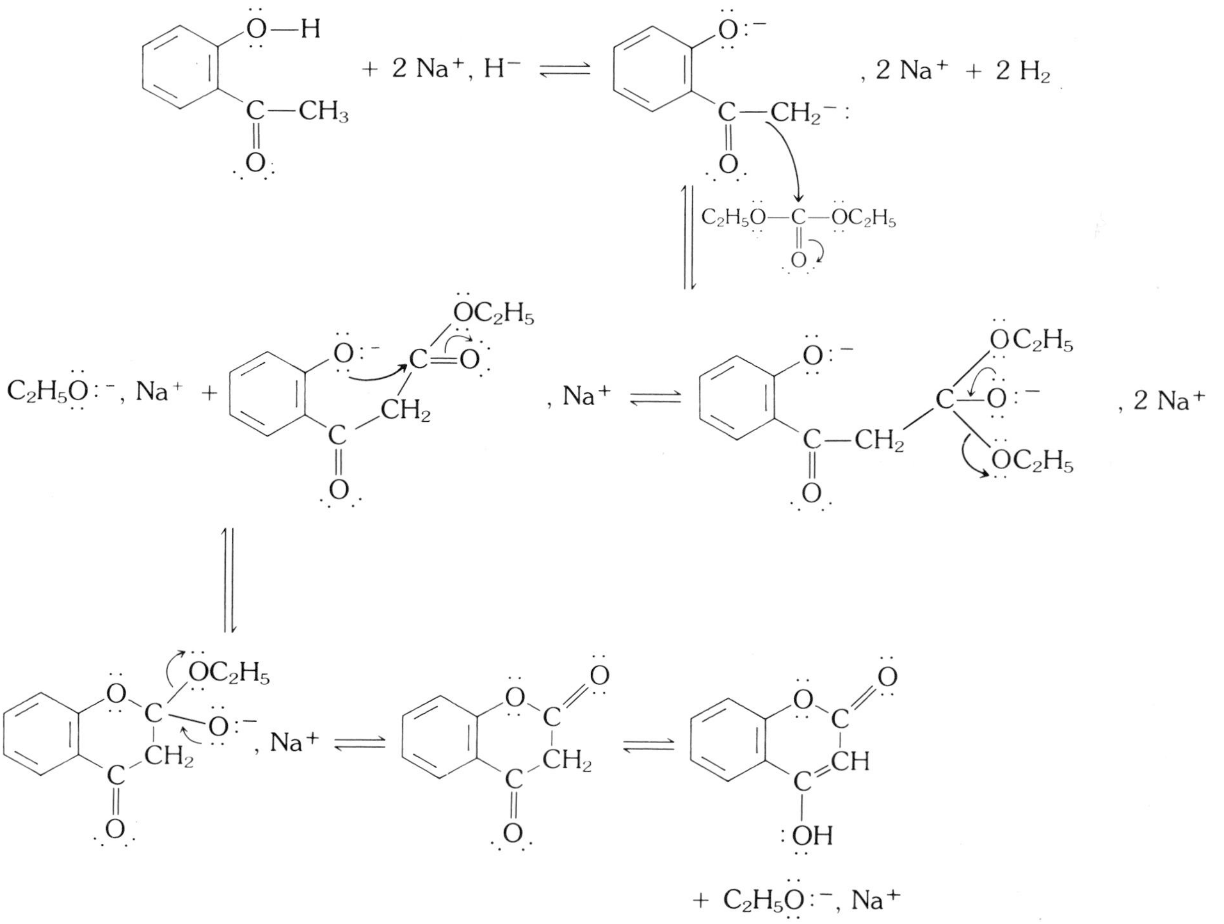

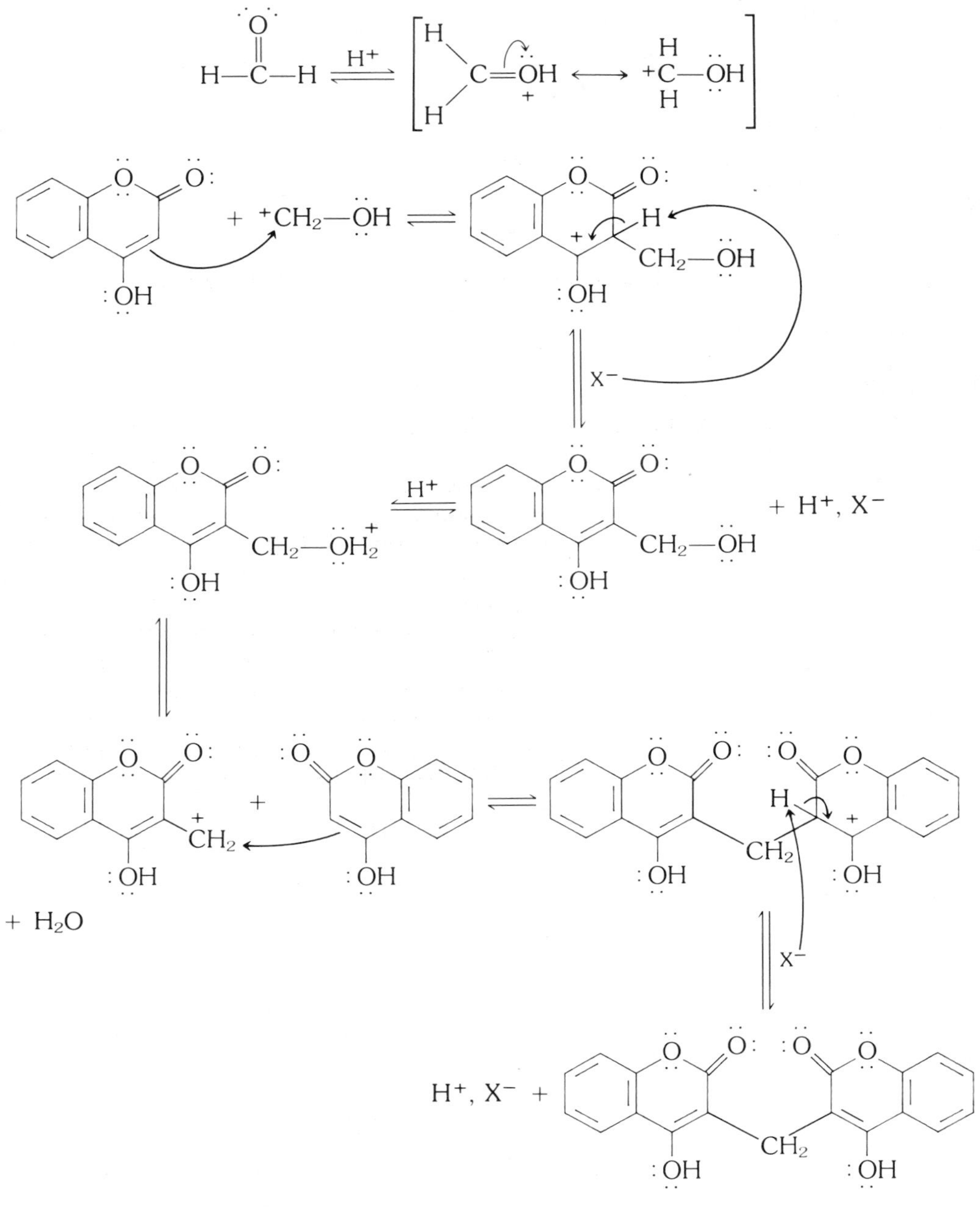

EXPERIMENTAL   PART A: Preparation of 4-Hydroxycoumarin

The reaction is shown above.

EXPERIMENTAL   Estimated time to complete the reaction: 4.0 hours.

Physical Properties of Reactants and Products

Compound	MW	Wt/Vol	mmol	mp(°C)	bp(°C)	Density	n_D
Sodium hydride (60% in oil dispersion)		85 mg	2.13				
Toluene	92.15	6.0 mL			111		
o-Hydroxyacetophenone	136.16	133 µL	1.1		218	1.13	1.5584
Diethyl carbonate	118.13	333 µL	2.75		126	0.96	1.3845
4-Hydroxycoumarin	162.15			213–214			

Reagents and Equipment

Important. *All equipment used in this reaction should be thoroughly dried in an oven at 110°C for half an hour just prior to use.*

In a 10-mL round-bottom flask containing a magnetic stirring bar and equipped with a Hickman condenser protected by a calcium chloride drying tube is placed 85 mg (2.13 mmol) of sodium hydride (60% in oil dispersion) and 3.0 mL of dry toluene. The Hickman still 14/10 ⊤ male joint should be wrapped with Teflon tape to prevent joint freeze-up.(■)

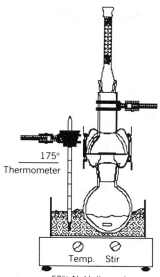

175°
Thermometer

60% NaH dispersion,
85 mg + $C_6H_5CH_3$, 6 mL +
o-$HOC_6H_4COCH_3$, 150 mg +
$(CH_3CH_2O)_2CO$, 324 mg

> **CAUTION:** *NaH is a flammable solid. Dispense in the* HOOD. *Toluene is distilled and stored over molecular sieves. This solvent is also dispensed in the* HOOD.

In rapid order place 133 µL (150 mg, 1.1 mmol) of o-hydroxyacetophenone, 3.0 mL of dry toluene, and 333 µL (324 mg, 2.75 mmol) of diethyl carbonate in a stoppered 10-mL Erlenmeyer flask.

Note. *The small volumes of the liquid reagents are dispensed using an automatic delivery pipet. Be sure to dry the removable plastic tips in the oven before use. A 10-mL graduated cylinder is used to measure the toluene.*

The drying tube is removed from the still head and the o-hydroxyaceto-phenone solution added with stirring to the reaction flask as rapidly as possible using a Pasteur pipet. The resulting solution turns yellow.

Reaction Conditions

The reassembled apparatus is placed in a sand bath and the temperature of the bath raised to about 175°C as rapidly as possible.

A distillate of ethanol and toluene (1.0 mL) is collected (2–3 fractions) in the collar of the condenser and the apparatus is then removed from the heat source. The reaction solution is allowed to cool and 3.0 mL of water added with stirring.

Isolation of Product

The resulting two-phase solution is transferred, using a Pasteur pipet, to a 12-mL centrifuge tube. The reaction flask is rinsed with an additional 2.0 mL of water and the rinse also added to the centrifuge tube. The toluene layer is separated using a Pasteur filter pipet and transferred to a second 12-mL test tube. This organic phase is then extracted with 3.0 mL of water, and the water extract is added to the original water phase. The combined aqueous layers are cooled in an ice bath and acidified by dropwise addition of concentrated HCl delivered from a Pasteur pipet. The solid product precipitates from the aqueous phase. The acid is added until the yellow color of the solution disappears (about 10 drops). The product is collected by vacuum filtration using a Hirsch funnel.(■)

Cyclized
product
collected
here

Acidified aqueous
phase, ~10 mL

Purification and Characterization

The crude 4-hydroxycoumarin is recrystallized from 50% ethanol using a Craig tube and dried on a porous clay plate.

Weigh the product and calculate the percentage yield. Determine the melting point and compare it with the literature value. Obtain the IR spectrum and compare it with that of an authentic sample.

PART B: 3,3'-Methylene Bis(4-hydroxycoumarin)

4-Hydroxycoumarin   Formaldehyde   Dicoumarol

EXPERIMENTAL   Estimated time to complete the reaction: 0.5 hour.

Physical Properties of Reactants and Products

Compound	MW	Wt/Vol	mmol	mp(°C)	bp(°C)
4-Hydroxycoumarin	162.15	50 mg	0.31	213–214	
Water		15 mL			100
37% Soln. formaldehyde		0.5 mL	6		
Dicoumarol	336.31			287–293	

Reagents and Equipment
In a 50-mL Erlenmeyer flask containing a boiling stone are placed 50 mg (0.31 mmol) of 4-hydroxycoumarin and 15 mL of water. The mixture is heated to boiling on a hot plate and to the resulting solution is added 0.5 mL (~200 mg, 6 mmol) of formaldehyde (37%).

> **WARNING:** *Formaldehyde is a cancer suspect agent. Dispense in the* HOOD.

Reaction Conditions
White crystals form immediately on addition of the formaldehyde solution. The flask is cooled in an ice bath.

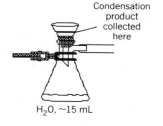

Condensation product collected here

H₂0, ~15 mL

Isolation of Product
The solid product is collected by vacuum filtration using a Hirsch funnel.(■)

Purification and Characterization
The material is recrystallized from a mixture of toluene-cyclohexanone (~2:1) using a Craig tube.

ENVIRONMENTAL DATA

Substance	Amount	TLV (mg/m³)	Emissions (mg)	Volume (m³)
Experiment 30A				
Toluene (SKIN) (Max., solv. distill.)	6 mL	375	5200	13.9
Ethanol (Max., Craig tube)	1.0 mL	1900	800	0.4
Diethyl carbonate	324 mg			
o-Hydroxyacetophenone	150 mg			

Substance	Amount	TLV (mg/m³)	Emissions (mg)	Volume (m³)
Experiment 30B				
Toluene (SKIN) (Max., Craig tube)	1.0 mL	375	870	2.3
Cyclohexanone (Max., Craig tube)	0.5 mL	100	500	5.0
Formaldehyde (Max.)	200 mg	1.5	200	133

TLV: Hydrochloric acid—5 ppm
TXDS: o-Hydroxyacetophenone—ipt-mus LD50: 100 mg/kg
Diethyl carbonate—suc-rat LD50: 8500 mg/kg
4-Hydroxycoumarin—ipr-mus LD50: 2000 mg/kg
Dicoumarol—orl-rat LD50: 710 mg/kg

QUESTIONS

6-77. In the first step of the above reaction, formation of 4-hydroxycoumarin, what is the purpose of using the Hickman still to remove the ethanol generated in the reaction?

6-78. The Dieckmann condensation is an intramolecular Claisen condensation. For example,

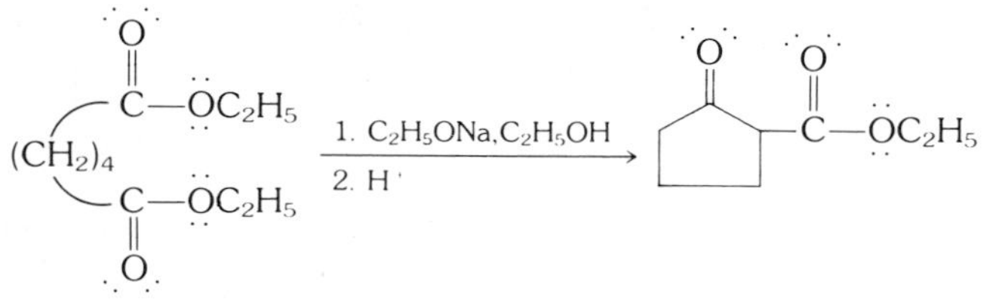

Suggest a suitable mechanism to account for the formation of the cyclic β-ketoester.

6-79. Predict the product formed in each of the following reactions:

   a. Acetophenone + diethyl carbonate $\xrightarrow[\text{2. H}^+]{\text{1. NaH,toluene}}$

   b. Acetophenone + ethyl formate $\xrightarrow[\text{2. H}^+]{\text{1. C}_2\text{H}_5\text{ONa,ethanol}}$

6-80. Suggest a synthesis for each of the following compounds utilizing the Claisen condensation. Any necessary organic or inorganic reagents may be used.

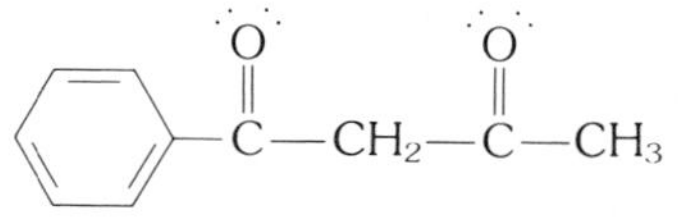

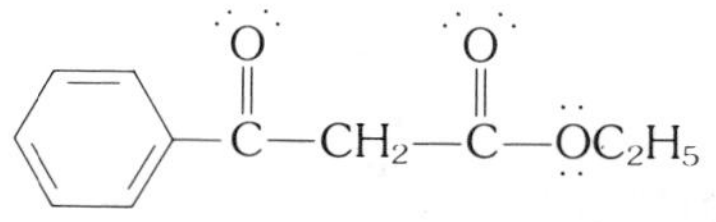

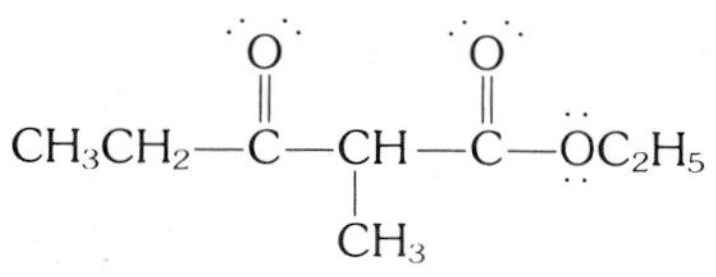

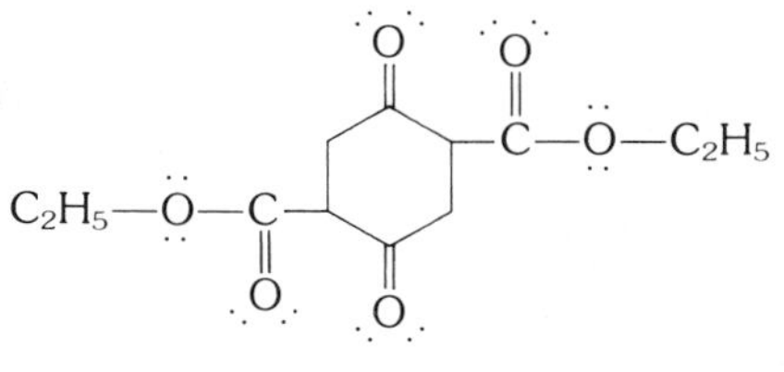

REFERENCES   1. A review of the Claisen condensation is given in "Organic Reactions": Hauser, C. R.; Swamer, F. W.; Adams, J. T. *Org. React.* **1954,** *8,* 59.
2. Selected references of Claisen condensations from "Organic Syntheses" include:
   a. Marvel, C. S.; Dreger, E. E. "Organic Syntheses"; Wiley: New York, 1941; Collect. Vol. I, p 238.
   b. Snyder, H. R.; Brooks, L. A.; Shapero, S. H. "Organic Syntheses"; Wiley: New York, 1943; Collect. Vol. II, p 531.
   c. Riegel, E. R.; Zwilgmeyer, F. *Ibid.,* p 126.
   d. Magnani, A.; McElvain, S. M. "Organic Syntheses"; Wiley: New York, 1955; Collect. Vol. III, p 251.
   e. Ainswoth, C. "Organic Syntheses"; Wiley: New York, 1963; Collect. Vol. IV, p 536.
   f. John, J. P.; Swaminathan, S.; Venkataramani, P. S. "Organic Syntheses"; Wiley: New York, 1973; Collect. Vol. V, p 747.

Experiment 31
Diazonium Coupling Reaction: Methyl Red

(azobenzene, 2-carboxy, 4'-*N*,*N*-dimethylamino-)

The generation of a diazonium salt and its use in the coupling reaction to produce an azo dye is demonstrated in this experiment.

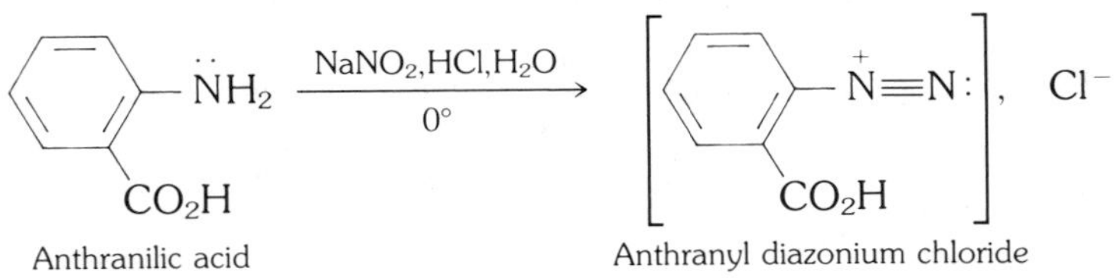

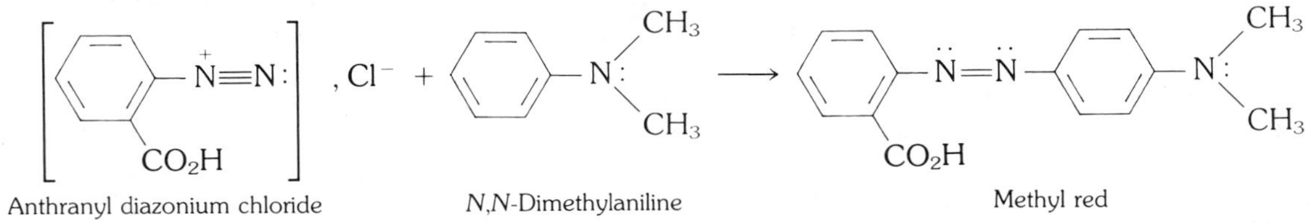

Anthranilic acid → Anthranyl diazonium chloride

Anthranyl diazonium chloride + *N*,*N*-Dimethylaniline → Methyl red

DISCUSSION   The coupling of a diazonium salt to a suitable aromatic substrate is an example of an electrophilic substitution reaction on an aromatic ring.

When aromatic primary amines are treated with nitrous acid (NaNO$_2$ + HCl), they are converted into diazonium salts. These salts are explosive when dry and therefore are generally not isolated.

Treatment of the diazonium salt with various aromatic molecules leads to the formation of azo derivatives by what is generally called a "coupling reaction." This is an important series of reactions since it has led to the synthesis of a wide variety of commercially important organic azo dyes.

The mechanistic sequence of the coupling reaction is given below.

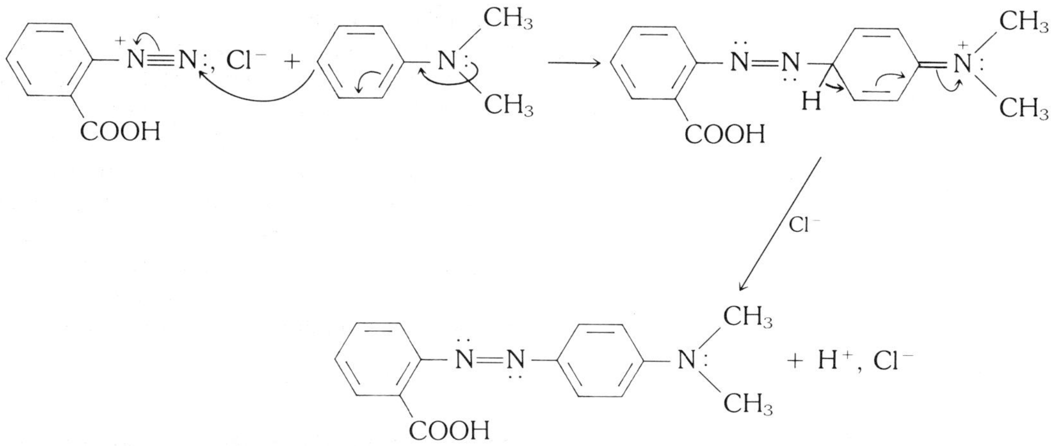

The reaction scheme at the top of the page shows the mechanism producing:

$$\text{N}=\text{N} - \text{C}_6\text{H}_4 - \text{N}(\text{CH}_3)_2 \quad + \text{ H}^+, \text{ Cl}^-$$

PART A: Anthranyl Diazonium Chloride

The reaction is shown above.

EXPERIMENTAL   Estimated time for the completion of the experiment: 1.0 hour.

Physical Properties of Reactants and Products

Compound	MW	Wt/Vol	mmol	mp(°C)
Anthranilic acid	137.14	65 mg	0.47	146–147
Conc. HCl		150 μL		
Water		600 μL		
Sodium nitrite	69.0	36 mg	0.52	271
Anthranyl diazonium chloride (not isolated)				

> **CAUTION:** *When dry, benzene diazonium 2-carboxylate detonates violently on being scraped or heated. It is strongly recommended that it be kept in solution at all times.*

Reagents and Equipment

Equip a 3.0-mL conical vial with a magnetic spin vane and an air condenser.(■) Add 65 mg (0.48 mmol) of anthranilic acid. A solution of 150 μL of concentrated HCl dissolved in 400 μL of water is then added to the vial, using a Pasteur pipet.

Note. *When preparing the acid solution, the acid is added to the water. These reagents are dispensed using automatic delivery pipets.*

If necessary, warm the mixture, with stirring, on a hot plate magnetic stirrer to obtain a homogeneous solution. The solution is then cooled in an ice bath, with stirring, for a period of 10 minutes.

In a 10 × 75 mm test tube or a small vial, prepare a solution of 36 mg (0.52 mmol) of sodium nitrite dissolved in 200 μL of water. This solution is also cooled in an ice bath.

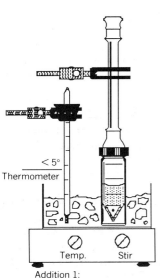

< 5°
Thermometer

Addition 1:
  Anthranilic acid, 65 mg
  + conc. HCl, 150 μL +
  H$_2$O, 600 μL +
  NaNO$_2$, 36 mg
Addition 2:
  C$_6$H$_5$N(CH$_3$)$_2$, 85 mg
Addition 3:
  CH$_3$CO$_2$Na, 68 mg +
  H$_2$O, 200 μL

Reaction Conditions

When both solutions in the ice bath are cooled to a temperature below 5°C, slowly add (dropwise) the nitrite solution to the stirred anthranilic acid solution while maintaining the temperature below 5°C. This transfer is accomplished using a Pasteur pipet. *The solution must be kept cool so that the diazonium salt will not hydrolyze to the corresponding phenol.*

After a period of 4–5 minutes, the clear solution of anthranyl diazonium chloride is checked for the presence of excess nitrous acid using potassium iodide-starch paper. If an excess is present, the test paper gives an immediate blue color. If no color is obtained, additional nitrite solution should be prepared and added until a positive test is observed.

PART B: The Coupling Reaction to Yield Methyl Red

The reaction is shown above.

EXPERIMENTAL   Estimated time for the completion of the experiment: 2.0 hours.

Physical Properties of Reactants and Products

Compound	MW	Wt/Vol	mmol	mp(°C)	bp(°C)	Density	n_D
Anthranyl diazonium chloride solution prepared in Part A							
N,N-Dimethylaniline	121.18	89 μL	0.7		194	0.96	1.5582
Sodium acetate	82.03	68 mg	.83	324			
10% NaOH solution		100 μL					
Methyl red	269.24			183			

Reaction Conditions

To the solution of anthranyl diazonium chloride prepared above, add fairly rapidly 89 μL (85 mg, 0.7 mmol) of N,N-dimethylaniline (automatic delivery pipet).

> **WARNING:** *This aniline compound is toxic and should be dispensed in the* HOOD. *The air condenser is removed when the addition is made; it is then reattached to the vial.*

The solution is stirred for an additional 15 minutes, keeping the temperature below 5°C.

A solution of 68 mg (0.83 mmol) of sodium acetate dissolved in 200 μL of water is prepared in a 10 × 75 mm test tube and transferred (Pasteur pipet) to the reaction mixture. This addition may be made without removing the air condenser. The resulting solution is maintained at 5°C, with stirring, for an additional 20 minutes.

The reaction vial is removed from the ice bath and allowed to warm to ambient temperature over a period of 15 minutes.

To the solution is added 100 μL of 10% aqueous NaOH solution (automatic delivery pipet). The reaction mixture is allowed to stand at room temperature for about 30 minutes. *The formation of the azo compound is a very slow reaction, but the rate is increased by raising the pH of the solution.*

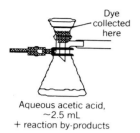

Dye
collected
here

Aqueous acetic acid,
~2.5 mL
+ reaction by-products

Isolation of Product

The precipitate of crude methyl red dye is collected by vacuum filtration using a Hirsch funnel.(■) The reaction flask is rinsed with 0.5 mL of water and this rinse used to wash the crystals. The crystals are then washed with 0.5 mL of dilute aqueous acetic acid, to remove unreacted N,N-dimethylaniline from the product, followed by 0.5 mL of water. This last wash is usually pale pink in color.

The small amounts of water and acetic acid are dispensed using a calibrated Pasteur pipet.

Purification and Characterization

The crude product is dissolved in 500 μL of methanol. If necessary, the mixture is warmed in a beaker of hot water to aid in the dissolution. The solution is cooled in an ice bath and the resulting crystals of methyl red are collected by vacuum filtration using a Hirsch funnel. They are then allowed to dry on a porous clay plate.

Weigh the product and calculate the percentage yield. Determine the melting point and compare it to the literature value. If further purification is desired, the material may be recrystallized from toluene using a Craig tube. Obtain an IR spectrum and compare it with that of an authentic sample.

ENVIRONMENTAL DATA

Substance	Amount	TLV (mg/m^3)	Emissions (mg)	Volume (m^3)
N,N-Dimethylaniline (SKIN) (transfer)	85 mg	25	2	<0.1
Methanol (SKIN) (Max., vacuum filtration)	0.5 mL	260	400	1.5

TLV: Hydrochloric acid—5 ppm
　　　Sodium hydroxide—2 mg/m^3
TXDS: Anthranilic acid—orl-rat LD50: 4549 mg/kg
　　　Methyl red—orl-rat TDLo: 12 g/kg/57W-C
　　　Sodium nitrite—orl-hum TDLo: 85 mg/kg/56W-C
　　　Sodium acetate—orl-rat LD50: 3530 mg/kg

QUESTIONS

6-81. In the experiment, a point is made that the formation of the azo compound is a slow reaction, but that the rate is increased by raising the pH of the solution. Why is this necessary? In other words, how does the acidity of the solution affect the reactivity of the N,N-dimethylaniline reagent?

6-82. In relation to question 6-81, diazonium salts couple with phenols in slightly alkaline solution. What effect does the alkalinity of the solution have on the reactivity of the phenol?

6-83. Starting with the appropriate aromatic amine and using any other organic or inorganic reagent, outline a synthetic sequence for the preparation of the azo dyes shown below (a,b,c).

a.

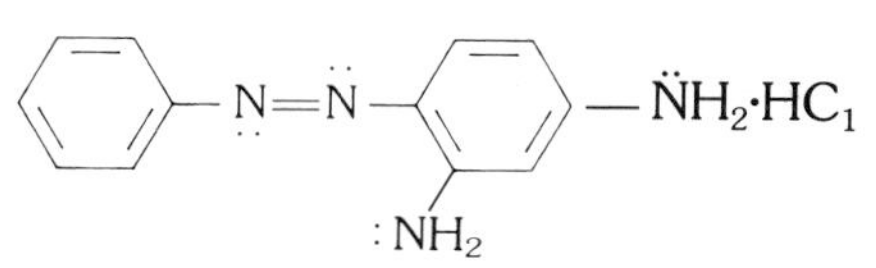

(Chrysoidine)

b.

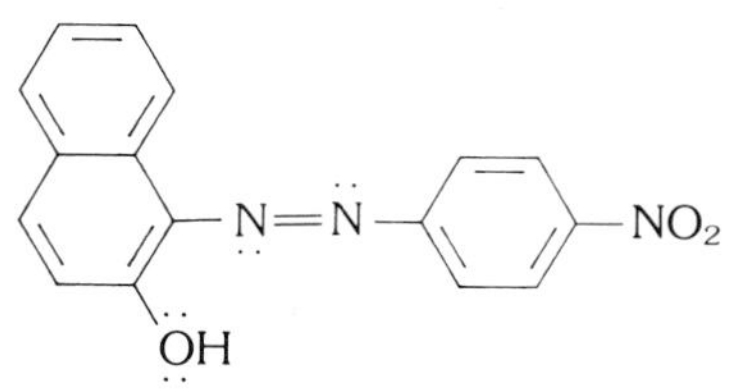

(Para red)

c.

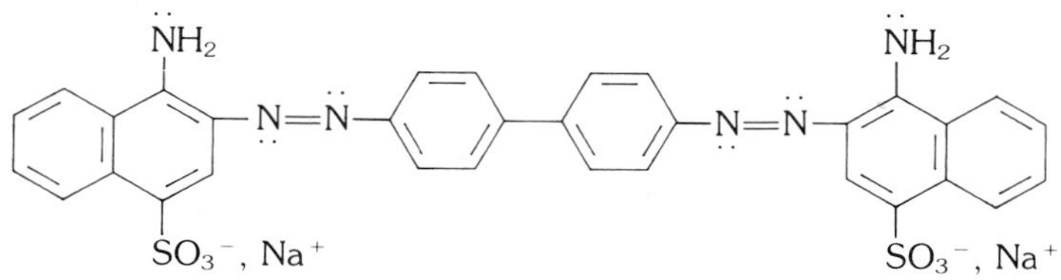

(Congo red)

6-84. What is the main feature of the azo dyes that cause them to be colored compounds?

6-85. Methyl orange is an acid–base indicator. In dilute solution at pH > 4.4, it is yellow.

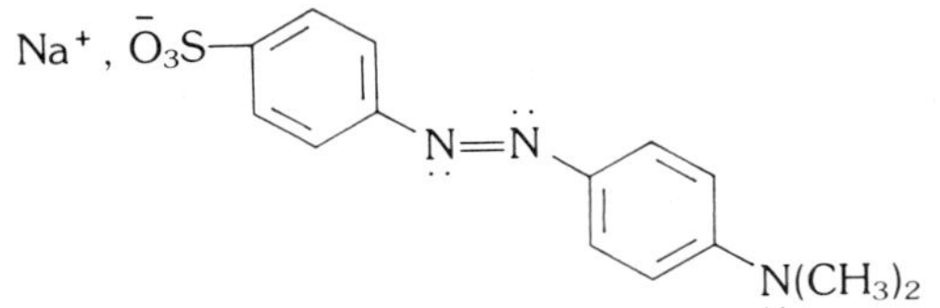

At a pH of 3.2 the solution appears red. Draw a structure of the species that is formed at the lower pH if the acid proton adds to the azo nitrogen atom adjacent to the aromatic ring containing the SO_3^- unit. Why does the proton add to this particular N atom when two other nitrogen atoms are available in the molecule?

REFERENCES   1. Selected coupling reactions with diazonium salts from "Organic Syntheses:"
   a. Fieser, L. F. "Organic Syntheses"; Wiley: New York, 1943; Collect. Vol. II, p 35.
   b. *Ibid.*, p 39.
   c. Hartwell, J. L.; Fieser, L. F. *Ibid.*, p 145.
   d. Conant, J. B.; Lutz, R. E.; Corson, B. B. "Organic Syntheses"; Wiley: New York, 1941; Collect. Vol. I, p 49.
   e. Santurri, P.; Robbins, F.; Stubbins, R. "Organic Syntheses"; Wiley: New York, 1973; Collect. Vol. V, p 341.
2. The synthesis of methyl red is also given in "Organic Syntheses:"
   Clarke, H. T.; Kirrer, W. R. "Organic Syntheses"; Wiley: New York, 1941; Collect. Vol. I, p 374.
3. The present experiment is an adaptation of that given in Vogel, A. I. "A Textbook of Practical Organic Chemistry", 4th ed.; Longman: London, 1978; p 716.

Experiments 32A and 32B
Friedel-Crafts Acetylation: Mono- and Diacetylferrocene

(iron, bis[η^5-cyclopentadienyl]-1-acetyl;
iron, bis[η^5-cyclopentadienyl]-1,1′-diacetyl)

This reaction illustrates the multiple-electrophilic acyl cation attack on an aromatic nucleus. The reaction can be carried out under a variety of conditions that lead to different product ratios. The highly colored acyl derivatives are easily separated and purified by both thin-layer and dry-column chromatography.

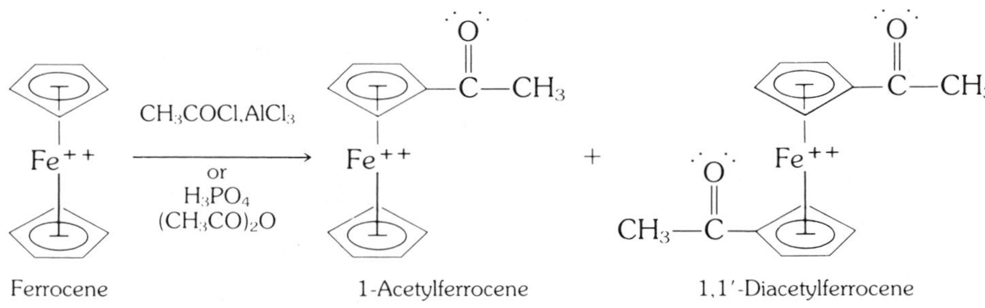

Ferrocene   1-Acetylferrocene   1,1′-Diacetylferrocene

DISCUSSION The generation of the appropriate carbocation, carbocation complex, or acylium ion in the presence of a fully delocalized ring system can lead to alkylation or acylation of an aromatic nucleus. This set of reactions, discovered by Charles Friedel and James Crafts in 1877, originally made exclusive use of aluminum chloride as the Lewis acid catalyst. The reaction now can be initiated by a range of Lewis acids other than aluminum chloride. In the present experiment 85% phosphoric acid is employed in Part A, while a more reactive Lewis acid, aluminum chloride, is utilized in Part B. Thus, the difference in reactivity among Lewis acid reagents is clearly observed by the variation in product ratios. In addition, product analysis by thin-layer chromatography demonstrates the great practical value of aliquot collection followed by subsequent TLC analysis.

The mechanism involves three steps: (1) ionization or partial ionization with complex formation of an alkyl halide or acyl halide; (2) attack by the formed carbocation (or complex) or acylium ion on the substrate delocalized ring; (3) loss of a proton by the arenium ion to return to a fully delocalized ring system. Ring deactivation on substitution can result in single product formation during acylation or, as in the case of ferrocene, in heteroannular substitution.

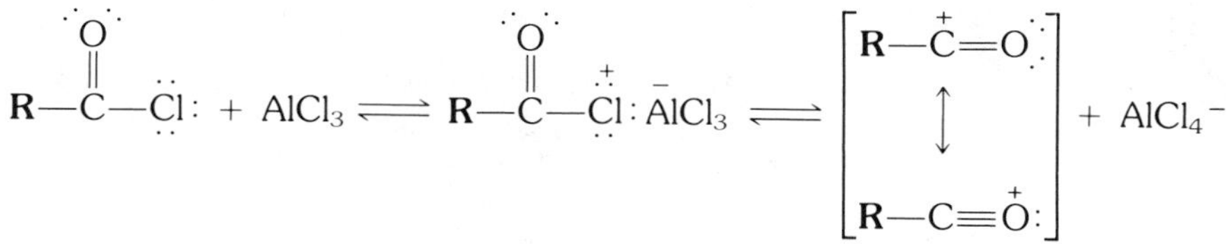

Acylium ion

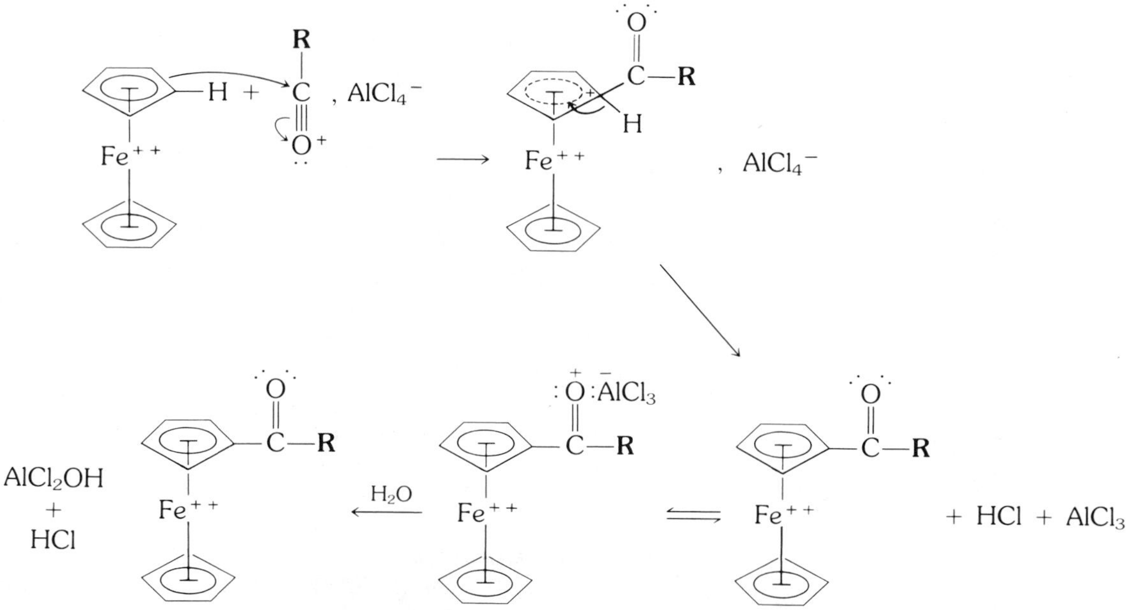

EXPERIMENTAL   Estimated time to complete experiment Parts A and B:
Week 1: 4 hours;
Week 2: 3.5 hours.

PART A: Acylation of Ferrocene:
Week 1 (Phosphoric Acid Catalyst)

Physical Properties of Reactants and Products

Compound	MW	Wt/Vol	mmol	mp(°C)	bp(°C)	Density	n_D
Ferrocene	186.04	100 mg	0.54	173			
Acetic anhydride	102.09	2.0 mL	21.2		140	1.08	1.3901
Phosphoric acid, 85%		0.3 mL					
Monoacetylferrocene	228.08			85–86			

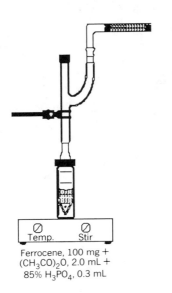

Ferrocene, 100 mg +
$(CH_3CO)_2O$, 2.0 mL +
85% H_3PO_4, 0.3 mL

HOOD

Reagents and Equipment
Place 100 mg (0.54 mmol) of ferrocene in a 5.0-mL conical vial containing a spin vane and equipped with a Claisen head protected by a calcium chloride drying tube.(■)

Note. *Prior to the addition of the ferrocene, the reaction system is flushed with a stream of dry nitrogen gas for several minutes while the glassware is gently warmed with a microburner.*

Using a calibrated Pasteur pipet, add 2.0 mL of reagent-grade acetic anhydride [HOOD] by removing the screwcap and immediately replacing it when the addition is complete.

Important. *The acetic anhydride must be pure and uncontaminated with hydrolysis products. Either freshly opened or distilled material should be employed.*

Dissolution occurs on gentle warming with stirring. After cooling the reaction mixture to room temperature, add 0.3 mL of 85% phosphoric acid dropwise through the Claisen head as above. The acid is delivered from a 1.0-mL graduated pipet. The mixture is stirred during this procedure and the time of the final addition is noted. Several drops of this mixture are removed by Pasteur pipet and placed in a capped vial. This is to be used for TLC analysis.

Instructions
The technique for obtaining a TLC aliquot is to remove a small amount of the reaction mixture by touching the open end of a Pasteur pipet to the surface of the solution. This is done by removing the cap from the Claisen head and inserting the pipet down the straight neck to the surface of the solution. Dissolve this aliquot in 10 drops of *cold* methylene chloride in a small capped vial.

Reaction Conditions
The unstirred reaction solution is allowed to stand for a period of 1.5 hours at room temperature.

Isolation of Product
The reaction is quenched by transfer of the solution by Pasteur pipet to a 12-mL capped centrifuge tube (cooled in an ice bath) containing 1.0 mL of ice water. The resulting mixture is neutralized by the dropwise addition (Pasteur pipet) of 3–4 mL of 25% aqueous sodium hydroxide solution.

Important. *Avoid an excess of base. Litmus or pH paper is used to confirm neutrality.*

The mixture is extracted with three 3-mL portions of methylene chloride. The capped tube is shaken, vented, and the layers allowed to separate. The bottom methylene chloride layer is then removed using a Pasteur filter pipet.

The methylene chloride extracts are combined in a 25-mL Erlenmeyer flask and dried over ~100 mg of granular anhydrous sodium sulfate for 20 minutes. The dried solution is transferred to a tared 10-mL Erlenmeyer flask using a **HOOD** Pasteur filter pipet in aliquots of 4 mL. This solution is concentrated [**HOOD**] under a stream of nitrogen gas after each transfer to a volume of approximately 0.5 mL. The drying agent is rinsed with an additional 1.0 mL of methylene chloride and the rinse combined with the concentrate. Several drops of this solution are removed by Pasteur pipet and placed in a capped vial containing 10 drops of *cold* methylene chloride. This is to be used for TLC analysis.

The remaining solvent is then removed. The crude yield of product is approximately 110 mg.

Option. *As an option, the combined extracts and wash may be left to evaporate* **HOOD** *[**HOOD**] in a 25-mL Erlenmeyer flask with the mouth covered by filter paper until the following week.*

Purification and Characterization
See Parts A and B, Week 2.

PART B: Acylation of Ferrocene: Week 1 (Aluminum Chloride Catalyst)

EXPERIMENTAL

Physical Properties of Reactants and Products

Compound	MW	Wt/Vol	mmol	mp(°C)	bp(°C)	Density	n_D
Aluminum chloride	133.34	150 mg	1.12	190			
Methylene chloride		4.0 mL			40		
Acetyl chloride	78.50	80 μL	1.12		51	1.11	1.3898
Ferrocene	186.04	100 mg	0.54	173			
Diacetylferrocene	270.11			130–131			

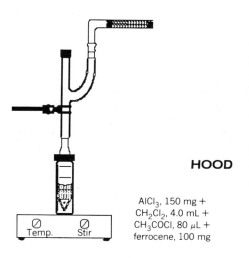

AlCl₃, 150 mg +
CH₂Cl₂, 4.0 mL +
CH₃COCl, 80 μL +
ferrocene, 100 mg

Temp. Stir

Reagents and Equipment
Equip a tared 5.0-mL conical vial with a magnetic spin vane and a Claisen head protected by a calcium chloride drying tube. Add 150 mg (1.12 mmol) of fresh, anhydrous aluminum chloride.(■)

Note. *Prior to the addition of the AlCl₃, the glassware is dried in an oven at 110°C for one-half hour.*

HOOD Using a calibrated Pasteur pipet, add 2.5 mL of methylene chloride to the reaction vial. With swirling, add 80 μL (1.12 mmol) of acetyl chloride from an automatic delivery pipet [**HOOD**]. To the resulting mixture is added a solution of 100 mg (0.54 mmol) of ferrocene dissolved in 1.5 mL of methylene chloride.

> **WARNING:** *It is important to minimize the exposure to moist air during these transfers. Both the aluminum chloride and the acetyl chloride are highly moisture sensitive so that rapid yet accurate manipulations are necessary to minimize deactivation of these reagents and hence variable results. In addition, both chemicals are irritants. Avoid breathing the vapors or allowing them to come in contact with skin. These reagents are dispensed in the* HOOD.

After the addition of the ferrocene solution, several drops of the resulting solution are removed by Pasteur pipet and placed in a capped vial. This is to be used for TLC analysis.

Instructions

The technique for obtaining a TLC aliquot is to remove a small amount of the reaction mixture by touching the open end of a Pasteur pipet to the surface of the solution. This is done by removing the cap from the straight neck of the Claisen head and inserting the pipet down the neck so as to touch the surface of the solution. Dissolve this aliquot in 10 drops of *cold* methylene chloride in a small capped vial.

Reaction Conditions

Following the addition of the ferrocene solution note the time, and initiate stirring. The reaction is allowed to proceed at room temperature for a period of fifteen minutes.

Isolation of Product

The reaction is quenched by transferring the mixture by Pasteur pipet to a 12-mL capped centrifuge tube containing 5.0 mL of ice water. The tube is cooled in an ice bath. The resulting solution is neutralized by the dropwise addition (calibrated Pasteur pipet) of ~0.5 mL of 25% aqueous sodium hydroxide solution.

Important. *Avoid an excess of base. Litmus or pH paper is used to confirm neutrality.*

The mixture is now extracted with three 3-mL portions of methylene chloride. The capped tube is shaken and vented, and the layers are allowed to separate. The bottom methylene chloride layer is then removed using a Pasteur filter pipet.

The methylene chloride extracts are combined in a 25-mL Erlenmeyer flask and dried over ~200 mg of granular anhydrous sodium sulfate for 20 minutes. The dried solution is transferred to a tared 10-mL Erlenmeyer flask using a Pasteur filter pipet in aliquots of 4 mL. After each transfer, the solution is con-

HOOD centrated [HOOD] under a stream of dry nitrogen gas to a volume of about 0.5 mL. The drying agent is rinsed with an additional 2.0 mL of methylene chloride and the rinse combined with the concentrate. Several drops of this solution are removed by Pasteur pipet and placed in a capped vial containing 10 drops of *cold* methylene chloride. This is to be used for TLC analysis.

HOOD The remaining solvent is then removed by warming in a sand bath [**HOOD**]. The crude yield of product is ~130 mg.

Option. *As an option, the combined extracts and wash may be left to evaporate* [HOOD] *in a 25-mL Erlenmeyer flask with the mouth covered by filter paper until the following week.*

Purification and Characterization

See Parts A and B, Week 2.

PARTS A and B: Thin-Layer Chromatography of Acylation Reactions: Week 1

The two samples saved above from both reaction mixtures immediately following addition of the acid catalysts and the two samples collected after the specified times of reaction (Part A, phosphoric acid catalyst; Part B, aluminum chloride catalyst) are analyzed by TLC. A standard mixture of the ferrocenes (supplied by the Instructor) is also analyzed. The developed TLC plates are used as a guide to determine the product mixtures obtained in the reactions and as an aid in determining the elution solvents required for separation of these mixtures by dry-column chromatography. (See section on Acylation of Ferrocene: Purification and Characterization, Parts A and B, Week 2.)

Information. *Thin-layer chromatography (TLC) in this experiment is carried out with Eastman Kodak silica gel-polyethylene terephthalate plates (#13179). The plates are activated at an oven temperature of 100°C for 30 minutes. They are then placed in a desiccator for cooling and stored until used. Development of the ferrocene derivatives is carried out with pure methylene chloride solvent. Visualization of unreacted ferrocene can be enhanced with iodine vapor. Each plate is ~1 × 4 in. Mark the plate lightly with a pencil 1.5 cm from the bottom. Using fine capillaries drawn from melting-point tubes, samples are applied to the plates by spotting the marked origin one to three times. The developing chamber is a screwcap jar or beaker (watch glass cover) in which 5 mL of methylene chloride gives a solvent depth of ~0.5 cm on the plate. See the Technique Section for the method of TLC analysis and determination of R_f values.*

PARTS A and B: Acylation of Ferrocene: Week 2

Purification and Characterization

The reaction products formed in Parts A and B are purified by dry-column chromatography. *See the Technique Section for a discussion of dry-column chromatography.*

The material obtained in Part A is dissolved in 0.5 mL (calibrated Pasteur pipet) of methylene chloride in a small vial. This solution is mixed with 300 mg of alumina (activity III) in a tared vial and the slurry evaporated under a stream of dry nitrogen [**HOOD**] to give a product/alumina mixture A. A chromatographic column is assembled in ascending order: prewashed cotton plug, 5 mm of sand, 60–80 mm of alumina (~5.0 g, activity III), *one half* of product/alumina mixture (A), and 10 mm of alumina.(■)

HOOD

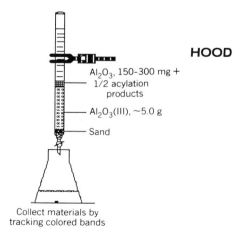

Al$_2$O$_3$, 150-300 mg + 1/2 acylation products

Al$_2$O$_3$(III), ~5.0 g

Sand

Collect materials by tracking colored bands

Information. *This procedure prevents overloading of the chromatographic column during the separation of the reaction products. If the yield of crude reaction products exceeds 75 mg (the usual case), half of the alumina admixture is introduced to the column. If the crude products, however, are obtained in quantities less than 75 mg, the entire admixture is added to the top of the column.*

If only half of the alumina-crude ferrocene acylation product mixture is on the column, it is important to reweigh the tared vial in order to establish a reasonably accurate estimate of overall yields obtained in the reaction.

Elution is commenced with pure hexane if TLC analysis indicated that unreacted ferrocene is present in the product mixture. The material will separate into two or three colored bands on the column during elution. Once the ferrocene band has been collected, the elution is continued with 1:1 CH_2Cl_2/hexane and then 9:1 to obtain the monosubstituted product. Further elution with 9:1 CH_2Cl_2/CH_3OH yields the disubstituted material. Collect and save each band

HOOD   separately. Remove the solvent [HOOD] under a stream of dry nitrogen gas. The volume of an eluted fraction should be in the range of 2–5 mL if the band is carefully tracked down the column. Solvent that elutes without color can be discarded into the solvent waste can. During concentration of the solvent, spot each fraction on a TLC plate to verify the separation and purity.

This procedure is repeated with the material isolated from Part B above.

Determine the melting point of the products and compare the results to the literature value. Obtain an IR spectrum of each material and compare the results with an authentic sample. Interpretation of the spectra allow an unambiguous determination of substitution based on the presence or absence of absorption in the 1100–900 cm^{-1} region of the spectrum.

Characterization of the fraction (total sample) isolated from Part A

Fraction 1, Ferrocene: mp _____ °C; _____ mg, _____ mmol.

Fraction 2, Monoacetylferrocene: mp _____ °C; _____ mg, _____ mmol.

    Total: _____ mmol, _____ % yield

Characterization of the fractions (total sample) isolated from Part B

Fraction 1, Monoacetylferrocene: mp _____ °C; _____ mg, _____ mmol.

Fraction 2, Diacetylferrocene: mp _____ °C; _____ mg, _____ mmol.

    Total: _____ mmol, _____ % yield

ENVIRONMENTAL DATA

Substance	Amount	TLV (mg/m³)	Emissions (mg)	Volume (m³)
Experiment 32A				
Acetic anhydride (reactant, solv., reflux, transfer)	2.0 mL	20	12	0.6
Methylene chloride (Max., evap.)	10 mL	350	13,350	38
Methylene chloride (Max., col. chromatog., evap.)	5.5 mL	350	7,340	21
Hexane (Max., col. chromatog., evap.)	5.0 mL	180	3,295	18
Methanol (Max., col. chromatog., evap.)	1.0 mL	260	790	3.0
Experiment 32B				
Acetyl chloride[a] (Max., reactant)	80 μL	10.3	88	8.6
Methylene chloride (Max., solv., evap.)	15 mL	350	20,025	57
Methylene chloride (Max., col. chromatog.)	5.5 mL	350	7,340	21
Hexane (Max., col. chromatog.)	5.0 mL	180	3,295	18
Methanol (Max., col. chromatog.)	1.0 mL	260	790	3.0

[a]Treated as HCl and acetic acid.

TLV: Ferrocene—10 mg/m³
      Sodium hydroxide—2 μg/m³
      Phosphoric acid—1 mg/m³

TXDS: Aluminum chloride—orl-rat LD50: 3700 mg/kg
       Sodium sulfate—orl-mus LD50: 5989 mg/kg
       Acetylferrocene—inv-mus LD50: 75 mg/kg

QUESTIONS

6-86. In the formation of diacetylferrocene, the product is always the one in which each ring is monoacetylated. Why does the second acetyl group become linked to the unsubstituted ring and not the ring already containing one acetyl unit?

6-87. Ferrocene cannot be nitrated using the conventional HNO_3-H_2SO_4 mixed acid conditions even though nitration is an electrophilic aromatic substitution reaction. Explain.

6-88. In contrast to nitration, ferrocene undergoes the acetylation and sulfonation reaction. Explain.

6-89. The bonding in ferrocene involves sharing of the six electrons from each cyclopentadienyl ring with the iron atom. Based on the electronic configuration of the iron species in the compound, show that a favorable 18-electron rare gas configuration is established.

6-90. In a manner similar to that in question 6-89, would you predict that ruthenocene and osmocene would be stable compounds containing two cyclopentadienyl rings? Explain.

6-91. Would you predict that bis(benzene)chromium (0) would be a stable compound? Give evidence to support your answer.

6.91a. Predict the major product(s) in each of the following Friedel-Crafts reactions. Name each product. If the reaction does not occur, offer a reasonable explanation for that fact.

a.

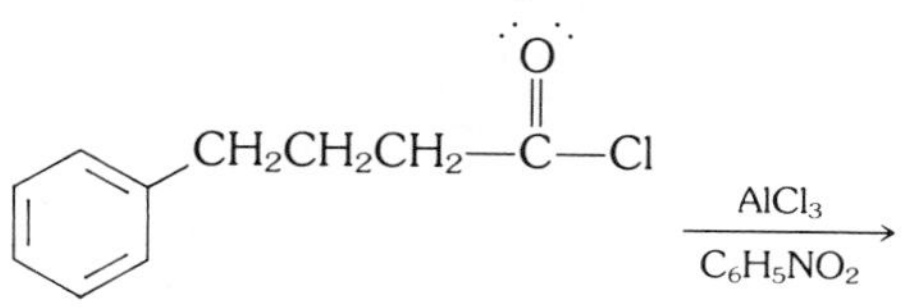

$$\text{C}_6\text{H}_5\text{—CH}_2\text{CH}_2\text{CH}_2\text{—C(=O)—Cl} \xrightarrow[\text{C}_6\text{H}_5\text{NO}_2]{\text{AlCl}_3}$$

b.

$$\text{C}_6\text{H}_6 + \text{C}_6\text{H}_5\text{Cl} \xrightarrow[\text{C}_6\text{H}_5\text{NO}_2]{\text{AlCl}_3}$$

c.

$$\text{C}_6\text{H}_5\text{—CH(CH}_3)_2 + \text{H—C(=O)—F} \xrightarrow[\text{CS}_2]{\text{BF}_3}$$

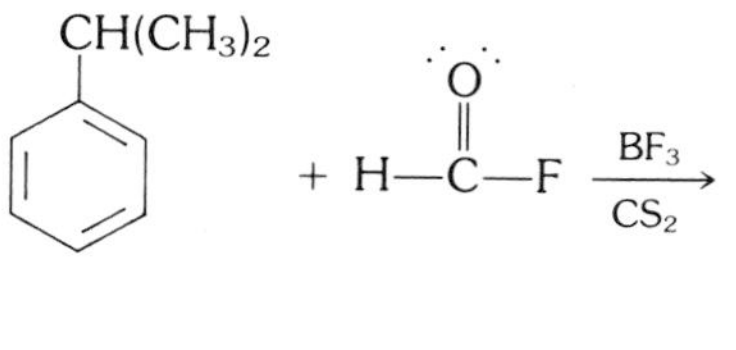

d.

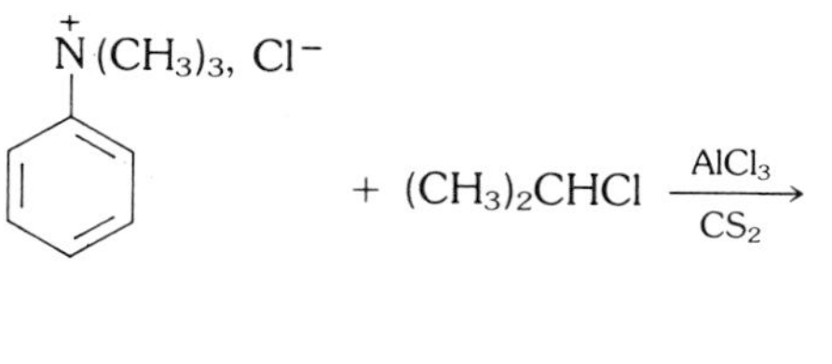

$$\text{C}_6\text{H}_5\text{—N}^+(\text{CH}_3)_3,\ \text{Cl}^- + (\text{CH}_3)_2\text{CHCl} \xrightarrow[\text{CS}_2]{\text{AlCl}_3}$$

e.

$$\text{C}_6\text{H}_5\text{—CH(CH}_3)_2 + \text{CH}_3\text{—C(=O)—Cl} \xrightarrow[\text{CS}_2]{\text{AlCl}_3}$$

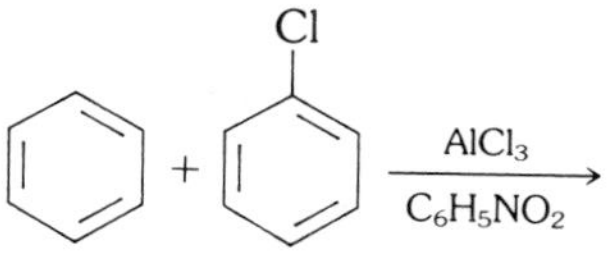

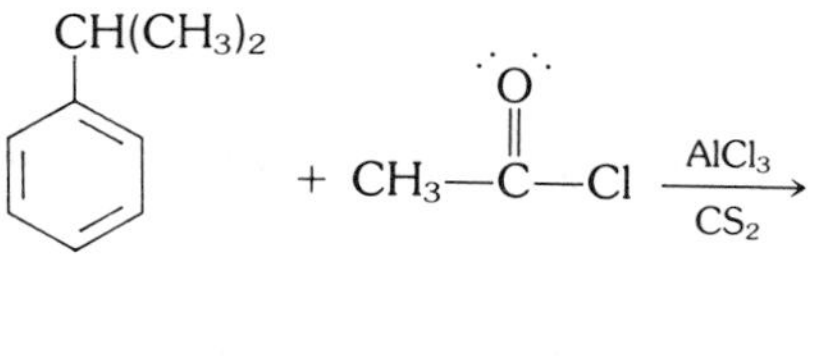

REFERENCES 1. The acetylation of ferrocene has been monitored using chromatographic techniques. Several references are listed:
    a. Bozak, R. E. *J. Chem. Educ.* **1966**, *43*, 73.
    b. Herz, J. E. *Ibid.*, 599.
    c. Bohen, J. M.; Joullie, M. M.; Kaplan, F. A. *J. Chem. Educ.* **1973**, *50*, 367.

2. Several reviews on the Friedel-Crafts reaction, selected from the many sited in the literature are listed:
    a. Gore, P. H. *Chem. Revs.* **1955**, *55*, 229.
    b. Johnson, W. S. *Org. React.* **1944**, *2*, 114.
    c. Berliner, E. *Org. React.* **1949**, *5*, 229.
    d. Olah, G.A., Ed. "Friedel-Crafts and Related Reaction"; Interscience: New York, 1963 to 1965; Vol. I–IV.

3. From the large number of Friedel-Crafts acetylation reactions given in "Organic Syntheses" several are listed:
    a. Adams, R.; Noller, C. R. "Organic Syntheses"; Wiley: New York, 1941; Collect. Vol. I, p 109.
    b. Marvel, C. S.; Sperry, W. M. *Ibid.*, p 95.
    c. Fieser, L. F. *Ibid.*, p 517.
    d. Fieser, L. F. "Organic Syntheses"; Wiley: New York, 1955; Collect. Vol. III, p 6.
    e. Lutz, R. E. *Ibid.*, p 248.

Experiment 33
Halogenation: 4-Bromoacetanilide

(acetamide, *N*-(4-bromophenyl)-)

In this experiment an electrophilic aromatic substitution is described. This type of reaction constitutes a very common route by which many aromatic compounds are prepared.

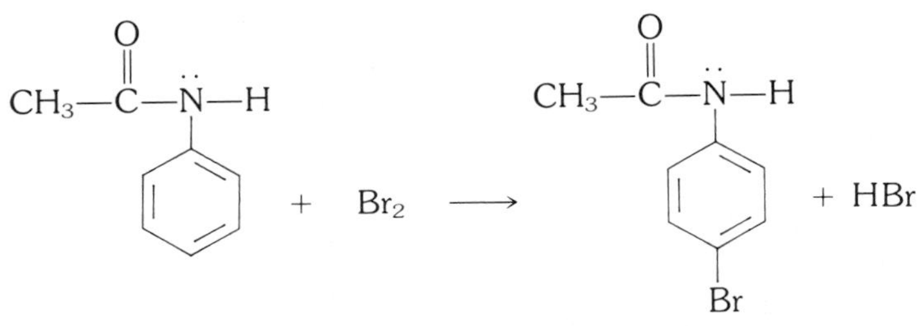

    Acetanilide   Bromine   4-Bromoacetanilide

DISCUSSION Aromatic compounds may be brominated by treatment with bromine in the presence of a Lewis acid catalyst such as ferric chloride. For very active substrates, such as amines, the reaction may proceed in the absence of a catalyst. In many cases, with amines or phenols, it is impossible to stop the bromination and all possible ortho and para positions are substituted. For this reason, primary aromatic amines are often converted to the corresponding acetanilide derivative, if the monosubstituted compound is desired. This effect is demonstrated in the present experiment. Furthermore, acetylation of the primary amine, aniline, effectively blocks the ortho positions due to steric hindrance. Electrophilic substitution by bromine is thus directed to the para position on the ring. The —NHCOCH$_3$ group is a less powerful *o,p* directing group than —NH$_2$ due to the presence of the carbonyl. Thus, substitution of the amino group renders the ring less nucleophilic. For these reasons only monosubstitution is observed.

The mechanism of the reaction is an illustration of the classic electrophilic substitution sequence on an aromatic ring. The mechanism shown below is presented as preceeding without the aid of a catalyst.

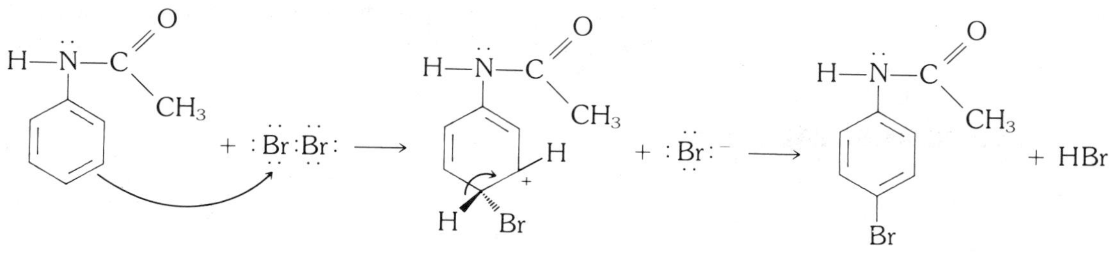

EXPERIMENTAL Estimated time for completion of the experiment: 1.5 hours.

Physical Properties of Reactants and Products

Compound	MW	Wt/Vol	mmol	mp(°C)	bp(°C)
Acetanilide	135.17	25 mg	0.19	114	
Glacial acetic acid	60.05	4 drops			118
Br$_2$-acetic acid reagent		3 drops			
4-Bromoacetanilide	214.08			168	

Reagents and Equipment

In a 3.0-mL conical vial fitted with a cap is placed 25 mg (0.19 mmol) of acetanilide to which is added four drops of glacial acetic acid using a medicine dropper. Stirring with a glass rod may be necessary to help dissolve the acetanilide. To the clear solution are added three drops of bromine-acetic acid reagent [**HOOD**]. The vial is immediately capped.

HOOD

> **WARNING:** *Bromine is a severe irritant. It is suggested that plastic gloves be worn since Br$_2$ burns require extended periods of time to heal. Dispense only in the HOOD. The reagent is prepared by mixing 2.5 mL of liquid bromine with 5.0 mL of glacial acetic acid.*

Reaction Conditions

The reddish-brown solution is allowed to stand at room temperature for 10 minutes with intermittent shaking. During this period yellow-orange colored crystals deposit from the solution.

Isolation of Product

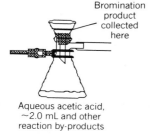

Bromination product collected here

Aqueous acetic acid, ~2.0 mL and other reaction by-products

Add 0.5 mL of water (calibrated Pasteur pipet) to the reaction mixture with swirling, followed by five drops of aqueous sodium bisulfite solution (33%). This treatment discharges the residual color, due to the presence of unreacted bromine, and results in the formation of white crystals. The reaction mixture is cooled in an ice bath for 10 minutes to maximize the product yield.

The white crystals of 4-bromoacetanilide are collected by vacuum filtration using a Hirsch funnel.(■) The filter cake is washed with three 0.25-mL portions of cold water (calibrated Pasteur pipet) and dried by drawing air through the crystals under reduced pressure for approximately 5 minutes.

Isolation and Characterization

The crude 4-bromoacetanilide is purified by recrystallization from 95% ethanol using the Craig tube.

Weigh the product and calculate the percentage yield. Determine the melting point and compare your result to the literature value. Obtain an IR spectrum of the material and compare it with that of an authentic sample.

ENVIRONMENTAL DATA

Substance	Amount	TLV (mg/m^3)	Emissions (mg)	Volume (m^3)
Acetic acid (transfer)	350 mg	25	22	0.9
Bromine (Max.)	170 mg	0.7	170	243
Ethanol (Max., Craig tube)	0.5 mL	1900	400	0.2

TLV: Sodium bisulfite—5 mg/m^3
TXDS: Acetanilide—orl-rat LD50: 800 mg/kg
 4-Bromoacetanilide—ipr-mus LD50: 250 mg/kg

QUESTIONS

6-92. Using resonance structures, show why the

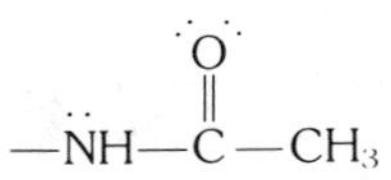

$$-NH-\overset{\overset{\displaystyle O}{\|}}{C}-CH_3$$

group is a less powerful o,p directing group than the $-\ddot{N}H_2$ unit.

6-93. Benzene is brominated in the presence of $FeBr_3$ catalyst.

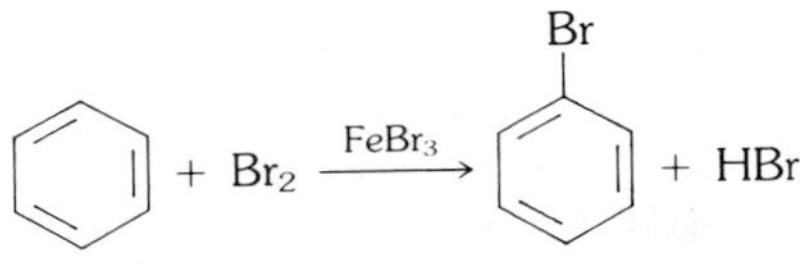

Suggest an appropriate mechanism for this reaction.

6-94. Draw the structure of the major monobromo product formed when each of the following compounds react with Br_2 in the presence of $FeBr_3$.

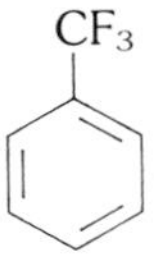

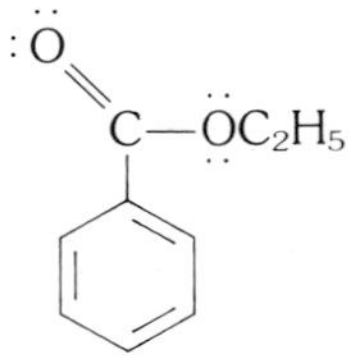

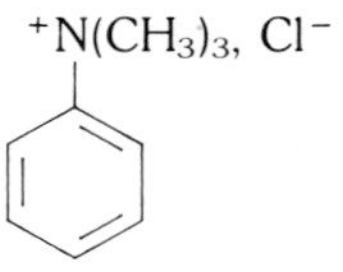

6-95. Arrange the compounds given below in order of increasing reactivity toward electrophilic substitution. Explain the reason(s) for your order.

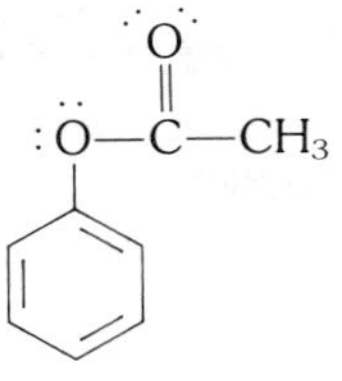

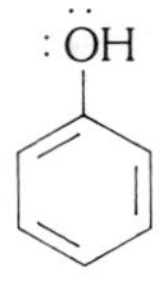

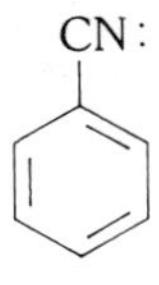

6-96. Explain why the bromination of m-xylene in acetic acid is accelerated by the addition of water.

6-97. In the experiment (Isolation of Product), sodium bisulfite solution is added to discharge the unreacted bromine reagent. What reaction is occurring in this step? Is HSO_3^- acting as an oxidizing or reducing agent? Write a balanced equation as part of your answer.

REFERENCES The references listed below are selected from a large number of examples given in "Organic Syntheses," which illustrate electrophilic aromatic substitution using bromine.
  a. Johnson, J. R.; Sandborn, L. T. "Organic Syntheses"; Wiley: New York, 1941; Collect. Vol. I, p 111.
  b. Langley, W. D. *Ibid.*, p 127.
  c. Adams, R.; Marvel, C. S. *Ibid.*, p 128.
  d. Smith, L. I. "Organic Syntheses"; Wiley: New York, 1943; Collect. Vol. II, p 95.
  e. Sandin, R. B.; McKee, R. A. *Ibid.*, p 100.
  f. Hartman, W. W.; Dickey, J. B. *Ibid.*, p 173.
  g. Coleman, C. H.; Talbot, W. F. *Ibid.*, p 592.
  h. Wilson, C. V. *Ibid.*, p 575.

Experiments 34A, 34B, 34C, and 34D

Nitration: 2,5-Dichloronitrobenzene; 2,5-Dibromonitrobenzene; N,N'-Diacetyl-2,3-dinitro-1,4-phenylenediamine; 5-Nitrosalicylic Acid

(benzene, 1,4-dichloro-2-nitro-; benzene, 1,4-dibromo-2-nitro-; acetamide, N,N'-(2,3-dinitro-p-phenylene)bis-; salicylic acid, 5-nitro-)

The reactions described below provide several examples of the nitration of aromatic compounds using 100% nitric acid.

DISCUSSION The nitration reactions described in this experiment demonstrate one of the classic electrophilic aromatic substitution reactions. Nitration has been used extensively as a means of introducing the NO_2 substituent on an aromatic ring, which by reduction, may be converted to the amino group. In the present case, 100% nitric acid is used in place of the usual HNO_3-H_2SO_4 mixture.

The nitro group is deactivating and thus nitration usually stops at the monosubstituted product. An exception occurs when activating groups are present. This fact can be seen by comparing the results of the nitration of 1,4-dichlorobenzene with that of N,N'-diacetyl-1,4-phenylenediamine. Because of the presence of the activating acetylamino groups, the dinitro derivative forms readily.

It is generally accepted that the nitronium ion, NO_2^+, is the electrophilic species that attacks the aromatic ring. The overall mechanism for nitration is given below.

$$HONO_2 + HONO_2 \rightleftharpoons H_2\overset{+}{O}-NO_2 + NO_3^-$$

$$H_2\overset{+}{O}-NO_2 + HONO_2 \rightleftharpoons H_3O^+ + NO_2^+ + NO_3^-$$

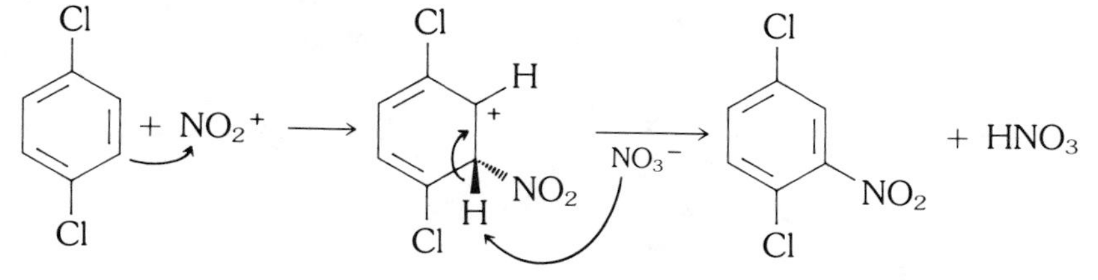

Preparation of 100% HNO₃

> **CAUTION:** *The reagents and the product of this preparation are highly corrosive. The distillation should be conducted in a HOOD. Prevent contact with eyes, skin, and clothing. Any spill should be neutralized using solid sodium carbonate or bicarbonate.*

EXPERIMENTAL

Estimated time of the preparation: 0.5 hour.

Important. *This acid reagent should be used immediately for the nitration experiments given below. The amount obtained is sufficient for the preparation of two or three of the nitro compounds.*

Physical Properties of Reactants and Products

Compound	MW	Wt/Vol	bp(°C)	Density
Conc. nitric acid (68%)		0.7 mL	120.5	1.41
Conc. sulfuric acid (96–98%)		1 mL	338	1.84
100% Nitric acid	63.01		83	1.40

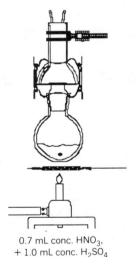

0.7 mL conc. HNO₃, + 1.0 mL conc. H₂SO₄

Reagents and Equipment

Add 0.7 mL of concentrated nitric acid to a 5-mL round-bottom flask containing a boiling stone and equipped with a Hickman still. Add 1.0 mL of concentrated sulfuric acid and swirl the assembly gently to mix the reagents.(■)

> **CAUTION:** *Sulfuric acid can cause severe burns. Nitric acid is a strong oxidizing agent. Prevent contact with eyes, skin, and clothing. A spill can be neutralized using sodium carbonate or bicarbonate. The acid additions are made using two clean, dry 1.0-mL graduated pipets.*

Reaction Conditions

The acid solution is now heated very gently with a microburner, keeping the microburner in constant motion, until approximately 0.2 mL of 100% nitric acid has been collected as the distillate in the collar of the still.

Isolation of Product

The 100% nitric acid is removed from the collar of the still (Pasteur pipet) and placed in a 1.0-mL conical vial fitted with a glass stopper. *It may be necessary to slightly bend the end of the pipet in a flame so that it can reach the collar of the still. The distillate is colorless or faintly yellow.*

Purification and Characterization

The 100% nitric acid is used as collected. No further purification is required.

Anhydrous nitric acid (white fuming) is a colorless liquid, bp 83°C. It is estimated that the purity of the nitric acid obtained in this preparation is 99.5–100%.

Part A: 2,5-Dichloronitrobenzene

Cl—⟨ ⟩—Cl + HNO₃ ⟶ Cl—⟨ ⟩—Cl
                                                NO₂

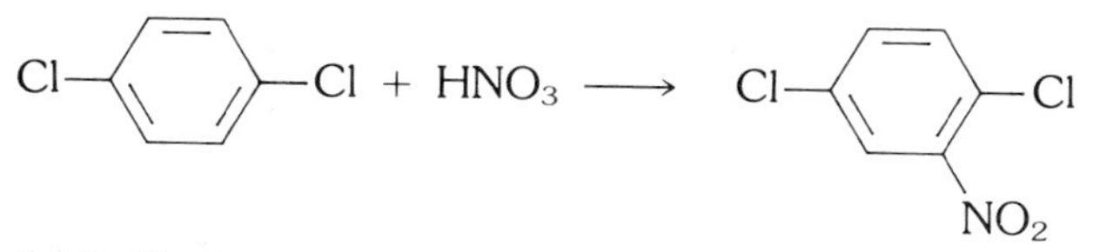

1,4-Dichlorobenzene   2,5-Dichloronitrobenzene

EXPERIMENTAL Estimated time to complete the experiment: 0.5 hour.

Physical Properties of Reactants and Products

Compound	MW	Wt/Vol	mmol	mp(°C)	bp(°C)	Density
1,4-Dichlorobenzene	147.01	38 mg	0.26	53		
100% Nitric acid	63.01	100 μL	2.4		83	1.50
2,5-Dichloronitrobenzene	192.00			56		

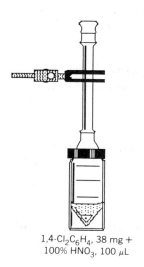

1,4-Cl₂C₆H₄, 38 mg +
100% HNO₃, 100 μL

Nitration
product
collected
here

Aqueous HNO₃, ~5 mL

Reagents and Equipment

Equip a 3.0-mL conical vial containing a magnetic spin vane with an air condenser.(■) Add 38 mg (0.26 mmol) of 1,4-dichlorobenzene followed by the addition of 100 μL of 100% nitric acid delivered from an automatic delivery pipet.

> **CAUTION:** *The nitric acid reagent is highly corrosive. Prevent contact with eyes, skin, and clothing. A spill is neutralized using solid sodium carbonate or bicarbonate.*

Reaction Conditions

The resulting solution is allowed to stand at room temperature for a period of 15 minutes. Next add, while stirring, 1.0 mL of water (calibrated Pasteur pipet), and then place the vial in an ice bath to cool.

Isolation of Product

After removal of the spin vane with forceps, the crystalline precipitate is collected by vacuum filtration using a Hirsch funnel.(■) Wash the filter cake with four 1.0-mL portions of water (calibrated Pasteur pipet) and then place it on a porous clay plate to dry.

Purification and Characterization

The product, consisting of fine, white needles, is sufficiently pure for characterization. It may be recrystallized from ethanol-water using a Craig tube if desired.

Weigh the 2,5-dichloronitrobenzene and calculate the percentage yield. Determine the melting point and mixed melting point with starting material. Compare your result to the literature value. Obtain an IR spectrum and compare it with that of an authentic sample.

PART B: 2,5-Dibromonitrobenzene

Br—⟨⟩—Br + HNO₃ ⟶ Br—⟨⟩—Br
                                          |
                                         NO₂

1,4-Dibromobenzene   2,5-Dibromonitrobenzene

EXPERIMENTAL   Estimated time to complete the experiment: 2.0 hour.

Physical Properties of Reactants and Products

Compound	MW	Wt/Vol	mmol	mp(°C)	bp(°C)	Density
1,4-Dibromobenzene	235.92	90 mg	0.38	87		
100% Nitric acid	63.01	168 μL	4.0		83	1.50
2,5-Dibromonitrobenzene	280.91			85–86		

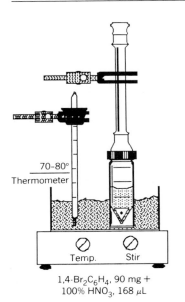

70–80°
Thermometer

Temp.   Stir

1,4-Br₂C₆H₄, 90 mg +
100% HNO₃, 168 μL

Nitration
product
collected
here

Aqueous HNO₃, ~5 mL

HOOD

Reagents and Equipment

To a 3.0-mL conical vial containing a magnetic spin vane and equipped with an air condenser, add 90 mg (0.38 mmol) of 1,4-dibromobenzene. (■) The vial is placed in a sand bath at a temperature of 70–80°C, and 168 μL of 100% nitric acid is added using an automatic delivery pipet.

> **CAUTION:** *The nitric acid reagent is highly corrosive. Prevent contact with eyes, skin, and clothing. A spill is neutralized using solid sodium carbonate or sodium bicarbonate.*

Reaction Conditions

Allow the vial to stand in the sand bath for 30 minutes. Then remove and cool to room temperature. While stirring, add 1.0 mL of water (calibrated Pasteur pipet) and place the vial in an ice bath.

Isolation of Product

The crystalline product is collected by vacuum filtration using a Hirsch funnel. (■) The filter cake is washed with four 1.0-mL portions of water (calibrated Pasteur pipet) and the product dried on a porous clay plate.

Weigh the crude 2,5-dibromonitrobenzene and calculate the crude percentage yield.

Purification and Characterization

The crude material is purified by column chromatography. A Pasteur filter pipet is packed with 750 mg of activated silica gel and the column wetted with pentane. The crude crystals are dissolved in the *minimum amount* of methylene chloride and transferred to the column using a Pasteur pipet. The product is eluted with 2.5 mL of pentane and the eluate collected in a 10-mL Erlenmeyer flask containing a boiling stone. The pentane solvent is evaporated **[HOOD]** by warming the flask in a sand bath. The yellow crystals are removed from the flask and dried on a porous clay plate.

Weigh the product and calculate the percentage yield. Determine the melting point and compare your value to that reported in the literature. Obtain an IR spectrum of your product and compare it with that of an authentic sample.

PART C:
N,N′-Diacetyl-2,3-dinitro-1,4-phenylenediamine

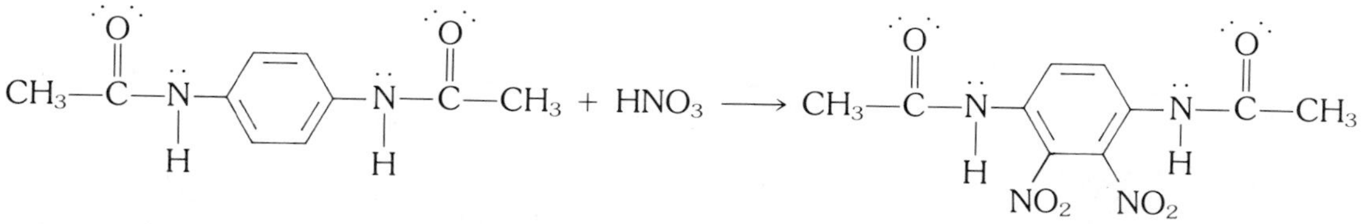

N,N′-Diacetyl-1,4-phenylenediamine

N,N′-Diacetyl-2,3-dinitro-
1,4-phenylenediamine

EXPERIMENTAL   Estimated time to complete the experiment: 0.5 hours.

Physical Properties of Reactants and Products

Compound	MW	Wt/Vol	mmol	mp(°C)	bp(°C)	Density
N,N′-Diacetyl-1,4-phenylenediamine	192	48 mg	0.25	312–315		
100% Nitric acid	63.01	100 μL	2.4		83	1.50
N,N′-Diacetyl-2,3-dinitro-1,4-phenylenediamine	282			257		

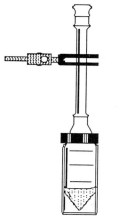

1,4-(CH₃CONH)₂C₆H₄, 48 mg +
100% HNO₃, 100 μL

Nitration
product
collected
here

Aqueous HNO₃, ~5 mL

Reagents and Equipment

To a 3.0-mL conical vial equipped with an air condenser add 48 mg (0.25 mmol) of *N,N′*-diacetyl-1,4-phenylenediamine followed by the dropwise addition **[CAUTION]** of 100 μL of 100% nitric acid from an automatic delivery pipet.(■) *The N,N′-diacetyl-1,4-phenylenediamine is prepared by the procedure outlined in Experiment 26B.*

> **CAUTION:** *The reaction is highly EXOTHERMIC. A vigorous reaction occurs if the acid is added too rapidly. The nitric acid reagent is highly corrosive; prevent contact with eyes, skin, and clothing. A spill is neutralized using sodium carbonate or bicarbonate.*

Reaction Conditions

The resulting solution is allowed to stand at room temperature for a period of 10 minutes. Water (1.0 mL) is added (calibrated Pasteur pipet) and the vial then placed in a ice bath to cool.

Isolation of Product

The resulting yellow precipitate is collected by vacuum filtration using a Hirsch funnel.(■) The filter cake is washed with four 1.0-mL portions of water (calibrated Pasteur pipet) and then dried on a porous clay plate.

Purification and Characterization

The product needs no further purification. Weigh the dried material and calculate the percentage yield. Determine the melting point and compare your result with the literature value. Obtain an IR spectrum of your product and compare it with that shown in Fig. 6.7 (*see over*).

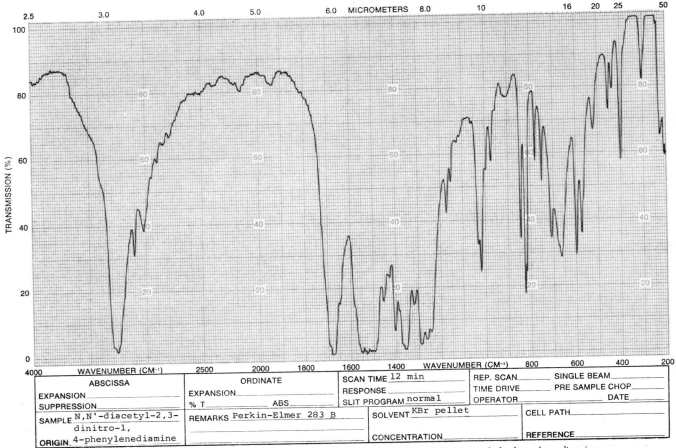

Fig. 6.7 IR spectrum: *N,N'-diacetyl-2,3-dinitro-1,4-phenylenediamine.*

PART D: 5-Nitrosalicylic Acid

Salicylic acid + HNO₃ ⟶ 5-Nitrosalicylic acid

EXPERIMENTAL Estimated time to complete the experiment: 0.5 hour.

Physical Properties of Reactants and Products

Compound	MW	Wt/Vol	mmol	mp(°C)	bp(°C)	Density
Salicylic acid	138.12	50 mg	0.36	159		
100% Nitric acid	63.01	100 μL	2.4		83	1.5
5-Nitrosalicylic acid	183.12			229–230		

This material may be used to prepare the caffeine 5-nitrosalicylate derivative (see Experiment 11).

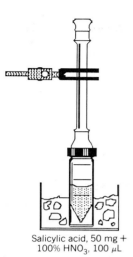

Salicylic acid, 50 mg +
100% HNO₃, 100 μL

CAUTION

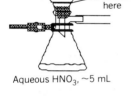

Nitration
product
collected
here

Aqueous HNO₃, ~5 mL

CAUTION: *This reaction should be conducted in a hood.*

Reagents and Equipment

To a 3.0-mL conical vial equipped with an air condenser, add 50 mg (0.36 mmol) of salicylic acid and place the vial in an ice bath to cool.(■) In addition, 100 μL of 100% nitric acid contained in a stoppered conical vial is also placed in an ice bath to cool.

CAUTION: *The nitric acid reagent is highly corrosive; prevent contact with eyes, skin, and clothing. A spill is neutralized using solid sodium carbonate or bicarbonate.*

Reaction Conditions

Add the cold nitric acid dropwise (Pasteur pipet) [**CAUTION**] to the salicylic acid. The vial containing the salicylic acid is kept in the ice bath during the addition.

CAUTION: *The reaction is highly EXOTHERMIC. A very vigorous reaction occurs if the acid is added too rapidly.*

The evolution of a red-brown gas (NO₂) is observed during the addition. Allow the vial to stand in the ice bath for an additional 20 minutes, after which 1.0 mL of distilled water is added (calibrated Pasteur pipet).

Isolation of Product

The orange-pink solid is collected by vacuum filtration using a Hirsch funnel.(■) The filter cake is washed with four 1.0-mL portions of cold water (calibrated Pasteur pipet) and dried on a porous clay plate.

Purification and Characterization

The product is recrystallized using a Craig tube by dissolving the material in the *minimum* amount of absolute ethanol, followed by the dropwise addition of water until precipitation occurs. Cool the mixture in an ice bath and separate the light-yellow crystals. They are dried on a porous clay plate.

Weigh the crystals and calculate the percentage yield. Determine the melting point and compare your value to that reported in the literature. Obtain an IR spectrum of the material and compare it with that of an authentic sample.

ENVIRONMENTAL DATA

Substance	Amount	TLV (mg/m³)	Emissions (mg)	Volume (m³)
Experiment 34				
Nitric acid (conc.) (30 minutes in open container)	0.7 mL	5	200	40
Sulfuric acid (conc.) (estimated)	1.0 mL	1	70	70
Experiment 34A				
Nitric acid (100%) (open container)	100 μL	5	90	18
Ethanol (Max., recryst.)	1.0 mL	1900	800	0.4
Experiment 34B				
Nitric acid (100% (open container)	168 μL	6	90	18
Pentane (Max., col. chromatog.)	3.5 mL	1800	2190	1.2
Methylene chloride (Max., col. chromatog.)	0.5 mL	350	670	1.9

ENVIRONMENTAL DATA
(continued)

Substance	Amount	TLV (mg/m³)	Emissions (mg)	Volume (m³)
Experiment 34C				
Nitric acid (100%) (open container)	100 μL	5	90	18
Experiment 34D				
Nitric acid (100%) (open container)	100 μL	5	90	18
Ethanol (Max., recryst.)	0.5 mL	1900	400	0.2

TLV: 1,4-Dichlorobenzene—75 ppm
TXDS: Nitric acid (white fuming)—inl-rat LC50: 244 ppm (NO_2)/30M, 5 mg/m³
Sulfuric acid (conc.)—unk-man LDLo: 135 mg/kg, 1 mg/m³
2,5-Dichloronitrobenzene—orl-rat LD50: 1210 mg/kg
Salicylic acid—orl-rat LD50: 891 mg/kg

QUESTIONS

6-98. Predict the position most likely to be taken by the incoming NO_2^+ ion in the mononitration of each of the following compounds. Explain the reasons for your choice.

a. (CF₃ benzene)   b. (CH(CH₃)₂ / CH₃ benzene)   c. (NO₂ / CH₃, CH₃ benzene)   d. (Br / benzene—C—OC₂H₅, O)

6-99. Write equations to show how nitronium ions might be formed using a mixture of nitric and sulfuric acids.

6-100. Which ring of phenyl benzoate would you expect to undergo nitration more readily? Explain.

Phenyl benzoate

6-101. Arrange the following compounds in order of increasing reactivity toward nitration. Given reasons for your order.

acetanilide   acetophenone   bromobenzene   toluene

6-102. Offer a reasonable explanation of why nitration of 1,4-dichlorobenzene yields the mononitro derivative, while N,N'-diacetyl-1,4-phenylenediamine forms the dinitro compound.

REFERENCES

1. Related to 100% nitric acid:
   a. Stern, S. A.; Mullhaupt, J. T.; Kay, W. B. *Chem. Revs.* **1960,** *60,* 185.
   b. The preparation used in this experiment is an adaptation of that reported by Cheronis, N. D.; Entrikin, J. B. "Semimicro Qualitative Organic Analysis"; Crowell: New York, 1947; p 258.
2. These references are selected from a large number of examples of nitration given in "Organic Syntheses." None use 100% nitric acid as the nitrating agent.
   a. Kamm, O.; Segur, J. B. " Organic Syntheses"; Wiley: New York, 1941; Collect. Vol. I, p 372.
   b. Robertson, G. R. *Ibid.*, p 376.
   c. Corson, B. B.; Hazen, R. K. "Organic Syntheses"; Wiley: New York, 1943; Collect. Vol. II, p 434.

d. Hartman, W. W.; Smith, L. A. *Ibid.*, p 438.

e. Huntress, E. H.; Shriner, R. L. *Ibid.*, p 459.

f. Kobe, K. A.; Doumani, T. F. "Organic Syntheses"; Wiley: New York, 1955; Collect. Vol. III, p 653.

g. Fitch, H. M. *Ibid.*, p 658.

h. Fanta, P. E.; Tarbell, D. S. *Ibid.*, p 661.

i. Howard, J. C. "Organic Syntheses"; Wiley: New York, 1963; Collect. Vol. IV, p 42.

j. Braum, C. E.; Cook, C. D. *Ibid.*, p 711.

k. Fetscher, C. A. *Ibid.*, p 735.

l. Mendenhall, G. D.; Smith, P. A. S. "Organic Syntheses"; Wiley: New York, 1973; Collect. Vol. V, p 829.

m. Newman, M. S.; Boden, H. *Ibid.*, p 1029.

Experiment 35
Quantitative Analysis of a Grignard Reagent: Phenylmagnesium Bromide

(magnesium, bromophenyl-)

This experiment outlines a technique for the analysis of Grignard reagents. It is similar in scope to Experiment 23.

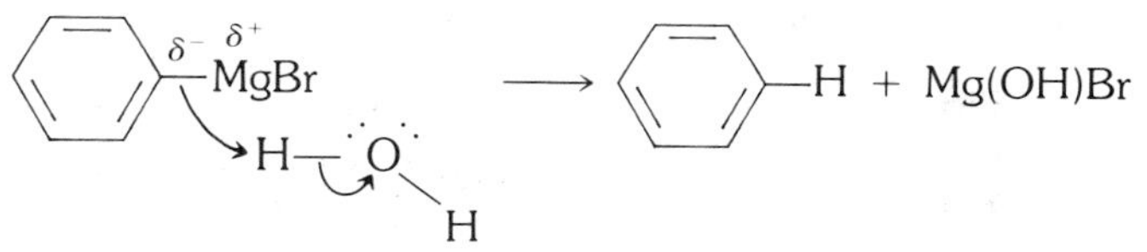

DISCUSSION For a discussion of the reaction see Experiment 23. Remember, however, that in the present experiment the hydrocarbon formed upon hydrolysis of phenylmagnesium bromide is benzene.

PART A: Phenylmagnesium Bromide

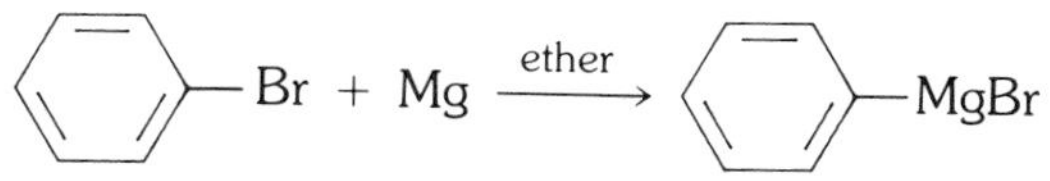

Bromobenzene   Phenylmagnesium bromide

EXPERIMENTAL Estimated time for the experiment: 1.0 hour.

Physical Properties of Reactants and Products

Compound	MW	Wt/Vol	mmol	mp(°C)	bp(°C)	Density	n_D
Bromobenzene	157.02	76 μL	0.72		156	1.50	1.5597
Diethyl ether	74.12	700 μL			35		
Magnesium	24.31	18 mg	0.73	649			
Iodine	253.81	1 crystal					
Benzene	78.12				80	0.88	1.5011

> **CAUTION:** *Ether is a flammable liquid and also a narcotic. All flames should be extinguished during the time of this experiment.*

Reagents and Equipment

This reagent is prepared exactly as described in Experiment 17. The reagents, amount of reagents, order of addition, workup manipulations, and precautions are the same. The equipment is also identical. (■)

The brown-gray Grignard reagent is cooled to room temperature and then analyzed by the titration method outlined in Experiment 23B.

ENVIRONMENTAL DATA

The Grignard preparation emissions are listed in Experiment 17.

TLV: Iodine—0.1 ppm
           Sodium hydroxide—2 mg/m^3

TXDS: Magnesium—orl-dog LDLo: 230 mg/kg
            Sulfuric acid—unk-man LDLo: 135 mg/kg

MTDS: Bromobenzene—dnr-esc 250 mg/L

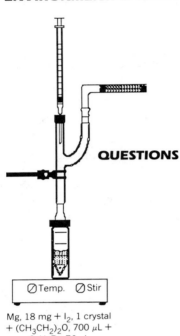

Mg, 18 mg + I$_2$, 1 crystal
+ (CH$_3$CH$_2$)$_2$O, 700 μL +
C$_6$H$_5$Br, 76 μL

QUESTIONS

6-103. What hydrocarbon would you expect to obtain by the action of water on each of the Grignard reagents listed below?
  a. Butylmagnesium bromide
  b. *sec*-Butylmagnesium bromide
  c. *iso*-Butylmagnesium bromide
  d. *t*-Butylmagnesium bromide

6-104. What product would each of the Grignard reagents in question 6-103 yield when treated with D$_2$O?

6-105. Consult your Quantitative Analysis text for the analysis of Mg^{2+} by the complexation titration method using EDTA (ethylenediaminetetraacetic acid). Do you think this technique might be adapted for the analysis of Grignard reagents? If so, outline a suitable procedure.

REFERENCES

For references relating to the preparation of Grignard reagents see Experiments 17 and 23.

Experiment 36
Grignard and Arene Halide Cross Coupling Reaction: 1-Methyl-2-(methyl-d$_3$)-benzene

(benzene, 1-methyl-(2-methyl-d$_3$))

This experiment illustrates the coupling reaction of a Grignard reagent with an alkyl halide to form a C—C bond. It also illustrates the technique of "labeling" particular hydrogen atoms for identification purposes. *The use of methyl iodide in the reaction produces o-xylene. This demonstrates only the coupling reaction.*

$$CD_3I + Mg \xrightarrow{\text{ether}} D_3CMgI$$

Methyl-d₃ iodide   Methyl-d₃-
                                   magnesium iodide

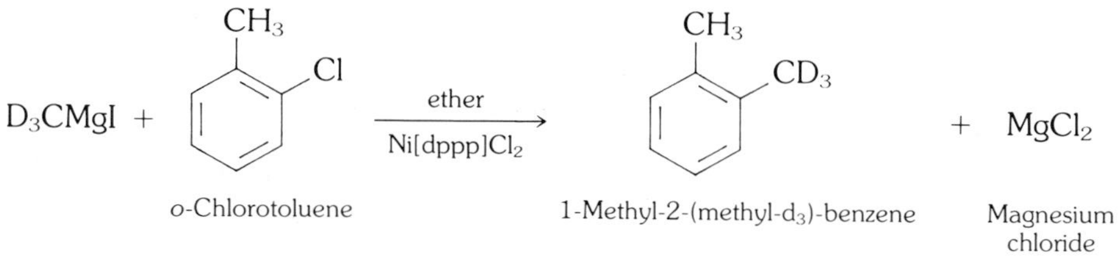

D₃CMgI +   o-Chlorotoluene $\xrightarrow[\text{Ni[dppp]Cl}_2]{\text{ether}}$ 1-Methyl-2-(methyl-d₃)-benzene   +   MgCl₂

Magnesium
chloride

DISCUSSION   The selective cross-coupling of a Grignard reagent with an organic halide to produce the desired C—C bond formation is demonstrated in this reaction. The reaction is catalyzed by a phosphine-nickel catalyst and has wide application for the synthesis of unsymmetrical alkanes and alkenes. Before the discovery of the nickel-phosphine activity, the cross-coupling of Grignard reagents with organic halides was seldom employed in synthetic practice. This was due to the formation of homocoupling products along with other side reactions.

The mechanism of the reaction is not fully understood at present. A discussion of the reaction is given in Tamao, K.; Sumitani, K.; Kumada, M. *J. Am. Chem. Soc.* **1972,** *94,* 4374.

EXPERIMENTAL   Estimated time to complete the experiment: 5.0 hours.

Physical Properties of Reactants and Products

Compound	MW	Wt/Vol	mmol	mp(°C)	bp(°C)	Density	n_D
Magnesium	24.31	65 mg	2.7	649			
Diethyl ether	74.12	700 μL			34.5		
Methyl-d₃-iodide	144.96	250 μL	3.93		42	2.28	1.5262
Ni[dppp]Cl₂	540.5	10 mg	0.02				
o-Chlorotoluene	126.59	150 μL	1.28		159	1.08	1.5268
1-Methyl-2-(methyl-d₃)-benzene	109.17				144		1.5055

Reagents and Equipment

Note. *All the glassware used in the experiment should be cleaned, dried in an oven at 110°C for at least 30 minutes, and then cooled in a desiccator before use.*

Equip a 5.0-mL conical vial containing a magnetic spin vane with a reflux condenser protected by a calcium chloride drying tube. Add 65 mg (2.7 mmol) of magnesium and 0.3 mL of anhydrous diethyl ether.(■)

A 4–5 in. piece of magnesium ribbon is scraped clean of oxide coating and cut into 1-mm length sections. This freshly cut material is handled only with forceps. The ether is dispensed in the **HOOD** *using an automatic delivery pipet.*

Prepare a solution of 250 μL of methyl-d₃ iodide in 200 μL of anhydrous diethyl ether in a capped vial. *These reagents may be dispensed using automatic delivery pipets.*

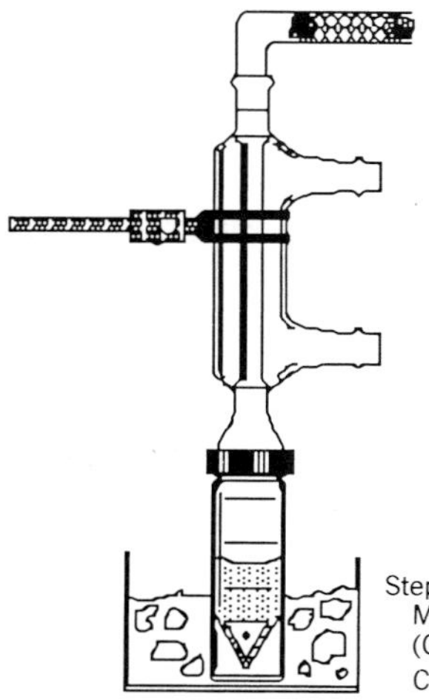

HOOD

Step 1:
Mg, 65 mg +
(CH₃CH₂)₂O, 0.5 mL +
CD₃I, 250 μL

CAUTION: *Methyl iodide is a cancer suspect agent.*

This solution is drawn into a Pasteur pipet and transferred to the reaction flask. This is done by removing the drying tube and inserting the pipet down the length of the condenser, allowing the pipet bulb to rest on the condenser lip.

Reaction Conditions

The reaction vial is placed in an ice bath and the methyl-d$_3$ iodide solution added dropwise with stirring. After the addition, the pipet is withdrawn, the drying tube reinstalled, and the mixture stirred for an additional 20 minutes.

Prepare a solution of 10 mg (0.02 mmol) of dichloro[1,3-bis(diphenyl-phosphino)propane]nickel(II) catalyst, 150 μL of o-chlorotoluene, and 200 μL of anhydrous diethyl ether in a capped vial. *The liquids are dispensed in the* **HOOD** *using automatic delivery pipets.*

Draw the solution into a Pasteur pipet, and in one portion add it to the reaction vial through the condenser as described above. The drying tube is reinserted as before. Place this mixture in a preheated sand bath at a temperature of 55°C and with stirring, heat for 2 hours.(■)

Upon addition of the catalyst, the solution turns green in color. After approximately half an hour of heating, the mixture becomes dark brown.

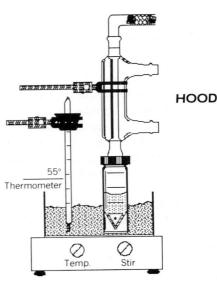

55°
Thermometer

Temp.   Stir

Step 2:
$(C_6H_5)_2P(CH_2)_3P(C_6H_5)_2 \cdot NiCl_2$,
10 mg
+ o-$CH_3C_6H_4Cl$, 150 μL
+ $(CH_3CH_2)_2O$, 200 μL

HOOD

Isolation of Product

Cool the reaction mixture in an ice bath and quench the reaction by the dropwise addition (calibrated Pasteur pipet) of 1.5 mL of 1.0M HCl solution. *The acid is added slowly since frothing occurs.*

The solution is extracted with 1 mL of ether. Using a Pasteur filter pipet, the aqueous layer is removed and discarded. The remaining ether phase is extracted with 1.0 mL of water followed by 1.0 mL of saturated sodium bicarbonate solution and then 1.0 mL of deionized water. This is followed by extraction with two 1.0-mL portions of 1.0N sodium thiosulfate solution and 1.0 mL of water.

The above amounts of extracting solutions are measured using calibrated Pasteur pipets. For each extraction operation, the vial is capped, shaken, and vented, and the layers are allowed to separate. The bottom aqueous phase is then removed and discarded.

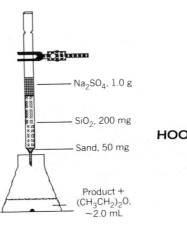

Na$_2$SO$_4$, 1.0 g

SiO$_2$, 200 mg

Sand, 50 mg

Product +
$(CH_3CH_2)_2O$,
~2.0 mL

HOOD

Purification and Characterization

The reaction product is isolated and purified using column chromatography. In a Pasteur filter pipet place 200 mg of activated silica gel (100 mesh) followed by 1.0 g of anhydrous sodium sulfate. The column is wetted with 0.5 mL of ether (calibrated Pasteur pipet).(■)

The ether solution of product is transferred to the column by Pasteur pipet and the material eluted from the column with two 0.5-mL portions of ether. The eluate is collected in a tared 10-mL Erlenmeyer flask containing a boiling stone. The solvent is evaporated by warming the flask in a sand bath [**HOOD**] to yield the "labeled" o-xylene.

Weigh the product and calculate the percentage yield. Determine the boiling point and refractive index and compare your values with those reported in the literature. Obtain an IR spectrum of the material and compare it with the spectrum shown in Fig. 6.8.

ENVIRONMENTAL DATA

Substance	Amount	TLV (mg/m³)	Emissions (mg)	Volume (m³)
Methyl iodide (SKIN) (Max., reactant, transfer, yield)	0.25 mL	10	450	47
Ether (Max., solv., col. chromatog., evap.)	3.2 mL	1200	2285	1.9
o-Chlorotoluene (Max., reactant, yield)	0.15 mL	250	49	0.2
Organonickel catalyst	1.0 mg			

TLV: o-Xylene—100 ppm
Hydrochloric acid—5 ppm
TXDS: Magnesium—orl-dog LDLo: 230 mg/kg
Sodium sulfate—orl-mus LD50: 5989 mg/kg
Sodium thiosulfate—scu-rbt LDLo: 4000 mg/kg
Sodium bicarbonate—orl-rat LD50: 4220 mg/kg
Silica gel—inv-mus LDLo: 343 mg/kg

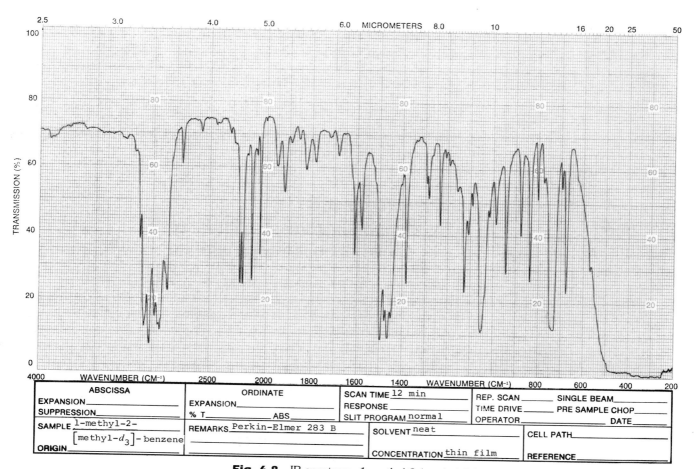

Fig. 6.8 IR spectrum: *1-methyl-2-(methyl-d₃)-benzene.*

QUESTIONS

a. ⬡—$(CH_2)_3CH_3$

b. [structure: benzene ring with $CH_2CH(CH_3)_2$ groups at 1,3 positions]

$CH_2CH(CH_3)_2$

$CH_2CH(CH_3)_2$

c. ⬡—CH_2—CH=CH_2

d. CH_3—[structure: 2,4,6-trimethylphenyl linked to 4-methylphenyl with CH_3 groups]—CH_3

6-106. Predict the reagents that could be used to prepare each of the compounds listed below by the Grignard coupling reaction. Give a suitable name to each reactant (*a,b,c,d*).

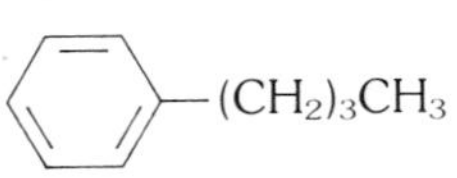

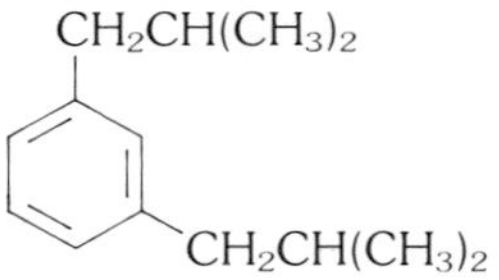

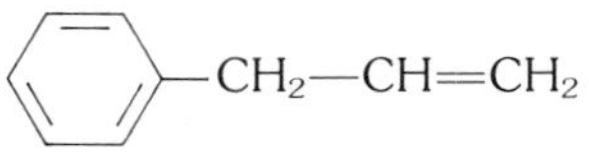

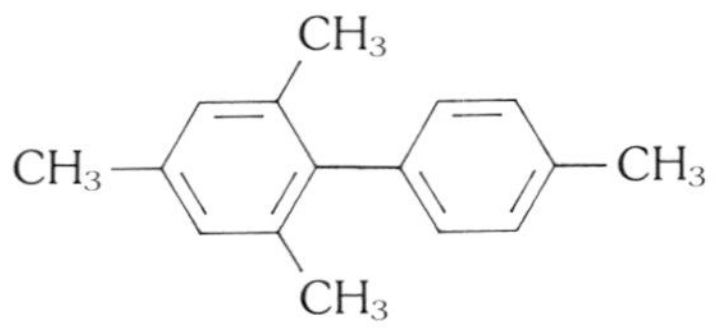

6-107. Write a structural formula for the dichloro[1,3-bis(diphenylphosphino)propane]-nickel(II) catalyst used in this reaction.

6-108. The Wurtz coupling reaction involves the treatment of haloalkanes with an active metal such as sodium. What products would be formed in the following reactions? Name them.

$$2\ \mathbf{R—X}\ +\ 2\ Na\ \xrightarrow[\text{solvent}]{\text{inert}}\ \mathbf{R—R}\ +\ 2\ NaX$$

a. Butylmagnesium bromide + Na ⟶

b. 1-Bromo-3-chlorocyclobutane + Na ⟶

6-109. In reference to question 6-108, explain why the coupling reaction of butylmagnesium bromide and propyl bromide under conditions of the Wurtz reaction is synthetically useless for the preparation of heptane.

REFERENCES

1. For general references on Grignard reagents refer to Experiment 17.
2. The present coupling reaction is based on the work reported by Kumada, M; Tamao, K; Sumitani, K. *Org. Synth.* **1978,** *58,* 127 and references found therein.
3. Selected Grignard coupling reactions presented in "Organic Syntheses" include:
   a. Lespieau, R.; Bourguel, M. "Organic Syntheses"; Wiley: New York, 1941; Collect. Vol. I, p 186.
   b. Gilman, H.; Catlin, W. E. *Ibid.,* p 471.
   c. Gilman, H.; Robinson, J. "Organic Syntheses"; Wiley: New York, 1943; Collect. Vol. II, p 47.
   d. Smith, L. I. *Ibid.,* p 360.
   e. Turk, A.; Chanan, H. "Organic Syntheses"; Wiley: New York, 1955; Collect. Vol. III, p 121.
4. Mayo, D. W.; Bellamy, L. J.; Merklin, G. T.; Hannah, R. W. *Spectrochim. Acta* **1985,** *41A,* 355.

Experiment 37
Nucleophilic Aromatic Substitution: 2,4-Dinitrophenylthiocyanate

(thiocyanic acid, 2,4-dinitrophenyl ester)

In this experiment an example of nucleophilic substitution of an activated aromatic ring is presented. In the reaction, a bromo substituent is replaced by a thiocyanate group. The rate of the reaction is increased by the presence of a phase transfer catalyst.

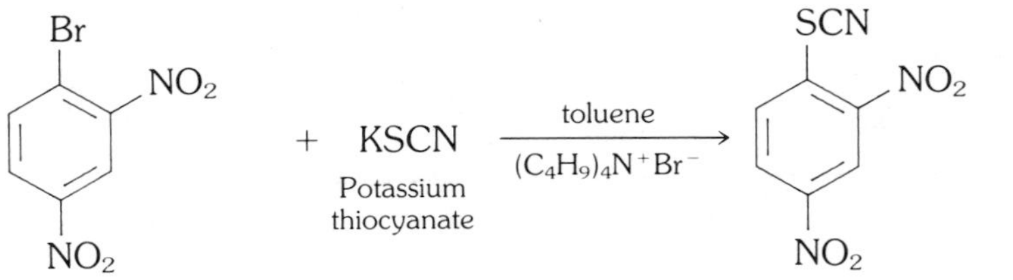

[reaction scheme: 2,4-Dinitrobromobenzene (Br, NO₂, NO₂ on benzene) + KSCN (Potassium thiocyanate) →(toluene, $(C_4H_9)_4N^+Br^-$)→ 2,4-Dinitrophenyl thiocyanate (SCN, NO₂, NO₂ on benzene)]

2,4-Dinitrobromobenzene

2,4-Dinitrophenyl thiocyanate

DISCUSSION Aromatic nucleophilic substitution generally takes place only if activating groups are present on the positions ortho and/or para to the leaving group. Activating groups are those that are electron-withdrawing. Groups such as

$$-NO_2, -SO_2CH_3, -\overset{+}{N}(CH_3)_3, -CF_3, -CN, -SO_3^-, -Br, -Cl, -I, -CO_2^-,$$

especially if another NO_2 group is present, are strongly activating. In the reaction illustrated in this experiment, two nitro groups are present on the ring, ortho and para to the departing bromo substituent. The reaction is hindered by electron-releasing groups.

The reaction proceeds in two steps, of which the first step is generally rate determining. A tetrahedral intermediate is formed by the attack of the nucleophilic reagent on the carbon atom to which the leaving group is attached. In the subsequent step, the leaving group then departs regenerating the aromatic nucleus. It is important to note that the reaction does not proceed by an S_N2 mechanism since an intermediate is formed. The sequence is outlined below.

EXPERIMENTAL Estimated time to complete the reaction: 2.0 hours.

Physical Properties of Reactants and Products

Compound	MW	Wt/Vol	mmol	mp(°C)	bp(°C)
2,4-Dinitrobromobenzene	247.01	50 mg	0.20	75	
Toluene	92.15	350 μL			111
Tetrabutylammonium bromide	322.38	5 mg	0.02	103–104	
Aqueous potassium thiocyanate (50%)		150 μL			
2,4-Dinitrophenylthiocyanate	244			138–139	

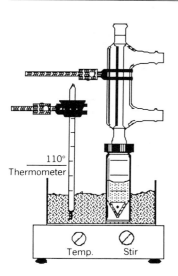

2,4-(NO$_2$)$_2$C$_6$H$_3$Br, 50 mg
+ C$_6$H$_5$CH$_3$, 350 μL +
Bu$_4$N$^+$Br$^-$, 5 mg +
50% aqueous KSCN, 150 μL

Reagents and Equipment

In a 3.0-mL conical vial containing a magnetic spin vane and equipped with a reflux condenser, place 50 mg (0.20 mmol) of 2,4-dinitrobromobenzene and 350 μL of toluene.(■) To this solution add 5.0 mg (0.02 mmol) of tetrabutylammonium bromide and 150 μL of a 50% aqueous potassium thiocyanate (wt/wt) solution.

CAUTION: *tetrabutylammonium bromide is an irritant. HANDLE WITH CARE! It is also hygroscopic and must be protected from moisture. The toluene solvent and the thiocyanate solution are dispensed in the HOOD using automatic delivery pipets.*

Reaction Conditions

The resulting mixture is heated at a sand bath temperature of 110°C with stirring for a period of 1.0 hour. It is then cooled to room temperature.

Isolation of Product

SAVE The toluene layer is separated from the aqueous layer [**SAVE**] by use of a Pasteur pipet and placed in a 3.0-mL capped vial. The toluene layer is then extracted with two 1.0-mL portions of water and the water extracts combined with the original water layer saved above. The combined aqueous layers are now extracted with two 0.5-mL portions of toluene and these toluene extracts combined with the original toluene layer.

The volumes of liquid used above are measured using calibrated Pasteur pipets. For each extraction, the mixture is shaken and vented, and the layers are allowed to separate. The transfers are made using Pasteur filter pipets.

The toluene solution is transferred by Pasteur pipet to a microcolumn (Pasteur filter pipet) containing 700 mg of anhydrous sodium sulfate previously wetted with toluene.(■) The dried toluene eluate, collected in a 10-mL Erlenmeyer flask containing a boiling stone, is concentrated by warming the solution in a sand **HOOD** bath under a gentle stream of nitrogen gas [**HOOD**]. A yellow, solid product is obtained.

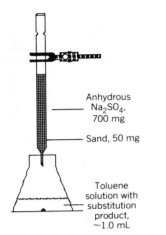

Anhydrous Na₂SO₄, 700 mg

Sand, 50 mg

Toluene solution with substitution product, ~1.0 mL

Purification and Characterization

The product is nearly pure but may be recrystallized from chloroform, using a Craig tube, if desired.

> **CAUTION:** *Chloroform is toxic! Dispense only in the* HOOD. *Do not breathe the vapors.*

Weigh the dried product and calculate the percentage yield. Determine the melting point and compare it with the literature value. Obtain an IR spectrum and compare your spectrum to that of an authentic sample.

ENVIRONMENTAL DATA

Substance	Amount	TLV (mg/m³)	Emissions (mg)	Volume (m³)
Toluene (SKIN) (Max., evap.)	1.35 mL	375	1214	3.2
Chloroform (Max., recryst.)	1.0 mL	50	1490	30

TXDS: Potassium thiocyanate—orl-hum LD50: 80 mg/kg
    Thiocyanic acid, 2,4-dinitrophenyl ester—orl-mus LD50: 2750 mg/kg
MTDS: Tetrabutylammonium bromide—dnd-mam:lyn 50 mmol/L

QUESTIONS   6-110. Explain the trend in the following reactions in terms of ease of reaction.

$$\text{C}_6\text{H}_5\text{Cl} \xrightarrow[\substack{350° \\ 31{,}000 \text{ kPa}}]{\text{NaOH(7\%)}} \text{C}_6\text{H}_5\text{OH} \;,\quad \text{4-Cl-C}_6\text{H}_4\text{NO}_2 \xrightarrow[160°]{\text{NaOH(15\%)}} \text{4-OH-C}_6\text{H}_4\text{NO}_2 \;,\quad \text{2-Cl-1,3,5-trinitrobenzene} \xrightarrow[\text{H}_2\text{O}]{\text{warm}} \text{2-OH-1,3,5-trinitrobenzene}$$

6-111. Complete each of the following reactions. Name the expected product.

a.

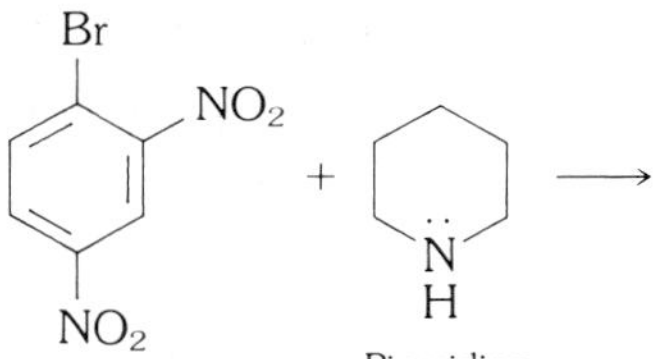

b. 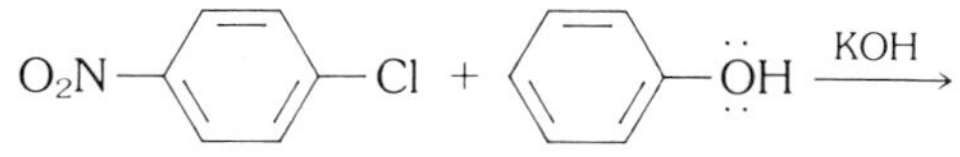 O_2N—⟨ ⟩—Cl + ⟨ ⟩—$\ddot{O}H$ $\xrightarrow{KOH}$

c. 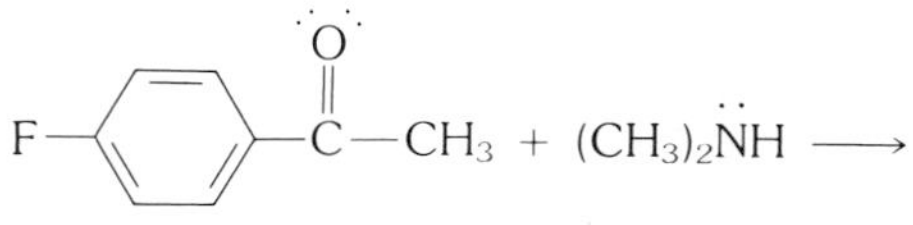 F—⟨ ⟩—$\overset{\overset{\ddot{O}}{\|}}{C}$—$CH_3$ + $(CH_3)_2\ddot{N}H$ $\longrightarrow$

6-112. Compare the potential energy diagram of an S_N2 substitution reaction on an aliphatic halide to that of the nucleophilic substitution reaction carried out in this experiment. Discuss the main differences in the diagrams in terms of the mechanisms.

6-113. Diagram the intermediate that is formed in the aromatic nucleophilic substitution reactions given below.

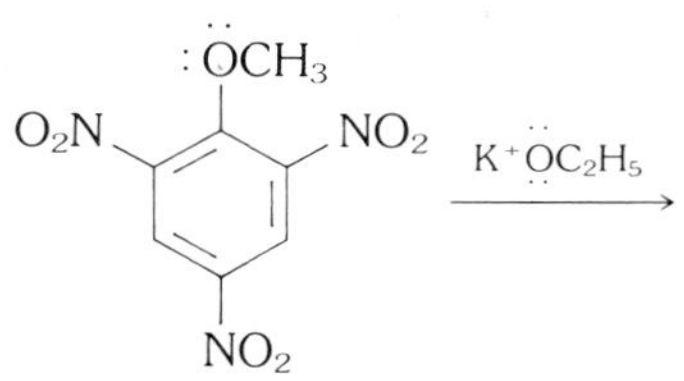

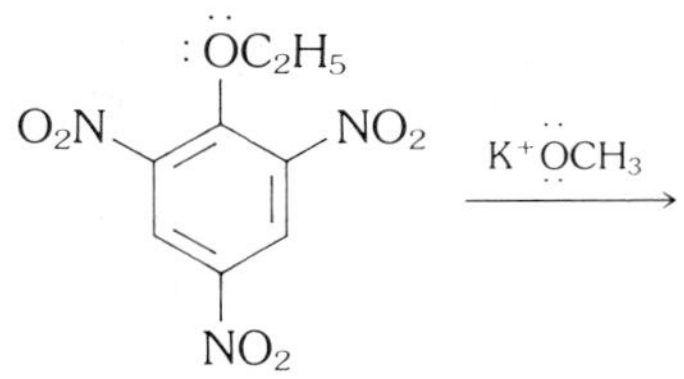

6-114. On workup of each of the reactions given in question 6-113, what product(s) would you expect to form? If more than one, indicate their ratio.

REFERENCES Selected examples of nucleophilic aromatic substitution reactions in "Organic Syntheses" include:

a. Hartman, W. W. "Organic Syntheses"; Wiley: New York, 1941; Collect. Vol. I, p 175.

b. Reverdin, F. *Ibid.*, p 219.

c. Hartman, W. W.; Byers, J. R.; Dickey, J. B. "Organic Syntheses"; Wiley: New York, 1943; Collect. Vol. II, p 451.

d. Brewster, R. Q.; Groening, T. *Ibid.*, 445.

e. Skorcz, J. A.; Kuminski, F. E. "Organic Syntheses"; Wiley: New York, 1973; Collect. Vol. V, p 263.

f. Kharasch, N.; Langford, R. B. *Ibid.*, p 474.

g. Bunnett, J. F.; Conner, R. M. *Ibid.*, p 478.

h. Sahyun, M. R. V.; Cram, D. J. *Ibid.*, p 926.

Experiment 38
Halogenation Using N-Bromosuccinimide: 9-Bromoanthracene

(Anthracene, 9-bromo-)

The use of N-bromosuccinimide as a highly specific brominating agent is demonstrated in this experiment.

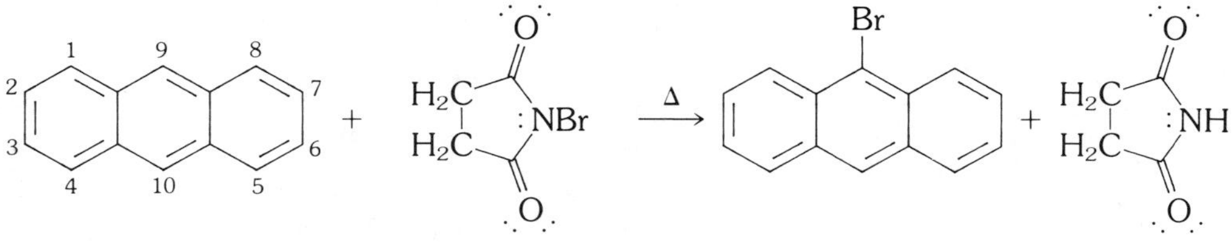

Anthracene   N-Bromosuccinimide   9-Bromoanthracene   Succinimide

DISCUSSION   N-Bromosuccinimide (NBS) is a highly specific brominating agent. Using this reagent, anthracene is brominated in the 9-position. NBS may also be used to brominate positions *alpha* to (1) a carbonyl group, (2) a triple bond, (3) an olefin (allylic position), (4) boron atoms, and (5) aromatic rings (benzylic position).

In the preparation of 9-bromoanthracene, the reaction is easily followed since N-bromosuccinimide (a reactant) and succinimide (a product) are both insoluble in carbon tetrachloride. The N-bromosuccinimide (NBS) reactant is more dense than the carbon tetrachloride solvent, and as reaction proceeds this solid disappears from the bottom of the reaction flask and the less dense succinimide forms and floats to the surface of the reaction solution. Other polynuclear hydrocarbons that have been brominated using NBS include naphthalene, phenanthrene, and acenaphthene.

Free radicals have been implicated in the mechanism of bromination using N-bromosuccinimide. In fact, the reaction proceeds only under photochemical conditions or in the presence of a free-radical initiator. The NBS reagent provides a source of Br_2 in the reaction mixture. This is a result of the rapid reaction of NBS with the hydrogen bromide that is formed in the substitution reaction. The generation of the bromine molecule initiates a chain reaction leading to the formation of 9-bromoanthracene.

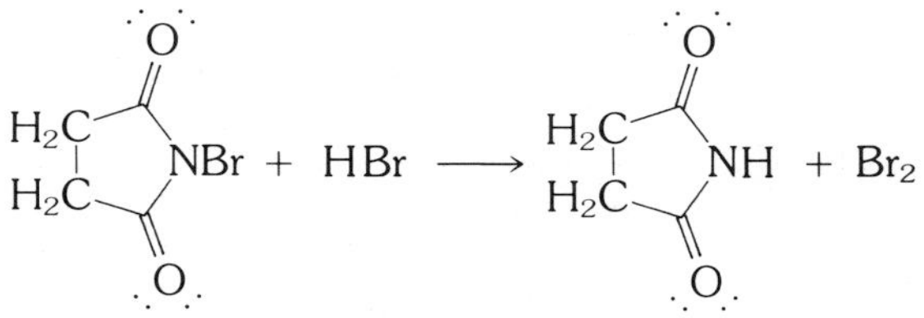

Inititation step: $: \ddot{Br} : \ddot{Br} : \xrightarrow{h\nu} 2 : \ddot{Br} \cdot$

Propagation step:

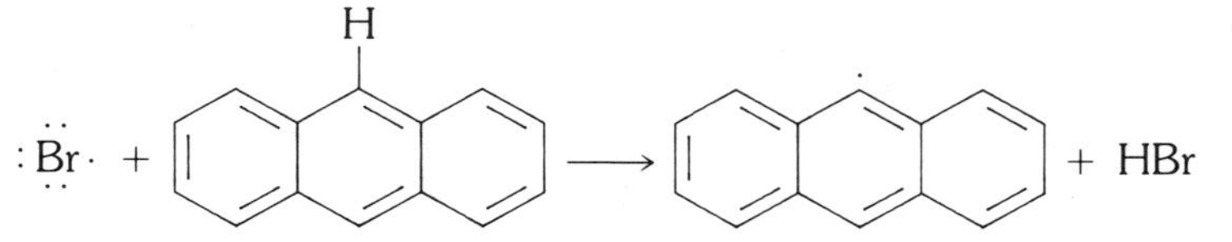

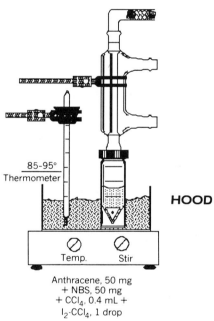

EXPERIMENTAL   Estimated time to complete the experiment: 2 hours.

Physical Properties of Reactants and Products

Compound	MW	Wt/Vol	mmol	mp(°C)	bp(°C)
Anthracene	178.24	50 mg	0.28	216	
N-Bromosuccinimide	177.99	50 mg	0.28	180–183	
Carbon tetrachloride	153.82	0.4 mL			77
Iodine-CCl$_4$ solution		1 drop			
9-Bromoanthracene	257.14			100–101	
Succinimide	99.09			126–127	

Reagents and Equipment

To a 1.0-mL conical vial containing a magnetic spin vane and equipped with a reflux condenser protected by a calcium chloride drying tube, add 50 mg (0.28 mmol) of anthracene and 50 mg (0.28 mmol) of N-bromosuccinimide.(■)

To this mixture add 0.4 mL of carbon tetrachloride followed by one drop of I$_2$-CCl$_4$ solution delivered from a Pasteur pipet.

> **WARNING:** CCl$_4$ *is a cancer suspect agent. It is dispensed in the* HOOD *using an automatic delivery pipet.*

Instructor Preparation of I$_2$-CCl$_4$ Solution

Iodine (0.2 g, 0.01 mol) is dissolved in 10 mL of carbon tetrachloride. The solution is placed in a [**HOOD**] for student use.

Reaction Conditions

The reaction mixture is heated, with stirring, to reflux in a sand bath (85–95°C) for a period of 1 hour. During this time the solution turns brown and crystals of succinimide appear at the surface of the reaction solution.

Isolation of Product

The succinimide product is collected by vacuum filtration of the warm solution using a Hirsch funnel.(■) The filter cake of succinimide is washed with three or four 0.5-mL portions of cyclohexane (calibrated Pasteur pipet). The washings are combined with the original filtrate.

The filtrate is concentrated to dryness [**HOOD**] under reduced pressure to give yellow-green crystals of 9-bromoanthracene. *Evaporation of the solvent can be accelerated by immersing the collection flask in warm water.*

Purification and Characterization

Weigh the air-dried succinimide and calculate the percentage yield. Determine the melting point and compare your value to that in the literature. Obtain an IR spectrum and compare your spectrum with that of an authentic sample.

The crude 9-bromoanthracene is purified by recrystallization from 95% ethanol using the Craig tube. Weigh the dried product and calculate the percentage yield. Determine the melting point and compare your result with the literature value. Obtain an IR spectrum and compare it with that of an authentic sample.

HOOD

HOOD

Anthracene, 50 mg
+ NBS, 50 mg
+ CCl$_4$, 0.4 mL +
I$_2$-CCl$_4$, 1 drop

85–95°
Thermometer

Temp.   Stir

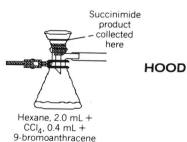

Succinimide
product —
collected
here

Hexane, 2.0 mL +
CCl$_4$, 0.4 mL +
9-bromoanthracene
product

ENVIRONMENTAL DATA

Substance	Amount	TLV (mg/m^3)	Emission (mg)	Volume (m^3)
Cyclohexane (Max., vac., evap.)	2.0 mL	1050	1560	1.5
Ethanol (Max., Craig tube)	1.0 mL	1900	800	0.4
Carbon tetrachloride (SKIN) (reflux, open transfer up to Max.)	0.4 mL	30	150–640	5–21

TXDS: Anthracene—orl-rat TDLo: 20 g/kg
  N-Bromosuccinimide—ipr-mus LDLo: 256 mg/kg
  Succinimide—orl-rat LD50: 14 g/kg

QUESTIONS

6-115. Predict and give a suitable name for the product(s) formed in the following reactions with NBS.

a. 1-Propene + NBS $\xrightarrow[h\nu]{CCl_4}$

b.  + NBS $\xrightarrow[h\nu]{CCl_4}$

c. $CH_3CH_2CH_2-CH=CH-CH_3$ $\xrightarrow[h\nu]{CCl_4}$ + NBS

d. + NBS $\xrightarrow[h\nu]{CCl_4}$

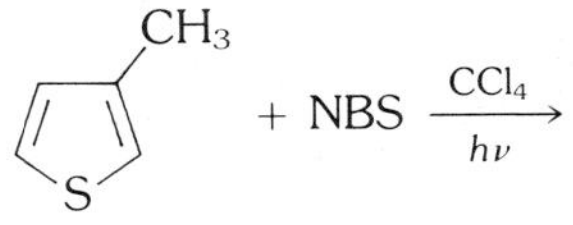

e. 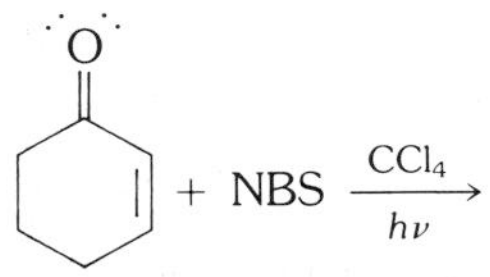 + NBS $\xrightarrow[h\nu]{CCl_4}$

6-116. When 1-octene is treated with NBS three monobromo straight-chain alkenes having molecular formula $C_9H_{17}Br$ are isolated from the reaction mixture. Identify these compounds and give each a suitable name.

6-117. Benzyl bromide, $C_6H_5CH_2Br$, can be prepared by treating toluene with NBS in the presence of a peroxide initiator. Suggest a suitable mechanism to account for this reaction.

6-118. The benzyl radical

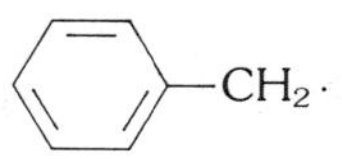

has unusual stability. Account for this fact by drawing appropriate resonance structures.

6-119. Suggest a suitable mechanism for the following reaction.

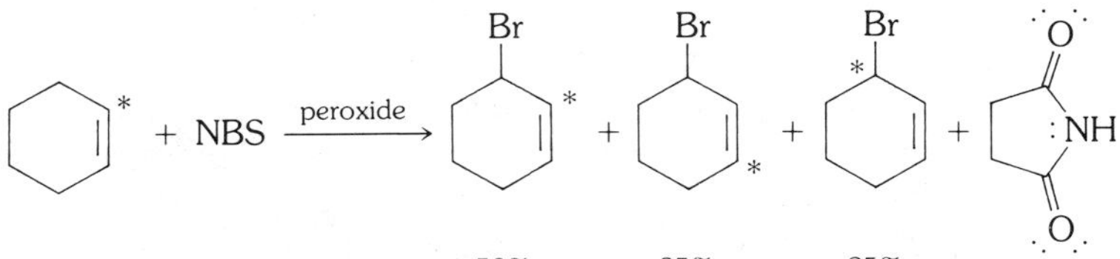

* = ^{13}c label   50%   25%   25%

6-120. The discussion indicates that other polynuclear hydrocarbons have been brominated using NBS. For the compounds shown below, give each a suitable name and draw the structure of the monosubstituted product that would form on reaction with NBS.

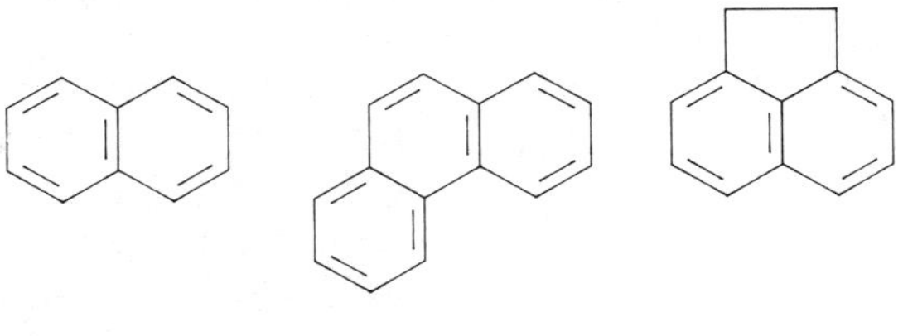

REFERENCES **1.** Review articles:

    a. Djerassi, C. *Chem. Revs.* **1948,** *43,* 271.

    b. Horner, L.; Winkelmann, E. H. *Angew. Chem.* **1959,** *71,* 349.

    c. Horner, L.; Winkelmann, E. H. *Newer Meth. Prep. Org. Chem.* **1964,** *3,* 151.

    d. Dauben, H. J. Jr.; McCoy, L. L. *J. Am. Chem. Soc.* **1959,** *81,* 4863.

   2. Brominations in "Organic Syntheses" using NBS:

    a. Greenwood, F. L.; Kellert, M. D.; Sedlak, J. "Organic Syntheses"; Wiley: New York, 1963; Collect. Vol. IV, p 108.

    b. Campaigne, E.; Tullar, B. F. *Ibid.,* p 921.

    c. Kalir, A. "Organic Syntheses"; Wiley: New York, 1973; Collect. Vol. V, p 825.

    d. Corbin, T. F.; Hahn, R. C.; Schechter, H. *Org. Synth.* **1964,** *44,* 30.

Experiment 39
Hypochlorite Oxidation of an Alcohol: Cyclohexanone

(cyclohexanone)

This experiment illustrates the oxidation of a secondary alcohol to a ketone.

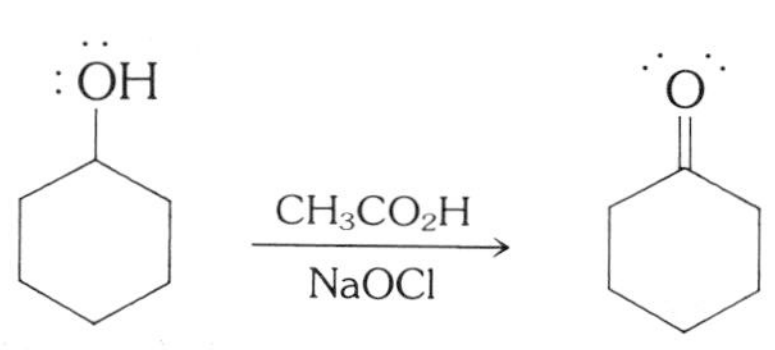

Cyclohexanol $\xrightarrow[\text{NaOCl}]{\text{CH}_3\text{CO}_2\text{H}}$ Cyclohexanone

DISCUSSION Sodium hypochlorite, or liquid bleach, contains 11.5–12.5% (1.8–2.0M) available chlorine. It is used extensively in swimming pool sanitation and as a bleach in the pulp and textile industries. A less concentrated product (5%) is used in laundries and around the household. The reaction described in this experiment illustrates the use of liquid bleach (11.5–12.5%) as an oxidizing agent in the organic laboratory.

    Sodium hypochlorite is prepared commercially by passing chlorine gas through a solution of aqueous sodium hydroxide.

$$\text{Cl}_2 + \text{Na}\ddot{\text{O}}\text{H} \rightleftharpoons \text{Na}\ddot{\text{O}}\text{Cl} + \text{NaCl}$$

The actual oxidizing agent in the present experiment is the chloronium ion, Cl^+ which is reduced in the reaction to chloride ion, Cl^-. The cyclohexanol acts as a reducing agent and thus becomes oxidized to cyclohexanone.

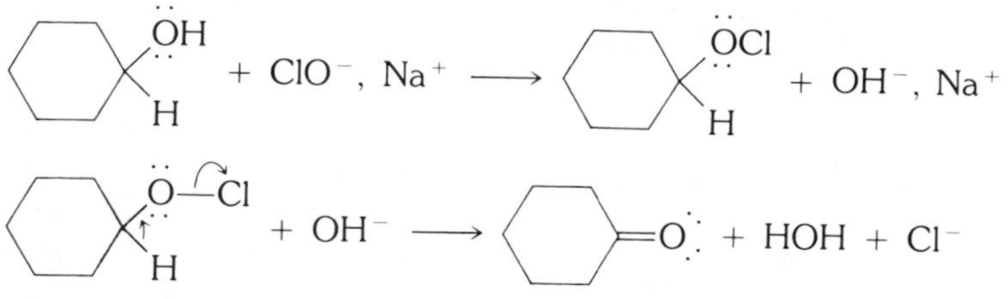

Complete details of the mechanism are not available to date.

EXPERIMENTAL   Estimated time to complete the experiment: 3.0 hours.

Physical Properties of Reactants and Products

Compound	MW	Wt/Vol	mmol	bp(°C)	Density	n_D
Cyclohexanol	100.16	100 mg	1.0	161	0.96	1.4641
Glacial acetic acid	60.05	250 μL		118		
Sodium hypochlorite solution (~12.5% aqueous)		2 mL				
Cyclohexanone	98.15			156	0.95	1.4507

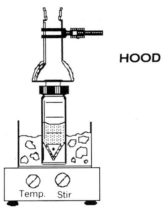

HOOD

Cyclohexanol, 100 mg +
CH₃COOH, 250 μL +
11.5–12.5% NaOCl, 2 mL

Reagents and Equipment

In a 5.0-mL conical vial containing a magnetic spin vane and equipped with a Hickman still, place 100 mg (1.0 mmol) of cyclohexanol and 250 μL of glacial acetic acid.(■) *The glacial acetic acid is dispensed in the* **HOOD** *by use of an automatic delivery pipet.*

The resulting solution is cooled in an ice bath. By use of a graduated pipet, there is now added dropwise with stirring, 2.0 mL of aqueous sodium hypochlorite solution (~12.5%, 1.8–2.0M). The ice bath is removed following the addition. *The NaOCl solution is added by inserting the pipet down the neck of the still just into the throat of the vial.*

Reaction Conditions

The resulting solution is stirred at room temperature for a period of 1 hour. *An excess of hypochlorite oxidizing agent should be maintained throughout the reaction period. The aqueous layer is monitored periodically using KI-starch paper. If a positive test is not obtained, additional sodium hypochlorite solution (1–3 drops) is added to ensure that an excess of the oxidizing agent is present. A Pasteur pipet inserted down the neck of the still is used to add the reagent and also to remove a few drops of solution for testing.*

120–130°C
Thermometer

Cyclohexanol, 100 mg +
CH₃COOH, 250 μL +
11.5–12.5% NaOCl, 2 mL +
NaHSO₃(sat. sol.),
~8–10 drops

Isolation of Product

Using a Pasteur pipet, add saturated, aqueous sodium bisulfite solution dropwise to the reaction mixture until the solution gives a negative KI-starch test.

The crude product is now distilled with stirring by heating the mixture at a sand bath temperature of 120–130°. (■) The first 0.5–1.0 mL of distillate is collected in the ring collar of the condenser. This material is then transferred to a 3.0-mL conical vial containing a spin vane using a Pasteur pipet. The condenser collar is rinsed with 0.5 mL of diethyl ether and the rinsing also transferred to the conical vial.

The distillate collected in the condenser collar consists of cyclohexanone, water, and acetic acid.

To neutralize the acetic acid present in the product mixture, anhydrous sodium carbonate (~100 mg) is added in small portions, with stirring, to the solution until evolution of CO_2 gas ceases. Add 50 mg of NaCl to the mixture. The resulting two-phase system is stirred until all the solid material dissolves.

The ether layer containing the cyclohexanone product is separated from the aqueous phase using a Pasteur filter pipet and transferred to a micro drying column. The column is assembled with a Pasteur filter pipet packed first with 300 mg of alumina (activity I) followed by 200 mg of anhydrous magnesium sulfate.(■) The column is wetted with diethyl ether before the transfer.

The eluate is collected in a tared 3.0-mL conical vial containing a boiling stone. The aqueous layer remaining in the conical vial is extracted with three 0.5 mL portions of ether and each extract is also passed through the column and combined with the original eluate.

The 3.0-mL conical vial is fitted with an air condenser and the ether removed by gentle warming in a sand bath in the **HOOD**.

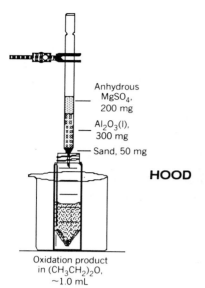

Anhydrous
MgSO₄,
200 mg

Al₂O₃(I),
300 mg

Sand, 50 mg

HOOD

Oxidation product
in (CH₃CH₂)₂O,
~1.0 mL

Purification and Characterization

The liquid residue of cyclohexanone isolated on evaporation of the ether solvent is sufficiently pure for characterization. Weigh the product and calculate the percentage yield. Determine the boiling point and compare your result with the literature value. Obtain an IR spectrum and compare it with that of an authentic sample.

ENVIRONMENTAL DATA

Substance	Amount	TLV (mg/m^3)	Emissions (mg)	Volume (m^3)
Cyclohexanol (Max., reactant)	100 mg	200	100	0.5
Cyclohexanone (SKIN) (Max., product)	77 mg	100	77	0.8
Acetic acid (Max., solv., distilled)	0.25 mL	25	262	10.5
Sodium hypochlorite solution (Max., treated as chlorine)	2.0 mL	3	250	83
Diethyl ether (Max., extract., evap.)	2.75 mL	1200	1610	1.4

TLV: Sodium hydrogen sulfite—5 mg/m^3
      Alumina—10 mg/m^3
TXDS: Sodium carbonate—orl-rat LDLo: 4000 mg/kg
       Magnesium sulfate—orl-mus LDLo: 5000 mg/kg
MTDS: Sodium hypochlorite—dnr-esc: 20 nL/disc

QUESTIONS

6-121. In the experiment, why is a solution of sodium bisulfite added to the reaction product mixture (Isolation of Product)? Write a reaction to account for what is happening. Is the bisulfite ion acting as an oxidizing or reducing agent?

6-122. In the isolation of the cyclohexanone product, 50 mg of sodium chloride is added to the water-cyclohexanone-diethyl ether mixture. Explain how the addition of sodium chloride aids in isolation of the cyclohexanone product.

6-123. Predict the product(s) for each of the following oxidation reactions. Give a suitable name for each reactant and product.

a. [structure: cyclooctatetraene with C(=O)—CH₃ group] $\xrightarrow[\text{2. H}^+]{\text{1. NaOBr}}$

c. $CH_3—CH—CH_3$ with $:OH$ below $\xrightarrow[\text{2. H}^+]{\text{1. NaOBr}}$

b. $CH_3CH_2\overset{..}{\underset{..}{O}}H$ $\xrightarrow[\text{2. H}^+]{\text{1. NaOI}}$

REFERENCES As a general reference for the use of sodium hypochlorite as an oxidizing agent see Fieser, L. F; Fieser, M. "Reagents for Organic Synthesis"; Wiley: New York, 1967; Vol. I, p 1084.

The above experiment is based on work reported by Zuczek, N. M.; Furth, P. S. *J. Chem. Educ.* **1981**, *58*, 824. See also Stevens, R. V.; Chapman, K. T.; Weller, H. N. *J. Org. Chem.* **1980**, *45*, 2030.

Experiment 40A, 40B
Chromium Trioxide-resin Oxidation of an Alcohol: 9-Fluorenone; Conversion to the 2,4-Dinitrophenylhydrazone

(fluoren-9-one)

In this experiment the oxidation of a secondary (2°) alcohol to a ketone using a polymer-bound chromium trioxide oxidizing agent is demonstrated. The progress of the reaction is followed by thin-layer chromatography (TLC). The product ketone is characterized by formation of its 2,4-dinitrophenylhydrazone derivative.

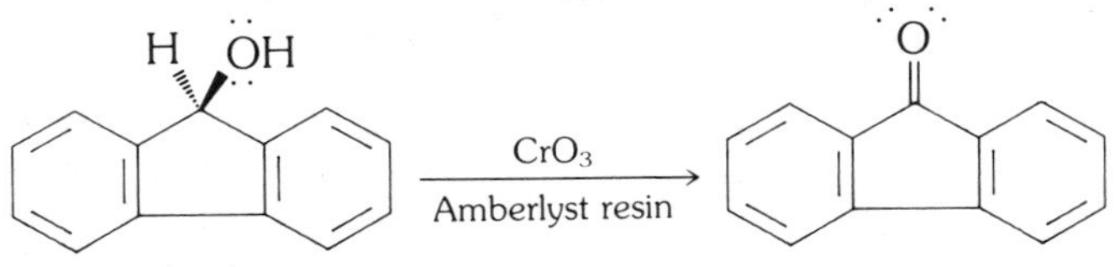

9-Fluorenol   9-Fluorenone

DISCUSSION This experiment illustrates the oxidation of a 2° alcohol to a ketone. The oxidizing agent commonly employed for this purpose is sodium dichromate or chromic oxide in sulfuric acid. In the present case, a convenient and advantageous polymer-bound chromium trioxide reagent is used. It is not only easy to prepare but is also easy to separate from the product mixture and can be recycled. Today, the use of polymeric reagents in organic synthesis is developing at a rapid pace. The mechanism of the oxidation is outlined below.

It should be noted that a solution of chromic oxide in aqueous sulfuric acid is used as a test reagent for 1° and 2° alcohols. A positive test is observed when the clear orange test reagent gives a greenish opaque solution upon addition of the alcohol.

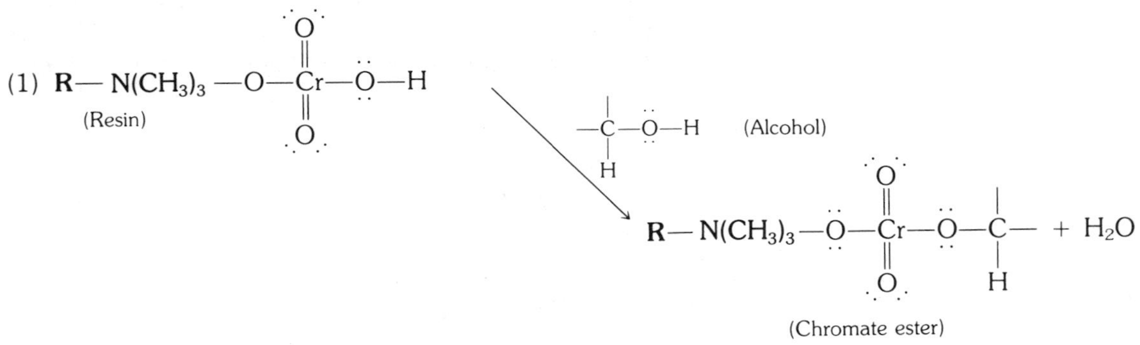

(2) $\text{R}\!-\!\text{N(CH}_3)_3\!-\!\ddot{\text{O}}\!-\!\overset{\overset{\displaystyle O}{\|}}{\underset{\underset{\displaystyle \ddot{O}}{\|}}{\text{Cr}}}\!-\!\ddot{\text{O}}\!-\!\overset{\displaystyle |}{\underset{\underset{\displaystyle H}{|}}{\text{C}}}\!-\!\longrightarrow\ \text{R}\!-\!\text{N(CH}_3)_3\!-\!\ddot{\text{O}}\!-\!\underset{\underset{\displaystyle \ddot{O}}{\|}}{\overset{\overset{\displaystyle \ddot{O}:^-}{\|}}{\text{Cr}}}\ ,\ \text{H}_3\text{O}^+\ +\ :\!\ddot{\text{O}}\!=\!\text{C}$

(Chromate ester)

$\underset{H}{\overset{\displaystyle\ddot{O}}{}}\underset{H}{}$

(Ketone)

PART A: 9-Fluorenone

EXPERIMENTAL   Estimated time to complete the experiment: 3–4 hours.

Physical Properties of Reactants and Products

Compound	MW	Wt/Vol	mmol	mp(°C)	bp(°C)
9-Fluorenol	182.23	100 mg	0.55	154	
Chromic oxide-resin		500 mg			
Toluene	92.15	3.5 mL			111
9-Fluorenone	180.22			84	

Reagents and Equipment

To a 10-mL round-bottom flask containing a magnetic stirring bar and equipped with a reflux condenser, add 100 mg (0.55 mmol) of 9-fluorenol, 500 mg of oxidizing resin, and 3.5 mL of toluene.(■) The resulting solution is sampled for TLC analysis (see the Technique section for an introduction to TLC).

Instructor Preparation

The CrO$_3$ resin is prepared by adding 35 g of Amberlyst A-26 resin to a solution of 15 g of CrO$_3$ in 100 mL of water. The mixture is stirred for 30 minutes at room temperature. The resin is then collected by vacuum filtration and successively rinsed with water and acetone. It is partially dried on the Buchner funnel by drawing air through the resin under vacuum for 1 hour and is then allowed to air-dry overnight.

Reaction Conditions

The contents of the flask are heated to reflux, with stirring, using a sand bath temperature of approximately 130°C.

Sample the solution for TLC analysis after a period of 5 minutes and every additional 15–20 minutes. In this manner, the progress of the reaction may be monitored until the conversion is complete. This point is reached when the TLC analysis shows that the 9-fluorenol has been completely consumed.

TLC conditions. *Eastman Kodak Fluorescent silica gel sheets (1.0 × 7 cm) are used. They are developed with 30% acetone in hexane and visualized by UV light. Reference R$_f$ values for 9-fluorenol and 9-fluorenone are determined using known compounds under the conditions cited above.*

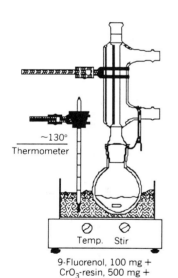

~130°
Thermometer

Temp.   Stir

9-Fluorenol, 100 mg +
CrO$_3$-resin, 500 mg +
CH$_3$C$_6$H$_5$, 3.5 mL

Isolation of Product

After cooling the reaction mixture, the resin is removed by gravity filtration through a cotton plug placed in a small funnel. The solution is transferred to the filter funnel using a Pasteur pipet, and the reaction flask and resin are rinsed with two 1.0-mL portions of methylene chloride (calibrated Pasteur pipet). The filtrate is collected in a tared 10-mL Erlenmeyer flask containing a boiling stone. The rinse is combined with the original filtrate.

HOOD The solvent is now removed [**HOOD**] from the filtrate under a stream of nitrogen by warming in a sand bath to yield a residue of the crude 9-fluorenone product.

Purification and Characterization

Obtain the weight of the crude product and calculate the percentage yield.

Recrystallize a 50-mg portion of the 9-fluorenone from hexane (about 1.0 mL of hexane/50 mg of ketone) using a Craig tube. Determine the melting point of this material and also calculate the percent recovery on recrystallization.

PART B: Fluorenone 2,4-Dinitrophenylhydrazone

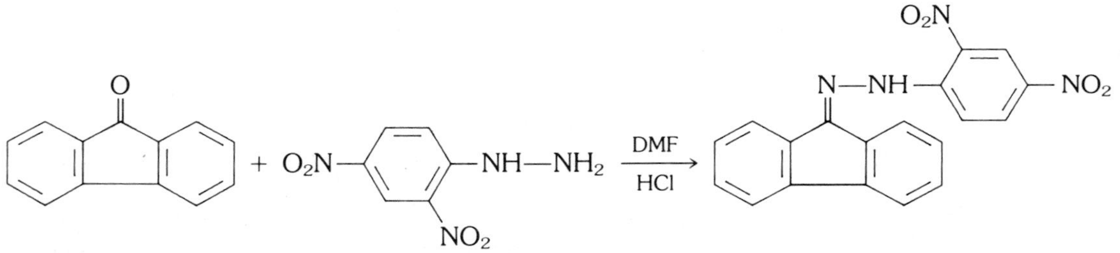

| 9-Fluorenone | 2,4-Dinitrophenylhydrazine | 2,4-Dinitrophenylhydrazone |

Physical Properties of Reactants and Products

Compound	MW	Wt/Vol	mmol	mp(°C)	bp(°C)
9-Fluorenone	180.22	38 mg	0.21	84	
2,4-Dinitrophenylhydrazine	198.14	50 mg	0.25	194	
Dimethylformamide	73.09	1.0 mL			149–150
2,4-Dinitrophenylhydrazone derivative	360.56			283	

The 9-fluorenone product may be further characterized by the preparation of its 2,4-dinitrophenylhydrazone derivative.

A solution of 50 mg (0.25 mmol) of 2,4-dinitrophenylhydrazine in 1.0 mL of dimethylformamide (DMF) is prepared in a 10 × 75 mm test tube. To this solution add 38 mg (0.21 mmol) of 9-fluorenone. After shaking the test tube so that all solids are in solution, two drops (Pasteur pipet) of concentrated HCl are added. The resulting mixture is allowed to stand at room temperature for a

2,4-DNP product collected here

DMF, 1 mL + 2*M* HCl, 1 mL + H$_2$O, 2 mL + reaction by-products

period of 5 minutes and then placed in an ice bath for an additional 5 minutes to complete precipitation of the derivative.

The crystals that deposit are collected by vacuum filtration using a Hirsch funnel. (■) They are washed with 1.0 mL of 2*M* HCl (dropwise) to remove unreacted 2,4-dinitrophenylhydrazine and DMF, followed by two 1.0-mL portions of water to remove any residual acid. The product is dried in a desiccator.

Weigh the derivative and calculate the percentage yield. Determine the melting point and compare it to the literature value.

ENVIRONMENTAL DATA

Substance	Amount	TLV (mg/m^3)	Emissions (mg)	Volume (m^3)
Toluene (SKIN) (Max., solv., evap.)	3.5 mL	375	3035	8.1
Methylene chloride (Max., wash, evap.)	2.0 mL	350	2670	7.6
Hexane (Max., Craig tube)	2.0 mL	180	1320	7.3
Hexane (Max., TLC solv.)	4.2 mL	180	2770	15
Acetone (Max., TLC solv.)	1.8 mL	1780	1425	0.8
Dimethylformamide (Max., deriv., solv., vac. filter)	1.0 mL	30	945	31.5

TLV: Concentrated HCl—5 ppm
TXDS: 9-Fluorenone—scu-rat TDLo: 300 mg/kg/26W-1
CrO$_3$—ihl-hmn TCLo: 100 μg/m^3 TFX:CAR

QUESTIONS

6-124. Suggest a suitable mechanism for the reaction of 9-fluorenone with 2,4-dinitrophenylhydrazine to form the corresponding 2,4-dinitrophenylhydrazone.

6-125. It is also possible to characterize 9-fluorenone by preparation of an oxime or semicarbazone. Formulate equations showing clearly the formation of these two derivatives and name each reagent used in the preparation.

6-126. As indicated in the discussion, a solution of chromic oxide in aqueous sulfuric acid is used as a test reagent for 1° and 2° alcohols.

a. What is this test (consult Chapter 7)?

b. Predict which of the alcohols listed below will give a positive test with the chromic oxide reagent. Give the structure for each of the alcohols:

1-heptanol   2,2,3-trimethyl-3-pentanol   cholesterol
3-methyl-2-butanol   4-*t*-butylcyclohexanol

REFERENCES

1. The present experiment is an adaptation of that reported by Wade, L. G., Jr.; Stell, L. M. *J. Chem. Educ.* **1980**, *57*, 438.
2. Introduction to the use of polymer-bound reagents:
   a. Leznoff, C. C. *Accounts Chem. Res.* **1978**, *1*, 327.
   b. Hodge, P. *Chem. in Britain* **1978**, *14*, 237.
3. Selected chromate oxidations reported in ''Organic Syntheses'' include:
   a. Conant, J. B.; Quayle, O. R. ''Organic Syntheses''; Wiley: New York, 1941; Collect. Vol. I, p 211.
   b. Sandborn, L. T. *Ibid.*, p 340.
   c. Bruce, W. F. ''Organic Syntheses''; Wiley: New York, 1943; Collect. Vol. II, p 139.
   d. Fieser, L. F. ''Organic Syntheses''; Wiley: New York, 1963; Collect. Vol. IV, p 189.
   e. *Ibid.*, p 195.
   f. Eisenbraun, E. J. ''Organic Syntheses''; Wiley: New York, 1973; Collect. Vol. V, p 310.

Experiment 41
Copper(II) Ion Oxidation of Benzoin: Benzil

(ethanedione, diphenyl-)

This experiment illustrates the oxidation of a secondary alcohol to a ketone using a dissolved metal ion catalyst.

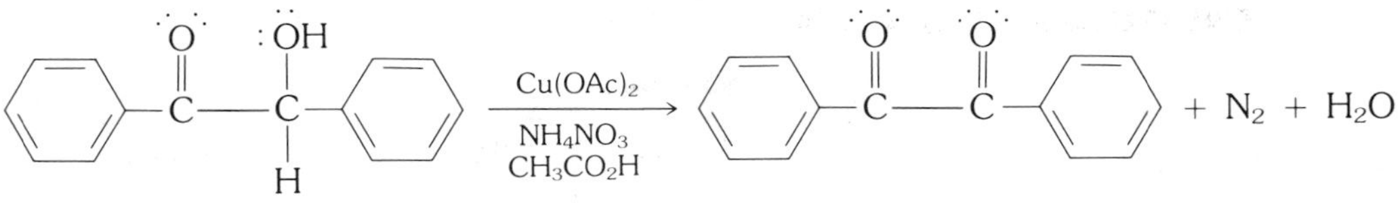

DISCUSSION   Benzil, a diketone, is obtained by the catalytic oxidation of benzoin using Cu^{2+} ion as the catalytic oxidant. The reaction is general for α-hydroxy ketones and is the basis of Fehling's test for reducing sugars. The mechanism of the oxidation shows the catalytic effect of the Cu^{2+} ion as it is continuously reduced and oxidized in the sequence outlined below. A key ingredient is the nitrate ion, which oxidizes the Cu^+ ion to the Cu^{2+} state and is in turn reduced to nitrite ion, which decomposes to yield nitrogen gas and water.

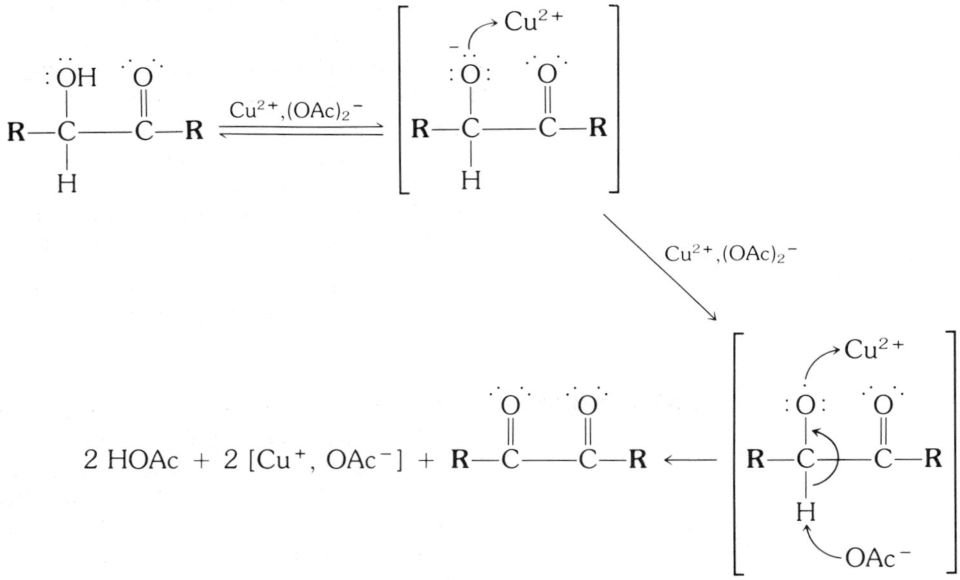

$$2\ Cu^+ + 2\ H^+ + NO_3^- \longrightarrow 2\ Cu^{2+} + NO_2^- + H_2O$$

$$2\ H^+ + NO_3^- \longrightarrow 2\ e^- + NO_2^- + H_2O$$

$$NH_4NO_2 \xrightarrow{\ H^+\ } N_2 + 2\ H_2O$$

EXPERIMENTAL Estimated time of the experiment: 2.5 hours.

Physical Properties of Reactants and Products

Compound	MW	Wt/Vol	mmol	mp(°C)
Benzoin	212.25	100 mg	0.47	137
Cupric acetate solution		350 μL		
Benzil	210.23			95

Instructor Preparation

The catalyst solution is prepared by dissolving 0.1 g of cupric acetate and 5 g of ammonium nitrate in 7.0 mL of deionized water (may require warming) followed by addition of 28 mL of glacial acetic acid. The container is placed in the **HOOD** and the solution dispensed by use of an automatic delivery pipet.

HOOD

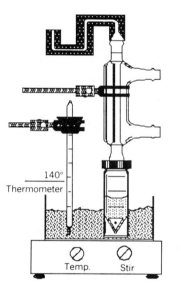

140°
Thermometer

Temp. Stir

Benzoin, 100 mg +
oxidation catalyst, 350 μL

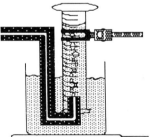

Benzil
collected
here

Aqueous acetic acid,
~1.5 mL and other
reaction by-products

Reagents and Equipment

A 1.0-mL conical vial containing a magnetic spin vane is equipped with a reflux condenser to which is attached a gas exit delivery tube. The system is charged with 100 mg (0.47 mmol) of benzoin and 350 μL of cupric acetate catalyst solution.(■) The gas delivery tube is led into the neck of an inverted 25-mL graduated cylinder that is immersed in a beaker of water.(■) This facilitates the measurement of the nitrogen gas evolved during the course of the reaction.

Note. *A calibrated test tube or a small buret may be used in place of the graduated cylinder. It is absolutely necessary that all connections be tight to prevent leakage of the gas evolved. The 7/10 T joint on the gas delivery tube is lightly greased.*

Reaction Conditions

The reaction mixture is heated with stirring at a sand bath temperature of ~140°C for a period of 1 hour or until the collected gas volume remains constant. *As the benzoin dissolves, the reaction mixture turns green and evolution of nitrogen gas commences. The theoretical volume of gas from 100 mg of benzoin is 10.6 mL (STP).*

Isolation of Product

If the gas delivery tube is used, *disconnect* it from the top of the condenser *before* removing the reaction vial from the heat source.

The reaction mixture is cooled to room temperature, 0.5 mL of cold water (calibrated Pasteur pipet) added, and the reaction vial placed in an ice bath for 10 minutes. The yellow crystals of benzil are collected by vacuum filtration using a Hirsch funnel.(■) The reaction vial and crystals are rinsed with an additional two 0.5-mL portions of cold water.

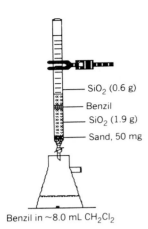

— SiO₂ (0.6 g)
— Benzil
— SiO₂ (1.9 g)
— Sand, 50 mg

Benzil in ~8.0 mL CH₂Cl₂

Purification and Characterization

The crude product is purified by recrystallization from methanol or 95% ethanol using the Craig tube. The yellow crystals of benzil are dried on a clay plate. *The benzil obtained after recrystallization often contains a small amount of benzoin impurity. It may be purified by chromatography on silica gel.*

A slurry of activated silica gel in methylene chloride is packed in a 1.0-cm diameter column to a height of 4.5 cm of the gel. (■) The sample of benzil is then introduced to the column followed by an additional 1.5 cm of silica gel (total silica gel amounts to 2.5 g). Approximately 8.0 mL of methylene chloride are used to elute the benzil, which is easily seen because of the yellow color. Concentration of the eluate gives pure benzil.

Weigh the dried product and calculate the percentage yield. Determine the melting point and compare your value to that reported in the literature. Obtain an IR spectrum of the product and compare it with that of the starting material and to an authentic sample of benzil.

ENVIRONMENTAL DATA

Substance	Amount	TLV (mg/m³)	Emissions (mg)	Volume (m³)
Acetic acid, 80% (v/v)	0.35 mL	25	112	4.5
Methanol (SKIN; Max., Craig tube)	1.0 mL	260	800	3.1
Ethanol (Max., Craig tube)	1.0 mL	1900	800	0.4

TXDS: Benzoin—orl-rat TDLo: 5460 mg/kg
Copper(II) acetate—orl-rat LD50: 595 mg/kg
Benzil—orl-rat LD50: 2710 mg/kg
AQTX: Ammonium nitrate—TLm 96: over 1000-100

QUESTIONS

6-127. In the directions given for the experiment, it is emphasized that the gas delivery tube must be disconnected from the top of the condenser before removing the reaction vial from the heat source. Why is this necessary?

6-128. What qualitative chemical tests would you perform to distinguish between benzoin and benzil? (See Chapter 7).

6-129. 1,2-Dicarbonyl compounds, such as benzil, can be characterized by reaction with 1,2-phenylenediamine to form a substituted quinoxaline.

a. Write the structure for the derivative obtained when benzil is the reactant. Do you think this compound would be colored? If so, why?
b. Suggest a suitable mechanism for the formation of the quinoxaline compounds based on the reaction scheme shown above.
c. What reagent would you react with 1,2-phenylenediamine to prepare the unsubstituted compound quinoxaline? Show a reaction scheme giving the structure of reactants and products.

REFERENCES

1. The synthesis of benzil is reported in "Organic Syntheses":
   a. Adams, R.; Marvel, C. S. *Org. Synth.* **1921,** *1,* 25.
   b. Clarke, H. T.; Dreger, E. E. "Organic Syntheses"; Wiley: New York, 1941; Collect. Vol. I, p 87.
2. For further information on the oxidation see: Weiss, M.; Appel, M. *J. Am. Chem. Soc.* **1948,** *70,* 3666.

Experiment 42
Ferric Chloride Oxidative Coupling of 2-Naphthol: 2,2'-Dihydroxydinaphthyl-1,1'

([1,1'-binaphthalene]-2,2'-diol)

The coupling reaction that aromatic phenols undergo in the presence of transition metal oxidants is demonstrated in this reaction. It mimics the biogenetic process that occurs in nature.

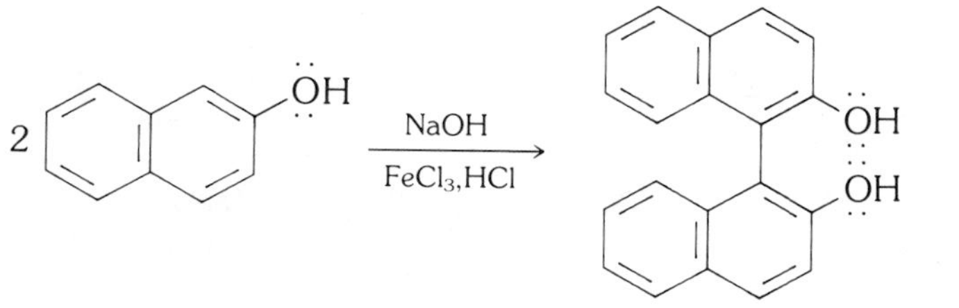

2-Naphthol   [1,1'-Binaphthalene]-2,2'-diol

DISCUSSION This reaction is important since it illustrates oxidative coupling of phenolic type species, which is an important biogenetic pathway in nature leading to the formation of natural products.

The coupling reaction involves oxidation of 2-naphthol by electron transfer to give an aryloxy radical, which then dimerizes to yield the product. The mechanism is shown below.

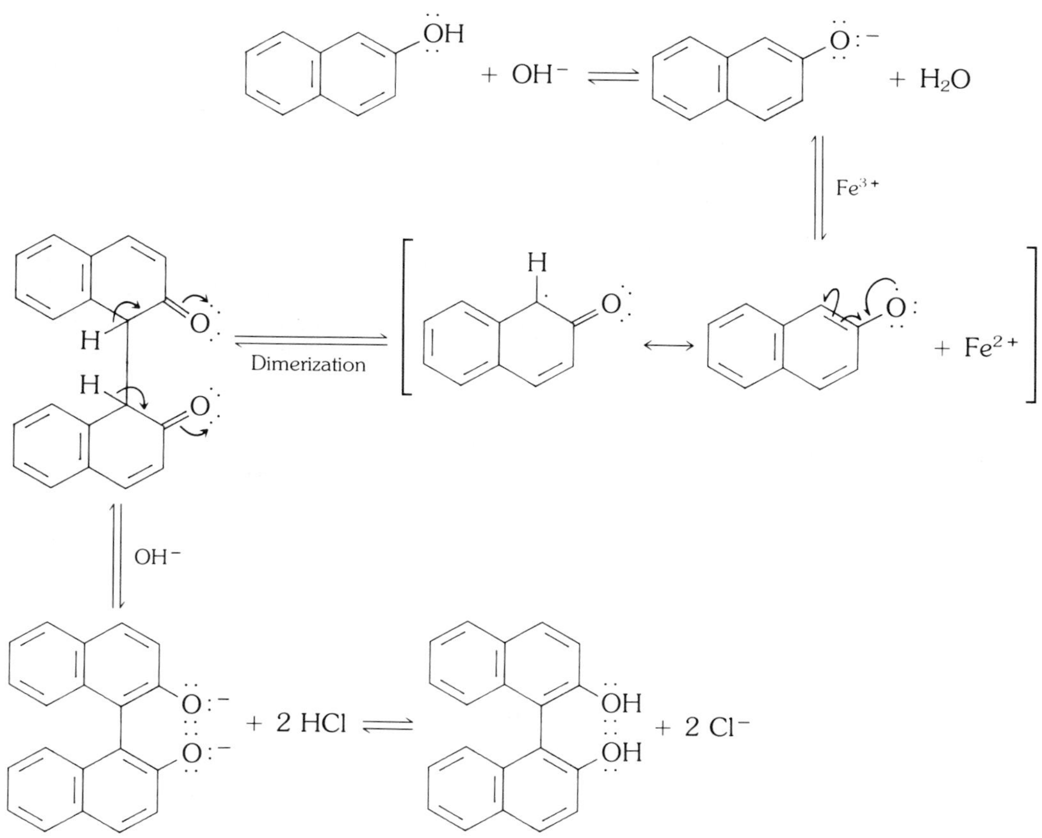

EXPERIMENTAL   Estimated time to complete the experiment: 2.0 hours.

Physical Properties of Reactants and Products

Compound	MW	Wt/Vol	mmol	mp(°C)
2-Naphthol	144.19	100 mg	0.69	123–124
Sodium hydroxide	40.00	30 mg	0.75	318.4
Ferric chloride · 6H$_2$O	270.30	297 mg	1.1	37
Conc. HCl	36.46	100 µL		
Dinaphthol	286.33			208–210

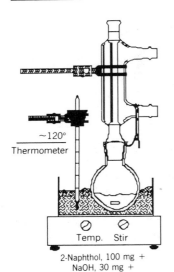

~120°
Thermometer

Temp.   Stir

2-Naphthol, 100 mg +
NaOH, 30 mg +
H$_2$O, 4.2 mL +
FeCl$_3$·6H$_2$O, 297 mg +
conc. HCl, 100 µl

Reagents and Equipment

To a 10-mL round-bottom flask containing a stirring bar and equipped with a reflux condenser, add 100 mg (0.69 mmol) of 2-naphthol, 30 mg (0.75 mmol) of sodium hydroxide, and 3.0 mL of water.(■)

> **CAUTION:** *Sodium hydroxide is a corrosive and toxic chemical. Do not allow it to touch the skin or to come in contact with the eyes.*

The reaction mixture is heated to reflux, with stirring, using a sand bath temperature of 120°C.

In a 10-mL Erlenmeyer flask prepare a solution of 180 mg (1.1 mmol) of anhydrous ferric chloride (MW162) or 297 mg of ferric chloride hexahydrate (MW270), 1.0 mL of water (calibrated Pasteur pipet), and 100 µL of concentrated hydrochloric acid.

> **WARNING:** *BE CAREFUL when mixing acid with water, especially if the acid is sulfuric. Add the acid TO the water. Avoid contact with the skin. The acid is dispensed using an automatic delivery pipet.*

Using a Pasteur pipet, transfer the ferric chloride solution, through the top of the condenser, to the reaction flask. The Erlenmeyer flask is rinsed with 200 µL of water, and the rinse is also added to the reaction flask as before.

Reaction Conditions

The resulting mixture is heated at reflux, with stirring, for a period of 45–60 minutes using a sand bath temperature of 120°C. It is allowed to cool to room temperature and then placed in an ice bath to complete the crystallization of the product.

Isolation of Product

The solid product is collected by vacuum filtration using a Hirsch funnel and the filter cake washed with two 1.0-mL portions of cold water (calibrated Pasteur pipet).(■)

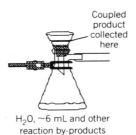

Coupled
product
collected
here

H$_2$O, ~6 mL and other
reaction by-products

Purification and Characterization

The crude product is recrystallized from 95% ethanol using a Craig tube, and the resulting crystals are dried on a porous clay plate.

Weigh the crystals and calculate the percentage yield. Determine the melting point and compare it with the literature value. Obtain an IR spectrum and compare your spectrum to that of an authentic sample.

ENVIRONMENTAL DATA

Substance	Amount	TLV (mg/m^3)	Emissions (mg)	Volume (m^3)
Hydrochloric acid, conc. (Max.)	100 μL	5	44	8.8
Ethanol (Max., Craig tube)	1.0 mL	1900	800	0.4

TLV: Ferric chloride—1 mg/m^3 as (Fe)
Sodium hydroxide—2 μg/m^3
TXDS: 2-Naphthol—orl-rat LD50: 2420 mg/kg

QUESTIONS

6-130. In the oxidative coupling reaction of 1-naphthol, it is possible to obtain three products.

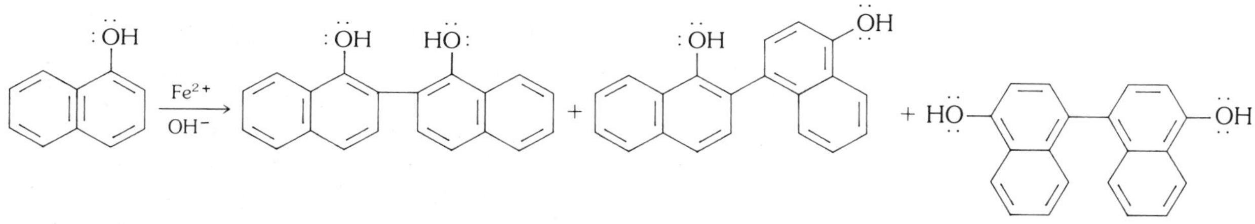

Account for the formation of this mixture by suggesting a mechanistic sequence similar to that presented in the discussion section of this experiment.

6-131. Predict the structure of the diphenylquinone product, $C_{28}H_{40}O_2$, formed by the oxidative coupling of 2,6-di-t-butylphenol with oxygen in the presence of alkali.

6-132. Substituted phenols, such as BHT, are used as antioxidants in processed foods.

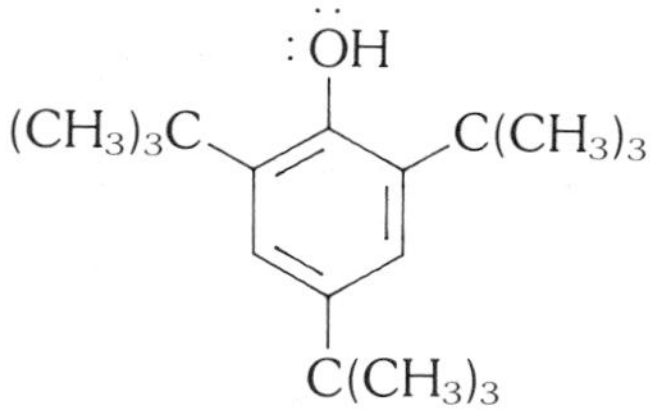

BHT

The role of the antioxidant is to stop spoilage caused by free radical reactions brought about by reaction of oxygen with compounds containing C=C bonds.
   a. Give a suitable chemical name for BHT.
   b. Can you suggest why this compound is an effective antioxidant?

REFERENCES

For references related to the oxidative coupling of phenolic derivatives, see:
   a. Dewar, M. J. S.; Nakaya, T. *J. Am. Chem. Soc.* **1968**, *90*, 7134.
   b. Scott, A. I. *Quart. Revs.* **1965**, *19*, 1.

Experiments 43A and 43B
Hypochlorite Oxidation of Methyl Ketones by the Haloform Reaction:
Benzoic Acid; p-Methoxybenzoic Acid

(benzoic acid; benzoic acid, p-methoxy-)

The well-known haloform reaction is often used to determine the presence of a methyl ketone. The reaction is also useful synthetically as a route to the preparation of organic acids.

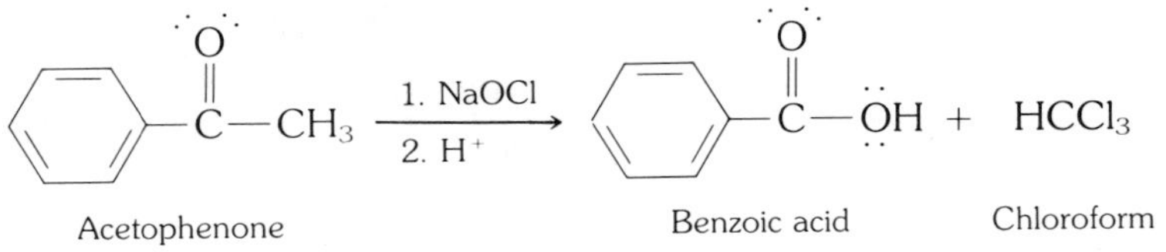

Acetophenone   Benzoic acid   Chloroform

DISCUSSION   The reaction of methyl ketones with a halogen in alkaline medium is known as the haloform reaction. In this experiment, halogen and base are present because of the equilibrium reaction:

$$H_2O + NaOCl + NaCl \longrightarrow 2\ NaOH + Cl_2$$

In the haloform reaction, two products are formed: (1) a haloform ($CHCl_3$, $CHBr_3$, or CHI_3), depending on the halogen employed; and (2) the carboxylic acid having one less carbon atom than the starting ketone. It is the formation of the carboxylic acid that gives the reaction synthetic utility.

If I_2 in a basic solution is used as the reactant, iodoform (CHI_3) is generated in the reaction. This compound is a solid and precipitates from the reaction medium. This observation has been used extensively as a chemical test for methyl ketones (see Chapter 7).

The reaction takes place in two stages. In the first stage, the methyl group is trihalogenated. In the second, base attacks the trihaloketone to generate the haloform and the alkali metal salt of the acid. Acidification yields the carboxylic acid. The sequence is shown below.

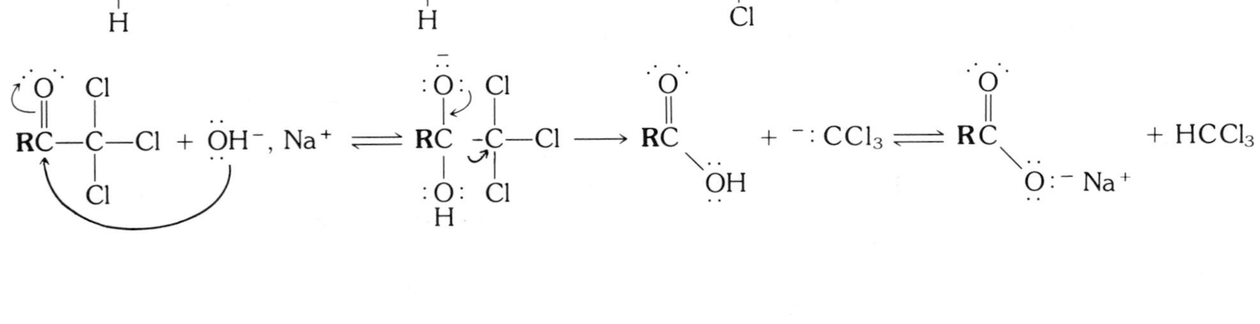

PART A: Benzoic Acid

The reaction is shown on page 262.

EXPERIMENTAL   Estimated time to complete the experiment: 1.5 hours.

Physical Properties of Reactants and Products

Compound	MW	Wt/Vol	mmol	mp(°C)	bp(°C)	Density	n_D
Acetophenone	120.16	20 µL	0.17	20.7	202.6	1.03	1.5372
Bleach (5%)		700 µL					
Benzoic acid	122.12			122–23			

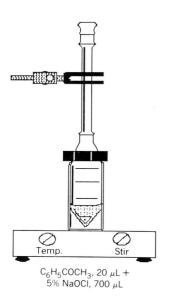

$C_6H_5COCH_3$, 20 µL +
5% NaOCl, 700 µL

Reagents and Equipment

To a 3.0-mL conical vial containing a magnetic spin vane and equipped with an air condenser, add 20 µL (21 mg, 0.17 mmol) of acetophenone and 700 µL of household bleach (5%).(■)

CAUTION: *Both reagents are irritants and should be dispensed in the* HOOD *using automatic delivery pipets.*

Reaction Conditions

The mixture is stirred at room temperature for a period of 0.5 hour.

Isolation of Product

Sodium sulfite (5 mg) is added to destroy any unreacted bleach and the reaction mixture stirred briefly. The resulting mixture is extracted with two 0.5-mL portions of ether (calibrated Pasteur filter pipet). Each ether extract portion is separated using a Pasteur filter pipet and discarded.

Oxidation
product
collected
here

Aqueous HCl, ~2 mL +
reaction by-products

Be careful. *The ether layer is the top layer; the lower alkaline layer contains the acid product. The ether extraction removes the chloroform generated in the reaction and any unreacted acetophenone. Dispose of the ether extracts in a suitable receptacle.*

The aqueous layer is made acidic (check with pH paper) by the dropwise addition of 3M HCl from a Pasteur pipet. A thick, white precipitate of benzoic acid then appears. The solid is collected by vacuum filtration using a Hirsch funnel and the filter cake washed with three 0.5-mL portions of water (calibrated Pasteur pipet).(■) The product is dried on a porous clay plate.

Purification and Characterization

The product is of sufficient purity for characterization. Weigh the material and calculate the percentage yield. Determine the melting point and compare your result with the literature value. Obtain an IR spectrum and compare it with an authentic sample of benzoic acid.

PART B: *p*-Methoxybenzoic Acid

$$CH_3\overset{..}{\underset{..}{O}}-\left\langle\right\rangle-\overset{\overset{\overset{..}{O}}{\parallel}}{C}-CH_3 \xrightarrow[\text{2. H}^+]{\text{1. NaOCl}} CH_3\overset{..}{\underset{..}{O}}-\left\langle\right\rangle-\overset{\overset{\overset{..}{O}}{\parallel}}{C}-\overset{..}{\underset{..}{O}}H + HCCl_3$$

p-Methoxyacetophenone   *p*-Methoxybenzoic acid   Chloroform

EXPERIMENTAL

Estimated time to complete the experiment: 1.5 hours.

Physical Properties of Reactants and Products

Compound	MW	Wt/Vol	mmol	mp(°C)	bp(°C)
p-Methoxyacetophenone	150.8	29 mg	0.19	38–39	258
Bleach (5%)		700 μL			
p-Methoxybenzoic acid	152.16			185	

Note. *The starting p-methoxyacetophenone reagent may be prepared in Experiment 52A.*

Reagents and Equipment

To a 3.0-mL conical vial containing a magnetic spin vane and equipped with an air condenser, add 29 mg (0.19 mmol) of *p*-methoxyacetophenone and 700 μL of household bleach (5%).(■)

> **CAUTION:** *The bleach is an irritant to skin and eyes. It is dispensed using an automatic delivery pipet in the* HOOD.

Reaction Conditions

The mixture is stirred for 0.5 hour with very gentle heating. *The lowest possible setting on a hot plate-magnetic stirrer is used.*

Isolation of Product

Sodium sulfite (5 mg) is now added; the reaction medium is stirred briefly and then cooled to room temperature. The resulting mixture is extracted with two 0.5-mL portions of ether (calibrated Pasteur pipet). The ether extract portions are separated using a Pasteur filter pipet and discarded.

Be careful. *The ether layer is the top layer; the bottom aqueous layer contains the product. The ether extraction removes the chloroform generated in the reaction and any unreacted p-methoxyacetophenone. Dispose of the ether extracts in a suitable receptacle.*

The aqueous layer is made acidic by the dropwise addition (check with pH paper) of 3*M* HCl from a Pasteur pipet. A thick, white precipitate of *p*-methoxybenzoic acid is formed. The mixture is cooled in an ice bath for 5 minutes and the solid collected by vacuum filtration using a Hirsch funnel.(■) The filter cake is washed with three 0.5-mL portions of cold water (calibrated Pasteur pipet) and the product dried on a porous clay plate.

Purification and Characterization

The product is reasonably pure but may be recrystallized from ethanol-water, if desired, using a Craig tube.

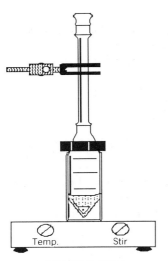

1,4-CH$_3$OC$_6$H$_4$COCH$_3$,
29 mg +
5% NaOCl, 700 μL

Oxidation product collected here

Aqueous HCl, ~2 mL + reaction by-products

Weigh the product and calculate the percentage yield. Determine the melting point and compare it with the literature value. Obtain an IR spectrum and compare it with that of an authentic sample.

ENVIRONMENTAL DATA

Substance	Amount	TLV (mg/m^3)	Emissions (mg)	Volume (m^3)
Household bleach (as Cl$_2$)^a	18 mg	3	18	6
Diethyl ether (Max.)	1.0 mL	1200	714	0.6
Chloroform (Max.)	—	50	21	0.4

^{a}Assuming all elemental chlorine escapes as chlorine gas.
TXDS: Acetophenone—orl-rat LD50: 900 mg/kg
Benzoic acid—orl-rat LD50: 2530 mg/kg
p-Methoxyacetophenone—orl-rat LD50: 1720 mg/kg
Sodium sulfite—inv-rat LD50: 115 mg/kg

QUESTIONS

6-133. In the haloform reaction, once the first α-hydrogen atom is replaced by a halogen atom, each successive hydrogen is more easily substituted until the trihalo species is obtained. Explain.

6-134. The haloform reaction using I$_2$ and NaOH is referred to as the "iodoform" test for methyl ketones (see Chapter 7). The test also gives positive results for compounds containing the —CH(OH)CH$_3$ grouping. This is due to the oxidation of the alcohol to the methyl ketone. Write a balanced equation for the conversion of this 2° alcohol unit to a methyl ketone in the presence of I$_2$ and NaOH. Identify which species is being oxidized and reduced.

6-135. If you were carrying out an industrial scale synthesis in which one step involved a haloform reaction to convert a methyl ketone into the corresponding acid having one less carbon atom, would you use NaOH and Cl$_2$, NaOH and Br$_2$, or NaOH and I$_2$ as the reagent? Give reasons for your choice.

6-136. Can you explain the fact that even though dibenzoylmethane is not a methyl ketone, it gives a positive iodoform test when treated with the NaOH and I$_2$ reagent?

REFERENCES

Selected examples from "Organic Syntheses" of the Haloform reaction include:
a. Sanborn, L. T.; Bousquet, E. W. "Organic Syntheses"; Wiley: New York, 1941; Collect. Vol. I, p 526.
b. Newman, M. S.; Holmes, H. L. "Organic Syntheses"; Wiley: New York, 1943; Collect. Vol. II, p 428.
c. Smith, L. I.; Prichard, W. W.; Spillane, L. J. "Organic Syntheses"; Wiley: New York, 1955; Collect. Vol. III, p 302.
d. Smith, W. T.; McLeod, G. L. "Organic Syntheses"; Wiley: New York, 1963; Collect. Vol. IV, p 345.
e. Staunton, J.; Eisenbraun, E. J. "Organic Syntheses"; Wiley: New York, 1973; Collect. Vol. V, p 8.

Experiment 44A
Air Oxidation of Aniline: Azobenzene

(diazene, diphenyl)

This reaction illustrates the base-promoted autoxidation of aniline to azobenzene.

DISCUSSION Primary aromatic amines have been oxidized to azo compounds using a variety of oxidizing agents. This experiment illustrates room-temperature, base-promoted autoxidation of aniline in dimethyl sulfoxide solution. A variety of o-, m-, and p-substituted anilines undergo this reaction, and it is of interest to note that these compounds are stable toward oxygen in the absence of the dimethyl sulfoxide solvent.

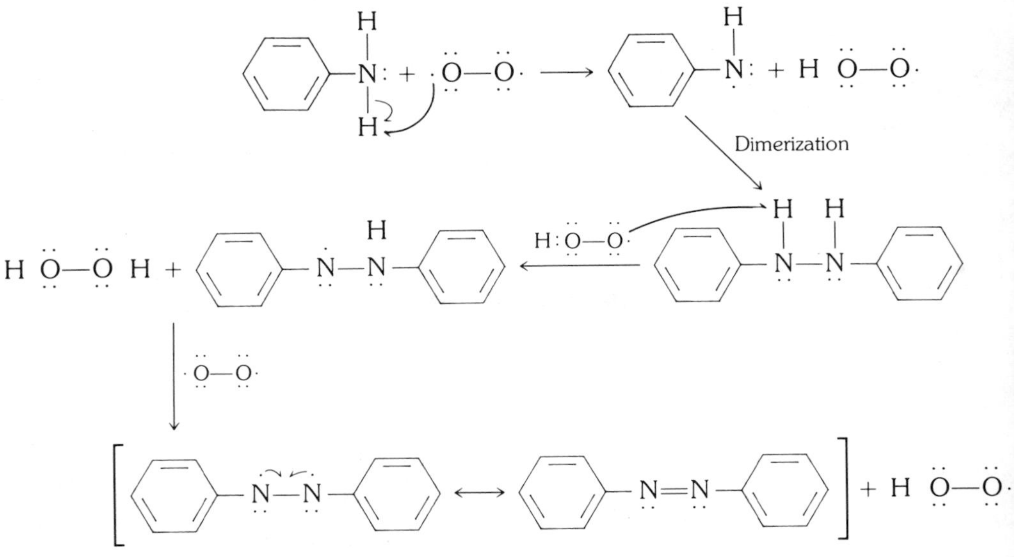

EXPERIMENTAL Estimated time to complete the reaction: 2.0 hours.

Physical Properties of Reactants and Products

Compound	MW	Wt/Vol	mmol	mp(°C)	bp(°c)	Density	n_D
Potassium t-butoxide	122.22	480 mg	3.9	256–258			
DMSO-t-butyl alcohol (80%)		8 mL					
Aniline	93.13	100 μL	1.08		184	1.02	1.5863
trans-Azobenzene	182.23			68.5			

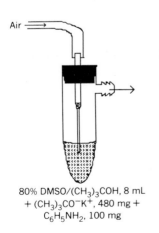

Air ⇌

80% DMSO/(CH$_3$)$_3$COH, 8 mL
+ (CH$_3$)$_3$CO$^-$K$^+$, 480 mg +
C$_6$H$_5$NH$_2$, 100 mg

Reagents and Equipment

To a 16 × 150 mm side-armed test tube fitted with an air inlet, add 480 mg (3.9 mmol) of potassium *tert*-butoxide and 8.0 mL of 80% dimethyl sulfoxide/ *tert*-butyl alcohol solution.(■) Using an automatic delivery pipet, add 100 μL (100 mg, 1.08 mmol) of aniline to the reaction mixture.

> **CAUTION:** *Potassium tert-butoxide is a moisture-sensitive, flammable solid; dimethyl sulfoxide is hygroscopic; aniline is highly toxic and a cancer suspect agent. These reagents are dispensed in the HOOD.*

Reaction Conditions

GENTLY   Air is now **GENTLY** bubbled through the solution at room temperature for a period of 1 hour. During this time the reaction mixture turns red and then quickly changes to brown.

Isolation of Product

The addition of 5–10 mL of water, followed by cooling of the mixture in an ice bath, yields crude, red azobenzene. The product is collected by filtration under reduced pressure using a Hirsch funnel and the filter cake washed with three 0.5-mL portions of cold water (calibrated Pasteur pipet).(■)

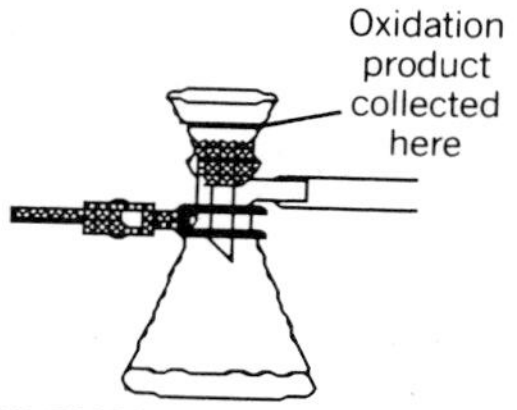

> **WARNING.** *Azobenzene is a cancer suspect agent.*

Purification and Characterization

The crude material is recrystallized, using a Craig tube, from 12:1 95% ethanol-water solution and dried on a porous clay plate.

Weigh the azobenzene and calculate the percentage yield. Determine the melting point and compare your value with that reported in the literature. Obtain an IR spectrum and compare it with that of an authentic sample.

80% DMSO/(CH₃)₃COH, 8 mL + aqueous KOH, 4.5 mL + (CH₃)₃COH, 0.5 mL + other reaction by-products

Oxidation product collected here

ENVIRONMENTAL DATA

Substance	Amount	TLV (mg/m³)	Emissions (mg)	Volume (m³)
Dimethyl sulfoxide, 80% (relatively involatile)	8.0 mL			
Aniline (SKIN) (Max., from yield)	100 mg	19	65	3.4
Ethanol (Max., Craig tube)	1.0 mL	1900	800	0.4

TLV: *t*-Butyl alcohol—100 ppm
TXDS: Dimethyl sulfoxide—orl-rat LD50: 19700 mg/kg
Azobenzene—orl-rat LD50: 1000 mg/kg

QUESTIONS

6-137. Azobenzene can also be prepared by the reduction of nitrobenzene using hot methanolic-sodium hydroxide solution with zinc dust. The stages of the reaction are outlined below.

nitrobenzene → nitrosobenzene → phenylhydroxylamine → azoxybenzene → azobenzene

Write a structural formula for each of the species involved in the transformation.

6-138. Azo compounds constitute a very useful group of initiators for free-radical polymerization reactions. The formation of a molecule of nitrogen on dissociation of the azo compound provides a strong driving force to generate an initial radical.

$$\text{R—N}\overset{..}{=}\overset{..}{\text{N}}\text{—R} \longrightarrow 2\ \text{R·} + \text{N}_2$$

Careful studies of the decomposition of AIBN, azoisobutyronitrile, have shown that the compound is less than 100% efficient in the production of radicals. That is, instead of producing two radicals per mole of AIBN, fewer are formed due to recombination of the radicals.

a. Give the structure for AIBN.

b. One of the compounds produced on dissociation of AIBN is a tetramethyl succinonitrile. Give a structure for this compound and account for its formation.

6-139. Based on the discussion in question 6-138, what product would be formed when

$$\text{D}_3\text{C—N}\overset{..}{=}\overset{..}{\text{N}}\text{—CD}_3$$

is allowed to decompose in isooctane solvent?

6-140. If the two azo compounds shown below are decomposed in the same reaction mixture, three products are formed in addition to the nitrogen gas. Give the structure of the expected compounds.

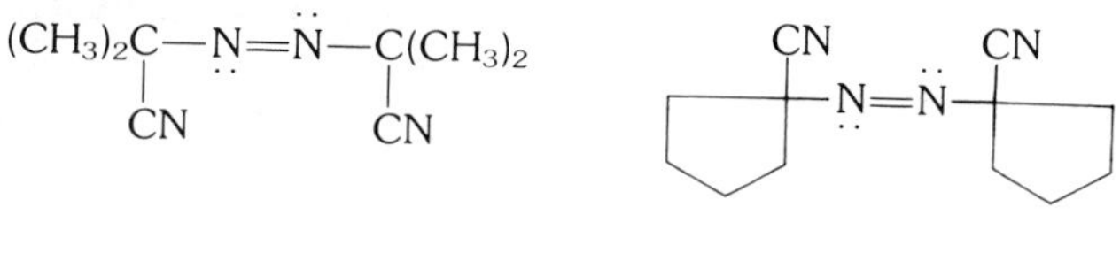

$$(\text{CH}_3)_2\underset{\underset{\text{CN}}{|}}{\text{C}}\text{—N}\overset{..}{=}\overset{..}{\text{N}}\text{—}\underset{\underset{\text{CN}}{|}}{\text{C}}(\text{CH}_3)_2$$

REFERENCES

1. The preparation of azobenzene is reported in Bigelow, H. E.; Robinson, D. B. "Organic Syntheses"; Wiley: New York, 1955; Collect. Vol. III, p 103.
2. Directions for the oxidation of a substituted aniline to prepare the corresponding azobenzene derivative is also given in Santuri, P.; Robbins. F.; Stubbings, R. "Organic Syntheses"; Wiley: New York, 1973; Collect. Vol. V, p 341.
3. The present experiment is based on the work of Russell, G. A.; Janzen, E. G.; Becker, H.; Smentowski, F. J. *J. Am. Chem. Soc.* **1962**, *84*, 2652.

Experiment 44B
Photochemical Isomerization: cis-Azobenzene

(diazene, diphenyl-)

The isomerization of *trans*-azobenzene to *cis*-azobenzene under photochemical conditions is demonstrated in this experiment. The course of the reaction is followed using thin-layer chromatography.

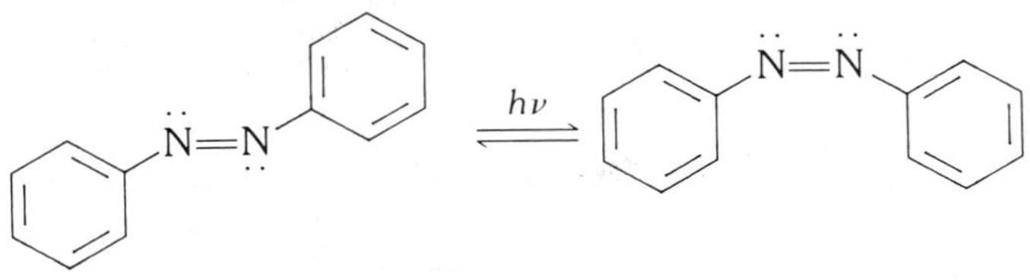

| *trans*-Azobenzene | *cis*-Azobenzene |

DISCUSSION

The photochemical conversion of *trans*- to *cis*-azobenzene proceeds by way of an excited electronic state of the trans isomer. Since both compounds are colored and have different R_f values, the conversion is conveniently followed using TLC.

Azo compounds, because of their intense color, are used commercially as synthetic dyes. Methyl red, prepared in Experiment 31, is an example of this class of materials.

EXPERIMENTAL

Estimated time to complete the experiment: 1.5 hours.

Physical Properties of Reactants and Products

Compound	MW	Wt/Vol	mmol	mp(°C)	bp(°C)
trans-Azobenzene	182.23	25 mg	0.14	68.5	
Methylene chloride	84.94	1.0 mL			40
cis-Azobenzene	182.23			71	

Reagents and Equipment
In a 3-mL conical vial place 25 mg (0.14 mmol) of azobenzene followed by 1.0 mL of methylene chloride.

> **CAUTION:** *Azobenzene is a cancer suspect agent. The azobenzene used may be prepared in Experiment 44A. The methylene chloride is dispensed in the HOOD using a graduated pipet.*

Determination of R_f Values

Using a capillary prepared from a melting-point tube, spot an activated silica thin-layer chromatographic plate (1.5 × 7 cm) with one drop of the above solution.

Refer to the Chromatographic Technique section for directions on setting up and carrying out a TLC experiment.

The chromatogram is developed in a 4 × 2 in. screwcap bottle using 5 mL of hexane. *Two spots are observed on the plate:*

trans-azobenzene ($R_f = 0.6$)

cis-azobenzene ($R_f = 0.1$)

The Isomerization Reaction

A second thin-layer plate is spotted with the same solution used above. The plate is now placed under a sunlamp (one foot away) for a period of 1 hour. The resulting plate is developed, as above, using hexane solvent.

Characterization

After development of the chromatogram, only one spot near the origin should be observed.

Determine the R_f value for this material and compare it to the control sample developed previously. The results should show that all the trans isomer present in the starting sample has been converted to the cis isomer.

ENVIRONMENTAL DATA

TLV: Hexane—50 ppm

TXDS: Azobenzene—orl-rat LD50: 1000 mg/kg
Silica gel—inv-mus LDLo: 234 mg/kg

QUESTIONS

6-141. Which of the two isomers of azobenzene would you expect to be the most thermodynamically stable? Explain.

6-142. Over the years the two isomers of azobenzene have been designated by various terms. Cis-trans, syn-anti, and **E-Z**. Using each set of terms, assign them to the isomers of azobenzene.

6-143. Listed below are the activation energies for the decomposition of a series of azo-compounds.

Compound	Activation energy (kjoule/mole)
$CH_3-N{=}\ddot{N}-CH_3$	210
$(CH_3)_2CH-N{=}\ddot{N}-CH(CH_3)_2$	170
$(C_6H_5)_2CH-N{=}\ddot{N}-CH(C_6H_5)_2$	110

Explain the trend in the table.

REFERENCES

This experiment is adapted from the work of Helmkamp, G. K.; Johnson, H. W., Jr. "Selected Experiments in Organic Chemistry", 2nd ed.; Freeman: San Francisco, 1968; p 32.

Experiment 45
Enzymatic Resolution of Optically Active Enantiomers: L-Alanine

(L-alanine)

The resolution of an optically active pair of enantiomers using a specific enzyme is demonstrated in this experiment.

$$CH_3-\underset{\underset{O}{\parallel}}{\underset{HN-C-CH_3}{\overset{H}{\underset{|}{C}}}}-\underset{\overset{O}{\parallel}}{C}-OH \xrightarrow[\text{acylase}]{\text{enzyme}} CH_3-\underset{\underset{O}{\parallel}}{\underset{HN-C-CH_3}{\overset{H}{\underset{|}{C}}}}-\underset{\overset{O}{\parallel}}{C}-OH + CH_3-\underset{NH_2}{\overset{H}{\underset{|}{C}}}-\underset{\overset{O}{\parallel}}{C}-OH$$

N-Acetyl-D,L-alanine   N-Acetyl-D-alanine   L-Alanine

DISCUSSION

This experiment illustrates the enzymatic resolution of a racemic mixture of N-acetyl-D-, and N-acetyl-L-alanine. The separation of the enantiomeric pair of isomers is possible because enzymatic hydrolysis of the N-acetyl derivative of the L-isomer occurs at a much faster rate than that of the corresponding N-acetyl-D-isomer. The reason for this difference in rate is that the enzyme is itself chiral and thus one enzyme–substrate interaction is preferred over the other. The L-alanine product is soluble in ethanol while the unreacted N-acetyl-D-isomer is not. This allows the L-alanine to be readily isolated.

EXPERIMENTAL

Estimated time to complete the experiment: 3.0 hours.

Physical Properties of Reactants and Products

Compound	MW	Wt/Vol	mmol	mp(°C)	bp(°C)
N-Acetyl-D,L-alanine	131.13	100 mg	0.76	137	
Water		1.0 mL			100
Ammonium hydroxide (6M)		170 μL			
L-Alanine	89.10			314d	

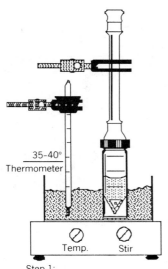

35–40°
Thermometer

Step 1:
CH₃CONHCHCO₂HCH₃
100 mg +
H₂O, 1 mL +
NH₄OH, 170 μL
+ acylase, 10 mg
Step 2:
6M CH₃CO₂H, 1.0 mL

Reagents and Equipment

To a 5.0-mL conical vial containing a magnetic spin vane and equipped with an air condenser, add 100 mg (0.76 mmol) of N-acetyl-D,L-alanine followed by 1.0 mL of water (calibrated Pasteur pipet).(■) A homogeneous solution is obtained by gentle warming in a sand bath. With stirring, add 170 μL of 6M NH₄OH solution and check the pH of the resulting solution using pH paper. If the pH is less than 8, an additional 10 μL of the ammonia solution is added and the pH again measured. *The pH of the solution must be above 8.*

HOOD

Note. *The ammonia solution is dispensed from an automatic delivery pipet in the* **HOOD.**

Add 10 mg of the enzyme acylase (grade I from porcine kidney).

Reaction Conditions

The homogeneous solution is heated in a sand bath at a temperature of 35–40°C for a period of 1 hour.

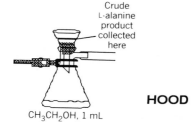

Crude
L-alanine
product
collected
here

CH₃CH₂OH, 1 mL

Isolation of Product

The reaction is quenched by the addition of 1.0 mL of 6M acetic acid solution (calibrated Pasteur pipet). *The addition of the acid denatures the protein.*

Using a Pasteur pipet, transfer the resulting solution to a 25-mL filter flask, add a boiling stone, and fit the flask with a Hirsch funnel containing filter paper. The solvent is removed under reduced pressure by swirling the flask on a hot plate [**HOOD**].

HOOD

Using a calibrated Pasteur pipet, add 1.0 mL of 95% ethanol. The mixture is warmed to ~80°C in a sand bath and cooled to room temperature, and then the crude solid product is collected by filtration under reduced pressure using a Hirsch funnel. (■)

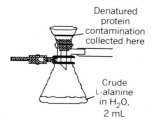

Denatured
protein
contamination
collected here

Crude
L-alanine
in H₂O,
2 mL

The collected material is transferred to a 10-mL Erlenmeyer flask and redissolved in 2.0 mL of water by heating the mixture to ~95°C in a sand bath. The hot solution is then filtered under reduced pressure using a Hirsch funnel. (■) The filtrate is collected in a 25-mL filter flask. *This operation is performed to remove any remaining protein.*

The filtrate is concentrated under reduced pressure as above. The resulting crude amino acid is scraped from the flask and placed in a Craig tube.

Purification and Characterization

The crude material is recrystallized by adding 100 μL of water, heating the mixture to ~90°C in a sand bath, and then, after allowing the solution to cool slightly, adding 200 μL of 95% ethanol. The solution is cooled to room temperature; the solid product is isolated and dried on a porous clay plate. *The water and alcohol are dispensed using automatic delivery pipets.*

Weigh the amino acid and calculate the percentage yield. Determine the melting point and compare your result with the literature value. Obtain an IR spectrum and compare it with that of an authentic sample. L-*Alanine has a very high melting point (over 300°C). It is recommended that a Fisher-Johns melting-point apparatus be used if the melting point is determined.*

ENVIRONMENTAL DATA

Substance	Amount	TLV (mg/m³)	Emissions (mg)	Volume (m³)
Acetic acid, 6M (Max. removal of excess under vacuum)	1.0 mL	25	360	14.4
Ethanol, 95% (Max. vac. filter)	1.2 mL	1900	950	0.5
Ammonium hydroxide, 6M (Max, pH control. Treat as NH₃)	170 μL	18	17	1.0
N-Acetyl-D,L-alanine (reactant, involatile)	100 mg			

QUESTIONS   **6-144.** In this experiment, the reaction is carried out using a specific enzyme. Several thousand different enzymes are known, each of which catalyzes a specific reaction. An enzyme is a complex protein.
    a. What is a protein?
    b. Draw the structural linkages that make up the protein polymer.
    c. What is the general name given to these linkages?

6-145. Enzymes function in our bodies at 37°C and nearly neutral pH. In this experiment you were cautioned not to warm the reaction mixture above 40°C. What process occurs if the temperature is raised above 40°C, and what is the general term used to describe this process?

6-146. Each of the known enzymes specifically catalyzes one and only one reaction. Can you offer a reasonable explanation to account for this fact?

6-147. All enzymes are "globular" proteins. What is meant by this term?

6-148. Humans can separate enantiomers biologically. For example, if racemic alanine is eaten, the **S**-isomer is metabolized and the **R**-isomer is excreted in the urine. Why do you think this is so?

6-149. The conventional method of separating enantiomers is to convert them into compounds that are diastereomers. What does this process accomplish as an aid in separating the original mixture of enantiomers?

6-150. Chromatographic techniques will separate compounds belonging to different functional group classes, members of the same functional group class having different molecular structures and also diastereomers such as **E**- and **Z**-alkenes. Do you think it might be possible to design a chromatographic packing material that would accomplish the separation of a mixture of enantiomers by this technique? If so, what specific features might this column packing necessarily contain?

REFERENCES
1. For a review of resolution techniques see Wilen, S. H. *Top. Stereochem.* **1971**, *6*, 107.
2. This experiment is a modification of the work reported by Greenstein and co-workers. Birnbaum, S. M.; Levintow, L.; Kingsley, R. B.; Greenstein, J. P. *J. Biol. Chem.* **1952**, *194*, 455.

Experiment 46
Beckmann Rearrangement: Benzanilide

(benzamide, *N*-phenyl-)

This experiment illustrates the rearrangement of a ketoxime (prepared *in situ*) to the corresponding amide.

Benzophenone   Hydroxylamine hydrochloride   Benzophenone oxime   Benzanilide

DISCUSSION   The Beckmann rearrangement was discovered in 1886 by E. Beckmann and is the reaction of a ketoxime in the presence of acid to yield the corresponding amide or amides. In the present case, only one amide is formed since benzophenone is a symmetrical ketone. Because the ketoximes are prepared from ketones, the reaction is often used to determine the structure of the starting ketone. This is done by identification of the acid and amine obtained on hydrolysis of the amide formed by the rearrangement.

The proposed mechanism involves an intramolecular (S_N2) migration. There are two significant points concerning the stereochemistry of the reaction: (1) the group that migrates is *anti* to the hydroxyl group and (2) the stereochemical features of the migrating group are retained.

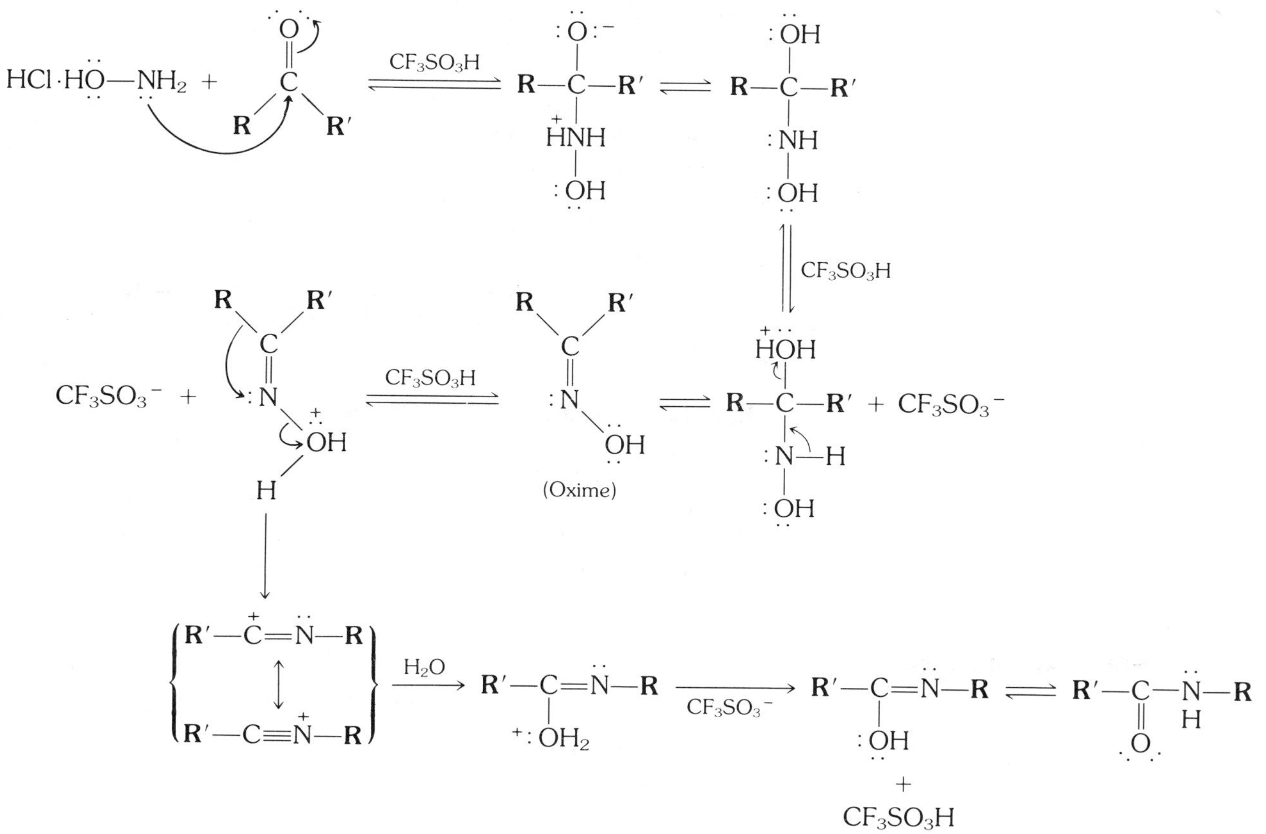

EXPERIMENTAL

Estimated time to complete the experiment: 2.5 hours.

Physical Properties of Reactants and Products

Compound	MW	Wt/Vol	mmol	mp(°C)
Benzophenone	182.21	100 mg	0.55	48.1
Hydroxylamine hydrochloride	69.49	51 mg	0.73	155–57
Triflic acid—formic acid solution		500 μL		
Benzanilide	197.24			162–64

Reagents and Equipment

In a 5.0-mL conical vial containing a magnetic spin vane and fitted with a reflux condenser, place 100 mg (0.55 mmol) of benzophenone and 51 mg (0.73 mmol) of hydroxylamine hydrochloride.(■) Using an automatic delivery pipet, add 500 μL of triflic acid-formic acid solution [**HOOD**] to this mixture.

HOOD

> **CAUTION:** *Triflic acid (trifluoromethane sulfonic acid) is one of the strongest acids known. It is very corrosive and toxic!*

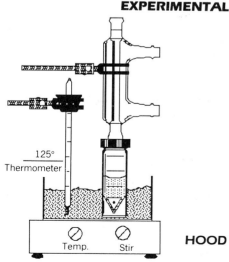

125°
Thermometer

Temp.   Stir

$(C_6H_5)_2CO$, 100 mg +
$HCl \cdot H_2NOH$, 51 mg +
$CF_3SO_3H \cdot HCO_2H$, 0.5 mL

Instructor Preparation
The acid solution is prepared by adding two drops of triflic acid to 5.0 mL of 90% formic acid.

Reaction Conditions
With stirring, heat the reaction mixture at reflux for 1.0 hour in a sand bath (125°C). The resulting solution is then cooled to room temperature.

Isolation of Product
To the cooled reaction solution add 1.0 mL of water (calibrated Pasteur pipet), and extract the resulting mixture with three 1.0-mL portions of methylene chloride. The organic phase is separated using a Pasteur filter pipet and transferred to a 10-mL Erlenmeyer flask. For each extraction, after addition of the methylene chloride, the vial is capped, shaken, and vented, and the layers are allowed to separate.

The combined methylene chloride extracts are dried over granular, anhydrous sodium sulfate. Using a Pasteur filter pipet, the dried solution is transferred to a clean, dry 10-mL Erlenmeyer flask containing a boiling stone.

HOOD   The solvent is evaporated [**HOOD**] under a gentle stream of nitrogen or in a warm sand bath.

Characterization and Purification
The crude benzanilide is removed from the Erlenmeyer flask and recrystallized from 95% ethanol using a Craig tube.

Weigh the product and calculate the percentage yield. Determine the melting point and compare it to the literature value. Obtain an IR spectrum and compare it to that of an authentic sample.

ENVIRONMENTAL DATA

Substance	Amount	TLV (mg/m³)	Emissions (mg)	Volume (m³)
Triflic acid-formic acid (For formic acid only, transfer, reflux)	0.5 mL	9	13	1.4
Methylene chloride (Max., less in hood)	3 mL	350	4000	11.4
Ethanol (Max., Craig tube)	1.0 mL	1900	800	0.4

TXDS: Benzophenone—orl-mus LD50: 2895 mg/kg
Hydroxylamine hydrochloride—orl-mus LD50: 400 mg/kg

QUESTIONS   **6-151.** Draw the structure of the product expected in each of the Beckmann rearrangements presented below. Give a suitable name for each product.

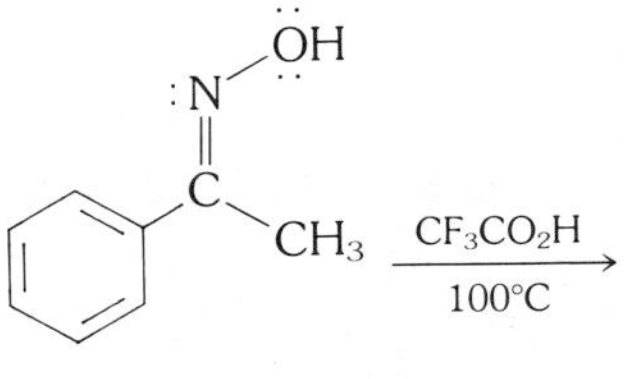

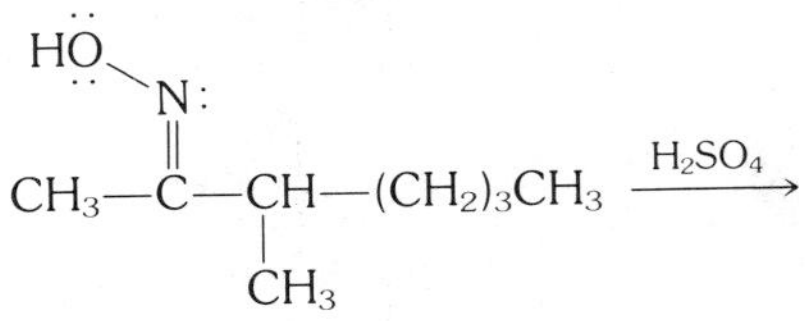

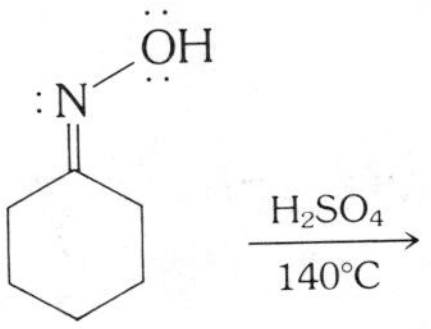

6-152. The picryl esters of oximes undergo the Beckmann rearrangement without an acid catalyst.

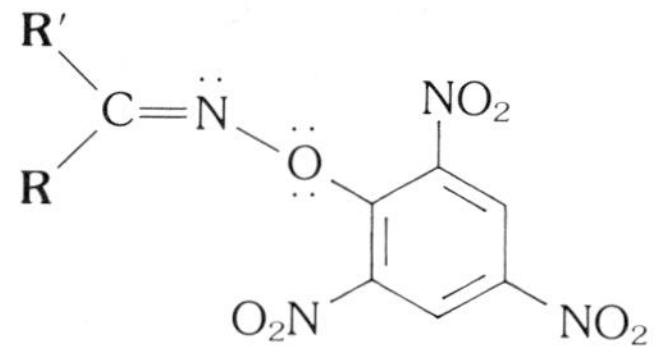

(Picryl ester)

Explain why a catalyst is not required in this rearrangement.

6-153. Compounds A and B undergo the Beckmann rearrangement with gentle heating of the solid compounds.

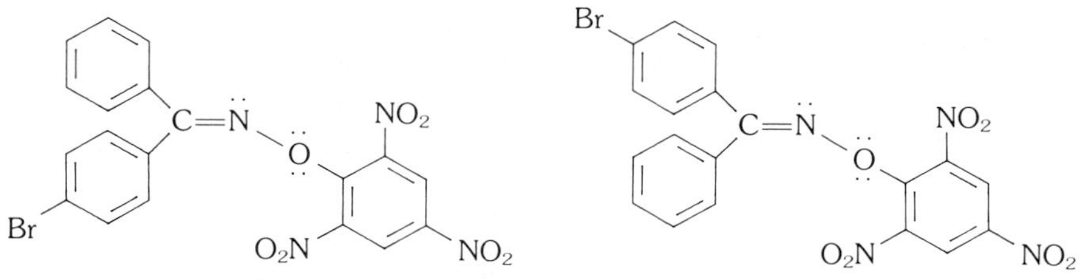

A

B

a. Write the structure of the products expected.

b. If a mixture of the two esters, A and B, are heated in the same reaction flask, what products would be formed? Explain.

REFERENCES

1. For reviews on the Beckmann rearrangement see:
   a. Blatt, A. H. *Chem. Revs.* **1933,** *12,* 215.
   b. Jones, B. *Ibid.* **1944,** *35,* 335.
   c. Donaruma, L. G.; Heldt, W. Z. *Org. React.* **1960,** *11,* 1.
2. An example of the Beckmann rearrangement using different reagents is given in "Organic Syntheses": Ohno, M.; Naruse, N.; Terasawa, I. "Organic Syntheses"; Wiley: New York, 1973; Collect. Vol. V, p 266.
3. The present reaction was adapted from the work of Ganboa, I.; Palomo, C. *Synth. Comm.* **1983,** *13,* 941.

SEQUENTIAL EXPERIMENTS

The synthesis of organic molecules in academic and industrial laboratories has been one of the great achievements of the organic chemist. A vast array of compounds that have had a profound effect on our way of life have been devised, synthesized, and applied to many of those areas that we all too often take for granted. Just over the last 30 years, new advances in pharmaceuticals, textiles, surfactants, plastics, synthetic oils, and the like have made a great impact on our society. All this due to the ingenuity, imagination, and knowledge of the chemist!

In practice, the synthesis of a new material usually requires a number of sequential steps. For example, the synthesis of the biologically active material cortisone requires 33 steps. Sequences of this length are rare, but syntheses requiring three to six steps are common. The objective of the organic chemist is to devise a multistep outline whereby the desired compound can be prepared efficiently, using inexpensive, readily available starting materials. For each individual step, the yield of product should be high with the minimum number of side reactions. In addition, the overall cost of the proposed project must be considered, including the time involved, type of equipment required, plus safety and environmental considerations.

The yield of product from each individual step has a great effect on the overall yield of the final product obtained from a multistep sequence. This is because chemical transformations always take place with some loss of material and in less than quantitative yields. *In a multistep synthesis, the overall yield is the mathematical product of the individual steps.* For example, if we assume that in a five-step sequence for the preparation of a new organosilicone dye, each step takes place in 85% yield, the overall yield would be $(0.85)^5 \times 100$ or 44%. This result points out why a synthetic route devised for a particular compound must be well planned so as to minimize losses at each stage of the chosen route.

Experiments 47A–C; 48A–D; 41, 49, 50A–B, 51; and 25C, 52A–B, 43B are designed as examples of multistep syntheses. When performing one or all of these multistep sequences, you will begin to understand the reason why laboratory technique is so important, since the product of one reaction is used as the starting material to prepare the next compound in the sequence. When you finally isolate the *target* compound, you will feel the excitement that has come to those who have made meaningful advances in the world of chemistry.

SEQUENCE A[2]: **Piperonal Reactions: The Synthesis of Piperonylonitrile from Piperonyl Alcohol (Experiments 47A–C)**

Experiment 47A
Piperonal

(benzaldehyde, 3,4-(methylenedioxy-)

This experiment illustrates the selective oxidation of a primary (1°) alcohol to an aldehyde using polymer-bound chromium trioxide as the oxidizing agent.

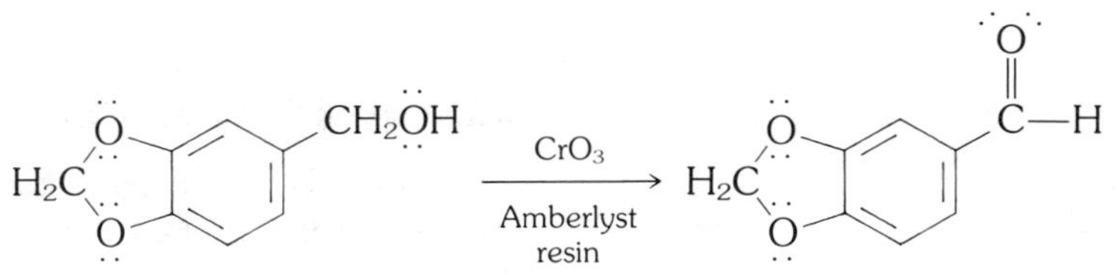

Piperonyl alcohol   Piperonal

DISCUSSION The selective oxidation of a 1° alcohol to an aldehyde using a chromium trioxide polymer-bound oxidizing agent is demonstrated in this experiment. Normally, treatment of 1° alcohols with aqueous sodium dichromate-sulfuric acid mixtures lead to the corresponding carboxylic acids. Another selective oxidizing agent

[2]Portions of Sequence A were previously published: Mayo, D. W.; Butcher, S. S.; Pike, R. M.; Foote, C. M.; Hotham, J. R.; Page, D. S. *J. Chem. Educ.* **1985,** *62,* 149.

that is frequently used is pyridinium chlorochromate. With both of these selective oxidizing agents, the reaction ceases at the aldehyde stage since the oxidations are conducted in nonaqueous media.

The mechanism of the oxidation is similar to that shown in Experiment 40A.

EXPERIMENTAL   Estimated time to complete the experiment: 3.5 hours.

Physical Properties of Reactants and Products

Compound	MW	Wt/Vol	mmol	mp(°C)	bp(°C)
Piperonyl alcohol	152.16	100 mg	0.65	58	
Chromic oxide resin		750 mg			
Dioxane	88.12	1.0 mL			101
Piperonal	150.14			37	263

Reagents and Equipment
In a 5-mL round-bottom flask containing a magnetic stirring bar and equipped with a reflux condenser, place 100 mg (0.65 mmol) of piperonyl alcohol, 750 mg of the oxidizing resin, and 1.0 mL of dioxane (calibrated Pasteur pipet).

Instructor Preparation
The CrO_3 resin is prepared by adding 35 g of Amberlyst A-26 to a solution of 15 g of CrO_3 in 100 mL of water. The mixture is stirred for 30 minutes at room temperature. The resin is then collected by filtration under reduced pressure and successively rinsed with water and acetone. It is partially dried on the Buchner funnel by drawing air through the resin for 1 hour and then allowed to air-dry overnight.

Reaction Conditions
With stirring, heat the reaction mixture at reflux for a period of 1.0 hour using a sand bath temperature of 125°C.

Isolation of Product
The reaction product is cooled to room temperature and the solution transferred by Pasteur filter pipet to a funnel containing a loose cotton plug covered with 500 mg of anhydrous sodium sulfate. The filtrate is collected in a 10-mL Erlenmeyer flask. The reaction flask and resin (include the walls of the condenser) are rinsed with three 0.5-mL portions of methylene chloride solvent (calibrated Pasteur pipet). Each rinse is transferred by Pasteur filter pipet to the funnel and the sodium sulfate is finally rinsed with an additional 0.1 mL of methylene chloride. Each rinse filtrate is combined with the original filtrate.

HOOD   A boiling stone is placed in the flask and the solvent removed [**HOOD**] using a sand bath maintained at a temperature of 125°C. A stream of nitrogen aids in this process. The crude piperonal is obtained as an oil and is purified by column chromatography.

Purification and Characterization
A short buret column is packed with approximately 1.75 g of silica gel and the crude product transferred by Pasteur pipet to the column. The flask is rinsed with 1.0 mL of methylene chloride (calibrated Pasteur pipet) and the rinse also transferred to the column.

Methylene chloride is now added to the column, 2.0 mL at a time (calibrated

Pasteur pipet), and the eluate collected in three tared 10-mL Erlenmeyer flasks. The first 4.0 mL of eluate are set aside. The next 6.0 mL are collected in one flask; this fraction contains the bulk of the product. One additional fraction of 3.0 mL is also collected in the third flask.

HOOD   The second fraction (6.0 mL) is concentrated [**HOOD**] using a warm sand bath and a slow stream of nitrogen to assist solvent removal. *Do not forget to add a boiling stone to the flask.*

Weigh the flask and contents. Reheat for 1 minute, cool, and weigh again. If the two weights are within 2.0 mg, the product is quite pure. If not, repeat the evaporation process until a constant weight is obtained.

Cool the product. If it does not solidify, place it in an ice bath and scratch the sides and bottom with a glass rod to induce crystallization. Determine the melting point and compare your value to that found in the literature. Obtain an IR spectrum and compare yours with an authentic sample and to the starting alcohol.

ENVIRONMENTAL DATA

Substance	Amount	TLV (mg/m^3)	Emissions (mg)	Volume (m^3)
Dioxane (SKIN) (Max.)	1.0 mL	90	1,030	11.4
Methylene chloride (Max., evap.)	1.6 mL	350	2,140	6.1
Methylene chloride (Max., col. chromatog., evap.)	14 mL	350	18,690	53

TXDS: Piperonal—orl-rat LD50: 1700 mg/kg
Silica gel—inv-mus LDLo: 234 mg/kg
Chromium trioxide—ihl-hmn TCLo: 100 $\mu g/m^3$ TFX:CAR
Sodium sulfate—orl-mus LD50: 5989 mg/kg

QUESTIONS   **6-154.** Primary alcohols can be oxidized to aldehydes and carboxylic acids. Often it is difficult to stop at the aldehyde stage. One technique frequently used to accomplish this (for those that boil below 100°C) is to remove the product from the reaction mixture as it is formed. This method is based on the fact that aldehydes have a lower boiling point than their corresponding alcohols. Explain this difference in boiling point.

6-155. A specific oxidizing agent for the conversion of primary alcohols to aldehydes is the pyridine-chlorochromite complex. Generally the oxidation is run in methylene chloride solvent. For example,

$$CH_3-\overset{\overset{\displaystyle CH_3}{|}}{C}=CHCH_2CH_2CH_2\ddot{O}H \xrightarrow[\substack{CH_2Cl_2 \\ RT}]{py \cdot CrO_3{}^+Cl^-} CH_3-\overset{\overset{\displaystyle CH_3}{|}}{C}=CHCH_2CH_2CHO$$

Oxidation of this alcohol with the conventional $Na_2Cr_2O_7$-H_2SO_4-water system produces the carboxylic acid. Offer an explanation for the difference in these results.

6-156. In reference to question 6-155, can you see another advantage of the pyridine chlorochromite reagent over that of the conventional conditions?

REFERENCES   1. The present procedure is based on the work reported by Cainelli, G.; Cardillo, G.; Orena, M.; Sandri, S. *J. Am. Chem. Soc.* **1978,** *98,* 6737.
2. The oxidation of a 1° alcohol to an aldehyde is given in "Organic Syntheses": Hurd, C. D.; Meinert, R. N. "Organic Syntheses"; Wiley: New York, 1943; Collect. Vol. II, p 541.

Experiments 47B and 47C
Piperonal O-(2,4-Dinitrophenyl)oxime;
Piperonylonitrile

(piperonylonitrile)

Experiments 47B and 47C illustrate the conversion of an aromatic aldehyde to an aromatic nitrile. It is an alternative to the conversion of an aromatic amine to an aromatic nitrile via the Sandmeyer reaction.

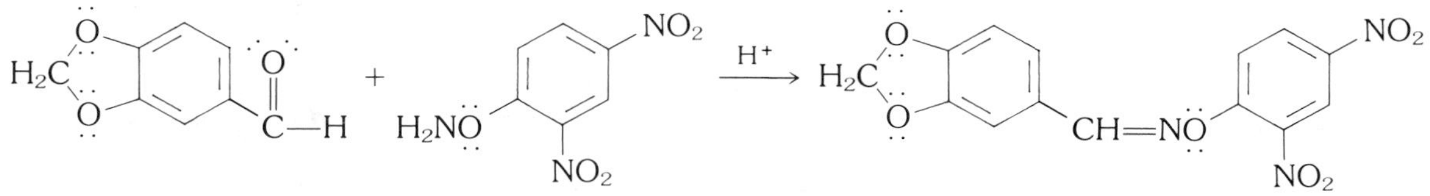

| Piperonal | O-(2,4-Dinitrophenyl)hydroxylamine | | Piperonal O-(2,4-dinitrophenyl)oxime |

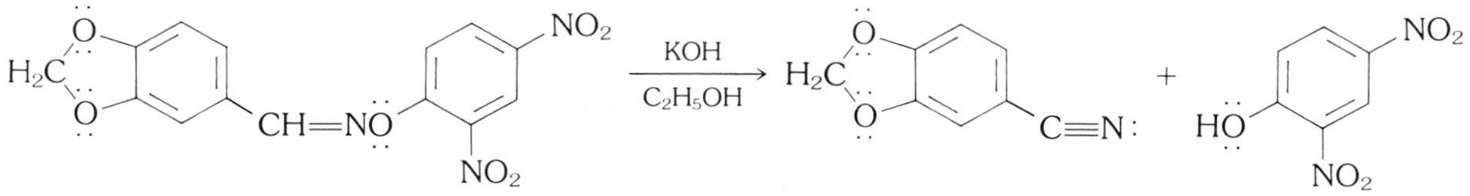

Piperonal O-(2,4-dinitrophenyl)oxime   Piperonylonitrile   2,4-Dinitrophenol

DISCUSSION An alternate method for the preparation of aromatic nitriles is demonstrated in this experiment. These compounds are usually prepared by the diazotization of the corresponding aromatic amine followed by treatment with copper(I) cyanide. This sequence is known as the Sandmeyer reaction. The present reaction converts an aromatic aldehyde to an O-phenylated oxime, which on treatment with alcoholic base yields the nitrile by an elimination reaction. The mechanistic sequence is outlined below. (The sequence for the formation of an oxime is outlined in Experiment 46.)

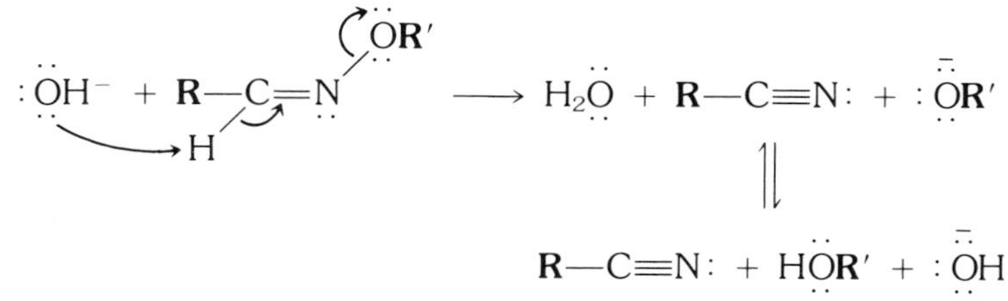

EXPERIMENTAL Estimated time to complete the experiment: 2 laboratory periods.

Physical Properties of Reactants and Products

Compound	MW	Wt/Vol	mmol	mp(°C)	bp(°C)
Piperonal	150.14	30 mg	0.20	37	263
O-(2,4-Dinitrophenyl)hydroxylamine	199.12	40 mg	0.20	111–112	
Ethanol	46.07	3.0 mL			
HCl (12M)		2 drops			
Piperonal O-(2,4-dinitrophenyl)oxime				194–195	

EXPERIMENT 47B Piperonal O-(2,4-dinitrophenyl)oxime

Reagents and Equipment

To a 5.0-mL conical vial containing a spin vane and equipped with a reflux condenser protected by a calcium chloride drying tube, add 40 mg (0.20 mmol) of O-(2,4-dinitrophenyl)hydroxylamine and 3.0 mL of absolute ethanol. *The preparation of O-(2,4-dinitrophenyl)-hydroxylamine is given in the Instructor's Manual.*

The vial is warmed gently with stirring in a sand bath to facilitate dissolution of the solid. The vial is detached from the condenser, 30 mg (0.20 mmol) of piperonal added, and the vial reattached to the condenser.

Following dissolution of the aldehyde, two drops of 12M HCl are introduced through the top of the reflux condenser by removing the drying tube. A Pasteur pipet is used for this addition. *As the HCl is delivered, the tip of the pipet should be held just above the surface of the solution.*

The drying tube is then reattached, and the contents are mixed by swirling.

Reaction Conditions

The oxime forms immediately. Complete precipitation is accomplished by cooling the reaction vial in an ice bath.

Isolation of Product

The solid precipitate is collected by filtration under reduced pressure using a Hirsch funnel. The crystals are washed with three 0.5-mL portions of cold absolute ethanol (calibrated Pasteur pipet) and then air dried on a porous clay plate.

Note. *Refrigeration of the filtrate for at least 24 hours produces another crop of oxime crystals. This second crop, collected by the same technique, may be combined with the initial product if its melting point is above 180°C. The collection of the second crop is generally not required to proceed with the synthesis. This portion, however, may be used for characterization of the oxime.*

Purification and Characterization

The oxime is of sufficient purity if the melting point of the material is 187°C or greater. A minimum of 50 mg of the oxime is necessary to proceed to Experiment 47C.

Obtain an IR spectrum of the oxime and compare it to that of the reference standard shown in Fig. 6.9. If there is a close spectral match, proceed to Experiment 47C.

Questions and references are at the end of Experiment 47C.

EXPERIMENT 47C Piperonylonitrile

Physical Properties of Reactants and Products

Compound	MW	Wt/Vol	mmol	mp(°C)	bp(°C)
Piperonal O-(2,4-dinitrophenyl)oxime	330.24	50 mg	0.15	194–195	
Ethanol	46.07	5 mL			78.5
Potassium hydroxide (0.2N)		2 mL			
Piperonylonitrile	147.13			92–93	

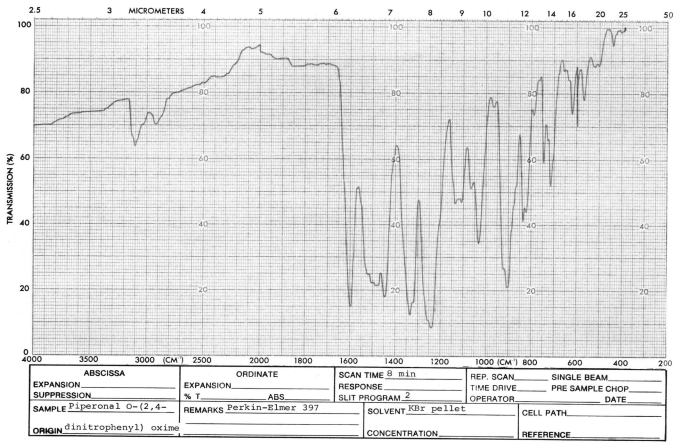

Fig. 6.9 IR spectrum: *piperonal O-(2,4-dinitrophenyl) oxime.*

Reagents and Equipment

To a 10-mL round-bottom flask containing a magnetic stirring bar and equipped with a reflux condenser, add 50 mg (0.15 mmol) of piperonal O-(2,4-dinitrophenyl) oxime, 5.0 mL of 95% ethanol, and 2.0 mL of 0.2N ethanolic KOH.

Note. *The oxime is prepared in Experiment 47B. The 0.2N ethanolic-KOH solution is prepared using 95% ethanol.*

Reaction Conditions

The reaction mixture is slowly heated to reflux by use of a sand bath 100–110°C and maintained at this temperature (mild reflux) for a period of 1 hour. During the initial warming period, the solution turns a deep yellow.

Note. *After the heating is terminated, the solution may be cooled, and the reaction vial removed, capped, and stored until the next laboratory period.*

Isolation of Product

The reaction flask is removed and the product mixture concentrated to a volume of 0.5 mL or less with a gentle stream of nitrogen gas and/or warming in a sand bath [**HOOD**]. *The concentration process takes a considerable length of time.*

A rinse solution is prepared by diluting 5% aqueous NaOH (1.0 mL) with distilled water (5.0 mL). This alkaline solution is used in three 2-mL portions to rinse the above reaction residue. The rinsings are transferred to a 12-mL capped centrifuge tube by use of a Pasteur pipet.

The resulting suspension is extracted with four 2-mL portions of methylene chloride (calibrated Pasteur pipet). The methylene chloride extract (bottom layer) is removed using a Pasteur filter pipet and the combined fractions placed in a 10-mL Erlenmeyer flask and dried over granular anhydrous sodium sulfate (0.5 g).

HOOD

By use of a Pasteur filter pipet, the dried solution is transferred to a 25-mL Erlenmeyer flask containing a boiling stone. The drying agent is rinsed with two 1-mL portions of methylene chloride, and the rinse solutions are combined with the original extract. The solvent is removed [**HOOD**] under a gentle stream of nitrogen and/or by warming in a sand bath to obtain the crude piperonylonitrile.

HOOD

Purification and Characterization

The crude product is purified by column chromatography using a Pasteur filter pipet filled with 300 mg of alumina (neutral, activity 1). The column is wetted with 1.0 mL of 1:1 methylene chloride/hexane solution.

The residue of crude nitrile isolated above is dissolved in the minimum amount of 1:1 methylene chloride/hexane solvent and the resulting solution transferred by Pasteur pipet to the column. The nitrile is eluted from the column with 2.0 mL of the CH_2Cl_2/hexane solvent and the eluate collected in a 10-mL Erlenmeyer flask containing a boiling stone.

The solvent is evaporated under a gentle stream of nitrogen or by warming in a sand bath [**HOOD**]. The white needles of piperonylonitrile may be dried on a porous clay plate.

HOOD

Weigh the product and calculate the percentage yield. Determine the melting point and compare it with the literature value. Obtain an IR spectrum and compare it with that of an authentic sample.

ENVIRONMENTAL DATA

Substance	Amount	TLV (mg/m³)	Emissions (mg)	Volume (m³)
Experiment 47B				
Ethanol (Max., solv., vacuum filter., cryst., wash)	4.5 mL	1,900	3,550	1.9
Experiment 47C				
Ethanol, 95% (Max., solv., evap.)	7.0 mL	1,900	5,250	2.8
Hexane (Max., col. chromatog.)	1.5 mL	180	990	5.5
Methylene chloride (Max., extraction, evap.)	10 mL	350	13,350	38
Methylene chloride (Max., col. chromatog.)	1.5 mL	350	2,000	5.7

TLV: Potassium hydroxide—2 mg/m³
TXDS: Piperonal—orl-rat LD50: 2700 mg/kg
      Piperonylonitrile—inv-mus LD50: 18 mg/kg
      1-Chloro-2,4-dinitrobenzene—orl-rat LD50: 1070 mg/kg

QUESTIONS

6-157. Suggest a suitable mechanism for the formation of piperonal O-(2,4-dinitrophenyl)oxime.

6-158. The Sandmeyer reaction is based on the replacement of the diazonium unit in aryldiazonium salts by chloro-, bromo-, or cyano- groups. Copper salt reagents are used.

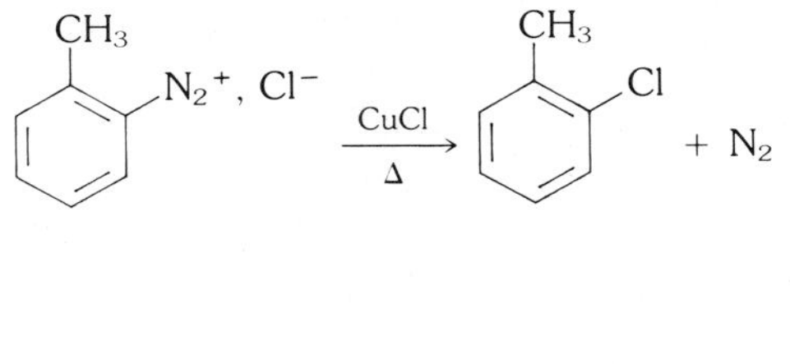

Carry out the following transformations using the Sandmeyer reaction.

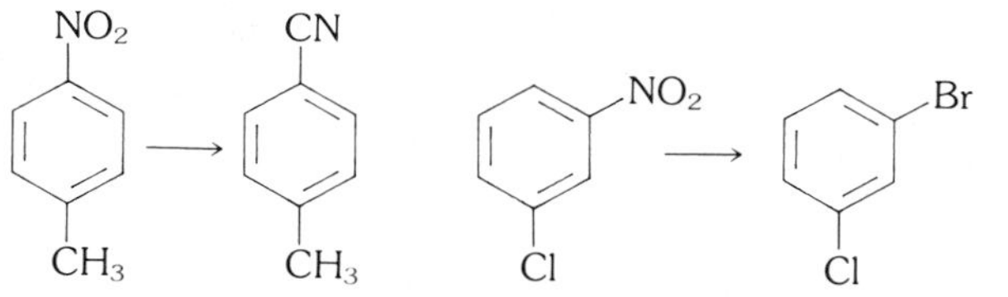

6-159. When CuCN is used in the Sandmeyer reaction, the preparation is generally carried out in a neutral medium. Can you offer an explanation of why this is done?

6-160. Outline a synthetic route for the preparation of nitriles using a carboxylic acid as the starting material.

REFERENCES The procedure outlined above for the preparation of piperonylonitrile is based on the work of Miller, M. J.; Loudon, G. M. *J. Org. Chem.* **1975,** *40,* 126.

SEQUENCE B: Use of the Ketal Protecting Group to Synthesize 1,1-Diphenyl-1-buten-3-one from Ethyl Acetoacetate (Experiments 48A–D)

Due to the fact that acetals and ketals hydrolyze under very mild conditions to regenerate the starting carbonyl unit, these groups are used to block or protect the carbonyl when performing a series of synthetic sequences. The series of reactions outlined in Experiments 48A–D demonstrate the use of a ketal protective group.

Experiment 48A
Ethyl Acetoacetate Ethylene Ketal

(1,3-dioxolane-2-acetic acid, 2-methyl-, ethyl ester)

The reaction of an alcohol (ethylene glycol) with a ketone in the presence of an acid catalyst forms the corresponding ketal, as presented in this experiment.

Ethyl acetoacetate   Ethylene glycol   Ethyl acetoacetate ethylene ketal

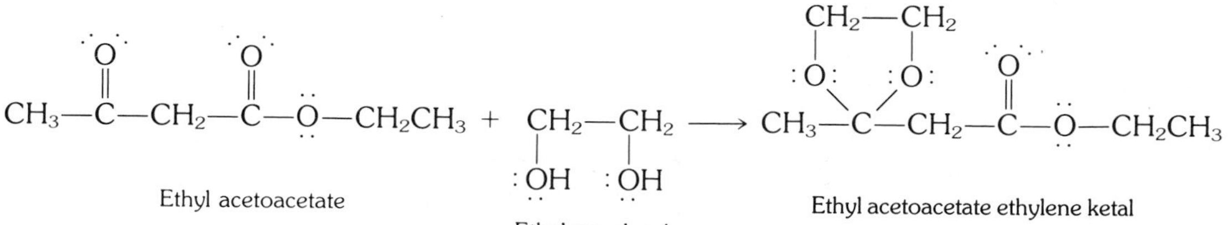

DISCUSSION This reaction represents a classic example of the placement of a protective group in a molecule. The formation of a cyclic ketal using ethylene glycol (a 1,2-diol) is often used to protect an aldehyde or ketone moiety.

To qualify as a good protecting group, certain qualifications must be met: (1) the reaction to form the group must proceed readily and in high yield, (2) it must possess stability under the conditions of further modification of the molecule, and (3) it must be readily removed when its protective function has been served.

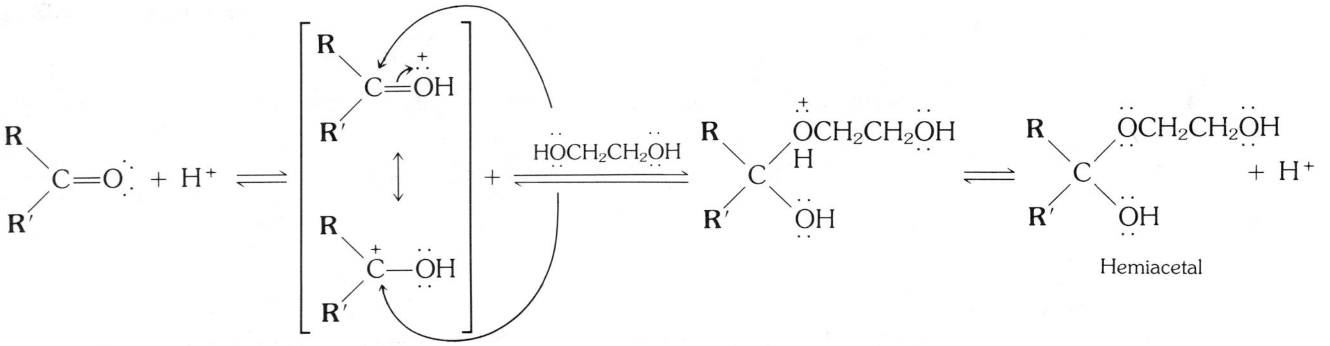

(1) Carbonyl + H$^+$ ⇌ [oxocarbenium resonance] + HOCH$_2$CH$_2$OH ⇌ hemiacetal (protonated) ⇌ Hemiacetal + H$^+$

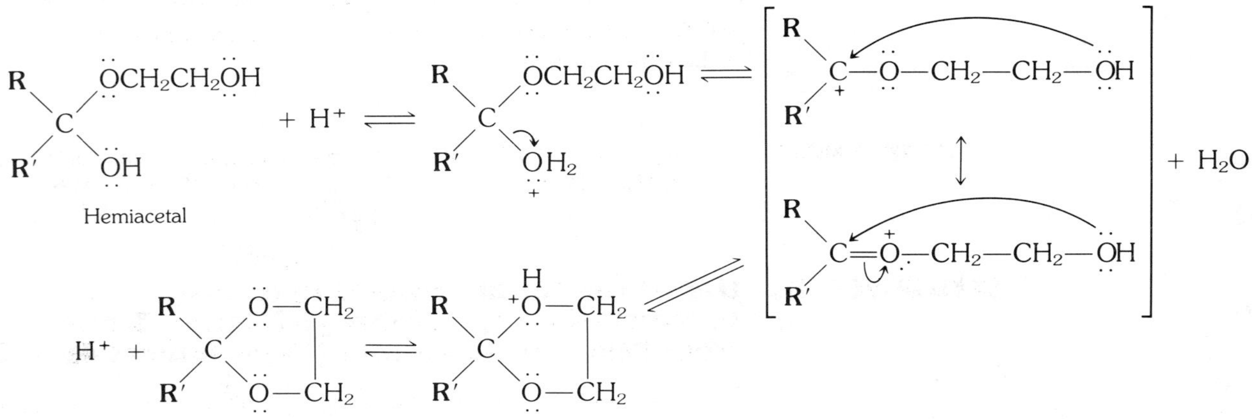

(2) Hemiacetal + H$^+$ ⇌ protonated ⇌ [oxocarbenium resonance] + H$_2$O

H$^+$ + ketal ⇌ protonated ketal ⇌ ...

As a protective unit, the ketal group is resistant to basic and nucleophilic reagents, but is readily hydrolyzed upon treatment with aqueous acid. The acid catalyzed mechanistic sequence is outlined above.

EXPERIMENTAL   Estimated time to complete the reaction: 3.0 hours.

Physical Properties of Reactants and Products

Compound	MW	Wt/Vol	mmol	mp(°C)	bp(°C)	Density	n_D
p-Toluenesulfonic acid monohydrate	190.22	15 mg	0.09	106			
Ethyl acetoacetate	130.15	200 µL	1.6		180	1.03	1.4194
Ethylene glycol	62.07	100 µL	1.8		198	1.11	1.4318
Toluene	92.15	2.0 mL			111		
Ethyl acetoacetate ethylene ketal	174				109 (17 torr)		1.4326

Reagents and Equipment

A 5-mL round-bottom flask containing a magnetic stirring bar and equipped with a Hickman still is charged with 15 mg (0.09 mmol) of p-toluenesulfonic acid monohydrate catalyst, 200 µL (206 mg, 1.6 mmol) of ethyl acetoacetate, 100 µL (111 mg, 1.8 mmol) of ethylene glycol, and 2.0 mL of anhydrous toluene. *It is convenient to dispense the ester and glycol using automatic delivery pipets. A graduated pipet or cylinder is sufficient to measure the toluene.*

Reaction Conditions

Using a sand bath (140–150°C), the solution is heated under vigorous reflux for a period of 1 hour.

As the reaction progresses, the water formed is separated from the reaction solution by azeotropic distillation with the toluene. The water then separates from the toluene on cooling in the upper portion of the Hickman still and collects on the walls and in the collar while the toluene returns to the distillation flask.

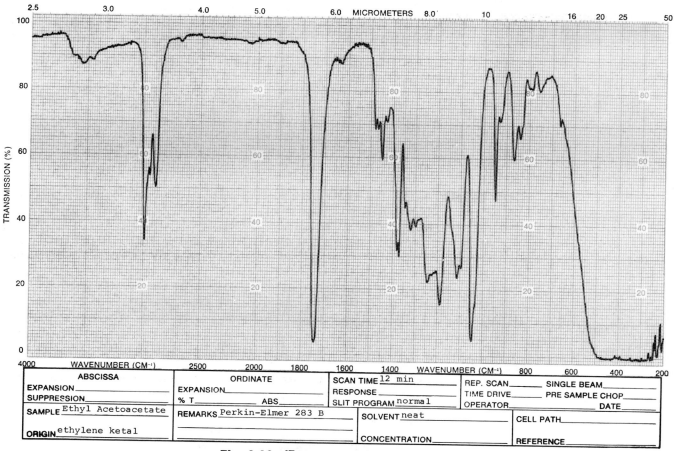

ABSCISSA		**ORDINATE**	SCAN TIME <u>12 min</u>	REP. SCAN_____ SINGLE BEAM_____
EXPANSION_____		EXPANSION_____	RESPONSE_____	TIME DRIVE_____ PRE SAMPLE CHOP_____
SUPPRESSION_____		% T_____ ABS_____	SLIT PROGRAM <u>normal</u>	OPERATOR_____ DATE_____
SAMPLE <u>Ethyl Acetoacetate</u>		REMARKS <u>Perkin-Elmer 283 B</u>	SOLVENT <u>neat</u>	CELL PATH_____
ORIGIN <u>ethylene ketal</u>			CONCENTRATION_____	REFERENCE_____

Fig. 6.10 IR spectrum: *ethyl acetoacetate ethylene ketal.*

Isolation of Product

The flask is cooled to room temperature and the toluene solution extracted with 1.0 mL of 1*M* aqueous NaOH solution followed by two 1-mL portions of water. During each extraction, the aqueous solution is added and the vial capped and then gently shaken. After separation of the layers, the cap is removed and the lower aqueous layer separated using a Pasteur filter pipet and discarded.

The remaining wet toluene solution is now dried over 500 mg of granular, anhydrous sodium sulfate.

The dried solution is transferred by Pasteur filter pipet to a clean, dry 5.0-mL conical vial equipped with a Hickman still. The drying agent and vial are rinsed with an additional 1 mL of toluene, and the rinse is combined with the original toluene solution using the Pasteur filter pipet.

Using a sand bath as the heat source, the toluene solvent is removed by distillation. As the toluene distillate collects in the collar of the still, it is removed with a Pasteur pipet. The high-boiling liquid residue that remains in the pot is the ethyl acetoacetate ethylene ketal product. *The toluene should distill slowly at a sand bath temperature of ~115–125°C. It is advisable not to heat the liquid residue above a sand bath temperature of 160°C.*

Purification and Characterization

The ketal is sufficiently pure for characterization.

Determine the refractive index of the product and compare your value to that in the literature. Obtain an IR spectrum and compare it with that shown in Fig. 6.10.

ENVIRONMENTAL DATA

Substance	Amount	TLV (mg/m³)	Emission (mg)	Volume (m³)
Ethylene glycol (Max., low vapor pressure)	111 mg	125	111	0.9
Toluene (Max., distilled)	3.0 mL	375	2600	6.9

TLV: Sodium hydroxide—2 $\mu g/m^3$
TXDS: Ethyl acetoacetate—orl-rat LD50: 3980 mg/kg
    p-Toluene sulfonic acid—orl-rat LD50: 2480 mg/kg
    Sodium sulfate—orl-mus LD50: 5989 mg/kg

QUESTIONS

6-160. What is meant by the term *azeotropic distillation*? How is this technique used to advantage in this experiment?

6-161. Cyclic ketals are important protecting groups because they are stable in aqueous base. That is, a carbonyl unit can be protected in alkaline solutions and regenerated (hydrolyzed) in aqueous acid. Based on this fact, carry out the following transformation.

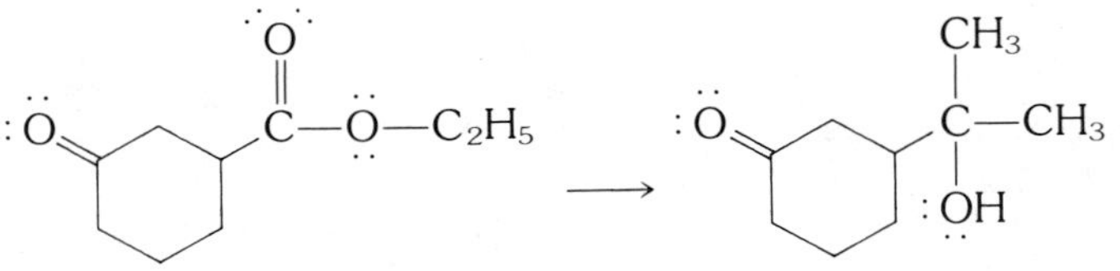

6-162. Aldehydes and ketones can also form cyclic thioketals when $HSCH_2CH_2SH$ is used in place of $HOCH_2CH_2OH$. These thioketals are useful because they undergo reduction with hydrogen in the presence of Raney nickel (Ni(Al)) catalyst to yield hydrocarbons.

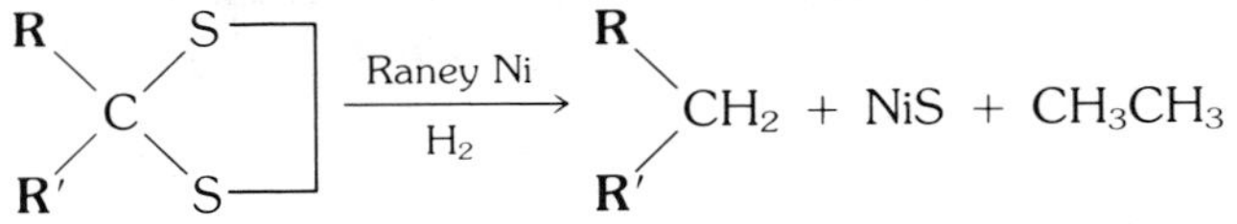

Knowing this fact, carry out the following transformations.

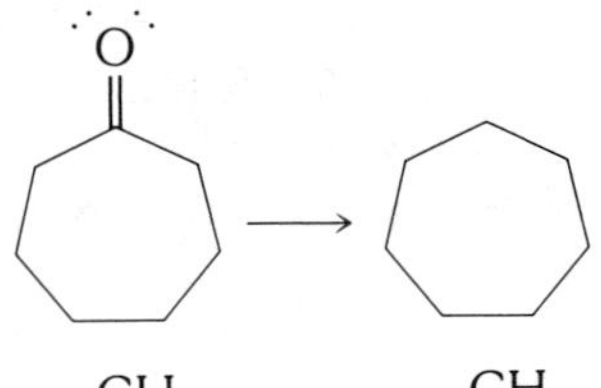

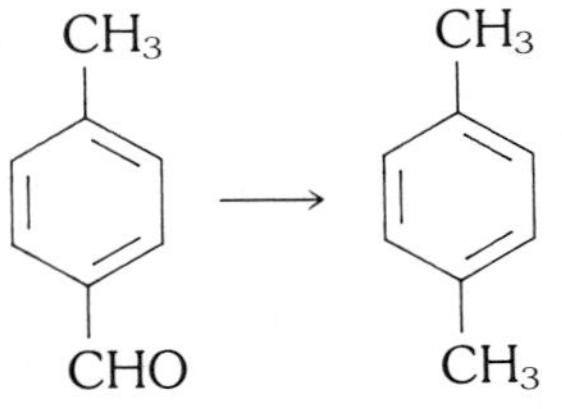

REFERENCES

1. This experiment is adapted from the following work:
   a. Paulson, D. R.; Hartwig, A. L.; Morgan, G. F. *J. Chem. Educ.* **1973,** *50,* 216.
   b. Rivett, D. E. A. *Ibid.* **1980,** *57,* 751.
2. References for the preparation of ketals appear in "Organic Syntheses":
   a. Renoll, M.; Newman, M. S. "Organic Syntheses"; Wiley: New York, 1955; Collect. Vol. III, p 502.
   b. Issidorides, C.H.; Gulen, R. "Organic Syntheses"; Wiley: New York, 1963; Collect. Vol. IV, p 679.
   c. Daignault, R. A.; Eliel, E. L. "Organic Syntheses"; Wiley: New York, 1973; Collect. Vol. V, p 303.

Experiment 48B
1,1-Diphenyl-1-hydroxy-3-butanone Ethylene Ketal

(1,3-dioxolane, 4-hydroxy-4,4-diphenyl-2-methyl-)

The addition of a Grignard reagent to an ester is a classic method for the synthesis of *tert*-alcohols. This experiment demonstrates the reaction.

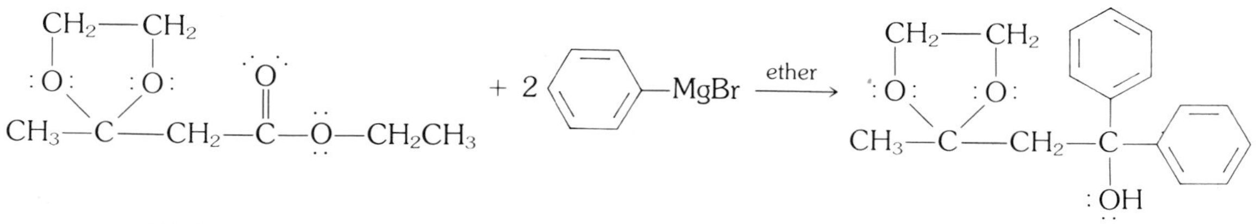

Ethyl acetoacetate
ethylene ketal

Phenylmagnesium
bromide

1,1-Diphenyl-1-hydroxy-3-
butanone ethylene ketal

DISCUSSION   This reaction is the second stage in this sequence. Note that the ketone group is protected by the cyclic ketal unit and thus is not vulnerable to attack by the Grignard reagent. For discussion of the reaction of Grignard reagents with carbonyl compounds, see Experiments 17 and 18. The mechanism is outlined below.

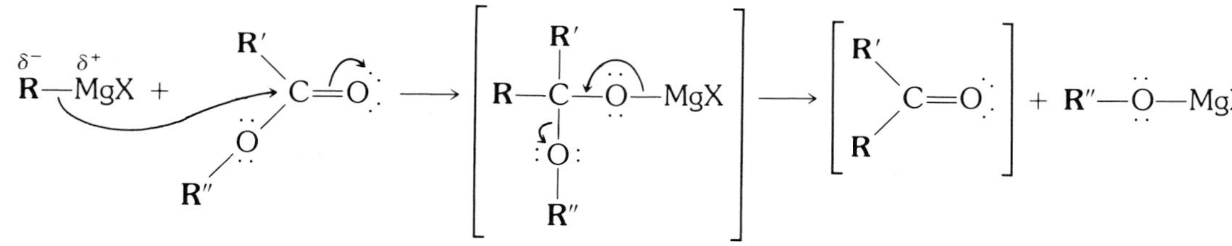

The intermediate ketone reacts with another equivalent of Grignard reagent to form the *tert*-alcohol. The mechanism of this stage is outlined in Experiment 17.

EXPERIMENTAL   Estimated time to complete the experiment: 3.5 hours.

Physical Properties of Reactants and Products

Compound	MW	Wt/Vol	mmol	mp(°C)	bp(°C)	Density	n_D
Bromobenzene	157.02	155 μL	1.4		156	1.50	1.5597
Diethyl ether	74.12	1.1 mL			34.5		
Iodine	253.81	1 crystal					
Magnesium	24.31	35 mg	1.4	649			
Ethyl acetoacetate ethylene ketal	174	100 mg	0.58		109 (17 torr)		1.4326
1,1-Diphenyl-1-hydroxy-3-butanone ethylene ketal	284			90–91			

Reagents and Equipment

Preparation of Phenylmagnesium Bromide

This reagent is prepared exactly as described in Experiment 17. The order of addition, manipulations, and precautions are the same. The equipment is also identical.

The following reagents and amounts are used:

1. Magnesium—35 mg (1.4 mmol)
2. Iodine—one small crystal
3. Anhydrous diethyl ether—700 μL
4. Bromobenzene—225 mg (1.4 mmol, 155 μL)

The Ketal-Ester Reagent

A solution of 100 mg (0.58 mmol) of the ketal-ester, prepared in Experiment 48A, dissolved in 300 μL of diethyl ether is prepared in a dry 1-mL conical vial. The solution is immediately drawn into a 1.0-cm^3 syringe, and the syringe is then inserted into the rubber septum on the Claisen head. An additional 100 μL of ether is placed in the vial as a rinse to be added later.

HOOD   **Note.** *The ether is measured in the* **HOOD** *using an automatic-delivery pipet. The ketal-ester reagent must be free of any acidic impurities. An ether solution of the material may be passed through a column of alumina (500 mg in a Pasteur filter pipet) if necessary.*

Reaction Conditions

With stirring, add the ether solution of the ketal-ester dropwise to the warm Grignard solution so as to keep the ether solvent under steady reflux. The rinse saved above is then added in like manner in one portion. The resulting mixture is then stirred for a period of 15 minutes while the reflux conditions are maintained by warming the reaction mixture with a sand bath.

The resulting mixture is then allowed to cool to room temperature. The conical vial is finally removed and capped.

Isolation of Product

CAREFUL   The alkoxide magnesium salt is hydrolyzed by the **CAREFUL**, dropwise addition of 1 mL of saturated ammonium chloride solution using a Pasteur pipet.

> **CAUTION:** *The addition of the dilute acid solution is accompanied by the evolution of heat. An ice bath should be kept handy to cool the solution if necessary. A two-layer reaction solution (ether-water) forms as the solid dissolves.*

The spin vane is now removed with forceps and set aside to be rinsed with an ether wash. The vial is capped tightly, shaken, and vented, and the layers are allowed to separate.

Using a Pasteur filter pipet, transfer the bottom aqueous layer to a clean 5.0-mL conical vial.

The ether layer is SAVED; it contains the crude reaction product.

This aqueous layer transferred to the 5.0-mL vial is now extracted with 0.5 mL of diethyl ether, rinsing the spin vane with the ether as it is added to the vial by means of a Pasteur filter pipet. The vial is capped, shaken, and vented, and the layers are allowed to separate. Using a Pasteur filter pipet, the bottom aqueous layer is removed and discarded. The ether layer is then combined with the ether extracts previously saved.

The combined ether solution is partitioned (using the technique just described) with several 1.0-mL portions of saturated ammonium chloride solution until the aqueous extracts are no longer alkaline to litmus. The wet ether solution is dried by passing the solution through a Pasteur filter pipet packed with 500 mg of anhydrous sodium sulfate. The vial is rinsed with an additional 0.5-mL portion of ether and this also is transferred to the drying column. The eluate is collected in a tared 10-mL Erlenmeyer flask containing a boiling stone.

HOOD   The ether is evaporated by gently warming the Erlenmeyer flask in a sand bath in the **HOOD** to yield a light yellow oil. Cooling this oil in an ice bath, while scratching the walls of the flask with a small glass rod, produces a yellow solid. The flask may be weighed to obtain a crude weight of product.

Purification and Characterization

The crude, yellow product is recrystallized from methanol using the Craig tube. The light yellow, cubic crystals are dried on a clay plate.

Weigh the crystals and calculate the percentage yield. Determine the melting point and compare it to the literature value. Obtain an IR spectrum and compare it with that in Fig. 6.11.

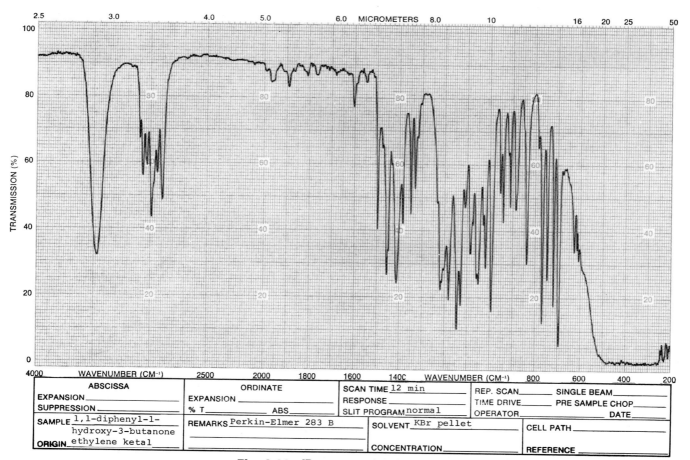

Fig. 6.11 IR spectrum: *1,1-diphenyl-1-hydroxy-3-butanone ethylene ketal.*

ENVIRONMENTAL DATA

Substance	Amount	TLV (mg/m³)	Emissions (mg)	Volume (m³)
Diethyl ether (Max., evap.)	2.1 mL	1200	1500	1.25
Methanol (SKIN) (Max., recryst. Craig tube)	2.0 mL	260	1580	6.1

TLV: Iodine—0.1 ppm
     Ammonium chloride—10 mg/m³
TXDS: Magnesium—orl-dog LDLo: 230 mg/kg
     Sodium sulfate—orl-mus LD50: 5989 mg/kg
TMDX: Bromobenzene—dnr-esc 250 mg/L

QUESTIONS

6-164. In the reaction of Grignard reagents with esters, a ketone is formed in the first stage of the reaction. Since ketones are more reactive toward Grignard reagents than esters, the ketone reacts immediately with additional Grignard reagent to yield the tertiary alcohol after hydrolysis. Offer an explanation of why ketones are more reactive than esters toward Grignard reagents.

6-165. Can you suggest a reason that Grignard reagents do not add to the carbonyl group of amides?

6-166. Give the structure and name of the product formed by reaction of propylmagnesium bromide with ethyl formate in ether solvent followed by hydrolysis using ammonium chloride solution.

REFERENCES

1. This experiment was adapted from the following work:
    a. Paulson, D. R.; Hartwig, A. L.; Morgan, G. F. *J. Chem. Educ.* **1973**, *50*, 216.
    b. Rivett, D. E. A. *Ibid.*, **1980**, *57*, 751.
2. For a general reference on the reaction of Grignard reagents with esters see Kharasch, M. S.; Reinmuth, O. "Grignard Reactions of Non-metallic Substances"; Prentice-Hall: New York, 1954; p 557.

Experiment 48C
1,1-Diphenyl-1-hydroxy-3-butanone

(3-butanone, 1-hydroxy-1,1-diphenyl-)

The acid hydrolysis of a ketal protecting group to regenerate the original ketone carbonyl unit is demonstrated in this experiment.

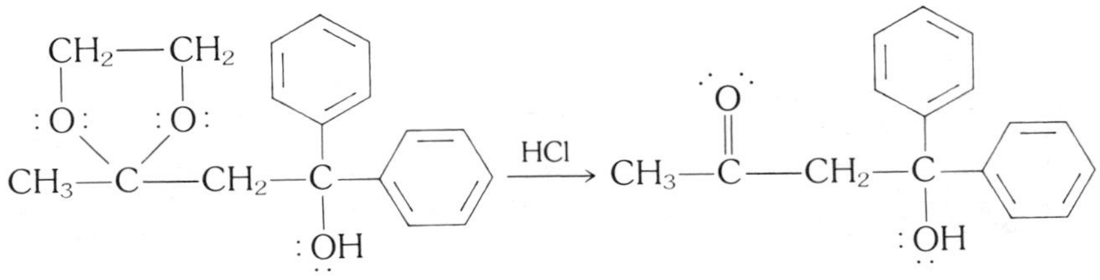

1,1-Diphenyl-1-hydroxy-3-butanone ethylene ketal   1,1-Diphenyl-1-hydroxy-3-butanone

DISCUSSION This reaction illustrates the removal of the ketal protecting group from a carbonyl unit. This is the third stage in this sequence of reactions. If the Grignard reaction conducted in Experiment 48B were used without first protecting the keto function, it too would have reacted with the Grignard reagent, since a ketone carbonyl is more reactive than an ester toward nucleophilic attack. The formation of the ketal and its hydrolysis constitutes an equilibrium reaction. The mechanism for the hydrolysis reaction is shown in Experiment 48A.

EXPERIMENTAL Estimated time to complete the experiment: 2.0 hours.

Physical Properties of Reactants and Products

Compound	MW	Wt/Vol	mmol	mp(°C)	bp(°C)
1,1-Diphenyl-1-hydroxy-3-butanone ethylene ketal	284	50 mg	0.19	90–91	
Acetone	58.08	2.0 mL			56
1M HCl solution		100 μL			
1,1-Diphenyl-1-hydroxy-3-butanone	238			85–86	

Reagents and Equipment

To a 5.0-mL conical vial containing a magnetic spin vane and equipped with an air condenser, add 50 mg (0.19 mmol) of the recrystallized hydroxy-ketal prepared in Experiment 48B, 2.0 mL of acetone, and 100 μL of 1M HCl solution. *The HCl is dispensed using an automatic-delivery pipet. A graduated cylinder* **HOOD** *or pipet may be used to deliver the acetone* [**HOOD**].

Reaction Conditions

The mixture is now heated at vigorous reflux for a period of 15 minutes using a sand bath maintained at a temperature of 80–90°C.

Isolation of Product

Saturated, aqueous sodium bicarbonate solution (1 mL) is added to the cooled reaction solution (calibrated Pasteur pipet) and the aqueous mixture extracted with three 1-mL portions of diethyl ether. For each extraction the ether is added, and the vial is capped, shaken, and vented. The ether layer is separated, using a Pasteur filter pipet and transferred to a 5.0-mL conical vial.

The combined ether extracts are partitioned with a 1-mL portion of saturated, aqueous sodium bicarbonate solution followed by a 1-mL portion of water. For each of these two extractions, the vial is capped, shaken, and vented, and the lower aqueous layer separated (Pasteur filter pipet) and discarded.

The wet ether layer is dried over ~500 mg of anhydrous sodium sulfate.

The dried ether solution is now transferred (Pasteur filter pipet) to a 10-mL Erlenmeyer flask. The drying agent is rinsed with an additional 1 mL of diethyl ether and the rinse solution also transferred to the Erlenmeyer flask.

HOOD The ether is evaporated [**HOOD**] using a filtered stream of air and/or by warming in a sand bath to yield the crude hydroxy-ketone.

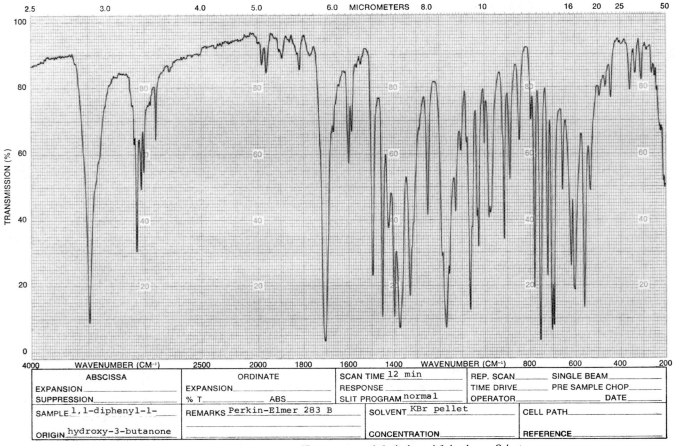

Fig. 6.12 IR spectrum: *1,1-diphenyl-1-hydroxy-3-butanone.*

Purification and Characterization

Recrystallization of the crude material from hexane using the Craig tube yields pure 1,1-diphenyl-1-hydroxy-3-butanone. The crystals are dried on a porous clay plate.

Weigh the product and calculate the percentage yield. Determine the melting point and compare your result with the literature value. Obtain the IR spectrum and compare it with Fig. 6.12.

ENVIRONMENTAL DATA

Substance	Amount	TLV (mg/m³)	Emissions (mg)	Volume (m³)
Hydrochloric acid (1*M*)	0.1 mL	7	3.7	0.5
Acetone (Max., solvent, water soluble, reflux)	2 mL	1780	1580	0.9
Diethyl ether (Max., extract, evap.)	4 mL	1200	2860	2.4
Hexane (recryst., Craig tube)	2 mL	180	1320	7.3

TLV: Sodium bicarbonate—5 mg/m³
TXDS: Sodium sulfate—orl-mus LD50: 5989 mg/kg

QUESTIONS

6-167. How could you distinguish between the starting reagent of this experiment and the isolated product by using chemical tests (see Chapter 7)?

6-168. Aldehydes and ketones form hemiacetals and hemiketals, respectively, when treated with alcohols in the presence of acid or base. Draw the structure of the cyclic hemiacetals generated in each of the following reactions.

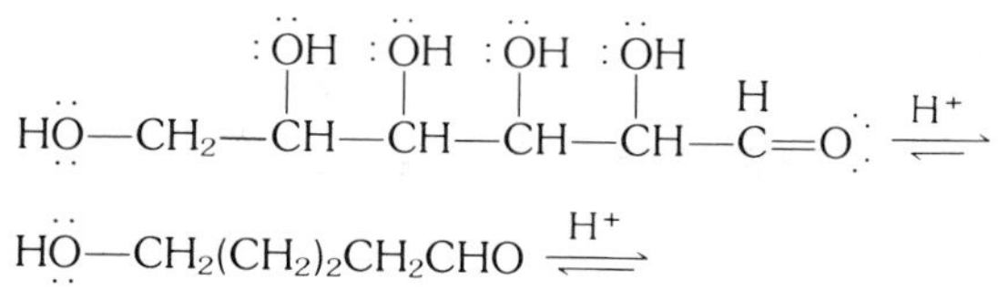

$$HO-CH_2-CH-CH-CH-CH-C=O \xrightleftharpoons{H^+}$$
(with OH groups and H)

$$HO-CH_2(CH_2)_2CH_2CHO \xrightleftharpoons{H^+}$$

6-169. Aldehydes and ketones form hydrates (gem-diols) upon addition of water under acid or base conditions.

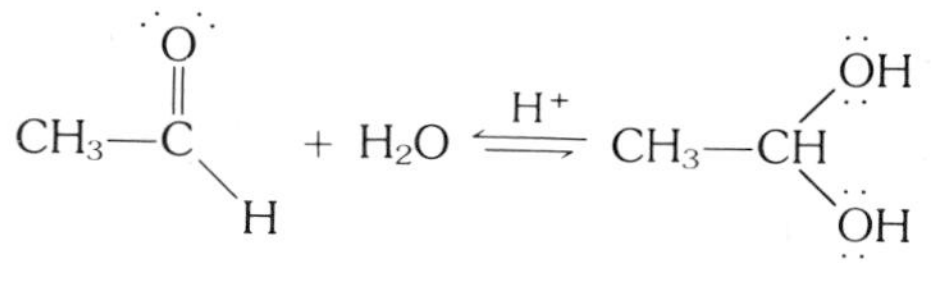

$$CH_3-\overset{O}{\overset{\|}{C}}{}_H + H_2O \xrightleftharpoons{H^+} CH_3-CH\overset{OH}{\underset{OH}{\diagup}}$$

For most aldehydes the equilibrium lies far to the left so the reaction is not synthetically important. Can you explain why each of the hydrates listed below is stable and can be isolated?

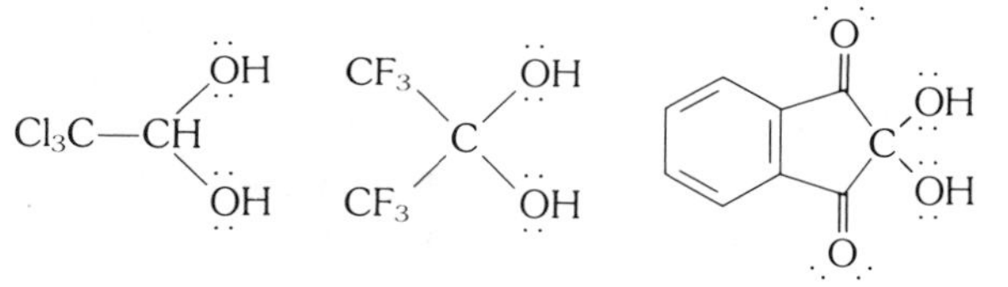

$$Cl_3C-CH\overset{OH}{\underset{OH}{\diagup}} \qquad \overset{CF_3}{\underset{CF_3}{\diagdown}}C\overset{OH}{\underset{OH}{\diagup}}$$

(and the indane-dione gem-diol structure with OH and OH)

REFERENCES

This experiment is adapted from the work reported by Rivett, E. A. *J. Chem. Educ.* **1980,** *57,* 751.

Experiment 48D
1,1-Diphenyl-1-buten-3-one

(1,1-diphenyl-1-buten-3-one)

The dehydration of a β-keto alcohol to give an α,β-unsaturated ketone is demonstrated in this experiment.

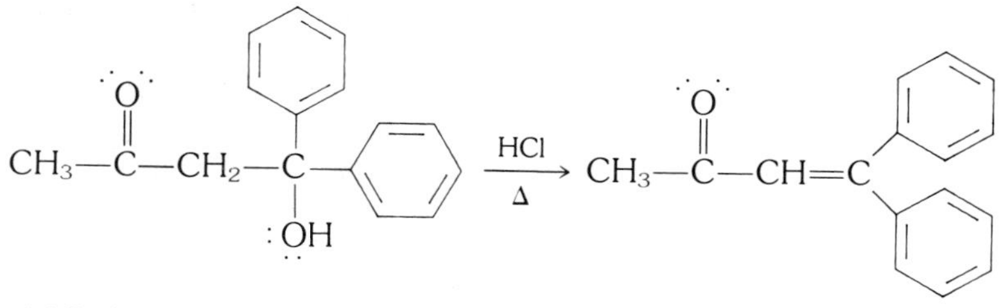

$$CH_3-\overset{O}{\overset{\|}{C}}-CH_2-\underset{OH}{\overset{}{C}}(C_6H_5)_2 \xrightarrow[\Delta]{HCl} CH_3-\overset{O}{\overset{\|}{C}}-CH=C(C_6H_5)_2$$

1,1-Diphenyl-1-hydroxy-3-butanone   1,1-Diphenyl-1-buten-3-one

DISCUSSION This experiment illustrates a very important reaction in organic chemistry; the dehydration of a 3° alcohol to yield an alkene. The reaction is classified as an acid-catalyzed elimination, and in the present case elimination may proceed in only one direction. Thus, only one olefin is produced. This completes the total sequence of reactions begun in Experiment 48A. The mechanism for the dehydration of an alcohol to an alkene is outlined in Experiment 9.

EXPERIMENTAL Estimated time to complete the reaction: 3.0 hours.

Physical Properties of Reactants and Products

Compound	MW	Wt/Vol	mmol	mp(°C)	bp(°C)
1,1-Diphenyl-1-hydroxy-3-butanone	238	35 mg	0.15	85–86	
Conc. HCl		100 μL			
Acetone	58.08	0.5 mL			56
1,1-Diphenyl-1-buten-3-one	220			34–36	192–194(15 torr)

Reagents and Equipment

In a 3.0-mL conical vial containing a magnetic spin vane and equipped with a reflux condenser are placed 35 mg (0.15 mmol) of 1,1-diphenyl-1-hydroxy-3-butanone (prepared in Experiment 48C), 100 μL of concentrated HCl, and 0.5 mL of acetone. *The reagents are dispensed in the* **HOOD.** *An automatic delivery pipet is used for the acid and a graduated pipet for the acetone.*

HOOD

Reaction Conditions

With stirring, the reaction mixture is heated at gentle reflux, using a sand bath temperature of ~70°C, for a period of 30 minutes. *The solution develops a deep red color about 5–10 minutes after the solution reaches the reflux temperature.*

Isolation of Product

The reaction solution is cooled to room temperature and diluted with 1 mL of water and then extracted with two 0.5-mL portions of diethyl ether. With each extraction after the addition of the ether, the vial is capped, shaken, and vented. The ether extracts are transferred to a 3.0-mL conical vial using a Pasteur filter pipet.

The ether solution is then extracted with two 0.5-mL portions of saturated sodium bicarbonate solution followed by one portion (0.5-mL) of water. The extraction procedure is identical to that outlined above. The resulting wet ether fraction is dried over 500 mg of anhydrous sodium sulfate.

The dried ether solution is transferred, using a Pasteur filter pipet, to a 10-mL Erlenmeyer flask containing a boiling stone. The drying agent is washed with an additional 0.5 mL of ether, which is also transferred to the Erlenmeyer flask.

HOOD The ether solvent is evaporated [**HOOD**] using a warm sand bath to yield the unsaturated ketone product as a yellow oil.

Purification and Characterization

The crude material is purified by column chromatography.

In a Pasteur filter pipet is placed 0.5 g of activated silica gel followed by 0.5 g of anhydrous sodium sulfate. The crude material isolated above is dissolved in 100 μL of diethyl ether and the solution then transferred by Pasteur pipet to the column. The material, which appears as a yellow band on the column, is eluted with toluene (~5.0 mL) and the eluate collected in a tared 25-mL filter flask.

HOOD   The toluene solvent is removed under reduced pressure while warming the flask on a sand bath [**HOOD**].

Weigh the pure 1,1-diphenyl-1-buten-3-one and calculate the percentage yield. Obtain an IR spectrum and compare it with the spectrum shown in Fig. 6.13.

ENVIRONMENTAL DATA

Substance	Amount	TLV (mg/m³)	Emissions (mg)	Volume (m³)
Hydrochloric acid (conc.)	0.1 mL	7	46	6.5
Acetone (Max., pipet, heat, dilute with water)	1.0 mL	1780	790	0.4
Diethyl ether (Max., extract, evap.)	1.6 mL	1200	1140	1.0
Toluene (Max., col. chromatog., evap.)	5.0 mL	375	4335	11.6

TLV: Sodium bicarbonate—5 mg/m³
TXDS: Sodium sulfate—orl-mus LD50: 5989 mg/kg
Silica—inv-mus LDLo: 234 mg/kg

QUESTIONS

6-170. Why is 1,1-diphenyl-1-buten-3-one colored?

6-171. The relative ease with which alcohols undergo dehydration is 3° > 2° > 1°. Offer a reasonable explanation for this order.

6-172. In relation to question 6-171, draw potential-energy diagrams that reflect the relative energy of activation for the dehydration of the alcohols.

6-173. Citric acid is converted to isocitric acid in the Krebs cycle through a dehydration–hydration sequence catalyzed by aconitase enzyme.

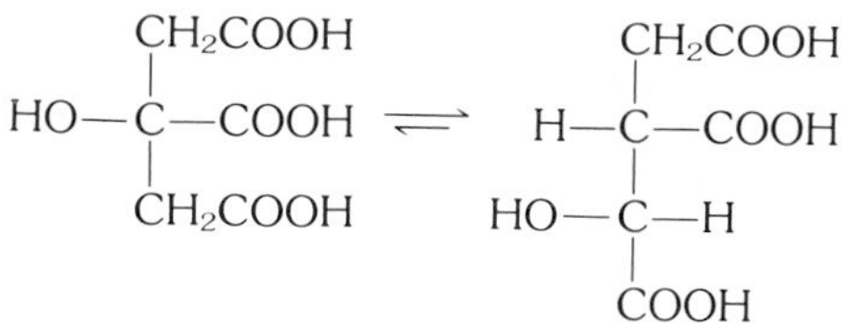

$$
\begin{array}{ccc}
CH_2COOH & & CH_2COOH \\
| & & | \\
HO-C-COOH & \rightleftharpoons & H-C-COOH \\
| & & | \\
CH_2COOH & & HO-C-H \\
& & | \\
& & COOH \\
\text{Citric acid} & & \text{Isocitric acid}
\end{array}
$$

The reaction proceeds by way of formation of an intermediate olefin, **Z**-aconitic acid. Give the structure of this acid intermediate.

REFERENCES

1. The above reaction is adapted from the work of the following individuals.
   a. Paulson, D. R.; Hartwig, A. L.; Moran, G. F. *J. Chem. Educ.* **1973**, *50*, 216.
   b. Rivett, D. E. A. *Ibid.* **1980**, *57*, 751.
2. Selected references from "Organic Syntheses" that illustrate the dehydration of alcohols to alkenes are listed below.
   a. Allen, C. F. H.; Converse, S. "Organic Syntheses"; Wiley: New York, 1941; Collect. Vol. I, p 226.
   b. Conant, J. B.; Tuttle, N. *Ibid.*, p 345.
   c. Norris, J. F. *Ibid.*, p 431.
   d. Adkins, H.; Zartman, W. "Organic Syntheses"; Wiley: New York, 1943; Collect. Vol. II, p 606.
   e. Wiley, R. H.; Waddey, W. E. "Organic Syntheses"; Wiley: New York, 1955; Collect. Vol. III, p 560.
   f. Grummitt, O.; Becker, E. I. "Organic Syntheses"; Wiley: New York, 1963; Collect. Vol. IV, p 771.

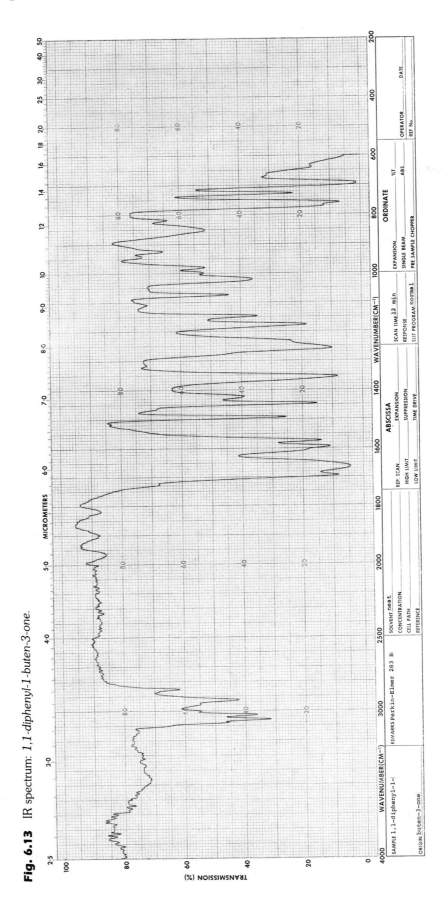

Fig. 6.13 IR spectrum: *1,1-diphenyl-1-buten-3-one.*

SEQUENCE C: **Synthesis of Hexaphenylbenzene from Benzoin and Z-Stilbene (Experiments 49, 50A, B, 51)**

The preparation of hexaphenylbenzene demonstrates the manner in which a variety of basic organic reactions can be integrated to prepare a desired end product. An outline of the overall scheme is presented below. (Experiment numbers are given in brackets.)

Benzoin $\longrightarrow$ Benzil $\longrightarrow$ Tetraphenylcyclopentadienone
[6]   [41]   [49]

$\searrow$ $\longrightarrow$ Hexaphenylbenzene   [51]

Z-Stilbene $\longrightarrow$ d,l-Stilbene $\longrightarrow$ Diphenylacetylene
dibromide [50A]   [50B]

As the flowchart illustrates, the total synthesis of hexaphenylbenzene involves an oxidation followed by an aldol condensation to prepare a diene and a bromination reaction followed by dehydrohalogenation to obtain a dienophile. A pericyclic Diels-Alder reaction using these end products completes the sequence.

Experiment 49
Tetraphenylcyclopentadienone

(2,4-cyclopentadien-1-one, 2,3,4,5-tetraphenyl-)

The use of the aldol condensation for the synthesis of a five-membered carbocyclic ring system is demonstrated in this experiment.

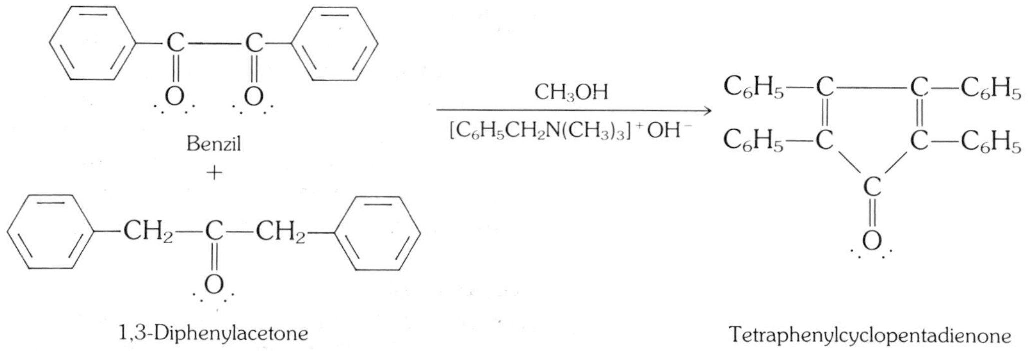

Benzil

+

1,3-Diphenylacetone

$\xrightarrow[\text{[C}_6\text{H}_5\text{CH}_2\text{N(CH}_3\text{)}_3]^+\text{OH}^-]{\text{CH}_3\text{OH}}$

Tetraphenylcyclopentadienone

DISCUSSION This experiment is a further example of the aldol condensation (see Experiment 22 for discussion). The reaction carried out in this sequence differs in that two ketones, one of which has no α-hydrogen atoms, are the reactants. It is unique because the selected reagents lead to the formation of a carbocyclic ring system. The aldol product initially formed undergoes an elimination reaction to yield a material that has a highly conjugated system of double bonds. In general, the more conjugation in a molecule, the less energy required to promote the

π-electrons to a higher energy level. In this case, energy in the visible region of the spectrum is absorbed, resulting in the product having a deep purple color.

The mechanism is a sequence of two aldol condensations. The first is intermolecular the second intramolecular. The mechanism is similar to that outlined in Experiment 22.

This product is the diene used for the preparation of hexaphenylbenzene (see Experiment 51).

EXPERIMENTAL   Estimated time to complete the experiment: 1.5 hours.

Physical Properties of Reactants and Products

Compound	MW	Wt/Vol	mmol	mp(°C)	bp(°C)
1,3-Diphenylacetone	210.28	50 mg	0.24	35	
Benzil	210.23	50 mg	0.24	95	
Triethylene glycol	150.18	0.25 mL			278
Benzyltrimethylammonium hydroxide (40% solution in methanol)		50 µL			
Tetraphenylcyclopentadienone	384.48			220–221	

Reagents and Equipment
In a 1.0-mL conical vial containing a magnetic spin vane and equipped with an air condenser, place 50 mg (0.24 mmol) of 1,3-diphenylacetone and 50 mg (0.24 mmol) of benzil followed by 0.25 mL of triethylene glycol.

Note. *The benzil used in this reaction must be free of benzoin impurity. If benzil is prepared according to Experiment 41, it should be purified by the chromatographic procedure cited therein.*

Reaction Conditions
The mixture is now heated with stirring in a sand bath maintained at a temperature of 155–165°C for a period of 10 minutes. The benzil dissolves during this time.

The reaction solution is removed from the bath and immediately 50 µL of a 40% benzyltrimethylammonium hydroxide-methanol solution is added (automatic-delivery pipet) to the hot reactants with gentle shaking. The vial and contents are again heated for 2–3 minutes at 150–160°C and then allowed to cool. As the cooling occurs, the appearance of dark purple-colored crystals of tetraphenylcyclopentadienone are evident. *Cooling may be accelerated by placing the vial under a stream of cold water.*

Isolation of Product
With stirring, 0.7 mL of cold methanol is added and the mixture cooled in an ice bath for 5–10 minutes. The dark crystals are collected by filtration under reduced pressure by use of a Hirsch funnel. The reaction vial and crystals are rinsed with several drops of cold methanol. The dropwise addition of cold methanol to the crystals is continued until the product appears purple and not brown in color. The crystalline product is dried on a porous clay plate.

Purification and Characterization
The product is of sufficient purity for use in the preparation of hexaphenylbenzene (Experiment 51.)

If a purer material is desired, it may be recrystallized from triethylene glycol. Weigh the tetraphenylcyclopentadienone product and calculate the percentage yield. Determine the melting point and compare it with the literature value. Obtain an IR spectrum of the material and compare it with that of an authentic sample.

ENVIRONMENTAL DATA

Substance	Amount	TLV (mg/m³)	Emissions (mg)	Volume (m³)
Methanol (SKIN) (Max., solv., recryst.)	2.0 mL	260	1580	6.1

TXDS: Benzyltrimethylammonium hydroxide—suc-mus LDLo: 35 mg/kg
Triethylene glycol—orl-rat LD50: 17 g/kg
Benzil—orl-rat LD50: 2710 mg/kg

QUESTIONS

6-174. Outline a complete mechanistic sequence to account for the formation of the tetraphenylcyclopentadienone compound.

6-175. Cyclopentadienone is unstable and rapidly undergoes the Diels-Alder reaction with itself. Write the structure for this Diels-Alder adduct.

6-176. The Diels-Alder adduct of question 6-175 undergoes a fragmentation reaction on heating to produce a bicyclotrienone compound plus carbon monoxide. Suggest a structure for this product.

6-177. Using the Huckel $4n + 2$ rule for aromaticity, predict which of the following species might be expected to show aromatic properties.

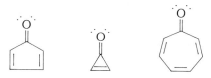

6-178. Based on questions 6-175 and 6-176, why is tetraphenylcyclopentadienone such a stable compound?

REFERENCES

1. For references on the aldol condensation refer to Experiment 22.
2. An "Organic Syntheses" preparation of tetraphenylcyclopentadienone is available: Johnson, J. R.; Grummitt, O. "Organic Syntheses"; Wiley, New York, 1955; Collect. Vol. III, p 80.

Experiment 50A
d,l-Stilbene Dibromide

(d,l-ethane, 1,2-dibromo-1,2-diphenyl-)

The bromination of an alkene is presented in this experiment. In the reaction, two chiral centers are generated yielding a mixture of optically active bromides. This demonstrates that the reaction is *stereospecific*.

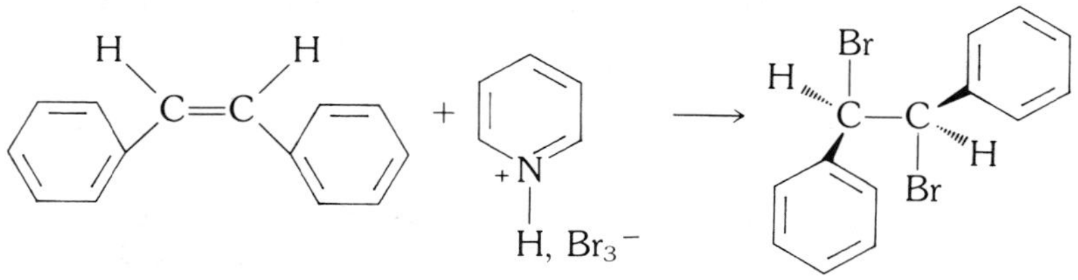

Z-Stilbene Pyridinium hydrobromide perbromide d,l-Stilbene dibromide

DISCUSSION Bromination of an alkene (cyclic or acyclic) is an example of an electrophilic addition reaction. The reaction proceeds in two stages. The first stage involves the formation of a cyclic *bromonium ion* intermediate. In a special case, this intermediate has actually been isolated as a tribromide salt. The second stage is a nucleophilic attack by a bromide ion on the intermediate, with inversion occurring at the carbon atom attacked, to yield a *vic*-dibromo derivative. This second stage is an S_N2 sequence. Cyclic alkenes provide evidence that the reaction is an anti addition, the bromine atoms being introduced trans to one another.

It is important to realize that if two different groups are present on one or both of the sp^2 carbon atoms of the alkene linkage, chiral carbon centers are generated on bromination of these carbon atoms. In the case with **Z**-stilbene, two chiral centers are generated, yielding a racemic mixture of enantiomeric bromides. The absence of the meso isomer confirms that the reaction is stereospecific.

Bromination of alkenes using a Br_2-CCl_4 solution (a red-brown color) is frequently used as a qualitative test for the presence of unsaturation in a compound. Rapid loss of color from the reagent solution is a positive test (see Chapter 7).

The mechanism of the bromination reaction is outlined below.

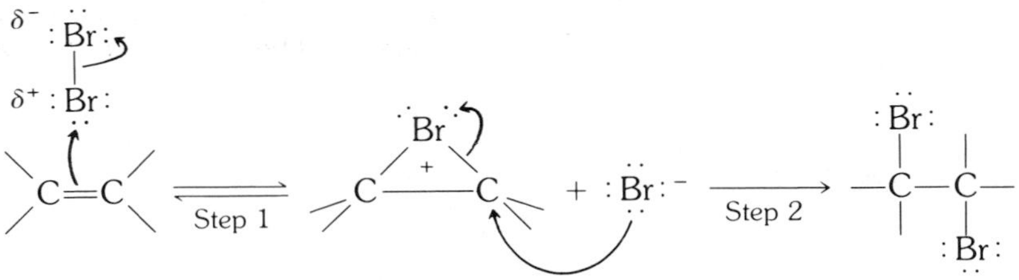

EXPERIMENTAL Estimated time to complete the reaction: 1.0 hour.

Physical Properties of Reactants and Products

Compound	MW	Wt/Vol	mmol	mp(°C)	bp(°C)	Density	n_D
Z-Stilbene	180.25	100 mg	0.55	5–6	141 (12 torr)	1.01	1.6130
Glacial acetic acid		2 mL			118		
Pyridinium hydrobromide perbromide	319.83	200 mg	0.63	205			
d,l-Stilbene dibromide	340.07			110			

Reagents and Equipment

In a 5.0-mL conical vial containing a magnetic spin vane and equipped with an air condenser, place 100 mg (0.55 mmol) of **Z**-stilbene. Next add 1 mL of glacial acetic acid and warm the mixture in a sand bath (or steam bath) to dissolve the liquid.

> **WARNING:** *Glacial acetic acid is corrosive and toxic. It is dispensed in the* HOOD *using an automatic pipet.*

To the cooled solution add 200 mg (0.63 mmol) of pyridinium hydrobromide perbromide.

> **WARNING:** *This compound is a lachrymator. It should be dispensed in the* HOOD.

Any perbromide that adheres to the side of the flask is washed down with an additional 1 mL of glacial acetic acid.

Reaction Conditions

With stirring, heat the reaction mixture in a sand bath at a temperature of 90–100°C until the yellow color of the perbromide disappears (about 5–10 minutes).

Isolation of Product

The vial is removed from the heat source and allowed to cool to approximately 40–50°C (water bath). Water (2 mL) is then added and the vial placed in an ice bath for a period of 15 minutes. The resulting crystalline solid is collected by vacuum filtration using a Hirsch funnel.

Purification and Characterization

The solid material is recrystallized from 95% ethanol using a Craig tube. The final product is dried on a porous clay plate.

Weigh the stilbene dibromide and calculate the percentage yield. Determine the melting point and compare your result with the literature value. Obtain an IR spectrum and compare it with that of an authentic sample.

ENVIRONMENTAL DATA

Substance	Amount	TLV (mg/m^3)	Emissions (mg)	Volume (m^3)
Acetic acid (Max., heat, vacuum filtration)	2.0 mL	25	2000	80
Ethanol (Max., Craig tube)	1.0 mL	1900	800	0.4

TXDS: Stilbene—ipr-mus LD50: 1150 mg/kg

QUESTIONS

6-179. Use Newman projections to show the addition of bromine to E-stilbene.

6-180. Are the results for the answer in question 6-179 different from those in this experiment? If so, how? What is the stereochemical relationship between the products formed in the two reactions?

6-181. Bromine undergoes addition to ethylene in the presence of a high concentration of Cl^- ion to give 1-bromo-2-chloroethane, as well as 1,2-dibromoethane. Chloride ion does not add to the C=C unless bromine is present. Suggest a suitable mechanism to explain these results. Is the rate of bromination significantly affected by the presence of the Cl^- ion?

REFERENCES

A large number of examples of the bromination of alkenes appear in "Organic Syntheses." Selected references are given below.
1. Allen, C. F. H.; Abell, R. D.; Normington, J. B. "Organic Syntheses"; Wiley: New York, 1941; Collect. Vol. I, p 205.
2. Snyder, H. R.; Brooks, L. A.; "Organic Syntheses"; Wiley: New York, 1943; Collect. Vol. II, p 171.
3. Rhinesmith, H. S. *Ibid.*, p 177.
4. Cromwell, N. H.; Benson, R. "Organic Syntheses"; Wiley: New York, 1955; Collect. Vol. III. p 105.
5. McElvain, S. M.; Kundiger, D. *Ibid.*, p 123.
6. Fieser, L. F. "Organic Syntheses"; Wiley: New York, 1963; Collect. Vol. IV, p 195.
7. Khan, N. A. *Ibid.*, p 969.
8. Paquette, L. A.; Barrett, J. H. "Organic Syntheses"; Wiley: New York, 1973; Collect. Vol. V, p 467.

Experiment 50B
Diphenylacetylene

(acetylene, diphenyl-)

This reaction is an illustration of dual dehydrohalogenation to yield an alkyne. The product is the dienophile used in the preparation of hexaphenylbenzene.

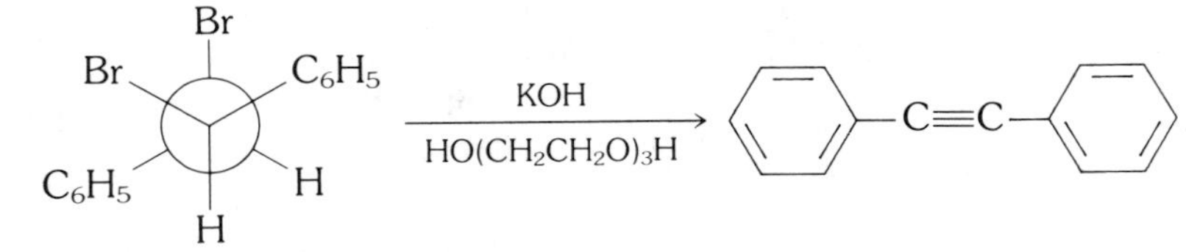

d,l-Stilbene dibromide   Diphenylacetylene

DISCUSSION   This reaction illustrates the dehydrohalogenation of a vic-dibromo compound to form an acetylenic linkage. It is a useful reaction for the generation of the $C\equiv C$ bond since the starting dibromides are readily available from alkenes. The reaction is usually run in the presence of a strong base and proceeds in two stages. In the first stage an intermediate bromoalkene is formed, which may be isolated if the conditions employed are not too drastic. In fact, it is a valuable route for the preparation of vinyl halides. The reaction involves the removal of the hydrogen on the carbon beta to the carbon to which the bromine is attached. The important E2 mechanism operates at this stage of the reaction where the hydrogen being removed is anti-periplanar to the departing bromo group.

The second stage of the reaction is more difficult, but at high temperatures and in the presence of a strong base, the bromoalkene undergoes further E2 elimination to form the alkyne linkage. The different stages of the mechanism are presented below.

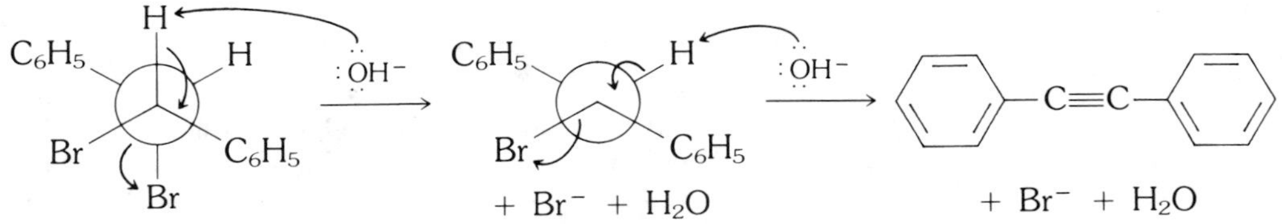

EXPERIMENTAL   Estimated time to complete the reaction: 1.0 hour.

Physical Properties of Reactants and Products

Compound	MW	Wt/Vol	mmol	mp(°C)	bp(°C)
d,l-Stilbene dibromide	340.07	80 mg	0.24	110	
Potassium hydroxide	56.11	75 mg	1.3	360	
Triethylene glycol	150.18	400 μL			278
Diphenylacetylene	178.23			61	

Reagents and Equipment

In a 3.0-mL conical vial containing a boiling stone and equipped with an air condenser, place 80 mg (0.24 mmol) of d,l-stilbene dibromide and 75 mg (1.3 mmol) of KOH flakes. Using an automatic-delivery pipet, add 400 μL of triethylene glycol to the vial.

Reaction Conditions

The reaction mixture is heated using a sand bath maintained at a temperature of 190°C for a period of 5 minutes.

Isolation of Product

The resulting dark reaction mixture is cooled to approximately 40–50°C (water bath) and 1.0 mL of water added. The vial is then placed in an ice bath for 15 minutes. The solid product is collected by filtration under reduced pressure using a Hirsch funnel.

Purification and Characterization

The solid diphenylacetylene is recrystallized from 95% ethanol (Craig tube) and dried on a porous clay plate.

Weigh the product and calculate the percentage yield. Determine the melting point and compare your result with the literature value. Obtain an IR spectrum and compare it with that of an authentic sample.

ENVIRONMENTAL DATA

Substance	Amount	TLV (mg/m³)	Emissions (mg)	Volume (m³)
Ethanol (Max., Craig tube)	1.0 mL	1900	800	0.4

TLV: Potassium hydroxide—2μg/m³
TXDS: Triethylene glycol—orl-rat LD50: 17 gm/kg

QUESTIONS

6-182. Both the E- and Z-2-chlorobutendioic acids dehydrochlorinate to give acetylene dicarboxylic acid.

$$\text{E,Z} \quad HOOC—C(Cl)=CH—COOH \longrightarrow HOOC—C\equiv C—COOH$$

The Z-acid reacts about 50 times faster than the E-acid. Explain.

6-183. Compounds containing the acetylenic bond undergo the Diels-Alder reaction. Formulate the product formed by the reaction of (1E,3E)-1,4-diphenyl-1,3-butadiene with diethyl acetylenedicarboxylate by heating at 140–150°C.

6-184. Acetylenes can be hydrated in the presence of acid and HgSO₄ by electrophilic addition of a molecule of water to the triple bond. The reaction proceeds by way of a carbocation intermediate. Hydration of acetylene produces acetaldehyde. Outline the steps that occur in this transformation.

6-185. In reference to question 6-184, explain why acetaldehyde is isolated instead of the vinyl alcohol.

REFERENCES

1. For a review on the preparation of alkynes *see* Jacobs, T. L. *Org. React.,* **1949,** *5,* 1.
2. A large number of elimination reactions leading to the formation of acetylenes appear in "Organic Syntheses." Selected references are given below.
   a. Allen, C. F. H.; Abell, R. D.; Normington, J. B. "Organic Syntheses"; Wiley: New York, 1941; Collect. Vol. I, p 205.
   b. Hessler, J. C. *Ibid.,* p 438.
   c. Abbott, W. T. "Organic Syntheses"; Wiley: New York, 1943; Collect. Vol. II, p 515.
   d. The synthesis of diphenylacetylene is reported: Smith, L. I.; Falkof, M. M. "Organic Syntheses"; Wiley: New York, 1955; Collect. Vol. III, p 350.
   e. Guha., P. C.; Sankaren, D. K. *Ibid.,* p 623.
   f. Khan, N. A. "Organic Syntheses"; Wiley: New York, 1963; Collect. Vol. IV, p 967.
   g. Campbell, K. N.; Campbell, B. K. *Ibid.,* p 763.

Experiment 51
Hexaphenylbenzene

(benzene, hexaphenyl)

This experiment illustrates the Diels-Alder reaction, which is used extensively as a method to form six-membered cyclic ring systems. The reaction is the final stage of Sequence C.

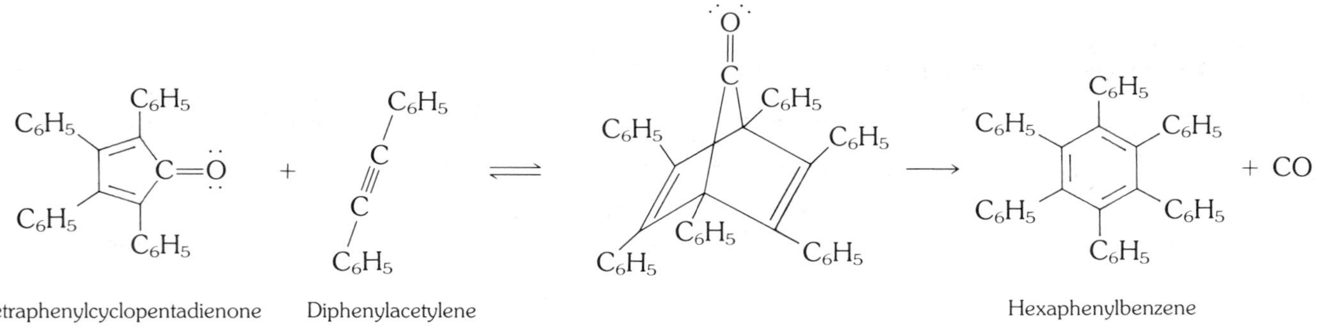

Tetraphenylcyclopentadienone   Diphenylacetylene   Hexaphenylbenzene

DISCUSSION   The Diels-Alder reaction is one of the most useful synthetic tools in organic chemistry. It is an example of a cycloaddition reaction between a conjugated diene and a dienophile (an -ene or -yne), which leads to the formation of six-membered cyclic rings.

By varying the nature of the diene and dienophile, a very large number of structures can be prepared using the Diels-Alder reaction. In the majority of cases, carbocyclic rings are generated, but ring closure can also occur with reactants containing heteroatoms. This leads to the synthesis of compounds containing heterocyclic rings.

The present reaction is unique in that the condensation of tetraphenylcyclopentadienone with diphenylacetylene leads to the formation of an aromatic ring with the evolution of carbon monoxide.

For further discussion on the Diels-Alder reaction see Experiment 14.

EXPERIMENTAL Estimated time to complete the experiment: 1.0 hour.

Physical Properties of Reactants and Products

Compound	MW	Wt/Vol	mmol	mp(°C)
Tetraphenylcyclopentadienone	384.48	20 mg	0.05	220–221
Diphenylacetylene	178.23	20 mg	0.11	61
Hexaphenylbenzene	534.66			465

Reagents and Equipment

In a 8 × 80 mm Pyrex tube sealed at one end, place 20 mg (0.05 mmol) of tetraphenylcyclopentadienone and 20 mg (0.11 mmol) of diphenylacetylene.

The tube is prepared from 8-mm Pyrex glass tubing. Tetraphenylcyclopenta-dienone is prepared in Experiment 49 and diphenylacetylene is prepared in Experiment 50B.

Reaction Conditions

The tube is clamped and the mixture heated with a microburner until a molten mass is obtained (See Fig. 6.14). Heating is continued for approximately 3–5 minutes during which time the dark purple color of the reactants gradually fades to white. *If the heating is too vigorous, the product will darken and decompose.*

Isolation of Product

The unreacted diphenylacetylene is separated from the mixture by clamping the tube at an angle and gently heating the contents with a microburner (Fig. 6.14).

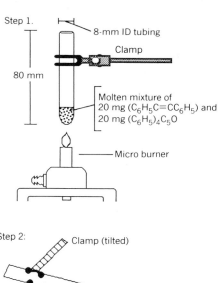

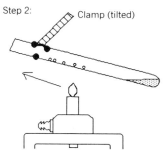

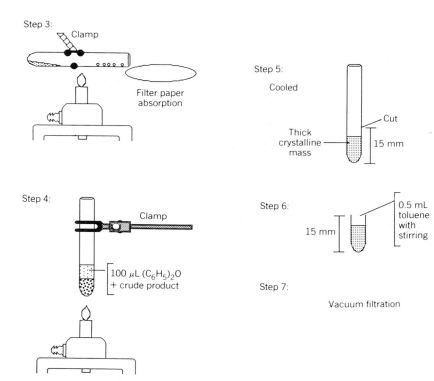

Fig. 6.14 *Step-by-step formation and isolation of hexaphenylbenzene.*

As the diphenylacetylene condenses on the walls of the tube, gentle heating is applied just below this liquid phase, forcing the acetylene compound to the top of the tube. The tube is tipped at an angle, and the liquid diphenylacetylene condensed at the top is removed by touching the end of the tube to a piece of filter paper (Fig. 6.14). *The product remains in the bottom of the tube as an off-white solid mass.*

Purification and Characterization

The tube is allowed to cool and 100 μL of diphenyl ether added to the crude product. The mixture is heated to dissolve the crystalline mass and the resulting solution then allowed to cool. *It may be necessary to add an additional 50 μL of diphenyl ether to dissolve the mass.*

The cooled tube is now cut approximately 2–3 cm above the thick crystalline mass (Fig. 6.14).

Directions

Hold the tube upright, score with a file, and break using a towel to protect the hands.

While stirring with a spatula, the mixture is diluted with 500 μL of cold toluene, and the crystals are collected by filtration under reduced pressure using a Hirsch funnel. The tube and crystals are then rinsed with an additional 0.5 mL of cold toluene and the product dried on a porous clay plate. Weigh the hexaphenyl-benzene product and calculate the percentage yield. Obtain an IR spectrum and compare it to that of an authentic sample.

ENVIRONMENTAL DATA

Substance	Amount	TLV (mg/m³)	Emissions (mg)	Volume (m³)
Diphenyl ether (Max., involatile)	100 μL	7	109	15.6
Toluene (SKIN) (Max., recryst.)	0.5 mL	375	2170	5.8
Toluene (SKIN) (Max., cryst., wash)	2.5 mL	50	3725	75

QUESTIONS   **6-186.** What starting materials would you use to prepare each of the following compounds by the Diels-Alder reaction?

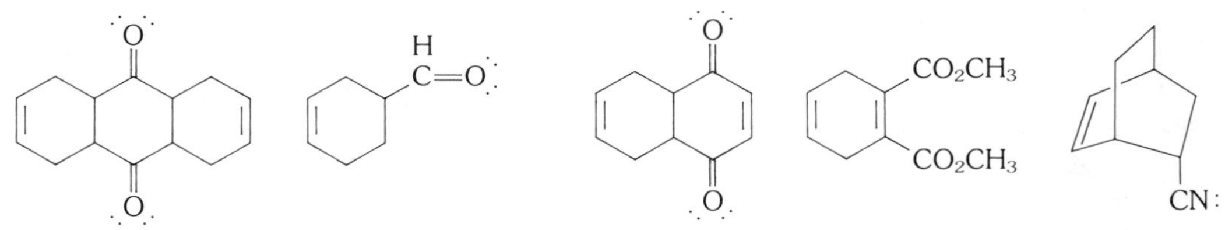

6-187. Diels-Alder reactions with benzene are rare. Two are shown below. Give the structures of the product produced in each reaction.

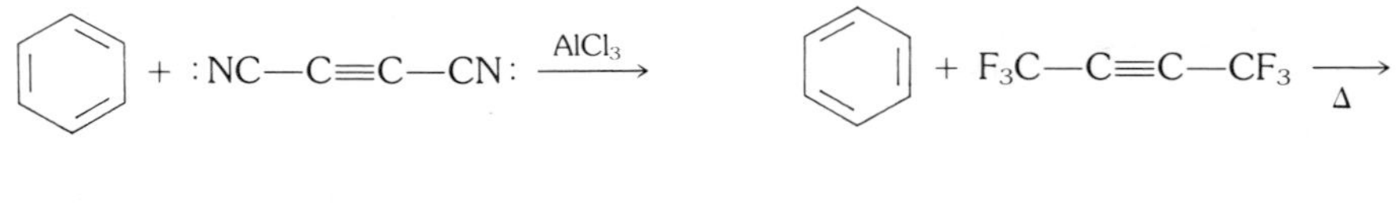

6-188. Below are shown two heteroatom compounds that undergo the Diels-Alder reaction. Formulate the product obtained in each reaction.

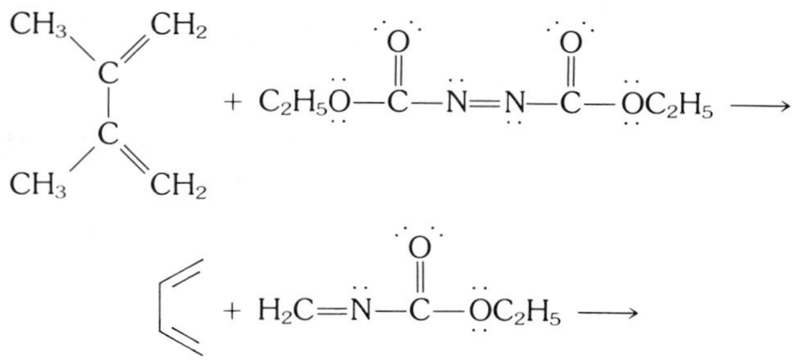

REFERENCES

1. Review articles:
   a. Norton, J. A. *Chem. Revs.* **1942,** *31,* 319.
   b. Kloetzel, M. C. *Org. React.* **1948,** *4,* 1.
   c. Holm, H. L. *Ibid.,* 60.
   d. Butz, L. W.; Rytina, A. W. *Org. React.* **1949,** *5,* 136.
   e. Sauer, J. *Angew. Chem. Int. Ed. Engl.* **1966,** *5,* 211.
   f. *Ibid.,* **1967,** *6,* 16.
2. An "Organic Syntheses" preparation using tetraphenylcyclopentadienone in a Diels-Alder reaction to obtain tetraphenylphthalic anhydride has been recorded.
      Grummitt, O. "Organic Syntheses"; Wiley: New York, 1955; Collect. Vol. III., p 807.

SEQUENCE D: Synthesis of 4-Methoxybenzoic Acid from 4-Ethylphenol

This series of reactions involves the synthesis of an ether using the Williamson reaction followed by two selective oxidations to obtain the final product.

4-Ethylphenol → 4-Ethylanisole →
[25C]
4-Methoxyacetophenone → 4-Methoxybenzoic acid
[52A]   [43B]

Experiments 52A and 52B
4-Methoxyacetophenone; 2,4-Dinitrophenylhydrazone

(acetophenone, 4-methoxy-)

This reaction illustrates the selective oxidation of a methylene group to a ketone function. Characterization of the resulting ketone product is achieved by formation of its 2,4-dinitrophenylhydrazone.

DISCUSSION   This reaction utilizes the silver/persulfate couple as an oxidizing agent to selectively oxidize substituted ethylbenzenes to the corresponding methyl ketone The reaction is carried out in aqueous media with a trace of silver ion. The overall mechanism is complicated, but a radical-ion species, SO_4^-, apparently is involved. The silver ion assists in the electron transfer process to generate this radical-ion. The complete details of the mechanism are not well understood.

The formation of 2,4-dinitrophenylhydrazones is one of the classical means used to characterize aldehydes and ketones (see Chapter 7 and Experiment 40B).

PART A: 4-Methoxyacetophenone

4-Ethylanisole	Ammonium peroxydisulfate	4-Methoxyacetophenone

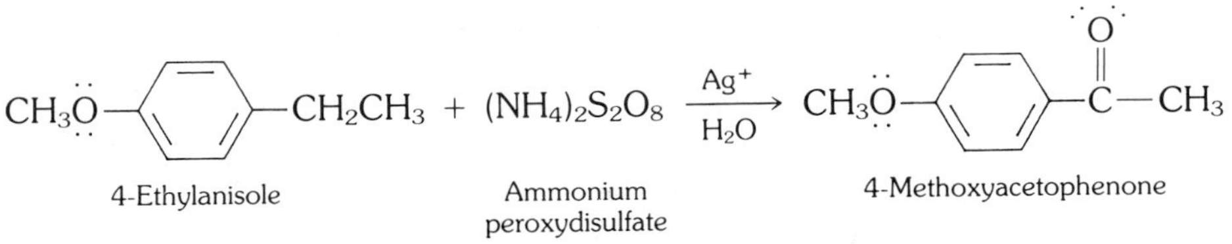

EXPERIMENTAL   Estimated time of the experiment: 5 hours (2 laboratory periods).

Physical Properties of Reactants and Products

Compound	MW	Wt/Vol	mmol	mp(°C)	bp(°C)	Density	n_D
4-Ethylanisole	136.20	50 mg	0.37		195–196	0.96	1.5120
Ammonium peroxydisulfate	228.18	167 mg	0.73	120			
Silver nitrate (0.2% aqueous solution)		750 μL					
4-Methoxyacetophenone	150.8			38–39	258		

Note. *A reaction time of 2.5–3.5 hours is required. It is recommended that another experiment be conducted during this time. The oxidation product can be isolated during the following laboratory period and the 2,4-dinitrophenyl-hydrazone prepared.*

Reagents and Equipment
In a 3.0-mL conical vial containing a magnetic spin vane and equipped with an air condenser, place 50 mg (0.37 mmol) of 4-ethylanisole, 167 mg (0.73 mmol) of ammonium peroxydisulfate, and 750 μL of 0.2% aqueous silver nitrate solution.

Note. *4-Ethylanisole is prepared in Experiment 25C and is of sufficient purity for use in this preparation. The silver nitrate solution is dispensed using an automatic-delivery pipet.*

Reaction Conditions
The reaction mixture is stirred at room temperature for 2.5–3.5 hours.

Isolation of Product
The resulting aqueous solution is extracted with three 0.5-mL portions of chloroform (calibrated Pasteur pipet). The wet chloroform extracts are separated with a Pasteur filter pipet and dried by transferring the solution to a Pasteur filter pipet packed with 1.0 g of anhydrous sodium sulfate.

To separate the two phases, draw them into the Pasteur filter pipet and gently transfer the bottom layer to the column. The aqueous layer is then returned to the original vial for the next extraction with additional chloroform.

The eluate is collected in a tared 5.0-mL conical vial containing a boiling stone. An additional 1.0 mL of chloroform is added to wash the column and this eluate combined with the original material collected from the column.

The solution is concentrated by placing the vial in a sand bath maintained at **HOOD** 75–85°C [**HOOD**]. A gentle stream of nitrogen gas impinging on the surface of the solution speeds up the evaporation process.

Purification and Characterization

The crude product is a brown liquid that may crystallize on standing.

Weigh the material and calculate the crude yield. Obtain an IR spectrum, as a liquid or melt, and compare your spectrum with that of an authentic sample.

PART B: 4-Methoxyacetophenone 2,4-Dinitrophenylhydrazone

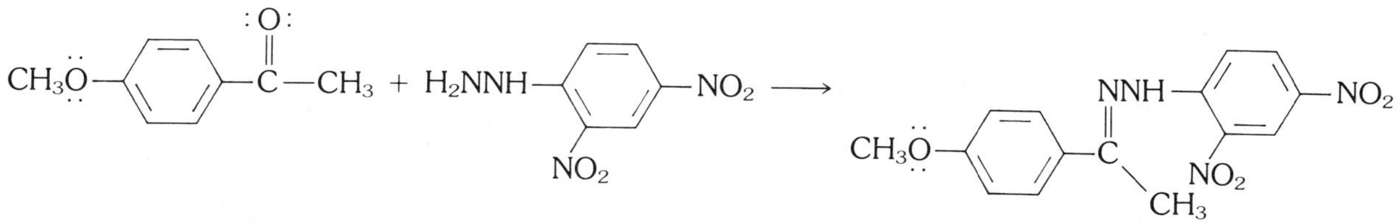

4-Methoxyacetophenone   2,4-Dinitrophenylhydrazine   4-Methoxyacetophenone
2,4-dinitrophenylhydrazone

EXPERIMENTAL   Estimated time of the experiment: 1.0 hour.

Physical Properties of Reactants and Products

Compound	MW	Wt/Vol	mmol	mp(°C)	bp(°C)
4-Methoxyacetophenone	150.8	30 mg	0.2	38–39	258
2,4-Dinitrophenylhydrazine	198.14	45 mg	0.23	145	
Dimethylformamide	73.09	450 μL			149–150
2,4-Dinitrophenylhydrazone	320.22			220	

Note. *This procedure is general for aldehydes and ketones. It is adapted from a method published by Howell, B. A., J. Chem. Educ.* **1984,** *61,* *176. For a further example see Experiment 40B, Fluorenone 2,4-dinitrophenylhydrazone.*

Reagents and Equipment

In a 1.0-mL conical vial, or a 10 × 75 mm test tube, place 30 mg (0.20 mmol) of 4-methoxyacetophenone. Add a solution of 45 mg (0.23 mmol) of 2,4-dinitrophenylhydrazine dissolved in 450 μL of dimethylformamide.

The total amount of crude 4-methoxyacetophenone prepared in Experiment 52A may be used to prepare the derivative. Adjust the reagent amounts accordingly. It may be convenient to reverse the order of addition. That is, add the ketone to the hydrazine-DMF solution.

The mixture is agitated to effect dissolution and then 2 drops of concentrated hydrochloric acid are added (Pasteur pipet).

Reaction Conditions

The resulting solution is allowed to stand at room temperature for a period of 15 minutes. It is then cooled in an ice bath. The red crystalline product should become evident during this period.

Isolation of Product

The crystals are collected by filtration under reduced pressure using a Hirsch funnel. The filter cake is washed with two 0.5-mL portions of *2M* hydrochloric acid, 1.0 mL of water, and 0.5 mL of cold 95% ethanol (calibrated Pasteur pipet).

Purification and Characterization

The crude, red crystalline derivative is then recrystallized from ethyl acetate using a Craig tube. After drying on a porous clay plate, weigh the material and calculate the percentage yield.

Determine the melting point and compare your results with those found in the literature.

ENVIRONMENTAL DATA

Substance	Amount	TLV (mg/m^3)	Emissions (mg)	Volume (m^3)
Chloroform (Max., extract, evap.)	2.5 mL	50	3725	75
Dimethylformamide (SKIN) (Max., solvent)	450 μL	30	425	14.2
Hydrochloric acid (Max., reactant)	100 μL	5	44	8.8
Ethanol (Max., cryst., wash)	0.5 mL	1900	400	0.2

TLV: Ammonium peroxydisulfate—2 mg/m^3
TXDS: 4-Methoxyacetophenone—orl-rat LD50: 1720 mg/kg
      2,4-Dinitrophenylhydrazine—orl-rat LD50: 654 mg/kg
      Silver nitrate—unk-man LDLo: 29 mg/kg
      Sodium sulfate—orl-mus LD50: 5989 mg/kg

QUESTIONS

6-189. Selenium dioxide is often used to effect oxidation of aldehydes and ketones to yield α-dicarbonyl compounds. Predict the product for each of the following reactions.

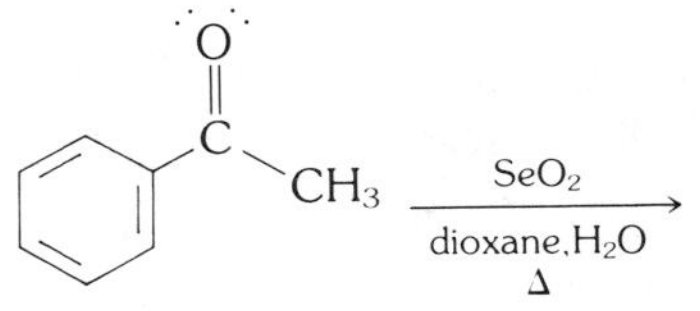

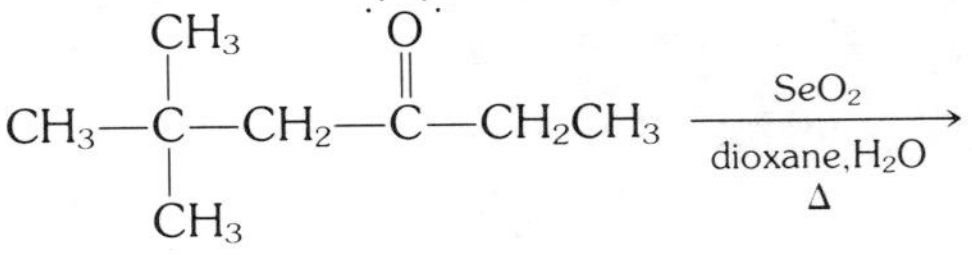

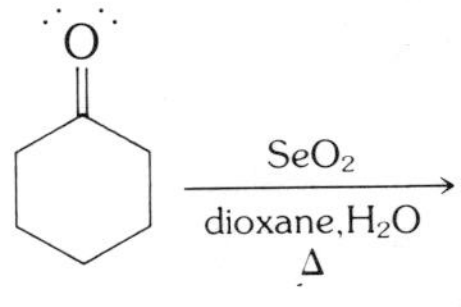

6-190. Using the above oxidation conditions with SeO$_2$, it was found that with unsymmetrical ketones oxidation occurred at the —CH$_2$— group that is most readily enolized. Based on this fact would you change any of the structures of the products you formulated in question 6-189?

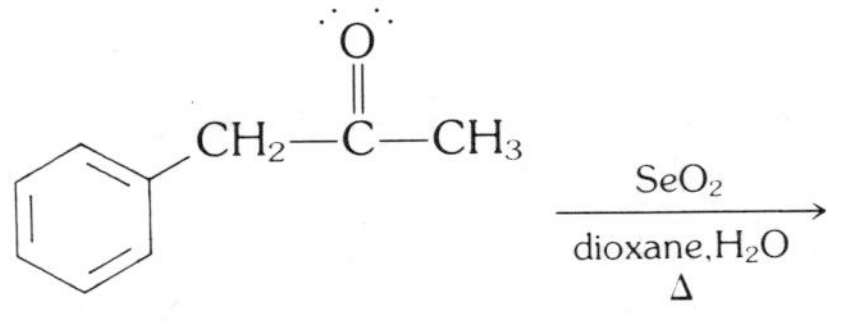

6-191. The Ag$^+$ ion acts to generate the SO$_4^{\cdot-}$ ion-radical from S$_2$O$_8^{2-}$. Propose an oxidation–reduction reaction sequence to account for the formation of the SO$_4^{\cdot-}$.

REFERENCES

1. Peroxydisulfate as an oxidizing agent:
   a. House, D. A. *Chem. Revs.* **1962,** *62,* 185.
   b. Daniher, F. A. *Org. Prep. and Proc.* **1970,** *2,* 207 and references therein.
2. Preparation of 2,4-dinitrophenylhydrazones:
   a. Howell, B. A. *J. Chem. Educ.* **1984,** *61,* 176.
   b. Shriner, R. L.; Fuson, R. C.; Curtin, D. Y.; Morrill, T. C. "The Systematic Identification of Organic Compounds", 6th ed.; Wiley: New York, 1980.

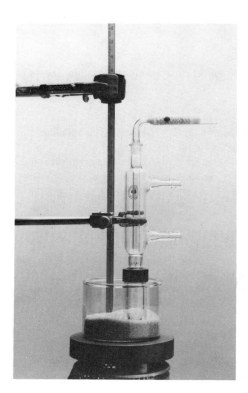

Chapter 7

Qualitative Identification of Organic Compounds
I. Organic Qualitative Analysis

Millions of organic compounds have been recorded in the literature. It may seem a bewildering task to attempt to identify one certain compound from this vast array. However, it is important to realize that the majority of these substances can be grouped into a comparatively small number of classes.

It is the object of organic qualitative analysis to place a given compound, through screening tests, into one of these classes, which in turn greatly simplifies the *identification* of the compound. This is usually done using a series of preliminary observations and chemical tests, in conjunction with the instrumentation that developments in the spectroscopic area of analysis have made available to the analyst. The advent of infrared, nuclear magnetic resonance, and mass spectroscopy has had a profound effect on the approach taken to identify a specific organic compound. Analysis of ultraviolet spectra may also be utilized to advantage with certain classes of materials.

The approach taken in this text follows the format outlined below.

1. Preliminary tests are performed to determine the physical nature of the compound.
2. Chemical tests, mainly to assist in identifying elements other than C, H, or O and also to detect common functional groups present in the molecule, are carried out. There is an added benefit especially in relation to the chemical detection of functional groups. This is, that an incredible amount of chemistry can be observed and learned in performing these tests. Later, as the spectroscopic techniques are developed, the number of chemical tests performed may be curtailed.
3. The spectroscopic method of analysis is utilized. As you develop further in your knowledge of chemistry, you will appreciate more and more the revolution that has taken place in chemical analysis over the past 25–30 years and what powerful methods are at your disposal for the identification of organic compounds.

A large number of texts have been published on organic qualitative analysis. Several are cited below.

REFERENCES

1. Shriner, R. L.; Fuson, R. C.; Curtin, D. Y.; Morrill, T. C. "The Systematic Identification of Organic Compounds", 6th ed.; Wiley: New York, 1980.

2. Schneider, F. L. "Qualitative Organic Microanalysis", in Vol. II of Monographien aus dem Gebiete der Qualitativen Mikroanalyse; Benedetti-Pichler, Ed.; Springer-Verlag: Wein, Austria, 1964.

3. Vogel, A. I. "Qualitative Organic Analysis", Part 2 of "Elementary Practical Organic Analysis"; Wiley: New York, 1966.

4. Cheronis, N. D.; Entrikin, J. B.; Hodnett, E. M. "Semimicro Qualitative Organic Analysis," 3rd ed.; Interscience: New York, 1965.

5. Cheronis, N. D.; Ma, T. S. "Organic Functional Group Analysis by Micro and Semimicro Methods"; Interscience: New York, 1964.

6. Feigl, F.; Anger, V. "Spot Tests in Organic Analysis"; Elsevier: New York, 1966.

7. Kamm, O. "Qualitative Organic Analysis," 2nd ed.; Wiley: New York, 1932.

PRELIMINARY TESTS

Overview

The objective of the preliminary tests is to provide a pathway or sequence to assist you in the route to follow in order ultimately to identify the unknown material at hand. It must be emphasized, however, that these tests frequently consume material. Given the amounts generally available at the micro- or semi-micro range, judicious selection of the tests to perform must be made; however, in some tests, the material may be recovered. You should always be aware of this fact. Each preliminary test that can be conducted with *little expenditure of time and material* can offer very valuable clues as to the class to which a given compound belongs.

It is important to realize that *negative* findings are often as important as *positive* results in identifying a given compound. Cultivate the habit of following a *systematic pathway or sequence* so that no clue or bit of information is lost or overlooked along the way. It is important also to develop the *attitude* and *habit* of planning ahead. Outline a logical plan of attack, depending on the nature of the unknown, and follow it. As you gain more and more experience in this type of investigative endeavor, the planning stage will become easier. At this initial phase of your development, the unknowns to be identified will be relatively pure materials. *Record all observations and results of the tests in your laboratory notebook.* Review these data as you execute the sequential phases of your plan. This serves to keep you on the straight and narrow path to success.

Nonchemical Tests

Physical State

If the material is a *solid*, a few milligrams of the sample may be viewed under a magnifying glass or microscope. This may give some indication as to the homogeneity of the material. Crystalline shape often is an aid to classify the compound.

Determine the melting point using a few milligrams of the solid material. If a narrow melting-point range (1–2°C) is observed, it is a good indication that the material is quite pure. If a broad range is observed, the compound must be recrystallized from a suitable solvent before proceeding. If the material undergoes decomposition on heating, it is worthwhile to try an evacuated melting point. If any evidence indicates that sublimation is occurring, a sealed-tube melting point should be run. Furthermore, this indicates that sublimation might be used to purify the compound, if necessary.

If the material is a *liquid*, the boiling point is determined by the ultramicro method. If sufficient material is on hand and the boiling point reveals that the material is relatively pure, the *density* and the *refractive index* add valuable information for identification purposes.

Color

Since the majority of organic compounds are colorless, examination of the color can occasionally provide a clue as to the nature of the sample. Use caution, however, since some impurities can give a substance color. Aniline is a classic example. When freshly distilled it is colorless, but on standing a small fraction oxidizes and turns the entire sample a reddish-brown color.

Colored organic compounds contain a *chromophoric group,* usually indicating extended conjugation in the molecule. For example, 1,2-dibenzoylethylene (Experiments 1A and 5) is light yellow; 5-nitrosalicylic acid (Experiment 34D) is light yellow; azobenzene (Experiment 44A) is red; tetraphenylcyclopentadienone (Experiment 49) is purple; crystal violet (Experiment 20) is green.

Can you identify the chromophore that causes these compounds to be colored? Note that a colorless liquid or white solid would *not contain* these units. Thus, compounds containing these groupings would be excluded from consideration as possible candidates.

Odor

Detection of a compound's odor can occasionally be of assistance, since the vast majority of organic compounds have no definitive odor. You should become familiar with the odors of the common compounds or classes. For example, aliphatic amines have a fishy smell; benzaldehyde (like nitrobenzene and benzonitrile) has an almond odor (Experiment 6); esters have fruity odors (Experiment 8A–D). Common solvents such as acetone, diethyl ether, and toluene all have distinctive odors. Butyric and caproic acids have rancid odors. In many cases, extremely small quantities of certain compounds can be detected by their odor. For example, a C_{16} unsaturated alcohol released by the female silk worm moth elicits a response from male moths of the same species at concentrations of 100 molecules/cm^3.

Odor detection involves your olfactory capabilities and thus can be considered a helpful lead, but very rarely can this property be used to strictly classify or identify a substance. As mentioned above, contamination by a small amount of an odorous substance is always a possibility.

> **CAUTION:** *You should be very cautious when detecting odors. Any odor of significance can be detected several inches from the nose. Do not place the container closer than this to your eyes, nose, or mouth. Open the container of the sample and gently waft the vapors toward you.*

Ignition Test[1]

Valuable information can be obtained by carefully noting the manner in which a given compound burns. The ignition test is carried out by placing 1–2 mg of the sample on a spatula followed by heating with a microburner. Do not hold the sample directly in the flame; heat the spatula about 1 cm from the flat end and move the sample slowly into the flame (see Fig. 7.1).

Important observations to be made concerning the ignition test are summarized in Table 7.1.

As the heating of the sample takes place you should make the following observations.

---

[1]For an extensive discussion on examination of ignition residues see Feigl, F.; Anger, V. "Spot Tests in Organic Analysis," 7th ed.; Elsevier: New York, 1966; p. 51.

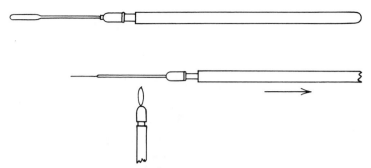

Fig. 7.1 *Heating on the microspatula. (Courtesy of Springer-Verlag, Wien, Austria).*

1. Any melting or evidence of sublimation: gives an approximate idea of the melting point by the temperature necessary to cause melting.
2. Color of the flame as the substance begins to burn (see Table 7.1).
3. Nature of the combustion (flash, quiet, or an explosion). Rapid, almost instantaneous combustion indicates high hydrogen content. Explosion indicates presence of nitrogen or nitrogen-oxygen–containing groups. For example, azo units (Experiment 44A) or nitro groups (Experiment 34).
4. Nature of a residue, if present, after ignition.
   a. If a black residue remains and disappears on further heating at higher temperature, the residue is carbon.
   b. If the residue undergoes swelling during formation, the presence of a carbohydrate or similar compound is indicated.
   c. If the residue is black initially but still remains after heating, an oxide of a heavy metal is indicated.
   d. If the residue is white, the presence of an alkali or alkaline earth carbonate or SiO_2 from a silane or silicone is indicated.

SEPARATION OF IMPURITIES   If the preliminary tests outlined above indicate that the unknown in question contains impurities, it may be necessary to carry out one of several purification

Table 7.1   Ignition Test Observations[a]

Type of Compound	Example	Observation
Aromatic compounds, unsaturated, or higher aliphatic compounds	Toluene	Yellow, sooty flame
Lower aliphatic compounds	Hexane	Yellow, almost nonsmoky flame
Compounds containing oxygen	Ethanol	Clear bluish flame
Halogen compounds	Methyl iodide	Smoky flame
Polyhalogen compounds	Chloroform	Generally do not ignite until burner flame applied directly to the substance
Sugars and proteins	Sucrose	Characteristic odor
Acid salts or organometallic compounds	Ferrocene	Residue

[a]Cheronis, N. D.; Entrikin, J. B. "Semimicro Qualitative Organic Analysis"; Interscience: New York, 1947; p 85.

steps. These techniques are discussed in earlier chapters and are summarized below for correlation purposes.

1. For a liquid, distillation is generally used (see Experiment 1).
2. For a solid, recrystallization is generally used (see Experiment 5).
3. Extraction is used especially if the impurity is insoluble in a solvent in which the compound itself is soluble (see Experiment 3).
4. Sublimation is a very efficient technique, if the compound sublimes (see Experiment 11).
5. Chromatography: gas, column, or TLC, is often used (see Experiments 2 and 8).

It should be realized that the above techniques may be applied to the separation of mixtures as well.

DETECTION OF ELEMENTS OTHER THAN CARBON, HYDROGEN, OR OXYGEN

The elements other than C, H, and O that are most often present in organic compounds are nitrogen, sulfur, and the halogens (F, Cl, Br, I). To detect the presence of these elements, the organic compound is generally fused with metallic sodium. This reaction converts them to the water-soluble inorganic compounds, NaCN, Na_2S and NaX. Inorganic qualitative analysis techniques enable the investigator to determine the presence of the corresponding anions.

$$\text{Organic compound containing} \begin{Bmatrix} C \\ H \\ O \\ N \\ S \\ X \end{Bmatrix} \xrightarrow[\Delta]{Na} \begin{Bmatrix} NaCN \\ Na_2S \\ NaX \end{Bmatrix}$$

Sodium Fusion: (Lassaigne's Test)[2]

The procedure recommended uses a sodium-lead alloy[3] (dri-Na from J.T. Baker Co.) in place of metallic sodium. This reagent is quite stable in air, is easily stored in a screwcapped bottle, and does not give a vigorous reaction in water. Any metallic residues obtained from running the fusion reaction should be disposed of properly to avoid lead pollution.

HOOD   **Important.** *The fusion reaction is carried out in the* **HOOD.**

In a small (10 × 75 mm) test tube supported in a transite board (see Fig. 7.2) is placed ~0.25 g of the sodium-lead alloy. The tube is heated with a flame until the alloy melts and sodium vapor is observed condensing on the walls of the tube.

Note. *Do NOT heat the alloy to redness.*

A small sample of your unknown compound is carefully added to the tube. If your unknown is a *liquid*, add 2–3 drops from a Pasteur pipet; if a solid, add ~5 mg using a spatula. Be careful not to get any sample on the sides of the tube.

If your sample is a volatile liquid (bp < 100°C) or if it contains nitrogen or sulfur, mix the material with ~20–30 mg of powdered sucrose prior to its addi-

[2]Lassaigne, J. L. *Ann.* **1843**, *48*, 367; also see Campbell, K. N.; Campbell, B. K. *J. Chem. Educ.* **1950**, *27*, 261.
[3]Vinson, J. A.; Grabowski, W. T. *J. Chem. Educ.* **1977**, *54*, 187.

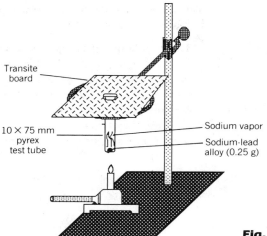

Fig. 7.2 *Apparatus for sodium fusion.*

tion to the test tube. This aids in the reduction of various nitrogen or sulfur compounds. Also, it absorbs volatile materials so that they may undergo the desired reaction before vaporization can occur.

The tube is heated gently, to initiate the reaction with sodium. Remove the flame until the reaction subsides and then heat to redness for 1–2 minutes.

A boiling stone is now added to the cooled test tube followed by 2.0 mL of distilled water. The resulting solution is reheated with stirring and filtered while warm through filter paper. The filtrate is collected in a 10-mL Erlenmeyer flask.

Note. *The filtration step may be eliminated if the aqueous solution is clear.*

If the filtration is performed, wash the filter paper with an additional 2.0 mL of water. This wash is combined with the original filtrate.

The fusion solution collected in the Erlenmeyer flask is used to test for the presence of CN^- (nitrogen), S^{2-} (sulfur), and X^- (halogens except F^-) as described below.

Sulfur

Place 2–3 drops (Pasteur pipet) of the fusion solution on a white spot plate followed by 2 drops of water. Now add 1 drop of dilute (2%) aqueous sodium nitroprusside solution. The formation of a deep blue-violet color is a positive test for sulfur.

$$Na_2S + Na_2Fe(CN)_5NO \longrightarrow Na_4[Fe(CN)_5NOS] + 2\,NaOH$$

Sodium nitroprusside   Blue-violet complex

Nitrogen[4]

The Reagents

1. 1.5% solution of *p*-nitrobenzaldehyde in 2-methoxyethanol.
2. 1.7% solution of *o*-dinitrobenzene in 2-methoxyethanol.
3. 2.0% solution of NaOH in distilled water.

All reagent drops are dispensed using Pasteur pipets.

On a white spot plate are placed 5 drops of reagent 1, 5 drops of reagent 2, and 2 drops of reagent 3. This mixture is gently stirred with a glass rod.

One drop of the fusion solution is now added. The formation of a deep-purple color is a positive test for the presence of CN^- ion; a yellow or tan

[4]Adapted from Guilbault, G. G.; Kramer, D. N. *Anal. Chem.* **1966**, *39*, 834. *Ibid., J. Org. Chem.* **1966**, *31*, 1103. See also Shriner, R. L.; Fuson, R. C.; Curtin, D. Y.; Morrill, T. C. "The Systematic Identification of Organic Compounds," 6th ed.; Wiley: New York, 1980; p 80.

coloration is negative. If a positive result is obtained, nitrogen is present in the sample.

The test is valid in the presence of halogens (NaX) or sulfur (Na$_2$S). It is much more sensitive than the traditional Prussian Blue test.[5]

The Halogens (Except Fluorine)

Using the Fusion Solution

In a 10×75 mm test tube containing a boiling stone is placed 0.5 mL (calibrated Pasteur pipet) of the fusion solution. This soluton is carefully acidified by the dropwise addition of dilute HNO$_3$ acid, delivered from a Pasteur pipet (test acidity with litmus paper). If a positive test for nitrogen or sulfur was obtained, the resulting solution is heated to a gentle boil for 1 minute over a microburner **[HOOD]** to expel any HCN or H$_2$S that might be present and then cooled to room temperature.

HOOD

To the fusion solution is now added 2 drops (Pasteur pipet) of aqueous 0.1M AgNO$_3$ solution.

A heavy curdy-type precipitate is a positive test for the presence of Cl$^-$, Br$^-$, or I$^-$ ion. A faint turbidity is a negative test.

AgCl precipitate is white.

AgBr precipitate is pale yellow.

AgI precipitate is yellow.

AgF is not detected by this test since it is relatively soluble in water.

The silver halides have different solubilities in dilute ammonium hydroxide solution.

Centrifuge the test tube and contents and remove the supernatant liquid using a Pasteur filter pipet. Add 0.5 mL (calibrated Pasteur pipet) of dilute ammonium hydroxide solution to the precipitate and stir with a glass rod to determine whether the solid is soluble.

AgCl is soluble in ammonium hydroxide due to the formation of the complex ion, [Ag(NH$_3$)$_2$]$^+$.

AgBr is slightly soluble in the reagent.

AgI is insoluble in this solution.

Further Test[6]

Once the presence of a halide ion has been established, a further test is also available to aid in distinguishing between Cl$^-$, Br$^-$, and I$^-$ ion.

As described above, acidify 0.5 mL of the fusion solution with dilute HNO$_3$. To this solution add 5 drops (Pasteur pipet) of a 1.0% aqueous KMnO$_4$ solution and shake the test tube for ~1 minute.

Now add 10–15 mg of oxalic acid, enough to decolorize the excess purple permanganate, followed by 0.5 mL of methylene chloride solvent. The test tube is stoppered, shaken, and vented, and the layers are allowed to separate. Observe the color of the CH$_2$Cl$_2$ (lower) layer.

A clear methylene chloride layer indicates Cl$^-$ ion.

A brown methylene chloride layer indicates Br$^-$ ion.

A purple methylene chloride layer indicates I$^-$ ion.

The colors may be faint and should be observed against a white background.

---

[5]See Vogel, A. I. "Elementary Practical Organic Chemistry," Part 2, 2nd ed.; Wiley: New York, 1966; p 37.

[6]For further tests to distinguish between the three halide ions see Shriner, R. L.; Fuson, R. C.; Curtin, D. Y.; Morrill, T. C. "The Systematic Identification of Organic Compounds," 6th ed.; Wiley: New York, 1980; p 81. Also see this reference (p 85) for a specific test for the F$^-$ ion.

The Beilstein Test[7]

Organic compounds that contain chlorine, bromine, or iodine and hydrogen are decomposed on ignition in the presence of copper oxide to yield the corresponding hydrogen halides. These gases react to form the volatile cupric halides which impart a green or blue-green color to a nonluminous flame. It is a very sensitive test, but some nitrogenous compounds and carboxylic acids also give positive results.

Pound the end of a copper wire to form a flat surface that can act as a spatula. The other end of the wire (~4 in. long) is stuck in a cork stopper to serve as a handle.

The flat tip of the wire is heated in a flame until coloration in the flame is negligible.

On the cooled flat surface is placed a drop (Pasteur pipet) of liquid unknown or a few milligrams of solid unknown. Gently heat the material in the flame. The carbon present in the compound will burn first, and thus the flame will be luminous, but then the characteristic green or blue-green color will be evident. It may be fleeting, so watch carefully.

It is recommended that a known compound containing a halogen be tested so that you become familiar with the appearance of the expected color.

F^- is not detected by this test since copper fluoride is not volatile.

SOLUBILITY CHARACTERISTICS

Determination of the solubility characteristics of an organic compound can often give valuable information as to its structural composition. It is especially useful when correlated with spectral analysis.

Several schemes have been proposed that place a substance in a definite group according to its solubility in various solvents. The scheme presented below is similar to that outlined in Shriner et al.[8]

There is no sharp dividing line between soluble and insoluble, and an arbitrary ratio of solute to solvent must be selected. We suggest that a compound be classified as soluble if its solubility is greater than 15 mg/500 μL of solvent. The solubility determinations are carried out at ambient temperature in 10 × 75 mm test tubes. The sample (15 mg) is placed in the test tube and a total of 0.5 mL of solvent added in three portions from a graduated or calibrated Pasteur pipet. Between addition of each portion, the sample is stirred vigorously with a glass stirring rod for 1.5 to 2 minutes. If the sample is water soluble, test the solution with litmus paper to assist in classification according to the solubility scheme on page 319.

To test with litmus paper, dip the end of a small glass rod into the solution and then gently touch the litmus paper with the rod. DO NOT DIP THE LITMUS PAPER INTO THE TEST SOLUTION.

In doing the solubility tests follow the scheme in the order given on page 319. *Keep a record of your observations.*

Step 1. Test for water solubility. If soluble, test with litmus paper.
Step 2. If water soluble, determine the solubility in diethyl ether. This test further classifies water-soluble materials.
Step 3. Water-insoluble compounds are now tested with 5% aqueous NaOH solution. If soluble, determine the solubility in 5% aqueous $NaHCO_3$. The use of the $NaHCO_3$ solution aids in distinguishing between strong (soluble) and weak (insoluble) acids.
Step 4. Compounds insoluble in 5% aqueous NaOH are tested with 5% HCl.

[7]Beilstein, F. *Ber.* **1872,** *5,* 620.
[8]Shriner, R. L.; Fuson, R. C.; Curtin, D. Y.; Morrill, T. C. "The Systematic Identification of Organic Compounds," 6th ed.; Wiley: New York, 1980.

Step 5. Compounds insoluble in 5% aqueous HCl are tested with concentrated H_2SO_4. If soluble, further differentiation is made using 85% H_3PO_4 as shown in the scheme.

Step 6. Miscellaneous neutral compounds containing sulfur or nitrogen are normally soluble in strong acid solution.

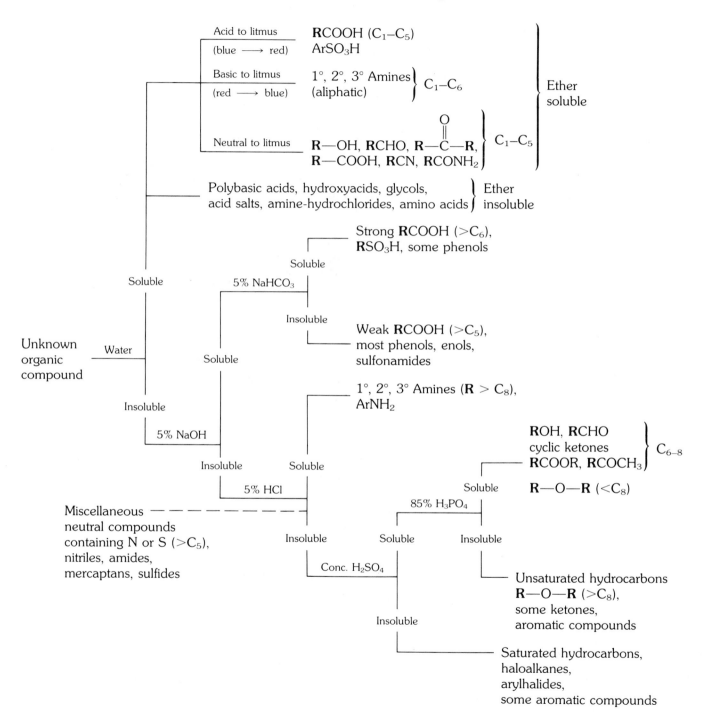

Note that to classify a given compound it may not be necessary to test its solubility in *every* solvent. *Do only those tests which are required to place the compound in one of the solubility groups.* Make your observations with care, and proceed in a logical sequence as you make the tests.

THE CLASSIFICATION TESTS[9]

Note. *For all tests given in this section, drops of reagents are measured using Pasteur pipets.*

Alcohols

Ceric Nitrate Test

Primary, secondary, and tertiary alcohols having fewer than 10 carbon atoms give a positive test as indicated by a change in color from *yellow* to *red*.

$$\underset{\text{Yellow}}{(NH_4)_2Ce(NO_3)_6} + \mathbf{RCH_2OH} \longrightarrow \underset{\text{Red complex}}{[\text{alcohol} + \text{reagent}]}$$

On a white spot plate is placed 5 drops of test reagent. (*The reagent is prepared by dissolving 4.0 g of ceric ammonium nitrate in 10 mL of 2N HNO$_3$. Warming may be necessary.*) Add 1–2 drops of the unknown sample (5 mg if a solid). Stir with a thin glass rod to mix the components and observe any color change.

1. If the alcohol is water insoluble, 3–5 drops of dioxane may be added, but run a blank to make sure the dioxane is pure. Efficient stirring gives positive results with most alcohols.
2. Phenols, if present, give a brown color or precipitate.

Chromic Anhydride Test—Jones Oxidation

The Jones oxidation test is a rapid method to distinguish primary and secondary alcohols from tertiary alcohols. A positive test is indicated by a color change from *orange* (the oxidizing agent, Cr^{6+}) to *blue-green* (Cr^{3+}).

$$\left.\begin{matrix}\mathbf{RCH_2OH} \\ \text{or} \\ \mathbf{R_2CHOH}\end{matrix}\right\} + \underset{\text{Orange}}{H_2Cr_2O_7} \xrightarrow{H_2SO_4} \underset{\text{Green}}{Cr_2(SO_4)_3} + \begin{matrix}\mathbf{RCO_2H} \\ \text{or} \\ \mathbf{R_2C{=}O}\end{matrix}$$

The test is based on oxidation of a primary alcohol to an aldehyde or acid and a secondary alcohol to a ketone.

On a white spot plate, place 1 drop of the liquid unknown (10 mg if a solid). Add 5–8 drops of acetone and stir the mixture with a thin glass rod. To the resulting solution is added 1 drop of the test reagent. Stir and observe any color change within a 2-second time period. (*The reagent is prepared by slowly adding a suspension of 5.0 g of CrO$_3$ in 5.0 mL of concentrated H$_2$SO$_4$ to 15 mL of water. Allow the solution to cool to room temperature before using.*)

1. Run a blank to make sure the acetone is pure.
2. Tertiary alcohols, unsaturated hydrocarbons, amines, ethers, and ketones give a negative test within the 2-second time frame for observing the color change. Aldehydes give a positive test.

The HCl/ZnCl$_2$ Test—The Lucas Test

The Lucas test is used to distinguish between primary, secondary, and tertiary monofunctional alcohols having fewer than 6 carbon atoms.

$$\underset{\text{Soluble}}{\mathbf{R{-}OH}} + H^+ \xrightarrow{ZnCl_2} \mathbf{R^+} + H_2O$$
$$\xrightarrow{Cl^-} \underset{\text{Insoluble}}{\mathbf{RCl}}$$

[9]For a detailed discussion of classification tests see Shriner, R. L.; Fuson, R. C.; Curtin, D. Y.; Morrill, T. C. "Systematic Identification of Organic Compounds," 6th ed.; Wiley: New York, 1980.

The test requires that the alcohol initially be in solution. As the reaction proceeds, the corresponding alkyl chloride is formed, which is insoluble in the reaction mixture. As a result, the solution becomes cloudy. In some cases a separate layer may be observed.

1. Tertiary alcohols react to give an immediate cloudiness to the solution. You may be able to see a separate layer of the alkyl chloride after a short time.
2. Secondary alcohols generally produce a cloudiness within 3–10 minutes. The solution may have to be heated to obtain a positive test.
3. Primary alcohols dissolve in the reagent but react very, very slowly.

In a 1.0-mL conical vial is placed 2 drops of the unknown (10 mg if a solid) followed by 10 drops of the Lucas reagent. (*The test reagent is prepared by dissolving 13.6 g of anhydrous $ZnCl_2$ in 10.5 g of concentrated HCl with cooling in an ice bath.*) Stir the mixture with a thin glass rod and allow the solution to stand. Observe the results. Based on the times given above, classify the alcohol.

1. Certain polyfunctional alcohols also give a positive test.
2. If an alcohol having 3 carbon atoms or fewer is expected, the reaction vial should be equipped with an air condenser. This is to prevent low-molecular-weight alkyl chlorides (volatile) from escaping and thus going undetected.

Aldehydes and Ketones

The 2,4-Dinitrophenylhydrazine Test

Aldehydes and ketones react rapidly with 2,4-dinitrophenylhydrazine to form 2,4-dinitrophenylhydrazones. These derivatives range in color from *yellow* to *red* depending on the degree of conjugation in the carbonyl compound.

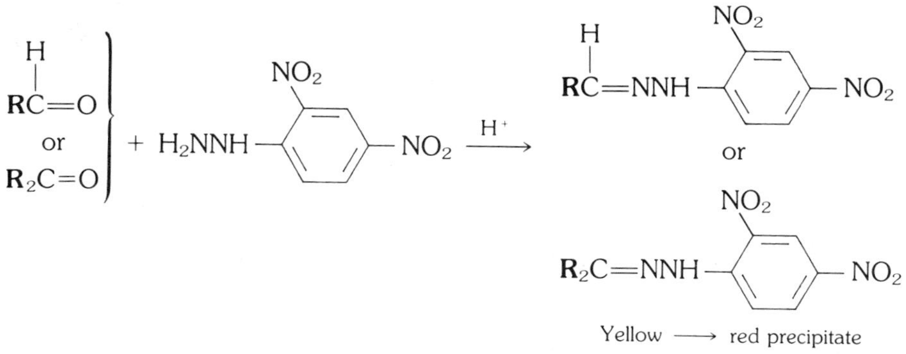

Yellow $\longrightarrow$ red precipitate

Examples of this reaction with ketones are demonstrated in Experiments 40B and 52B.

On a white spot plate is placed 7–8 drops of 2,4-dinitrophenylhydrazine reagent. (*The reagent is prepared by dissolving 1.0 g of 2,4-dinitrophenylhydrazine in 5.0 mL of concentrated sulfuric acid. This solution is slowly added, with stirring, to a mixture of 10 mL of water and 35 mL of 95% ethanol.*) Add 1 drop of a liquid unknown. If the unknown is a solid, 1 drop of a solution prepared by dissolving 10 mg of the material in 10 drops of ethanol is added. The mixture is stirred with a thin glass rod. The formation of a red to yellow precipitate is a positive test.

Esters and amides do not interfere with the test.

Silver Mirror Test for Aldehydes—Tollens Reagent

This reaction involves the oxidation of aldehydes to the corresponding carboxylic acid using an alcoholic solution of silver ammonium hydroxide. A positive test is the formation of a *silver* mirror or a black precipitate of finely divided silver.

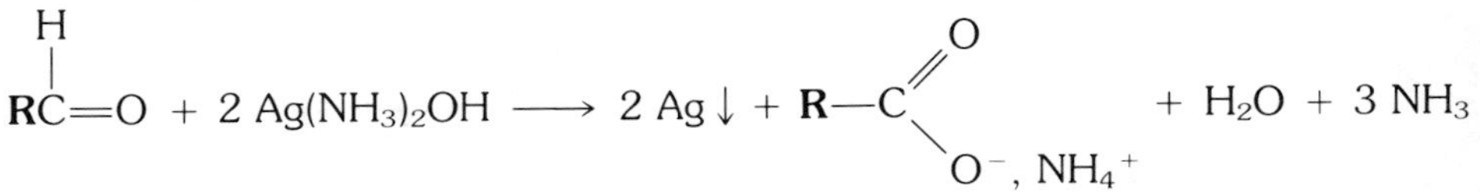

$$\overset{H}{\underset{|}{R}} C = O + 2\ Ag(NH_3)_2OH \longrightarrow 2\ Ag\downarrow + R-C\overset{O}{\underset{O^-,\ NH_4^+}{\diagup}} + H_2O + 3\ NH_3$$

In a 10 × 75 mm test tube is placed 1.0 mL of a 5% aqueous solution of $AgNO_3$ followed by 1 drop of aqueous 10% NaOH solution. Concentrated aqueous ammonia is now added drop by drop (2–4 drops), with shaking, until the precipitate of silver oxide just dissolves. Add 1 drop of the unknown (10 mg if a solid), with shaking, and allow the reaction mixture to stand for 10 minutes at room temperature. If no reaction has occurred, place the test tube in a sand bath at 40°C for 5 minutes. Observe the result.

1. Avoid a large excess of ammonia.
2. Reagents must be well mixed. Stirring with a thin glass rod is recommended.
3. *This reagent is freshly prepared for each test. It should not be stored since decomposition occurs with the formation of AgN_3, which is explosive.*
4. This oxidizing agent is very mild and thus alcohols are not oxidized under these conditions. Some sugars, acyloins, hydroxylamines, and substituted phenols do give a positive test.

Ammonium Salts, Amides, Nitriles

Ammonium salts, amides, and nitriles undergo hydrolysis in alkaline solution to form ammonia gas or an amine.

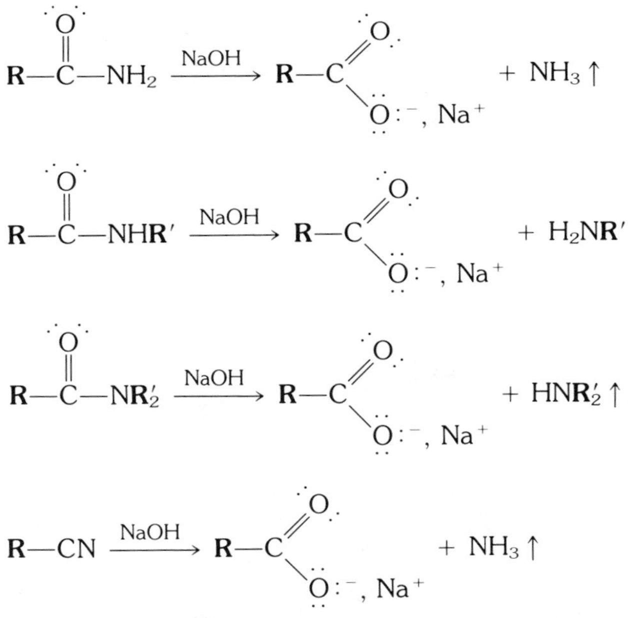

$$R-\overset{O}{\overset{||}{C}}-NH_2 \xrightarrow{\text{NaOH}} R-C\overset{O}{\underset{O:^-,\ Na^+}{\diagup}} + NH_3\uparrow$$

$$R-\overset{O}{\overset{||}{C}}-NHR' \xrightarrow{\text{NaOH}} R-C\overset{O}{\underset{O:^-,\ Na^+}{\diagup}} + H_2NR'\uparrow$$

$$R-\overset{O}{\overset{||}{C}}-NR_2' \xrightarrow{\text{NaOH}} R-C\overset{O}{\underset{O:^-,\ Na^+}{\diagup}} + HNR_2'\uparrow$$

$$R-CN \xrightarrow{\text{NaOH}} R-C\overset{O}{\underset{O:^-,\ Na^+}{\diagup}} + NH_3\uparrow$$

Detection of ammonia from ammonium salts, primary amides, and nitriles by use of a color test using copper sulfate solution can be carried out. The same test may also be used for secondary and tertiary amides that can generate low-molecular-weight (volatile) amines on hydrolysis.

In a 1.0-mL conical vial containing a boiling stone and equipped with an air condenser are placed 1–2 drops of the unknown liquid (10 mg if a solid) and

0.5 mL of 20% aqueous NaOH solution. This mixture is heated to *gentle* reflux on a sand bath. A strip of filter paper moistened with 2 drops of 10% aqueous copper sulfate solution is placed over the top of the condenser. Formation of a *blue* color (copper ammonia or amine complex) is a positive test.

The filter paper may be held in place using a small test tube holder or other suitable device.

Amines—The Hinsberg Test

The Hinsberg test is useful for distinguishing between primary, secondary, and tertiary amines. The reagent used is *p*-toluenesulfonyl chloride in alkaline solution.

Primary amines with fewer than seven carbons form a sulfonamide that is soluble in the alkaline solution. Acidification of the solution results in the precipitation of the insoluble sulfonamide.

$$CH_3-\bigcirc-SO_2Cl + \mathbf{R}-NH_2 \xrightarrow{NaOH}$$

$$CH_3-\bigcirc-SO_2NHR + NaCl + H_2O \underset{\substack{excess \\ acid}}{\overset{\substack{excess \\ base}}{\rightleftharpoons}} CH_3-\bigcirc-SO_2NR^-, Na^+$$

(Insoluble)   (Soluble)

Secondary amines form an insoluble sulfonamide in the alkaline solution.

$$CH_3-\bigcirc-SO_2Cl + \mathbf{R}_2NH \xrightarrow{NaOH} CH_3-\bigcirc-SO_2NR_2 + NaCl + H_2O \xrightarrow{\substack{excess \\ base}} \text{no change}$$

(Insoluble)

Tertiary amines normally give no reaction under these conditions.

$$CH_3-\bigcirc-SO_2Cl + \mathbf{R}_3N \xrightarrow{NaOH} CH_3-\bigcirc-SO_3^- + NR_3 + 2\,Na^+ + Cl^-$$

(Soluble)   (Oil)

HOOD In a 1.0-mL conical vial containing a boiling stone and equipped with an air condenser are placed 0.5 mL of 10% aqueous sodium hydroxide solution, 1 drop of the sample unknown (10 mg if a solid), followed by 30 mg of *p*-toluenesulfonyl chloride [**HOOD**]. The mixture is heated to reflux for 2–3 minutes on a sand bath and then cooled in an ice bath. Test the alkalinity of the solution using litmus paper. If it is not alkaline, add additional 10% aqueous NaOH dropwise.

SAVE Using a Pasteur filter pipet, separate the solution from any solid that may be present. Transfer the solution to a clean 1.0-mL conical vial [**SAVE**].

SAVE *Note. If an oily upper layer is obtained at this stage, remove the lower alkaline phase [SAVE] using a Pasteur filter pipet. To the remaining oil add 0.5 mL of cold water and stir vigorously to obtain a solid material.*

If a solid is obtained it may be (1) the sulfonamide of a secondary amine, (2) recovered tertiary amine if the original amine was a solid, or (3) the insoluble salt of a primary sulfonamide derivative (if the original amine had more than 6 carbon atoms).

1. If the solid is a tertiary amine, it is soluble in aqueous 10% HCl.
2. If the solid is a secondary sulfonamide, it is insoluble in aqueous 10% NaOH.
3. If no solid is present, acidify the alkaline solution by addition of 10% aqueous HCl. If the unknown amine is primary, the sulfonamide will precipitate.

Unsaturated Hydrocarbons—Alkenes and Alkynes

Bromine in Methylene Chloride

Unsaturated hydrocarbons readily add bromine. An example of this reaction is given in Experiment 50A.

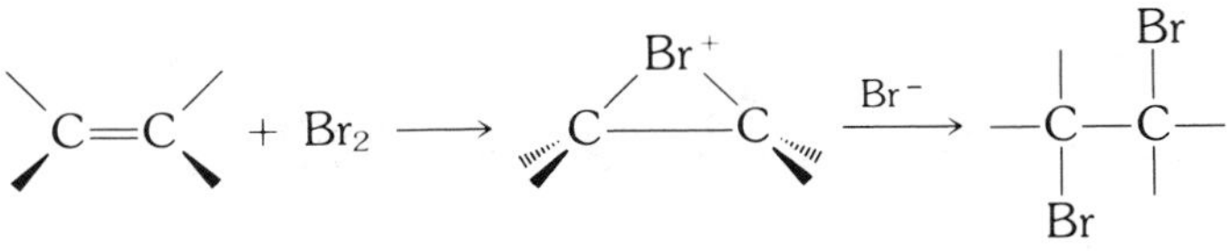

The test is based upon the decolorization of a red-brown bromine-methylene chloride solution.

In a 10×75 mm test tube is placed 2 drops of a liquid unknown (15 mg if a solid) followed by 0.5 mL of methylene chloride [**HOOD**]. Add dropwise, with shaking, a 2% solution of Br_2 in methylene chloride solvent [**HOOD**]. The presence of an unsaturated hydrocarbon will require 2–3 drops of the reagent before the reddish-brown color of bromine persists in the solution.

HOOD
HOOD

Caution: *Bromine is highly toxic and can cause burns.*

---

1. Methylene chloride is used in place of the usual carbon tetrachloride since it is less toxic.
2. Phenols, enols, amines, aldehydes, and ketones interfere with this test.

Permanganate Test—Baeyer Test for Unsaturation

Unsaturation in an organic compound can be detected by the decolorization of permanganate solution. The reaction involves the cis hydroxylation of the olefin to give a 1,2-diol (glycol).

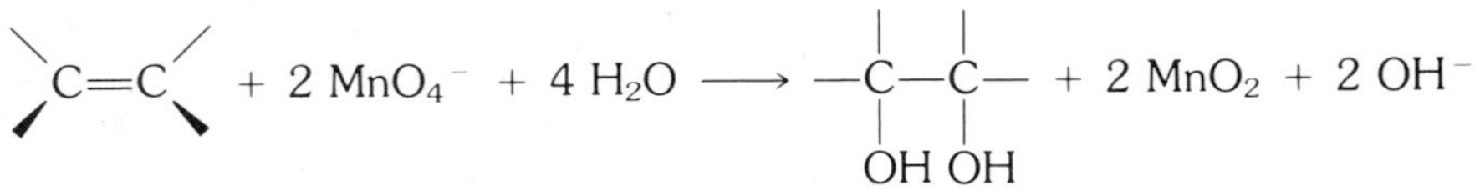

On a white spot plate is placed 0.5 mL of *alcohol-free* acetone followed by 2 drops of the unknown compound (15 mg if a solid). There is now added dropwise (2–3 drops), with stirring, a 1% aqueous solution of $KMnO_4$. A positive test is the discharge of purple permanganate color from the reagent and the precipitation of brown manganese oxides.

Any functional group that undergoes oxidation with permanganate interferes with the test (phenols, aryl amines, most aldehydes, primary and secondary alcohols, etc.).

Phenols and Enols

Most phenols and enols form colored complexes in the presence of ferric ion, Fe^{3+}.

$$6 \left[\begin{array}{c} :\ddot{O}H \\ \bigcirc \end{array} \right] + Fe^{3+} \rightleftharpoons \left[\left(\bigcirc - \ddot{\underset{..}{O}} \right)_6 Fe \right]^{3-} + 6\,H^+$$

Phenols give red, blue, purple, or green colors. Sterically hindered phenols give a negative test. Enols generally give a tan, red, or red-violet color.

On a white spot plate are placed 2 drops of water, or 1 drop of water plus 1 drop of ethanol, or 2 drops of ethanol, depending on the solubility characteristics of the unknown. To this solvent system is added 1 drop (10 mg if a solid) of the substance to be tested. Stir the mixture with a thin glass rod to complete dissolution. Add 1 drop of 2.5% aqueous ferric chloride solution (light yellow in color). Stir and observe any color formation. If necessary, a second drop of the $FeCl_3$ solution may be added.

1. The color developed may be fleeting or it may last for many hours. A slight excess of the ferric chloride solution may or may not destroy the color.
2. An alternate procedure using $FeCl_3$-CCl_4 solution in the presence of pyridine is available.[10]

Esters—Hydroxamate Test

Esters of carboxylic acids can be identified by conversion to hydroxamic acid salts. Acidification of this salt produces the corresponding hydroxamic acid, which is identified by formation of a red to purple color in the presence of Fe^{3+} ion.

$$(Ar, \mathbf{R})-\overset{\ddot{O}:}{\underset{OR'}{C}} + H_2N-OH \xrightarrow{KOH} (Ar, \mathbf{R})-\overset{\ddot{O}:}{\underset{NHO^-K^+}{C}} + \mathbf{R'}OH + H_2O$$

$$\left[\begin{array}{c} (Ar, \mathbf{R}) \\ \underset{H}{\diagdown} \overset{C=\ddot{O}}{\underset{N}{\diagup}} \diagdown \ddot{\underset{..}{O}} \end{array} Fe \right]_3 \xleftarrow{Fe^{3+}} (Ar, \mathbf{R})-\overset{\ddot{O}:}{\underset{NHOH}{C}} \xleftarrow{H^+}$$

Red-violet

In a 3.0-mL conical vial containing a boiling stone and equipped with an air condenser is placed 1 drop of the liquid unknown (10 mg if a solid) followed by 0.5 mL of 1.0M ethanolic hydroxylamine hydrochloride solution. To this solution is added dropwise 10% methanolic KOH until the resulting solution is pH ~ 10 (pH paper). This mixture is heated to reflux temperature using a sand bath for 5 minutes, cooled to room temperature, and acidified to pH 3–4 by dropwise addition of 5% aqueous HCl solution. Add 2 drops of 5% aqueous $FeCl_3$ solution. The formation of a red to purple color is a positive test.

1. It is suggested that a blank be run for comparison purposes.
2. Acid chlorides, anhydrides, lactones, and imides also give a positive test.

Methyl Ketones, Methyl Carbinols—Iodoform Test

The iodoform test involves hydrolysis and cleavage of methyl ketones to form a yellow precipitate of iodoform.

[10]Soloway, S; Wilen, S. H. *Anal. Chem.* **1952**, *4*, 979.

$$\text{R}\overset{\displaystyle \overset{..}{\underset{..}{O}}}{\overset{\|}{\text{C}}}\text{—CH}_3 + 3\text{I}_2 + 3\text{KOH} \longrightarrow \text{R}\overset{\displaystyle \overset{..}{\underset{..}{O}}}{\overset{\|}{\text{C}}}\text{—CI}_3 + 3\text{KI} + 3\text{H}_2\text{O}$$

$$\Big\downarrow \text{KOH}$$

$$\text{R}\overset{\displaystyle \overset{..}{\underset{..}{O}}}{\overset{\|}{\text{C}}}\text{—O}^-\ \text{K}^+ + \text{CHI}_3 \downarrow$$

Yellow

It is also a positive test for compounds that on oxidation generate methyl ketones under these reaction conditions. For example, methyl carbinols (secondary alcohols having at least one methyl group attached to the carbon atom to which the OH unit is linked), acetaldehyde, and ethanol give positive results.

In a 3.0-mL conical vial equipped with an air condenser is placed 2 drops of the unknown liquid (10 mg if a solid) followed by 5 drops of 10% aqueous KOH solution.

Note. *If the sample is insoluble in the aqueous phase, either mix vigorously or add dioxane or bis-(2-methoxyethyl) ether to obtain a homogeneous solution.*

The mixture is warmed on a sand bath to 50–60°C and the KI-I$_2$ reagent is added dropwise until the solution becomes dark brown in color (~1.0 mL). Additional 10% aqueous KOH is now added (dropwise) until the solution is again colorless. (This reagent is prepared by the instructor. Mix 3 g of KI and 1 g I$_2$ in 20 mL of water.)

CAUTION: *Iodine is highly toxic and can cause burns.*

After a period of 2 minutes, cool the solution and determine whether a yellow precipitate (HCI$_3$) has formed. If a precipitate is not observed, reheat as before for another 2 minutes. Cool and check again for the appearance of iodoform.

1. The iodoform test is reviewed elsewhere.[11]
2. An example of the general haloform reaction using bleach to oxidize a methyl ketone is given in Experiments 43A and 43B.

Alkyl Halides

Silver Nitrate Test

Alkyl halides that undergo the S$_N$1 substitution react with alcoholic silver nitrate to form a precipitate of the corresponding silver halide.

$$\text{RX} + \text{AgNO}_3 \longrightarrow \text{AgX} \downarrow + \text{RONO}_2$$

Secondary and primary halides react slowly or not at all at room temperature. However, they do react at elevated temperatures.

Tertiary halides react immediately at room temperature.

In a 1.0-mL conical vial are placed 0.5 mL of 2% ethanolic AgNO$_3$ solution and 1 drop of unknown (10 mg if a solid). A positive test is a precipitate within 5 minutes. If no reaction occurs, add a boiling stone and equip the vial with an air condenser. Heat the solution at *gentle* reflux for an additional 5 minutes using a sand bath. Cool the solution.

If a precipitate is formed, add 2 drops of dilute HNO$_3$. AgX will not dissolve in nitric acid solution.

---

[11]Fuson, R. C.; Bull, B. A. *Chem. Revs.* **1934,** *15,* 275.

1. The order of reactivity for **R** groups is allyl = benzyl > tertiary > secondary >>> primary. For the halide leaving groups: I > Br > Cl.
2. Acid halides, α-haloethers, and 1,2-dibromo compounds also give a positive test at room temperature. Only the activated aryl halides give a positive test at elevated temperatures.

Sodium Iodide in Acetone

Primary alkyl chlorides and bromides can be distinguished from aryl and vinylic halides by reaction with sodium iodide in acetone.

$$\textbf{R}-X + NaI \xrightarrow{\text{acetone}} \textbf{R}-I + NaX \downarrow$$

$$X = Cl, Br$$

Primary alkyl bromides undergo a S_N2 displacement reaction within 5 minutes at room temperature, primary alkyl chlorides only at 50°C.

In a 1.0-mL conical vial are placed 1 drop of a liquid unknown (10 mg if a solid) and 3 drops of acetone. To this solution is added 0.5 mL of NaI-acetone reagent. (*The reagent is prepared by dissolving 3 g of NaI in 25 mL of acetone*). A positive test is the appearance of a precipitate of NaX within 5 minutes. If no precipitate is observed, add a boiling stone and equip the vial with an air condenser. Warm the reaction mixture in a sand bath at ~50°C for 5 minutes. Cool to room temperature and determine whether a reaction has occurred.

1. Benzylic and allylic chlorides and bromides, acid chlorides and bromides, and α-halo ketones, esters, amides, and nitriles also give a positive test at room temperature.
2. Primary and secondary alkyl chlorides, and secondary and tertiary alkyl bromides, react at 50°C under these conditions.
3. If the solution turns red-brown in color, I_2 is being liberated.

Nitro Compounds

Many nitro compounds give a positive test based on the following reaction.

$$\textbf{R}-NO_2 + 4\,H_2O + 6\,Fe(OH)_2 \longrightarrow \textbf{R}-NH_2 + 6\,Fe(OH)_3 \downarrow$$

$$\text{Red-brown}$$

The nitro derivative oxidizes the iron(II)hydroxide to iron(III)hydroxide, the latter being a red-brown solid.

In a 1.0-mL conical vial is placed 5–10 mg of the unknown compound followed by 0.4 mL of freshly prepared 5% aqueous ferrous ammonium sulfate. After mixing, 1 drop of 3N sulfuric acid is added followed by 10 drops of methanolic 2N KOH. The vial is capped, shaken vigorously, vented, and allowed to stand over a 5-minute period. The formation of a red-brown precipitate, usually within 1 minute, is a positive test.

Acids

The presence of a carboxylic acid is detected by its solubility behavior. An aqueous solution of the acid will be acidic to litmus paper. Since a sulfonic acid would also give a positive test, the test for sulfur is used to distinguish between the two types of acids. A water-soluble phenol is acidic toward litmus paper but also would give a positive ferric chloride test.

PREPARATION OF DERIVATIVES

Based on the preliminary and classification tests carried out to this point, you should have established the type of functional group present (or lack of one) in the unknown organic sample. The next step is to consult a set of tables containing a listing of known organic compounds by functional group and/or physical properties or both. Using the physical property data for your compound, you can select a few possible candidates that appear to "fit" the data you have collected. On a chemical basis, the final step in the identification sequence is to prepare one or two *crystalline derivatives* of your compound. Selection of the specific compound and thus final confirmation of its identity can then be made from the extensive derivative tables that have been accumulated. With the advent of spectral analysis, the preparation of derivatives is often not necessary, but the wealth of chemistry that can be learned by the beginning student in carrying out these procedures is extensive and important. The preparation of selected derivatives for the most common functional groups are given below. Condensed tables of compounds and their derivates are summarized in Appendix D. For extensive tables see the References cited below.

Important. *In each of the procedures outlined below, drops of reagents are measured using Pasteur pipets.*

REFERENCES

1. Shriner, R. L.; Fuson, R. C.; Curtin; D. Y.; Morrill, T. C. "The Systematic Identification of Organic Compounds," 6th ed.; Wiley: New York, 1980.
2. Rappoport, Z. "Handbook of Tables for Organic Compound Identification," 3rd ed.; CRC Press: Boca Raton, FL, 1967.

ACIDS (Tables D.1, D.2)

Preparation of Acid Chlorides

$$R-\overset{O}{\underset{||}{C}}-\overset{..}{\underset{..}{O}}H + Cl-\overset{O}{\underset{||}{S}}-Cl \xrightarrow{DMF} R-\overset{O}{\underset{||}{C}}-Cl + HCl + SO_2$$

In a 1.0-mL conical vial containing a boiling stone and equipped with a reflux condenser are placed 20 mg of the unknown acid, 4 drops of thionyl chloride, and 1 drop of N,N-dimethylformamide.

> **CAUTION:** *This reaction is run in the* **HOOD** *since hydrogen chloride and sulfur dioxide are evolved. Thionyl chloride is an irritant and is harmful to breathe.*

After allowing the mixture to stand at room temperature for 10 minutes, heat it at gentle reflux in a sand bath for a period of 15 minutes and then cool to room temperature. It is then diluted with 5 drops of methylene chloride solvent.

The acid chloride is not isolated but is used directly in the preparations given below.

Amides

$$R-\overset{O}{\underset{||}{C}}-Cl + 2 NH_3 \longrightarrow R-\overset{O}{\underset{||}{C}}-NH_2 + NH_4Cl$$

HOOD In a 3.0-mL conical vial containing a magnetic spin vane and equipped with an air condenser is placed 10 drops of concentrated aqueous ammonia **[HOOD]**. The flask is cooled in an ice bath and the acid chloride solution prepared above transferred by Pasteur pipet, *dropwise* with stirring, to the ammonia solution. *It* is convenient to make this addition down the neck of the condenser. The amide may precipitate during this operation. After the addition is complete, the ice bath is removed and the mixture stirred for an additional 5 minutes. Methylene chloride (10 drops) is added and the mixture stirred to dissolve any precipitate. The methylene chloride layer is separated from the aqueous layer using a Pasteur filter pipet and transferred to a Pasteur filter pipet containing 200 mg of anhydrous sodium sulfate. The eluate is collected in a Craig tube containing a boiling stone. The aqueous phase is extracted with 0.5 mL of methylene chloride and the methylene chloride layer separated as before and also passed through the column. Both eluates are combined. The methylene chloride solution is now

HOOD   evaporated using a sand bath [**HOOD**]. The amide product is recrystallized from an ethanol-water mixture using the Craig tube. Dissolve the material in ~0.5 mL of ethanol, add water (dropwise) to the cloud point, cool in an ice bath, and collect the crystals in the usual manner. Dry the crystalline amide on a porous clay plate and determine the melting point.

Anilides

In a 3.0-mL conical vial containing a magnetic spin vane and equipped with an air condenser are placed 5 drops of aniline and 10 drops of methylene chloride. The solution is cooled in an ice bath and the acid chloride solution prepared above transferred by Pasteur pipet, *dropwise* with stirring, to the aniline solution

HOOD   [**HOOD**]. *It is convenient to make this addition down the neck of the condenser.* After the addition is complete, the ice bath is removed and the mixture stirred for an additional 10 minutes.

The methylene chloride layer is now transferred to a 10 × 75 mm test tube and washed with 0.5 mL of water, 0.5 mL of 5% aqueous HCl solution, 0.5 mL of 5% aqueous sodium hydroxide solution, and finally 0.5 mL of water. For each washing, the test tube is shaken and the top aqueous layer removed by Pasteur filter pipet. The wet methylene chloride layer is transferred to a Pasteur filter pipet containing 200 mg of anhydrous sodium sulfate and the eluate collected in a Craig tube containing a boiling stone. The original test tube is rinsed with an additional 10 drops of methylene chloride and this rinse also passed through the column. Both eluates are combined.

HOOD   The methylene chloride solvent is evaporated using a sand bath in the **HOOD**. The crude anilide is recrystallized from an ethanol–water mixture using the Craig tube. Dissolve the material in ~0.5 mL of ethanol, add water (dropwise) to the cloud point, cool in an ice bath, and collect the crystals in the usual manner. Dry the purified derivative product on a porous clay plate and determine the melting point.

Toluidides

The same procedure described for the preparation of anilides is used except that *p*-toluidine replaces the aniline.

ALCOHOLS (Table D.3) Phenyl- and α-Naphthylurethans

$$Ar—N=C=O + RO—H \longrightarrow Ar—\underset{H}{N}—C\overset{O}{\diagup}—OR$$

Isocyanate   Urethan

In a 1.0-mL conical vial containing a boiling stone and equipped with an air condenser protected by a calcium chloride drying tube is placed 15 mg of an anhydrous alcohol or phenol. The air condenser is removed from the vial and 2 drops of phenyl isocyanate or α-naphthyl isocyanate is added. The air condenser is replaced immediately. If the unknown is a phenol, one drop of pyridine is added in a similar manner.

> **CAUTION:** *This addition must be done in the* HOOD. *The isocyanates are lachrymatory! Pyridine has a strong odor of the amines.*

If a spontaneous reaction does not take place, the vial is heated at ~ 80–90°C, using a sand bath, for a period of 5 minutes and then cooled in an ice bath. It may be necessary to scratch the sides of the vial to induce crystallization. The solid product is collected by vacuum filtration using a Hirsch funnel and purified by recrystallization from ligroin. For this procedure the solid is placed in a 10 × 75 mm test tube and dissolved in 1.0 mL of warm (60°–80°C) ligroin. If diphenyl (or dinaphthyl) urea is present, which is formed by reaction of the isocyanate with water, it is insoluble in the solvent. The warm ligroin solution is transferred to a Craig tube using a Pasteur filter pipet. The solution is cooled in an ice bath and the resulting crystals collected in the usual manner. After drying the product on a porous clay plate, determine the melting point.

3,5-Dinitrobenzoates

3,5-Dinitrobenzoyl chloride

In a 1.0-mL conical vial containing a boiling stone and equipped with an air condenser protected by a calcium chloride drying tube are placed 100 mg of 3,5-dinitrobenzoyl chloride and 2 drops of the unknown alcohol. The mixture is then heated to ~ 10°C below the boiling point of the alcohol (but not over 100°C) on a sand bath for a period of 5 minutes. Water (0.3 mL) is added and the vial placed in an ice bath to cool. The solid ester is collected by vacuum filtration using a Hirsch funnel and the filter cake washed with three 0.5-mL portions of 2% aqueous sodium carbonate solution followed by 0.5 mL of water. The solid product is recrystallized from an ethanol-water mixture using a Craig tube. Dissolve the material in ~0.5 mL of ethanol. Add water (dropwise) to the cloud point, cool in an ice bath, and collect the crystals in the usual manner. After drying the product on a porous clay plate, determine the melting point.

ALDEHYDES AND KETONES (Tables D.4, D.5)   2,4-Dinitrophenylhydrazones

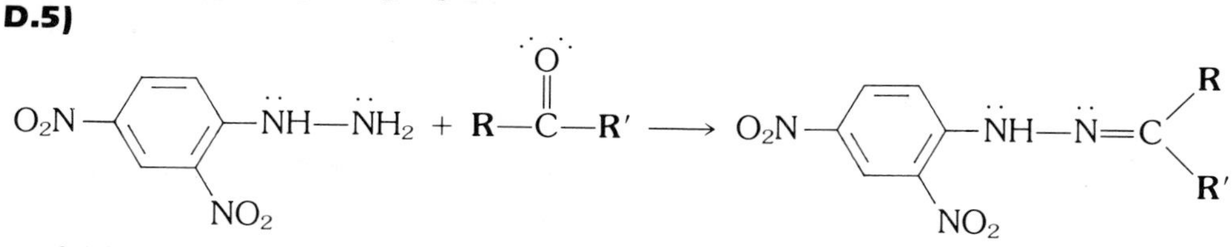

2,4-Dinitrophenylhydrazine   A 2,4-dinitrophenylhydrazone

The procedure outlined in the Classification Test Section for aldehydes and ketones (p. 321) is used. If necessary, the derivative can be recrystallized from 95% ethanol.

Semicarbazones

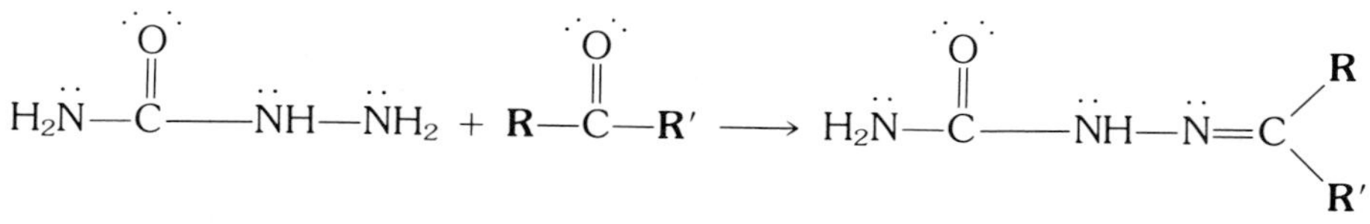

Semicarbazide   A semicarbazone

In a 1.0-mL conical vial are placed 12 mg of semicarbazide hydrochloride, 20 mg of sodium acetate, 10 drops of water, and 12 mg of the unknown carbonyl compound. Cap the vial, shake vigorously, vent, and allow the vial to stand at room temperature until crystallization is complete. Cool the vial in an ice bath if necessary. Collect the crystals by vacuum filtration using a Hirsch funnel and wash the filter cake with 0.2 mL of cold water. Dry the crystals on a porous clay plate. Determine the melting point.

AMINES (Tables D.6, D.7)   *Primary and Secondary Amines: Acetamides*

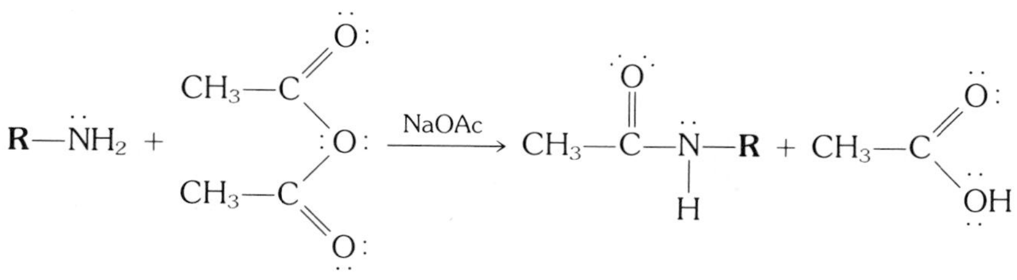

In a 1.0-mL conical vial equipped with an air condenser are placed 20 mg of the unknown amine, 5 drops of water, and 1 drop of concentrated HCl.

In a small test tube a solution of 40 mg of sodium acetate trihydrate dissolved in 5 drops of water is prepared. The stoppered solution is set aside for use in the next step.

The solution of amine hydrochloride is warmed to ~50°C in a sand bath. It **HOOD** is then cooled and 40 μL of acetic anhydride added in one portion **[HOOD]** through the condenser. This is followed *immediately* by the addition of the sodium acetate solution using a Pasteur pipet. Swirl the contents of the vial to ensure complete mixing.

The reaction mixture is allowed to stand at room temperature for ~5 minutes and then placed in an ice bath for an additional 5–10 minutes. The white crystals are collected by vacuum filtration using a Hirsch funnel and the filter cake washed with two 0.1-mL portions of water. The crystals are dried on a porous clay plate. Determine the melting point.

Primary and Secondary Amines: Benzamides

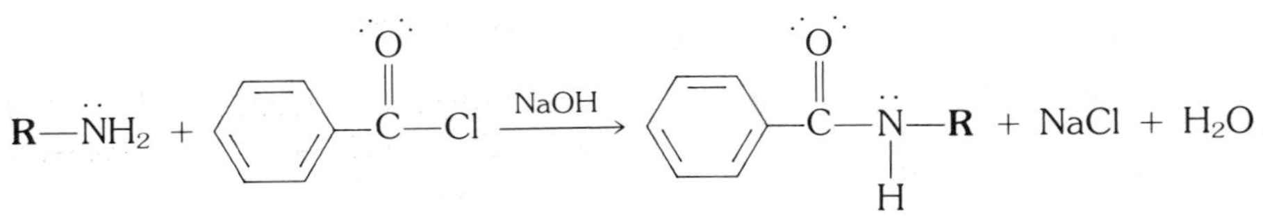

HOOD   In a 1.0-mL conical vial are placed 0.4 mL of 10% aqueous NaOH solution, 25 mg of the amine, and 2–3 drops of benzoyl chloride [**HOOD**]. The vial is capped and shaken over a period of ~10 minutes. It is vented periodically to release any pressure buildup.

The crystalline precipitate is collected by vacuum filtration using a Hirsch funnel and the filter cake washed with 0.1 mL of dilute HCl followed by 0.1 mL of water. It is generally necessary to recrystallize the material from methanol or aqueous ethanol using the Craig tube. Dry the product on a porous clay plate and determine the melting point.

Primary, Secondary, and Tertiary Amines: Picrates

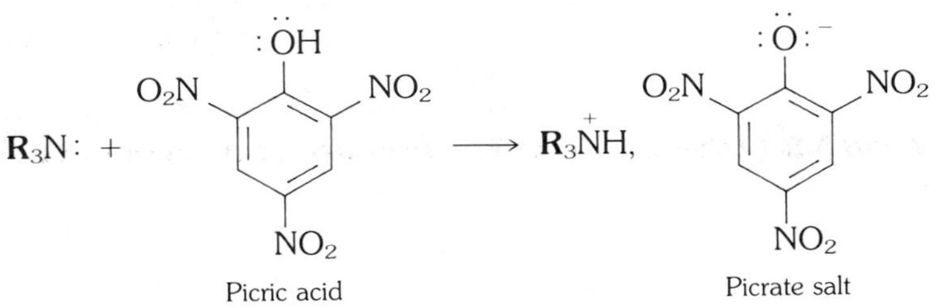

Picric acid   Picrate salt

In a 1.0-mL conical vial containing a boiling stone and equipped with an air condenser are placed 15 mg of the unknown amine and 0.3 mL of 95% ethanol.

Note. *If the amine is not soluble in the ethanol, shake the mixture to obtain a saturated solution and then transfer this solution, using a Pasteur filter pipet, to another vial.*

There is now added 0.3 mL of a saturated solution of picric acid in 95% ethanol.

> **CAUTION:** *Picric acid explodes when rapidly heated or by percussion.*

The mixture is heated at reflux, using a sand bath, for ~1 minute and then allowed to cool slowly to room temperature. The yellow crystals of the picrate are collected by vacuum filtration using a Hirsch funnel. The material is dried on a porous clay plate and the melting point determined.

ACID CHLORIDES AND ANHYDRIDES (Table D.8)

Amides

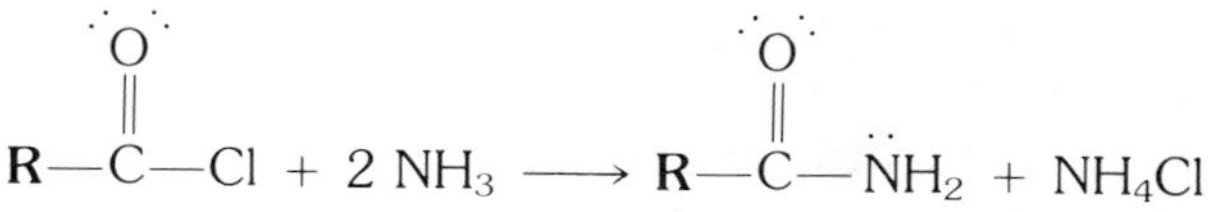

$$R-\overset{\overset{\displaystyle ..O..}{\|}}{C}-Cl + 2\,NH_3 \longrightarrow R-\overset{\overset{\displaystyle ..O..}{\|}}{C}-\overset{..}{N}H_2 + NH_4Cl$$

HOOD

In a 10 × 75 mm test tube is placed 0.4 mL of ice cold, concentrated ammonium hydroxide solution. To the solution is added slowly, [**HOOD**] with shaking, 15 mg of the unknown acid chloride or anhydride. Stopper the test tube and allow the reaction mixture to stand at room temperature for approximately 5 minutes. Collect the crystals by vacuum filtration using a Hirsch funnel, and wash the filter cake with 0.2 mL of ice cold water. Recrystallize the material using a Craig tube, from water or an ethanol-water mixture. Dry the purified crystals on a porous clay plate and determine the melting point.

AROMATIC HYDROCARBONS (Table D.9)

Picrates

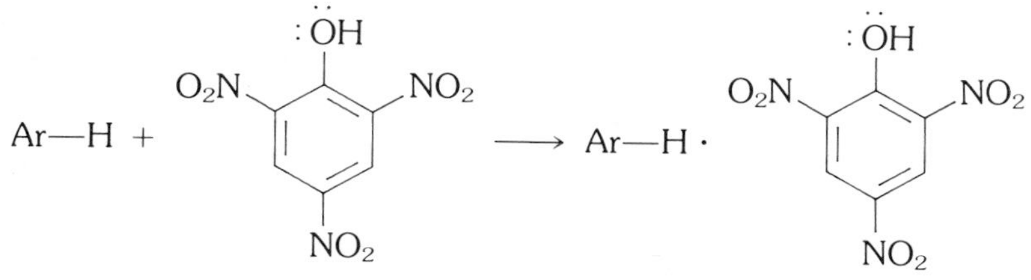

Picric acid

Picrate complex

The procedure outlined under Amines, Picrates is used to prepare these derivatives (p. 332).

PHENOLS (Table D.10)

α-Naphthylurethans

The procedure outlined under Alcohols, Phenyl- and α-Naphthylurethans, is used to prepare these derivatives (p. 330).

Bromo Derivatives

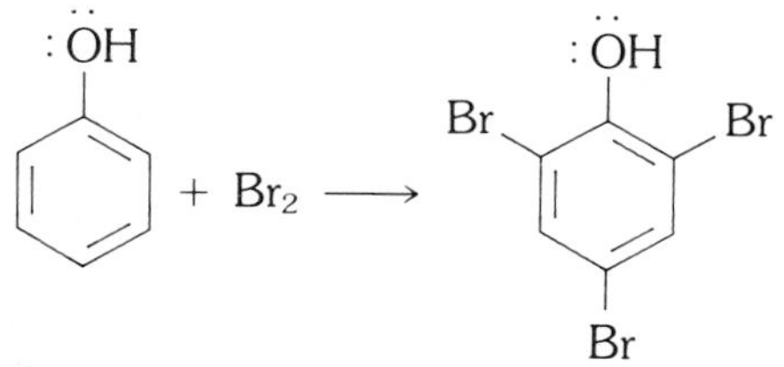

HOOD
(Instructor's Prep.)

In a 1.0-mL conical vial is placed 10 mg of the unknown phenol followed by 2 drops of methanol and 2 drops of water. To this solution is added 3 drops [**HOOD**] of brominating agent from a Pasteur pipet. (*The brominating reagent is prepared by adding 1.0 mL (3 g) of bromine* [**CAUTION! HOOD**] *to a solution of 4.0 g of KBr in 25 mL of water.*) The addition is continued (dropwise) until the yellow color of bromine persists. Water (4 drops) is now added, the vial capped, shaken, vented and then allowed to stand at room temperature for 10 minutes. The crystalline precipitate is collected by vacuum filtration using a Hirsch funnel and the filter cake washed with 0.5 mL of 5% aqueous sodium bisulfite solution. The solid is recrystallized from ethanol or an ethanol-water mixture using a Craig tube. Dissolve the material in ~0.5 mL of ethanol, add water to the cloud point, cool in an ice bath, and collect the crystals in the usual manner. The purified product is dried on a porous clay plate and the melting point determined.

**HYDROCARBONS,
HALOGENATED
HYDROCARBONS, NITRILES,
AMIDES, NITRO COMPOUNDS,
ETHERS, and ESTERS
(Tables D.11, D.12, D.13,
D.14, D.15, D.16, D.17)**

These compounds do not give derivatives directly. They are usually converted into another material that can then be derivatized. The procedures are for the most part lengthy and frequently give mixtures of products. It is recommended that compounds belonging to these classes be identified using spectral methods.

QUESTIONS

7-1. The following six substances have approximately the same boiling point and all are colorless liquids. Suppose you were given six unlabeled bottles, each of which contained one of these compounds. Explain how you would use simple chemical tests to identify the contents of each bottle.

propanoic acid	ethylbenzene
propyl butanoate	diisobutylamine
1-pentanol	styrene

7-2 A friend of yours, who is a graduate student attempting to establish the structure of a chemical species from field clover, isolated an alcohol that was found to have an optical rotation of $+49.5°$. Combustion analysis gave a molecular formula of $C_5H_{10}O$. It was also observed that this alcohol readily decolorized Br_2-methylene chloride solution. On this basis, the alcohol was subjected to catalytic hydrogenation and it was found to absorb 1 mol of hydrogen gas. The product of the reduction gave a positive ceric nitrate test, indicating that it too was an alcohol. However, the compound was optically inactive. Your friend has come to you for assistance in writing the structure of the two alcohols. What do you believe the structures are?

7-3. A colorless liquid is soluble in water and diethyl ether. It has a boiling point of 97–98°C, and the sodium fusion test gave negative results for the presence of halogens, nitrogen, and sulfur. The sample gave a positive ceric nitrate test and also decolorized potassium permanganate solution with precipitation of a brown solid. Treatment with 2,4-dinitrophenylhydrazine solution and the iodoform test gave negative results. Reaction with phenyl isocyanate produced a white solid product, which on purification melted at 70°C. Identify the liquid.

7-4 An unknown compound burned with a yellow, nonsmoky flame and was found to be insoluble in 5% aqueous sodium hydroxide solution but soluble in concentrated sulfuric acid. Measurement of its boiling point gave a range of 130–131°C. It was found to give a semicarbazone having a melting point of 204–206°C. However, it gave a negative result when treated with Tollens reagent and did not decolorize the Baeyer reagent. It also gave a negative haloform test. Identify the unknown compound.

7-5. An unknown acid of mp 139–141°C burned with a yellow-sooty flame, and the sodium fusion test showed that nitrogen was present. The sample did not react with *p*-toluene-sulfonyl chloride but did give a positive test when treated with 5% aqueous ferrous ammonium sulfate solution, acidified with $3N$ sulfuric acid and then followed by methanolic potassium hydroxide solution. A 200 mg sample of the acid neutralized 12.4 mL of $0.098N$ sodium hydroxide solution. Identify the acid. Does your compound agree with the experimental equivalent weight?

7-6. An unknown liquid was found to have a boiling point of 207–209°C and to burn with a yellow-sooty flame. The sodium fusion test showed that it contained a halogen and nitrogen but not sulfur. It gave a positive Beilstein test. The compound was insoluble in water and dilute alkali but dissolved in diethyl ether and dilute hydrochloric acid. It reacted with picric acid to form a yellow precipitate having mp 134–135°C. Identify the liquid.

7-7. A hydrocarbon, C_6H_{10}, burned with a yellow, almost nonsmoky flame. On catalytic hydrogenation over platinum catalyst, it absorbed 1 mol of hydrogen to form compound B. It also decolorized Br_2-methylene chloride solution to yield a dibromo derivative, C. Ozonolysis of the hydrocarbon gave only one compound, D. Compound D gave a positive haloform test when treated with I_2-NaOH solution. On treatment of compound D with an alcoholic solution of silver ammonium hydroxide, a silver mirror was formed within a few minutes. Identify the hydrocarbon and compounds B, C, and D.

7-8. A high-boiling liquid has bp 202–204°C and burns with a yellow-sooty flame. Sodium fusion indicates that halogens, nitrogen, and sulfur are not present. It is not soluble in water, dilute sodium bicarbonate solution, or dilute hydrochloric acid. However, it proved to be soluble in 5% aqueous sodium hydroxide solution. The compound gives a purple color with ferric chloride solution and a precipitate when reacted with bromine water. Treatment with hydroxylamine reagent did not give a reaction, but a white precipitate was obtained when the compound was treated with α-naphthyl isocyanate. On drying, this white, solid derivative had mp 127–129°C. Identify the original liquid and write a structure for the solid derivative. After identifying the unknown liquid, can you indicate what the structure of the precipitate obtained on reaction with bromine might be?

7-9. A colorless liquid has bp 199–201°C and burns with a yellow-sooty flame. The sodium fusion test proved negative for the presence of halogens, nitrogen, and sulfur. It was not soluble in water, 5% aqueous sodium hydroxide, or 5% hydrochloric acid. However, it dissolved in sulfuric acid with evolution of heat. It did not give a precipitate with 2,4-dinitrophenylhydrazine solution and did not decolorize bromine-methylene chloride solution. The unknown liquid did give a positive hydroxamate test and was found to have a saponification equivalent of 136. Identify the unknown liquid.

7-10. Your friend of question 7-2 still needs your help. A week later a low-melting solid A was isolated, which combustion analysis indicated had the composition $C_9H_{10}O$. The substance gave a precipitate when treated with 2,4-dinitrophenylhydrazine solution. Furthermore, when reacted with the odoform reagent, a yellow precipitate of CHI_3 was observed. Acidification of the alkaline solution from the odoform test produced a solid material, B.

Reduction of compound A with $LiAlH_4$ gave compound C, $C_9H_{12}O$. This material, C, also gave compound B when treated with odoform reagent.

Vigorous oxidation of A, B, or C with sodium dichromate-sulfuric acid solution gave an acid having mp 121–122°.

Your friend needs your assistance in writing the structures for compounds A, B, C. Can you identify the three compounds?

7-11. An unknown substance burned with a clear blue flame and on analysis was found to have the molecular formula $C_4H_{10}O_4$. It gave a positive ceric nitrate test and reacted with the Lucas reagent on gentle heating within 4–5 minutes. On acetylation with acetic anhydride, it produced a pleasant-smelling compound having the formula $C_{12}H_{18}O_8$. What is the structure of the compound?

7-12. An unknown compound, X, was soluble in ether but only slightly soluble in water. It burned with a clear blue flame, and analysis showed it to have the molecular formula $C_5H_{12}O$. It gave a positive test with the Jones reagent, producing a new compound, Z, having formula $C_5H_{10}O$. Compound Z gave a positive haloform test and formed a semicarbazone. Compound X on treatment with sulfuric acid produced a hydrocarbon, Y, of formula C_5H_{10}. Hydrocarbon Y readily decolorized Br_2-methylene chloride solution and on ozonolysis produced acetone as one of the products. Identify the structure of each of the lettered compounds.

II. The Interpretation of Infrared Spectra

INTRODUCTION TO GROUP FREQUENCIES

Studies of the vibrational spectra of thousands of molecules have revealed that many of the normal modes associated with particular atomic arrangements may be transferred from one molecule to another. These vibrational frequencies are associated with small groups of atoms that are essentially uncoupled from the rest of the molecule. The absorption bands that result from these modes, therefore, are characteristic of the small group of atoms regardless of the composition of other parts of the molecule. These vibrations are known as the *group frequencies*, and the interpretation of infrared spectra of complex molecules based on group frequency assignments is an extremely powerful aid in the elucidation of molecular structure.

Note. *An introduction to the theory of this effect is given in Appendix B.*

We will discuss group frequencies in the following sequence:

Part A. Group Frequencies of the Hydrocarbons

Alkanes

Alkenes

Alkynes

Arenes

Part B. Factors Affecting the Carbonyl Frequencies

Mass effects

Geometric effects

Electronic effects (inductive and conjugative effects)

Interaction effects

Part C. Group Frequencies of the Functional Groups

Hydrocarbons	Amines
Alcohols	Nitriles
Aldehydes	Amides, primary
Ketones	Amides, secondary
Esters	Isocyanates
Acid halides	Thiols
Carboxylic acids	Halogens
Anhydrides	Phenyl
Ethers	

Part D. Strategies for Interpreting Infrared Spectra

PART A: Group Frequencies of the Hydrocarbons

Characteristic Group Frequencies of Alkanes

The saturated hydrocarbons and the alkane section of mixed structures contain only C—C and C—H bonds. The fundamental modes derived from the hydrocarbon portion of these molecules, therefore, are limited to C—C and C—H stretching and bending vibrations. These two types of oscillators are good examples of structural units that give rise to both excellent and very poor group frequencies. We have established (Appendix B) that a change in dipole moment during the vibration is essential for the absorption process to occur in the infrared. The exact relationship between the vibrational displacements and the observed band intensities is rather complex. It is related to the slope of the curve of the variation of dipole moment with the normal coordinate at the equilibrium point. The intensity of the fundamental is proportional to the square of the derivative of the dipole moment with respect to the normal coordinates. The absorptivity of infrared bands, therefore, usually can be gauged from a rough estimate of the magnitude of the oscillating dipole moment.

The *C—C group* is an oscillator that, at best, will possess a very small dipole moment because of the symmetry inherent in the bond. Because of the low bond polarization, the absorption bands associated with this system can be expected to be quite weak and difficult to identify. In addition, the C—C oscillator, in most cases, will be directly connected to other C—C oscillators with similar or identical frequencies. In such an arrangement mechanical coupling effects are to be expected. This coupling will give rise to a very complex absorption pattern unique to a particular compound. Although these highly coupled vibrations have little value as group frequencies, they are the most powerful means of identifying organic materials by modern chemical instrumentation. The region of the infrared spectrum where these frequencies predominate, 1500 to 500 cm^{-1}, is often referred to as the "fingerprint region."

The *C—H oscillator* is at the other extreme from the C—C case. It gives rise to excellent group frequencies. The light terminal H atom, which is connected to a relatively massive carbon atom by a strong bond (large force constant), possesses a high natural frequency. Because of the separation of this frequency from other frequencies, the only coupling that can influence the oscillator is that of other C—H groups connected to same carbon. A rule of thumb to remember is: "coupling generates as many modes as there are coupled oscillators." The methyl group, CH$_3$—, has three stretching modes; the methylene group, —CH$_2$—, two stretching modes; and the methine group, —CH—, a single stretching mode. Since there is very little mechanical coupling of C—H oscillators beyond the local carbon atom, the natural frequencies of the methyl, methylene, and methine groups remain relatively constant when these groups are transferred from compound to compound or from group to group within the same system. The dipole moment of the C—H bond is not large, but its derivative is sufficient to give rise to reasonably identifiable absorption bands. Because many compounds will contain several C—H bonds of similar character, they will also possess an equal number of nearly equivalent C—H frequencies. These C—H stretching modes overlap, often to produce the most intense absorption bands observed in the infrared spectrum of a material.

The three *coupled C—H stretching modes of methyl groups* can be described in terms of two antisymmetric vibrations that are degenerate, or nearly degenerate, (Appendix B) depending on the symmetry of the system, plus a symmetric mode (Fig. 7.3). In most cases, the two antisymmetric methyl stretching modes,

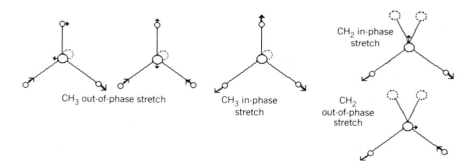

CH$_3$ out-of-phase stretch   CH$_3$ in-phase stretch   CH$_2$ in-phase stretch   CH$_2$ out-of-phase stretch

Fig. 7.3 *Stretching vibration of methyl and methylene groups.*

while not rigorously degenerate, will give rise to a pair of very close lying bands that are seldom resolved. The antisymmetric methyl modes are the highest frequency vibrations of the purely sp^3 hybridized C—H bonds. These fundamentals occur near 2960 cm^{-1}. The methyl symmetric stretching mode is found close to 2870 cm^{-1} (Fig. 7.4).

The two *coupled stretching modes of the methylene group* are very similar in displacement pattern to the fundamental vibrations of the water molecule (Fig. B.5), p. 392. The higher frequency mode, as in water, is the antisymmetric stretch. This fundamental occurs near 2925 cm^{-1} in hydrocarbons. The symmetric stretching mode (which is particularly sensitive to adjacent hetero-atoms bearing lone-pair electrons) in saturated hydrocarbons has the lowest frequency of the sp^3 hybridized coupled C—H oscillators. The symmetric stretch occurs close to 2850 cm^{-1} (Fig. 7.3).

The *methine group* has a *single uncoupled mode*. As relatively few groups of this type are normally present in a structure compared to methyl and methylene groups, this vibration gives rise to a weak fundamental, usually masked by the absorption of the other alkane groups. The absorption of the tertiary C—H bond occurs in the 2900 cm^{-1} region.

With very few exceptions the sp^3 hybridized C—H bonds have their fundamental modes in the 3000–2800 cm^{-1} region.

Since the hydrogen atom is much lighter than other atoms, it undergoes most of the displacement. The mass term in the expression for frequency of a diatomic molecule is actually a *reduced* mass. In the case of hydrogen and carbon this is defined by:

$$1/\mu \;=\; 1/m_H \;+\; 1/m_C \qquad \text{since } m_C \gg m_H, \qquad \mu \simeq m_H$$

It is possible, therefore, to express to a very good first approximation the vibrational frequency of this system by a simple Hooke's law relationship assuming an infinite mass for the carbon atom.

$$\bar{\nu} = \frac{1}{2\pi c}\sqrt{\frac{k}{m_H}}$$

Employing the above expression it is possible to predict a frequency shift on substitution by deuterium for hydrogen of $\bar{\nu}_H/\bar{\nu}_D = 1.41$. In practice these frequency shifts are somewhat less than the theoretical values, usually falling in the range 1.32–1.38 (Fig. 7.5). If the $\bar{\nu}_H/\bar{\nu}_D$ ratio departs significantly from this range, the result can be taken as an indication that one or possibly both of the

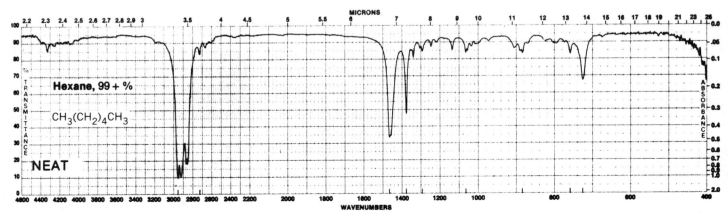

Fig. 7.4 *IR spectrum: hexane.*

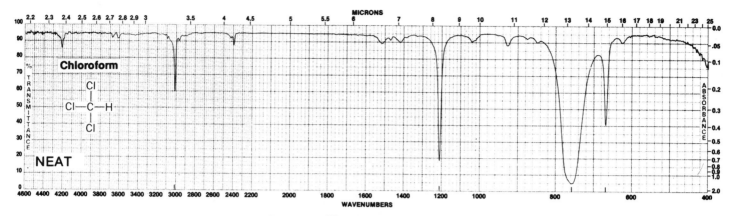

Fig. 7.5a *IR spectrum: HCCl₃.*

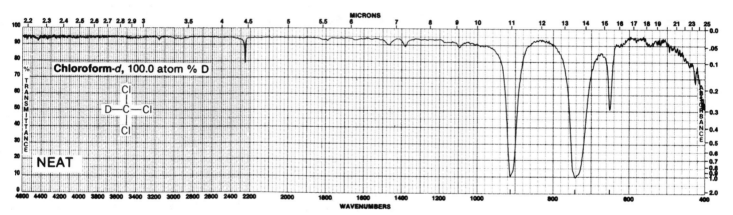

Fig. 7.5b *IR spectrum: DCCl₃.*

fundamentals are not behaving in harmonic fashion and that coupling is present. (See also Experiment 36 for an example of the shift of methyl frequencies on deuterium substitution.) Chloroform is one of the rare exceptions to the 3000 cm⁻¹ rule (see p. 338). The single uncoupled C—H stretching mode occurs at 3022 cm⁻¹. The rise in frequency in this case results from extensive substitution by the strongly electronegative chlorine atoms.

The *bending frequencies of the C—H oscillator* will now be discussed. As in the case of stretching modes of the alkyl groups, the bending fundamentals are coupled only with those oscillators directly bonded to the carbon. Since there are three independent H—C—H bond angles, there are *three deformation or bending vibrations* associated with the *methyl group* (Fig. 7.6). As in the stretching fundamentals, these modes can be described in terms of two antisymmetric degenerate or nearly degenerate vibrations that occur near 1460 cm⁻¹. The third vibration is a symmetric bending mode (umbrella) close to 1375 cm⁻¹ (Figs. 7.4 and 7.6), which is easily identified.

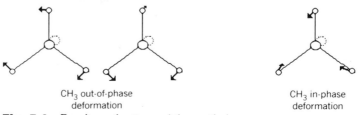

CH₃ out-of-phase deformation   CH₃ in-phase deformation

Fig. 7.6 *Bending vibrations of the methyl group.*

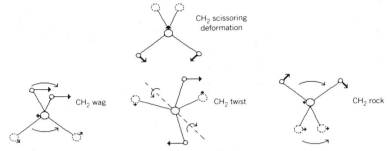

Fig. 7.7 *Bending vibrations of the methylene group.*

The *methylene group* has a single H—C—H bond angle, and a *single deformation mode* (scissoring, Fig. 7.7) directly analogous to that of the water molecule. The symmetric bend of the —CH_2— group occurs near 1450 cm^{-1} in hydrocarbons (Figs. 7.4 and 7.7).

Three other bending modes are available to the *methylene group* (Fig. 7.7). When the methylene unit is fused into the molecule, three vibrational degrees of freedom develop, which are related to rotational motion of the isolated system; for example, the rotational motion of the water molecule (see Appendix B). These fundamentals (wag, twist, and rock) are subject to significant coupling to adjacent methylene groups. The transitions of these modes are also rather weak in intensity. Thus, wag, twist, and rock are not useful group frequencies with the exception of a component of the rocking vibration (Fig. 7.7) in certain structures. In those molecules having four or more methylene groups in a row, the coupled mode corresponding to the *all in-phase* rocking vibration develops a significant dipole moment change, and a stable frequency. Thus, the in-phase rock that occurs near 720 cm^{-1} (Fig. 7.4) in the fingerprint region gives rise to an absorption band of sufficient intensity to allow for confident assignment.

The bending modes associated with the *methine hydrogen* are hard to identify in most hydrocarbons. (The in-plane bend can be particularly important, however, in certain hetero-atom systems; see discussion of the aldehyde functional group.)

The C—H vibrational modes of the alkanes (or mixed compounds containing alkyl groups) that are characteristic and reliable group frequencies can be summarized as in Table 7.2 and B.1.

Table 7.2 Alkane Normal Modes

C—H Vibrational Mode	$\bar{\nu} \pm 10$ (cm^{-1})
Methyl groups	
Antisymmetric (degenerate) stretch	2960
Symmetric stretch	2870
Antisymmetric (degenerate) deformation	1460
Symmetric (umbrella) deformation	1375
Methylene groups	
Antisymmetric stretch	2925
Symmetric stretch	2850
Symmetric deformation (scissor)	1450
Rocking mode (all-in-phase)	720

CHARACTERISTIC GROUP FREQUENCIES OF ALKENES

Alkenes (olefins) possess the carbon-carbon double bond, C=C. The group frequencies of these molecules will be those of the alkanes for the saturated portion of the molecule, plus those modes contributed by the unsaturated group.

C=C Stretching

The stretching fundamental of the C=C group is a useful group frequency. The increase in the force constant in going from the single bond to the double bond moves the frequency to sufficiently high values ($\bar{\nu}_{C=C}$ = 1616 cm^{-1}, ethylene) to decouple this mode from adjacent C—C vibrations. Substitution of carbon for hydrogen on the double bond tends to raise the C=C stretching frequency, as the effective force constant has been shown to include an increased compression term (Fig. 7.8). It is possible to classify open-chain unsaturated systems into two groups as in Table 7.3 and B.2.

The high-frequency group has quite weak bands unless conjugation occurs. Indeed, in the tetrasubstituted case, if the four groups are identical, the C=C band is formally forbidden in the infrared. When the alkyl groups are similar but not identical, the magnitude of the dipole moment is so small that it becomes difficult to detect an absorption band. In this situation the molecule is considered to have a "pseudo-center of symmetry." In the low-frequency set more intense absorption bands occur, with the average intensity falling in the medium-to-strong range.

Olefinic C—H

A number of the fundamental modes associated with the olefinic C—H groups are good group frequencies.

The C—H stretching frequencies occur above 3000 cm^{-1} and are localized in two regions as follows: (1) If two C—H groups are present on an sp^2 carbon, two coupled (antisymmetric and symmetric) vibrations occur near 3080 and 3010 cm^{-1}. (2) If a single C—H group is attached to an sp^2 hybridized carbon, it gives rise to a single mode near 3030 cm^{-1}. (3) A vinyl group will have both sets of bands, but the lower frequency modes are seldom resolved (Figs. 7.9 and 7.10).

The olefinic C—H bending frequencies fall into two categories: (1) There are bending modes that occur in the plane of the double bond. These fundamentals are not useful group frequencies. (2) There are bending modes that occur out of the plane of the double bond. These fundamentals are useful group frequencies.

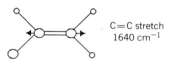

Fig. 7.8 *Double-bond stretching vibration.*

C=C stretch
1640 cm^{-1}

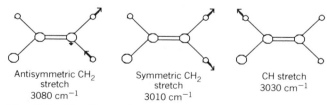

Antisymmetric CH$_2$
stretch
3080 cm^{-1}

Symmetric CH$_2$
stretch
3010 cm^{-1}

CH stretch
3030 cm^{-1}

Fig. 7.9 *C—H stretching vibrations of the vinyl group.*

Table 7.3 Substitution Classification of C=C Stretching Frequencies

C=C Normal Mode	$\bar{\nu}$ (cm^{-1})
trans-, tri-, tetra-substituted	1680–1665
cis-, vinylidene- (terminal-1,1-), vinyl-substituted	1660–1620

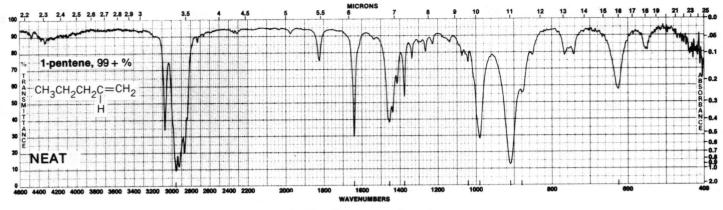

Fig. 7.10 *IR spectrum: 1-pentene.*

The Out-of-Plane Deformation Modes

Vinyl Groups

The vinyl group has three C—H bonds and therefore three out-of-plane bending modes. Two of these normal modes are good group frequencies. The fundamental with the two trans hydrogens bending in-phase occurs near 990 cm^{-1} (Fig. 7.11). A second vibration primarily involving the two hydrogens attached to the terminal carbon, bending in-phase together (wag), is located close to 910 cm^{-1} (Fig. 7.11). The absorption bands resulting from both of these fundamentals are strong and easily detected in the fingerprint region.

Vinylidene Groups

The 1,1-substituted system will have two bending frequencies. One of these bending modes is a good group frequency. When both hydrogens wag out of the C=C plane together, a strong dipole moment change develops and gives rise to an intense absorption band near 890 cm^{-1}. This mode is related to the low-frequency mode found in the vinyl group at 910 cm^{-1} (Figs. 7.11 and 7.12).

Trans Olefins

A trans olefin has two out-of-plane bending modes, but again only one gives rise to a good group frequency. The mode that involves both hydrogens moving in-phase together occurs close to 965 cm^{-1}. This mode is directly related to the high-frequency mode of the vinyl group near 990 cm^{-1} (Fig. 7.11).

Cis Olefins

A cis substituted C=C group does not possess a very good out-of-plane group frequency. The only mode with reasonable intensity involves the in-phase bend of the two hydrogens. This fundamental is derived from a rotational type motion that couples to the rest of the system (Fig. 7.13). Therefore, the cis mode is not localized, but occurs in a broad region near 700 cm^{-1}.

Trisubstituted Olefins

In the case of **R$_2$C=CHR** systems we have a single out-of-plane bending vibration that occurs near 820 cm^{-1}. The mode is uncoupled and gives rise to a medium-intensity band.

Tetrasubstituted Olefins

These groups have no C—H bending modes.

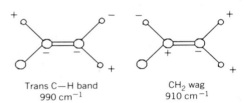

Fig. 7.11 *Out-of-plane C—H bending vibrations of the vinyl group.*

Trans C—H band
990 cm^{-1}

CH$_2$ wag
910 cm^{-1}

CH$_2$ wag
890 cm^{-1}

Fig. 7.12 *Out-of-plane C—H bending vibration of the vinylidene group.*

C—H bend
~700 cm^{-1}

Fig. 7.13 *Out of plane C—H bending vibration of cis-substituted HC=CH group.*

Table 7.4   Alkene Normal Modes

C—H Vibrational Mode	$\bar{\nu} \pm 10$ (cm^{-1})
Stretching modes	
Antisymmetric stretch (=CH$_2$)	3080
Symmetric stretch (=CH$_2$)	3020
Uncoupled stretch (=CH)	3030
Out-of-plane bending modes	
Vinyl group	
Trans-hydrogens (in-phase)	990
Terminal-hydrogens (wag)	910
Vinylidene group	
Terminal (wag)	890
Trans group	
Trans hydrogens (in-phase)	965
Cis group	
Cis hydrogens (in-phase)	~700
Trisubstituted group	
Uncoupled hydrogen	820
Tetrasubstituted group: no modes	

Overtones

The overtones of the fundamentals that involve wagging of the terminal hydrogen atoms in the vinyl and vinylidene groups occur with unusual intensity. These bands are observed at 1825 and 1785 cm^{-1}, slightly more than double the fundamental frequency. Thus, these harmonics exhibit negative anharmonicity in addition to unusual intensity for forbidden vibrations.

The group frequencies of the olefinic C—H modes can be summarized as in Tables 7.4 and B.3.

CHARACTERISTIC GROUP FREQUENCIES OF ALKYNES

C≡C Stretching Vibration

Triple bond formation further increases the force constant involved in the C—C stretching vibration. Thus, acetylenes possess the highest of all observed C—C stretching frequencies. The triple bond group frequency is located near 2120 cm^{-1} in monosubstituted alkynes. Compression effects on the force constant similar to those observed in substituted olefins raise the stretching mode into the 2225 cm^{-1} region in disubstituted acetylenes. The alkyne stretching vibration involves a relatively small dipole moment change. The resulting bands, therefore, are particularly weak in the infrared. In the disubstituted case, if the two groups are identical, the vibration is infrared inactive. Here, as in the olefins, a pseudo-center of symmetry can operate to significantly suppress the intensity of the mode. Even though these bands occur in a region that is essentially devoid of other absorptions their inherent weakness and variable intensity present significant problems in making confident band assignments. The high frequency of the mode effectively decouples the vibration from the rest of the system. Thus, triple bonds show little evidence of any first-order coupling. On the other hand, these vibrations are prone to second-order effects that complicate their interpretation. The triple bond assignments must be handled with care. These vibrations are a relatively difficult set of group frequencies to deal with, if they can be observed at all.

Alkyne C—H Vibrations

The stretching of acetylenic C—H bonds gives rise to the highest carbon–hydrogen vibrations observed. These are relatively intense uncoupled single

Table 7.5 Alkyne Normal Modes

C≡C, C—H Vibrational Mode	$\bar{\nu} \pm 10$ (cm^{-1})
Triple bond stretch (monosubstituted)	2120
Triple bond stretch (disubstituted)	2225
C—H bond stretch (monosubstituted)	3300

sharp modes that occur near 3300 cm^{-1}. Although this normal mode occurs in the same region as O—H and N—H fundamental vibrations, acetylenic C—H stretches usually can be distinguished by the sharpness of the band. They are highly reliable group frequencies.

The bending modes of the C—H acetylenic group do not give rise to reliable group frequencies.

The group frequencies of the alkynes are summarized in Tables 7.5 and B.4.

CHARACTERISTIC GROUP FREQUENCIES OF ARENES

Aromatic ring systems represent the final class of hydrocarbons to be considered in this section. The discussion will center on the benzene ring, but many of the more complicated systems have been examined in detail.

The infrared spectra of aromatic compounds possess many needle-sharp bands. This characteristic sets these spectra apart from the spectra of aliphatic compounds. It arises from the fact that aromatic systems are tightly bound rigid molecules having little opportunity for rotational isomerism. With aliphatic compounds the observed spectrum, in reality, often contains the spectrum of a complex mixture of rotamers. These isomers all exhibit very similar but not identical spectra that overlap and result in band broadening.

THE GROUP FREQUENCIES OF THE PHENYL GROUP

The group frequencies of the phenyl group can be classified as carbon-hydrogen vibrations consisting of stretching, and out-of-plane bending modes, plus carbon-carbon ring stretching and out-of-plane bending modes. The in-plane bending modes in both cases are not effective group frequencies.

C—H Stretching Modes
The C—H stretching vibrations occur as a series of weak bands in the region from 3100 to 3000 cm^{-1}. This is consistent with sp^2 hybridization of the carbon atom. These modes directly overlap the olefinic C—H stretching fundamentals. Substitution of hetero-atoms into the ring can significantly perturb these frequencies (oxygen raises the mode into the 3200 to 3100 cm^{-1} region). As the bands are generally weak, they may be masked by strong aliphatic absorption in mixed compounds if they lie close to 3000 cm^{-1}. Care must be taken in the assignment of these modes.

C=C Stretching Modes
The phenyl ring modes, which possess excellent group frequency properties, involve two pairs of closely related C=C stretching vibrations. These vibrations are related to degenerate fundamentals in unsubstituted benzene. On ring substitution the degeneracy is removed because of the lowered symmetry. The ring vibrations ν_{8a} and ν_{8b} (the numbering has been carried over from the benzene fundamental assignments) involve displacements in three carbon units at each end of the ring, which are analogous to the symmetric and antisymmetric stretching modes of water. The two modes result from the two sets of displacements, which are in-phase (Fig. 7.14).

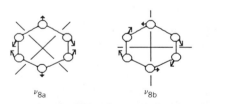

ν_{8a} ν_{8b}

Fig. 7.14 *Ring stretching vibrations of benzene, $\nu_{8a,b}$.*

ν_{19a} ν_{19b}

Fig. 7.15 *Ring stretching vibrations of benzene, $\nu_{19a,b}$.*

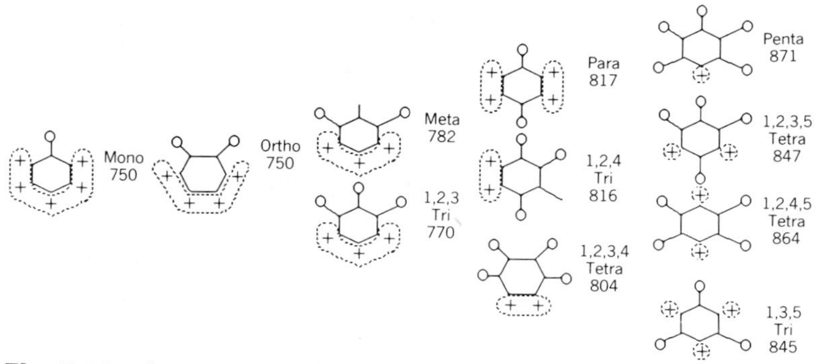

Fig. 7.16 *Arene out-of-plane C—H bending vibrations.*

These vibrations are degenerate and inactive in the infrared spectrum of the unsubstituted ring. In general, the higher frequency vibration, ν_{8a}, is the more intense of the pair. If a substituent is conjugated to the ring system, however, the lower frequency component gains in intensity and sometimes becomes the most intense member of the pair. These two modes are also substituent independent with para disubstitution; however, if the groups are identical or nearly identical, the modes are infrared forbidden or active with greatly suppressed intensity. These modes, ν_{8a} and ν_{8b} occur near 1600 and 1580 cm^{-1}, respectively. The second pair of vibrations corresponds to the identical displacements of the first pair of modes with the sets now out-of-phase (Fig. 7.15). Substitution on the ring removes the degeneracy, giving two bands corresponding to ν_{19a} and ν_{19b} in benzene.

These fundamentals, as with ν_{8a} and ν_{8b} are substituent independent on mono or para substitution. The dipole moment change associated with these normal modes generally gives rise to rather intense absorption bands. The high-frequency component, however, is sensitive to electron withdrawing substituents that can significantly suppress the intensity of this mode. These fundamentals occur near 1500 and 1450 cm^{-1}, respectively.

Carbon—Hydrogen Bending Vibrations

The C—H bending normal modes of group frequency value are the out-of-plane vibrations. These fundamentals are useful guides to the substitution pattern on the ring system. There will be as many out-of-plane vibrations as there are C—H groups. The modes of interest, however, are those fundamentals in which all the hydrogens move in-phase. These vibrations have substantial dipole moment changes and, thus, give rise to intense absorption bands (Fig. 7.16). The very strong intensity of the out-of-plane deformation modes plays a key role in our ability to make confident assignments for these fundamentals, as they fall in the heart of the fingerprint region. The five all-in-phase bending vibrations are as presented in Fig. 7.16 and Tables 7.6 and B.5.

While there is considerable overlap of the ranges, the uncertainty of the assignment can often be reduced by the identification of an additional strong

Table 7.6 Arene Out-of-Ring Plane C—H Deformation Modes

Number of Adjacent H's	cm^{-1} range
5	770–730
4	770–735
3	810–750
2	860–800
1	900–860

Table 7.7 Arene Group Frequencies

Arene Fundamental	$\bar{\nu}$ (cm^{-1})
C—H stretch	3100–3000
C=C ring stretch (ν_{8a})	1600 ± 10
C=C ring stretch (ν_{8b})	1580 ± 10
C=C ring stretch (ν_{19a})	1500 ± 10
C=C ring stretch (ν_{19b})	1450 ± 10
C—H out-of-plane bend (1H)	900–860
C—H out-of-plane bend (2H)	860–800
C—H out-of-plane bend (3H)	810–750
C—H out-of-plane bend (4H)	770–735
C—H out-of-plane bend (5H)	770–730
C—C ring out-of-plane bend (1; 1,3; 1,3,5)	690 ± 10
C—H out-of-plane bend sum tones	2000–1650

Fig. 7.17 *Benzene out-of-plane bending vibration, ν_4.*

band in the 690 cm^{-1} region. This band results from a carbon-carbon out-of-plane ring deformation, ν_4 of benzene (Fig. 7.17). This mode is substituent insensitive to mono-, meta-, and 1,3,5-substitution. Thus, a band will seldom occur in this region in ortho disubstituted systems.

Sum Tone Patterns

Some of the out-of-plane C—H bending modes can be excited simultaneously with other low-frequency fundamentals. As these *sum tone* transitions are formally forbidden, the resulting absorption bands are very weak. With optically thick samples, however, weak bands are observed in the 2000–1650 cm^{-1} region. The pattern of these bands is highly characteristic of the substitution arrangement on the ring, since they have their origin in the out-of-plane C—H bending frequencies. Unless carbonyl groups are present in the molecule, the 2000–1650 cm^{-1} region will be open for observation. These combination band patterns can be used to remove ambiguity about the ring substitution pattern based on assignments of the out-of-plane fundamentals (Fig. 7.18).

The group frequencies of the arenes can be summarized as in Tables 7.7 and B.6.

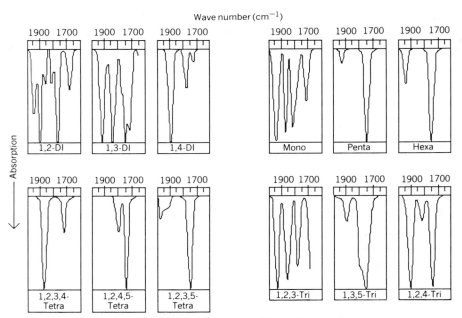

Fig. 7.18 *Combination band patterns in the 2000–1650 cm^{-1} region.*

PART B: The Characteristic Group Frequencies of the Carbonyl

The carbonyl unit is perhaps the single most important functional group in organic chemistry. It is certainly the most commonly occurring functionality. Infrared spectroscopy can play a powerful role in the characterization of the carbonyl because this group possesses all of the properties that give rise to an excellent group frequency. (1) The carbonyl group has a large dipole moment derivative, which gives rise to very intense absorption bands. (2) As a result of the large force constant, it has a stretching frequency that occurs at high values outside the fingerprint region. In addition, this portion of the spectrum is devoid of most other fundamentals; (3) its stretching fundamental occurs in a range that is reasonably narrow (little coupling), 1750 ± 150 cm^{-1}, but sensitive enough to the local environment to allow for considerable interpretation of the surrounding structure. (4) The range of frequencies is determined by a number of factors that are now well understood in terms of the effects outlined below.

MASS EFFECTS

The mode of principal interest is the stretching vibration. In this oscillator the C and O atoms undergo comparable displacements; thus, we must replace the simplified mass expression in the Hook's law approximation that was applied to the C—H stretching fundamentals by the reduced mass μ (where $\mu = m_C m_O/(m_C + m_O)$, see above). On substitution of the isotopes ^{13}C and ^{18}O, the predicted small frequency shifts ($\sim$30–40 cm^{-1}) are observed. These results are consistent with a relatively low degree of mechanical coupling to the rest of the system. This lack of coupling is expected, because the force constant of the multiple bond is significantly different from the values of the force constants of the bonds that connect the carbonyl to the rest of the molecule.

GEOMETRIC EFFECTS

Geometric effects can play a major role in determining the location of the carbonyl frequency within the 1750 cm^{-1} region. Although the displacement of the oxygen atom involves simply stretching or compressing the C=O bond, the displacement of the carbon atom is more complex. This latter movement also contains a compression component of the force constants in the two connecting single bonds as the carbonyl carbon is being stretched. In the opposite phase of the vibration, a stretching component of the force constants in the two connecting bonds is required, as the carbonyl carbon is being compressed. The magnitude of these additional force constant components is angle dependent. As the angle between the single bonds (C—CO—C angle) decreases, the contribution of the single bond components to the effective C=O stretching force constant will be raised. Since the frequency of the vibration is directly proportional to the square root of the force constant, a decrease in the internal carbonyl bond angle will raise the frequency (Table 7.8). Alternatively, an increase in the internal carbonyl bond angle will lower the carbonyl frequency.

RESONANCE AND INDUCTIVE EFFECTS

Resonance and inductive effects can profoundly influence the vibrational frequency of the carbonyl group. In the following discussion the carbonyl stretching mode for acetone ($\bar{\nu}_{C=O}$ 1715 cm^{-1}, liq.) will be used as a reference frequency representative of simple alkyl substitution on the carbonyl group. Effects that perturb this reference fundamental to either higher or lower values will be examined.

Table 7.8 Variation of Carbonyl Frequency vs. Bond Angle

Ring size	7	6	5	4	3	(2)
Lactones	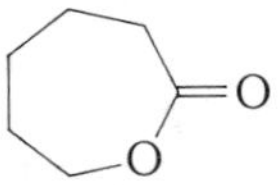 1727	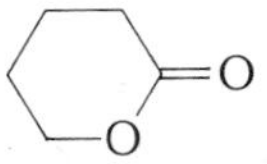 1740	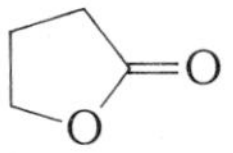 1775	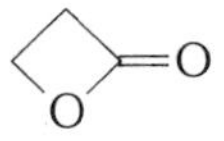 1832		
Ketones	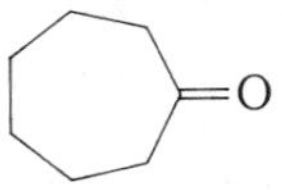 1699	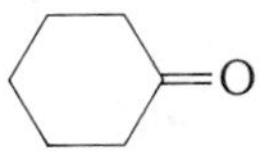 1710	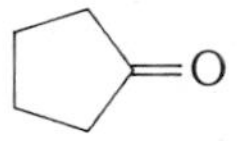 1744	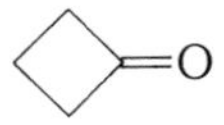 1782	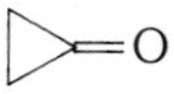 1906	2049
Lactams		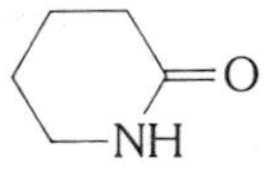 1670	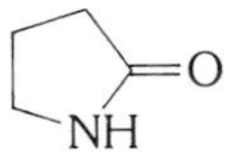 1695	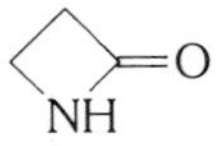 1750		

Electronic Effects That Raise the Carbonyl Frequency

When an alkyl substituent is replaced by a more electronegative system, the balance of contributing resonance forms in the carbonyl is slightly shifted away from dipolar forms by strong inductive effects. This shift results in a larger effective C=O force constant and higher frequencies. For example, in hexanoyl chloride, $\bar{\nu}_{C=O}$ = 1805 cm^{-1} (see Fig. 7.23).

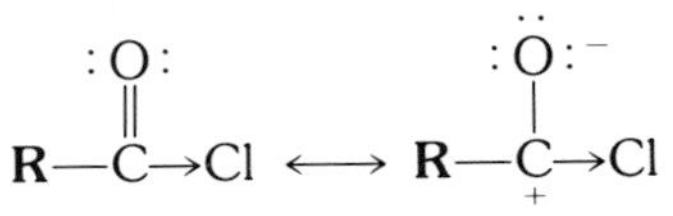

Electronic Effects That Lower the Carbonyl Frequency

Direct conjugation of the carbonyl via α,β-unsaturation will introduce new dipolar carbonyl resonance forms that lower the effective force constant values and thus, decrease the carbonyl stretching frequency. For example, consider cyclohexene methyl ketone (I), $\bar{\nu}_{C=O}$ = 1685 cm^{-1}, and acetophenone (II), $\bar{\nu}_{C=O}$ = 1687 cm^{-1}.

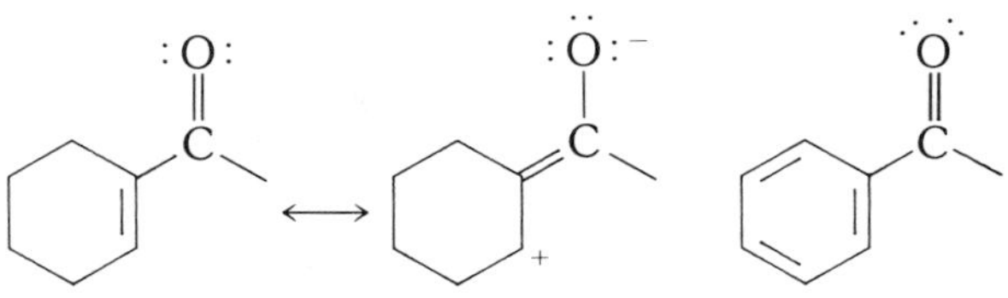

The Case of Ester Carbonyl Vibrations: Competing Inductive and Resonance Effects

The substitution of the more electronegative oxygen for carbon in going from ketones to esters will raise the carbonyl frequency in esters as a result of inductive influences (see Electronic Effects that Raise the Carbonyl Frequency, above). On the other hand, the lone-pair electrons present on the ether oxygen of the ester will be in direct conjugation with the carbonyl. This latter interaction will generate dipolar resonance forms that will tend to drop the C=O frequency. The balance between these two competing effects in esters, which might be difficult to anticipate, ends in favor of the inductive effects. The ester carbonyls commonly are located 20–40 cm^{-1} higher than the simple aliphatic ketones, in the range 1755–1735 cm^{-1}.

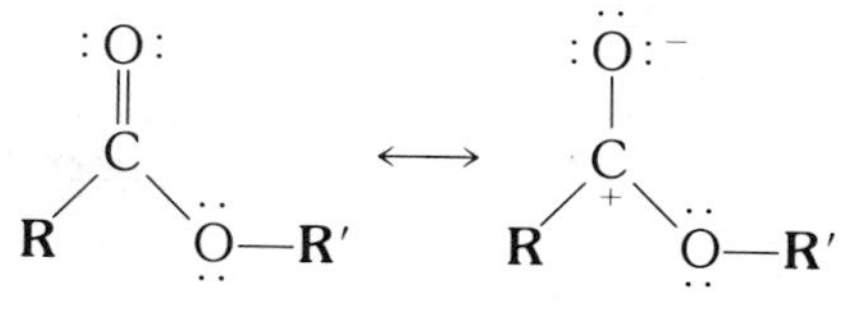

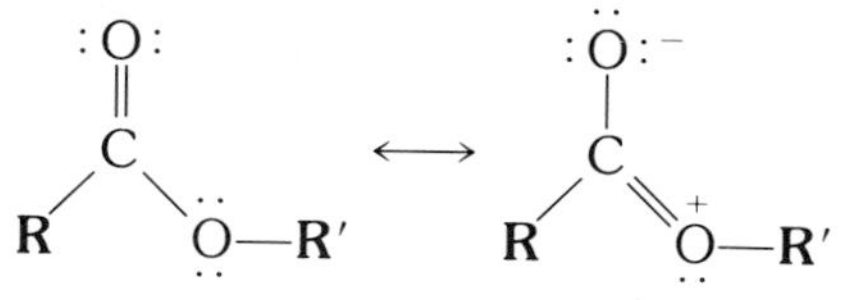

Now let us examine ester carbonyl frequencies in somewhat greater detail.

If the ester carbonyl is directly involved with α,β-unsaturation, the normal ester frequency is lowered by 20–30 cm^{-1}. Thus, unsaturated ester carbonyl frequencies occur very nearly in the same region as simple aliphatic ketone frequencies. For example, ethyl benzoate, $\bar{\nu}_{C=O} = 1720$ cm^{-1}.

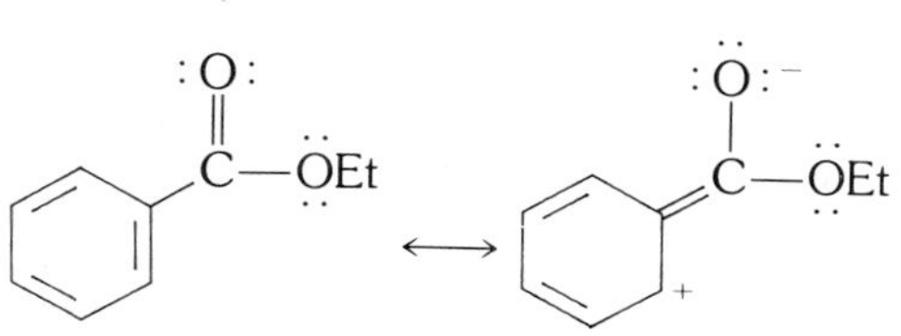

If the ester is conjugated, but the conjugation is located adjacent to the ether oxygen rather than alpha to the carbonyl group, then the carbonyl frequency is raised. The higher $\nu_{C=O}$ results from resonance competition for the lone-pair electrons of the ether oxygen by the carbonyl and the new conjugating group, for example, phenyl acetate, $\bar{\nu}_{C=O} = 1769$ cm^{-1}. These frequency shifts support the arguments concerning competition between inductive and resonance effects taking place within the ester ether group.

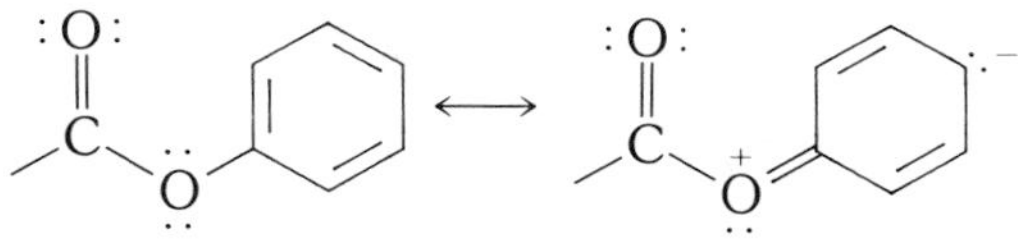

If the ester group is directly conjugated on both sides, then the resonance effects should cancel, and we would expect this type of system to exhibit near normal carbonyl frequencies, for example, phenyl benzoate, $\bar{\nu}_{C=O} = 1743$ cm^{-1} as compared to ethyl acetate, $\bar{\nu}_{C=O} = 1742$ cm^{-1}.

Interaction Effects Interaction effects vary from those that have a dramatic impact on the spectra to those that are barely detectable. Our understanding of these terms completes the discussion of the major factors affecting the carbonyl group frequencies. We can roughly divide this area of discussion into intramolecular and intermolecular type interactions.

Intramolecular Carbonyl Interactions

First-order coupling effects are rarely observed as this oscillator is rather effectively decoupled from the rest of the molecule by differences in frequency. The most spectacular example (see Appendix B, p. 394) is the case of carbon dioxide.

In CO_2 the two coupled oscillators are aligned for maximum interaction and possess identical frequencies. The splitting between the antisymmetric and symmetric levels is very large, approximately 1000 cm^{-1}. A second, much less dramatic, example is the case of anhydrides. In this instance the two oscillators are joined through a central oxygen. Delocalization across the connecting atom operates to maintain planarity of the system and thereby to increase the coupling. Even so, since the carbonyls are no longer held at the optimum angle and are vibrationally insulated by an intervening atom, the first-order coupling drops to ~70 cm^{-1}, less than 10% of the CO_2 value; for example, hexanoic anhydride, $\bar{\nu}_{C=O}$ = 1817, 1750 cm^{-1}. The uncoupled vibration would be expected to occur near 1770 cm^{-1}. This latter value is consistent with an oxygen substituted carbonyl in which conjugation of the lone-pair electrons on the ether oxygen has been nearly cancelled. The full inductive effect of the ether oxygen atom on the carbonyl stretching vibration can be inferred from these data. Equalized resonance competition by both carbonyl systems of the anhydride for the ether lone-pair electrons might be expected to bring about just such an effect.

Intramolecular H-bonding of carbonyls can be enhanced by resonance interactions and, thus, significantly perturb the stretching frequency. For example, let us consider the anthraquinone series shown below. In structure I, H-bonding is not present and a single stretching frequency for the quinone is observed at 1675 cm^{-1}. This value reflects direct conjugation with the aromatic ring, and with the methoxyl groups conjugated in equivalent fashion with both carbonyls. Little coupling between the carbonyls is observed through the ring. In structure II a phenol group replaces one of the methoxyl groups. Resonance forms operating through a tightly hydrogen-bonded six-membered ring act to reduce the effective force constant of the carbonyl with the result that the observed frequency, $\bar{\nu}_{C=O}$ = 1636 cm^{-1}, is lowered nearly 40 cm^{-1}. In structure III both methoxyl groups have been replaced by phenolic groups. This change results in both carbonyls undergoing strong H-bonding, which involves resonance forms similar to those found in structure II. Thus, a single band is observed at $\bar{\nu}_{C=O}$ = 1627 cm^{-1}. This downward shift approaching 50 cm^{-1} can be primarily ascribed to strong internal H-bonds present in the anthraquinone system.

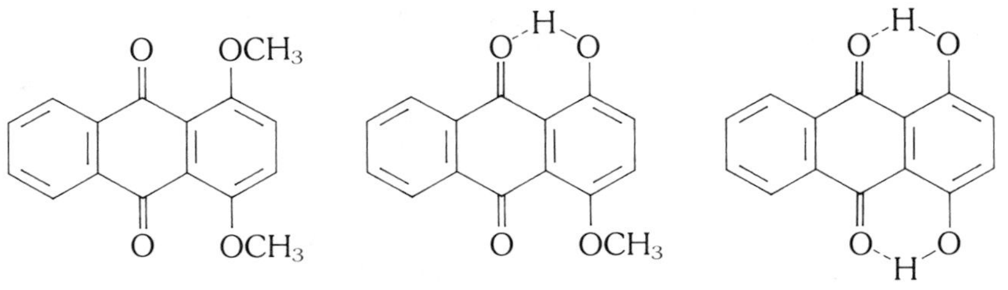

Anthraquinone derivatives

Several other examples of very strong intramolecular H-bonding are known. In tropolone, where the hydrogen appears to be essentially equidistant from the two carbonyls, the absorption occurs at 1605 cm^{-1}.

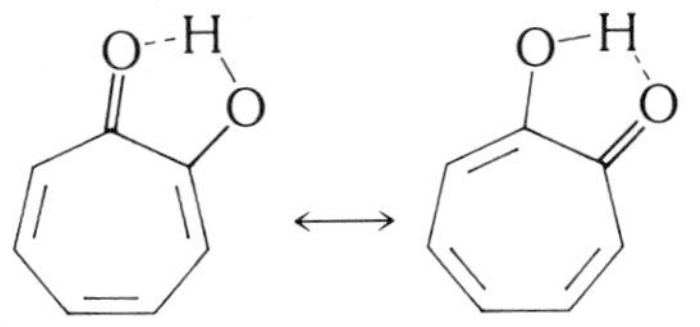

Tropolone (H is equidistant from oxygens)

Table 7.9 Field Effects in Chloroacetones

Compound	$\bar{\nu}_{C=O}$ (cm^{-1})
Acetone	1715
Chloroacetone	1752, 1726
1,1-Dichloroacetone	1743, 1724
1,1,1-Trichloroacetone	1729

Second-order coupling (Fermi resonance) often occurs with carbonyl frequencies in complex organic molecules. In the large majority of cases, the frequency match of overtone and fundamental is relatively poor so that the frequency of the fundamental is not affected. The main evidence for the interaction in these cases will be weak shoulders associated with the main carbonyl peak. An example of the coupling not being trivial is the case of cyclopentanone. The Fermi interaction involves an overtone or combination level as the splitting collapses to a singlet in 2,2,5,5-tetra-d$_4$-cyclopentanone.

Field effects will also perturb the carbonyl frequency. The classic case is that of the chloroacetones. In those rotamers, in which the chlorine atom is in the eclipsed position with respect to the oxygen, repulsive lone-pair interactions occur. These field effects result in suppression of the contribution of the dipolar carbonyl resonance form, and therefore result in a rise (~30 cm^{-1}) in the stretching frequency (Table 7.9).

These arguments are supported by the observation that two carbonyl frequencies are present in the mono- and dichloroacetones, and single frequencies in acetone and trichloroacetone. The small frequency rise observed in the low-frequency component is attributed to inductive effects, which can have only a minimal influence on the high-frequency component shift.

Transannular interactions occur when cyclic carbonyl groups are sterically positioned so that the carbon atom of the carbonyl is oriented toward an electron-rich center lying across the ring. This interaction can greatly enhance the dipolar resonance form of the carbonyl and result in a significant drop in the stretching frequency. The effect has a major impact on the carbonyl frequency of the alkaloid protopine ($\bar{\nu}_{C=O}$ = 1660 cm^{-1}).

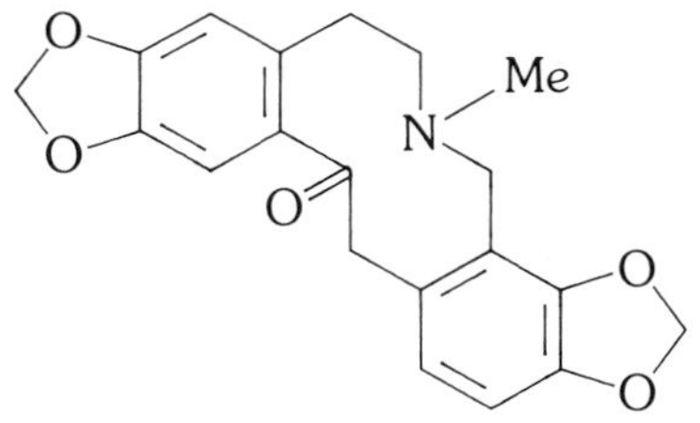

Protopine

Most interesting, however, are the results obtained from a number of excellent model compounds. For example, the cyclooctaaminoketone (I), $\bar{\nu}_{C=O}$ = 1666 cm^{-1}, and its perchlorate salt (II), which exhibits no carbonyl absorption band at all!

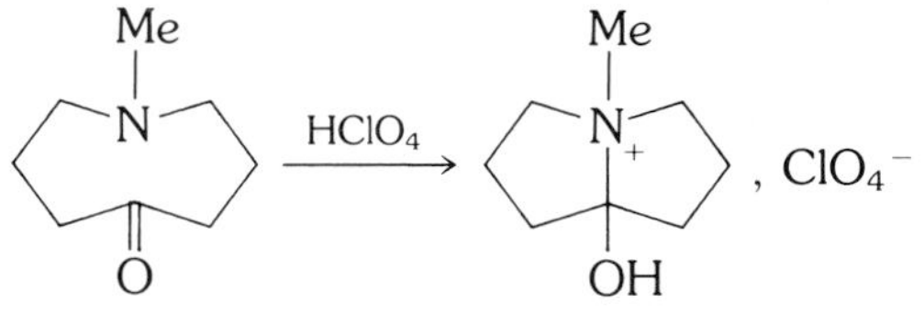

Intermolecular Carbonyl Interactions

Strong intermolecular H-bonding can significantly perturb carbonyl frequencies. It is known in the case of aliphatic carboxylic acids that these substances form strongly H-bonded dimers when neat or in highly concentrated solutions. Association through the carbonyl groups leads to the formation of a symmetric eight-membered ring containing two H-bonds. Coupling through the tightly bonded ring results in a splitting of the carbonyl levels of approximately the same magnitude as found in anhydrides ($\Delta\bar{\nu}_{C=O} = 70$ cm^{-1}). As the dimer possesses a center of symmetry, the in-phase mode will not be active in the infrared ($\bar{\nu}_{C=O} = \sim 1650$ cm^{-1}). The out-of-phase stretch of the carbonyls, however, will be active. The antisymmetric C=O stretch gives rise to a strong band in the infrared ($\bar{\nu}_{C=O} = \sim 1720$ cm^{-1}). In very dilute solution, it is sometimes possible to observe these systems in the monomeric state. Under these conditions the carbonyl frequencies return to expected values ($\bar{\nu}_{C=O} = \sim 1770$ cm^{-1}). As the discussion of group frequencies expands, we will see a number of other examples of the effect of intermolecular H-bonding on the carbonyl group frequency.

The interaction of weak H-bonds is relatively hard to detect in the infrared, as the shifts are measured in terms of a few wavenumbers. One of the better examples is the effect on the carbonyl stretch of acetone ($\bar{\nu}_{C=O} = 1722$ cm^{-1}) as measured in hexane solution. When the hydrocarbon solvent is replaced by chloroform, weak H-bonds (O $\cdots$ H—C) develop, and the carbonyl mode drops 12 wavenumbers to 1710 cm^{-1}.

Weak dipolar interactions between carbonyls can also be observed in the infrared. The frequency shifts caused by these interactions parallel the development of polarization in the carbonyl group, as can be judged by the data in Table 7.10.

The major factors perturbing carbonyl frequencies can be summarized as follows:

Factors that raise the C=O frequency

1. Electronegative substitution
2. Decrease in C—CO—C internal bond angle

Factors that lower the C=O frequency

1. Conjugation
2. Hydrogen bonding

As several of these factors may be operating simultaneously, careful judgment as to the contribution of each individual effect must be exercised in predicting carbonyl frequencies. This judgment develops rapidly with practice at interpretation.

This completes the discussion of factors affecting carbonyl group frequencies. A number of additional examples will be discussed in detail in Part C, in which a survey of the infrared spectra of functional groups is considered. Carbonyl frequencies are summarized in Tables 7.11 and B.7.

Table 7.10 Carbonyl Dipolar Interactions[a]

Compound	$\Delta\bar{\nu}_{C=O}$ (cm^{-1})
Acetyl chloride	15
Phosgene	13
Acetone	21
Acetaldehyde	23
Dimethylformamide	50

[a]Shift measured between dilute nonpolar solution and neat sample.

Table 7.11 Carbonyl Group Frequencies

Functional Group	$\bar{\nu}_{C=O}$ (cm^{-1})
Ketones, aliphatic, open chain (**R**$_2$CO)	1725–1700
Ketones, conjugated	1700–1675
Ketones, ring	(see Table 7.8)
Acid halides	>1800
Esters, aliphatic	1755–1735
Esters, conjugated	1735–1720
Esters (conjugated to oxygen)	1780–1760
Lactones	(see Table 7.8)
Anhydrides, aliphatic, open chain	1840–1810 and 1770–1740
Acids, aliphatic	1725–1710
Amides	(see Part C)
Lactams	(see Table 7.8)
Aldehydes	1735–1720

PART C: The Characteristic Frequencies of Functional Groups

Now that we have examined the major group frequencies associated with the common hydrocarbon platforms (platform = hydrocarbon structural unit supporting a functional group), and the principal parameters affecting the carbonyl group, let us consider the vibrations associated with the common functional groups that lead to good group frequency correlations. For the most part we will use, in these discussions, a series of infrared spectra derived from straight chain aliphatic C$_6$ compounds.

HEXANE The spectrum of normal hexane (Fig. 7.4) obtained with the pure liquid, as expected, contains simply the group frequencies of an aliphatic hydrocarbon. The antisymmetric and symmetric methyl stretching modes occur below 3000 cm^{-1} at $\bar{\nu}$ = 2960 and 2876 cm^{-1}. The antisymmetric and symmetric methylene stretching fundamentals occur near $\bar{\nu}$ = 2938 and 2860 cm^{-1}. The antisymmetric methyl deformation ($\bar{\nu}$ = 1467 cm^{-1}) overlaps the symmetric methylene scissoring vibration, which is found as a difficult-to-identify shoulder at $\bar{\nu}$ = 1455 cm^{-1}. The symmetric methyl bend (umbrella mode) is easily assigned to the sharp band at 1379 cm^{-1}. Finally, the all-in-phase rocking mode of a sequence of four or more methylene groups can be identified by its intensity in the fingerprint region near 725 cm^{-1}. We often will be able to identify this collection of platform group frequency bands as we progress through the infrared spectra of the following series of compounds (Fig. 7.4 and Table 7.2).

HEXANOL If an oxygen atom is inserted across one of the terminal C—H bonds, we obtain the alcohol, 1-hexanol. The change in the infrared spectrum obtained with a sample path length of less than half that used to obtain the spectrum of hexane is remarkable (Fig. 7.19). A very intense band appears at 3350 cm^{-1}, which is assigned to the stretching mode of the single O—H group (Table 7.13 and B.8). The very broad and intense properties of this absorption are characteristic of the stretching of H-bonded hydroxyl groups. The increase in intensity of this mode also reflects an increase in the polarity of the bond involved in the vibration over that of the C—H bond in hexane. A second strong band in the spectrum is located near 1058 cm^{-1}. This absorption has been identified as the C—O stretching mode. The vibrational displacements of this fundamental are similar

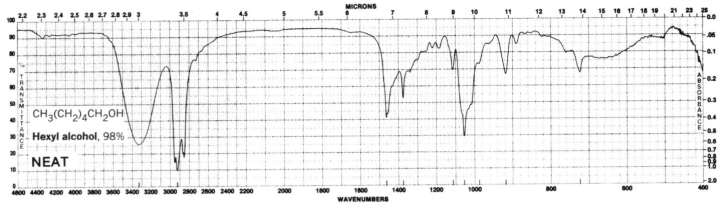

Fig. 7.19   *IR spectrum: 1-hexanol.*

to the antisymmetric stretch of water. Since the vibration involves significant displacement of the adjacent C—C oscillator, the vibration will be substitution sensitive. These latter shifts can be of value in determining the nature of the alcohol (primary, secondary, tertiary, see Table 7.12).

The only other new modes observed in going from hexane to 1-hexanol are the O—H bending vibrations. Two types of bending vibrations would be expected, the in-plane and out-of-plane displacements. The in-plane bend of the O—H oscillator is not a very good frequency because it is coupled to adjacent methylene group bending vibrations (wagging). It can be found because of its breadth (O—H bonding) as an underlying absorption running across the 1500–1300 cm^{-1} region (Table 7.13 or B.8). The out-of-plane O—H bending fundamental occurs at lower frequencies as a broad band (H—bonding) near 650 cm^{-1}. The group frequencies of the hydrocarbon portion of the molecule are easily identified, including the rocking fundamental (727 cm^{-1}), which is superimposed on the broad out-of-plane bending O—H mode. The normal modes of the hydroxyl group possess many of the characteristics that lead to excellent group frequency correlations (Fig. 7.19).

HEXANAL   If two terminal hydrogen atoms of hexane are replaced by a single oxygen atom, we have hexanal. The aldehyde functional group gives rise to several good group frequencies (Fig. 7.20 and Tables 7.14 or B.9). The system has strong bonds and a large dipole, and it is essentially decoupled from the rest of the molecule by the low frequency C—C connecting bond. A component of the

Table 7.12   Substitution Effects on C—O Stretch of Alcohols

Type of —OH Substitution	$\bar{\nu}_{C-O}$ (cm^{-1})
RCH$_2$—OH	1075–1000
R$_2$CH—OH	1150–1075
R$_3$C—OH	1200–1100
C$_6$H$_5$—OH	1260–1180

Table 7.13   Normal Modes of the Hydroxyl Group

$\bar{\nu}$ (cm^{-1})	Intensity	Mode Description
3500–3200	Very strong	O—H stretch (only strong when H-bonded)
1500–1300	Medium-strong	O—H in-plane bend (overlaps CH$_2$, & CH$_3$ bend)
1260–1000	Strong	C—C—O antisymmetric stretch
650	Medium	O—H out-of-plane bend

Table 7.14   Normal Modes of the Aldehyde Group

$\bar{\nu}$ (cm^{-1})	Intensity	Mode Description
2750–2720	Weak-medium	C—H stretch in Fermi resonance with C—H bend
1735–1720	Very strong	C=O stretch
1420–1405	Medium	CH$_2$ sym. bend, —CH$_2$— alpha to —CHO carbonyl
1405–1385	Medium	C—H in-plane bend

C—H stretching mode of the aldehyde group can be assigned to a band of weak to medium intensity at $\bar{\nu} = 2723$ cm^{-1}. The low frequency of this mode is interpreted on the basis of a Fermi resonance interaction. The aldehyde in-plane C—H bending fundamental found at 1390 cm^{-1} would be expected to generate an overtone very close to the aldehyde group C—H stretching mode, which must occur near 2775 cm^{-1}. The two levels interact and split to give two components. The more easily identified low-frequency component is very characteristic of the aldehyde group and is located near 2730 cm^{-1}. The higher frequency component often is masked by other aliphatic C—H stretching absorptions in the 2900 to 2800 cm^{-1} region. In the case of hexanal, the upper band is observed as a distinct shoulder occurring at 2823 cm^{-1}. The carbonyl stretching frequency is the most intense band in the spectrum and is located at 1728 cm^{-1}. The structural change in going from ketone to aldehyde will produce mass effects that will lower the frequency, and inductive and hyperconjugative effects that will raise the carbonyl frequency. The outcome of this competition is that aliphatic aldehyde carbonyl stretching modes generally occur at slightly higher values than do the saturated ketones. The only other identifiable group frequency associated with the aldehyde system is the C—H in-plane bending fundamental ($\bar{\nu} = 1390$ cm^{-1}) responsible for the overtone that undergoes Fermi resonance with the aldehyde C—H stretching mode. It should be noted that the aldehyde group perturbs one of the aliphatic chain group frequencies. Thus, the frequency of the symmetric deformation (scissoring) of the methylene group alpha to the carbonyl is lowered ($\bar{\nu} = 1408$ cm^{-1}) and intensified by hyperconjugation with the carbonyl. It is also evident that the vibrational modes of the aliphatic portion of the molecule now contribute a much smaller fraction of the overall absorption by the sample. Hence the spectrum is obtained with an even smaller path length than that of 1-hexanol.

3-HEPTANONE   If we insert a carbonyl group between the first two methylene groups of hexane, we have the ketone 3-heptanone. The only group frequency mode associated with this group is the stretching frequency ($\bar{\nu}_{C=O} = 1718$ cm^{-1}), which occurs

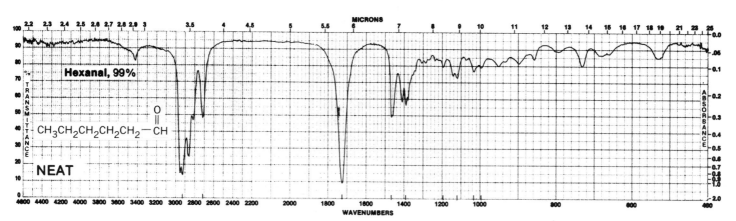

Fig. 7.20   *IR spectrum: hexanal.*

Table 7.15 Normal Modes of the Ketone Group

$\bar{\nu}$ (cm^{-1})	Intensity	Mode Description
3430–3410	Very weak	Not fundamental, overtone of carbonyl stretch
1725–1700	Very strong	C=O stretch
1430–1415	Medium	CH$_2$ sym. bend, —CH$_2$— alpha to ketone carbonyl

within the expected region for an aliphatic ketone (Fig. 7.21 and Table 7.15 or B.10). As in the case of aldehydes having α-methylene groups, the symmetric bending modes (scissoring) of the adjacent methylene groups at carbons 2 and 4 in 3-heptanone are perturbed to lower frequencies ($\bar{\nu} = 1425$ cm^{-1}) by hyperconjugation with the carbonyl. The shift of the methylene bending frequency is a useful indication of the substitution surrounding the ketone. The very weak band at 3425 cm^{-1} can be confidently assigned as the overtone of the carbonyl stretching frequency ($2\bar{\nu}_{C=O} = 3444$ cm^{-1}). The intensity drop from that of the fundamental and the frequency contraction are typical of these forbidden transitions. Note that in this molecule the sequence of methylene groups has dropped below four, and the rocking vibration ($\sim$720 cm^{-1}) is no longer detectable.

1-HEXYLACETATE   Replacing the O—H hydrogen of hexanol with CH$_3$CO— gives 1-hexylacetate. The very strong band found at 1743 cm^{-1} is typical of the carbonyl frequency of an aliphatic ester, particularly acetate esters (Fig. 7.22 and Table 7.16 or B.11). Two very intense bands occur in the spectra of acetate esters in the 1250–1000 cm^{-1} region. In primary acetates these bands are found near 1250 and 1050 cm^{-1} (hexyl acetate, 1242, 1042 cm^{-1}). The higher frequency mode is assigned to the antisymmetric C—CO—O stretch (similar to that in water), and the lower frequency to the antisymmetric O—CH$_2$—C stretch. While there is some coupling of these vibrations to the adjacent structure, resonance through the carbonyl group by the ether oxygen tends to localize the higher of the two vibrations. The lower mode, which would be expected to be more highly coupled, is in fact more subject to substitution effects. The upper mode is found in most saturated esters at frequencies slightly lower than those that occur in acetates (1210–1160 cm^{-1}). The only other absorption to note in acetate esters is the methyl symmetric bending mode. Here the umbrella deformation of the methyl adjacent to the carbonyl occurs at a slightly lower (hyperconjugation) frequency ($\bar{\nu} = 1366$ cm^{-1}), and the band is significantly intensified by the interaction with the carbonyl as compared to the deformation of the methyl at the end of the hexyl chain ($\bar{\nu} = 1384$ cm^{-1}).

HEXANOYL CHLORIDE   Hexanoyl chloride can be formed from hexane by exchanging three terminal hydrogens for an oxygen and a chlorine. The carbonyl stretching mode dominates the spectrum (Fig. 7.23 and Table 7.17 and B.12). It is an extremely intense band occurring near 1802 cm^{-1}. The high frequency and intensity result from inductive effects of chlorine substitution directly on the carbonyl group. The chlorine modes, on the other hand, are not easy to identify, and these vibrations do not develop good group frequencies. The symmetric deformation (scissoring) of the methylene group adjacent to the carbonyl group again shows the same frequency decrease resulting from hyperconjugation as observed with the ketones and aldehydes. In this case the mode ($\bar{\nu} = 1408$ cm^{-1}) gains considerable intensity via interaction with the highly polarized carbonyl group.

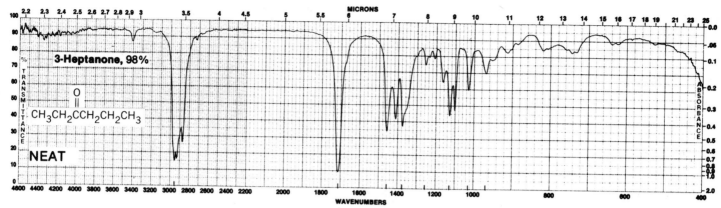

Fig. 7.21 *IR spectrum: 3-heptanone.*

Table 7.16 Normal Modes of the Ester Group

$\bar{\nu}$ (cm^{-1})	Intensity	Mode Description
1755–1735	Very strong	C=O stretch
1370–1360	Medium	CH$_3$ symmetric bend alpha to ester carbonyl
1260–1230	Very strong	C—CO—O antisymmetric stretch—acetates
1220–1160	Very strong	C—CO—O antisymmetric stretch—higher esters
1060–1030	Very strong	O—CH$_2$—C antisymmetric stretch primary acetates
1100–980	Very strong	O—CH$_2$—C antisymmetric stretch higher esters (may overlap with upper band)

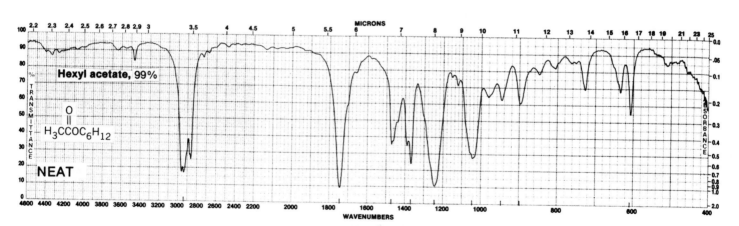

Fig. 7.22 *IR spectrum: 1-hexylacetate.*

Table 7.17 Normal Modes of the Acid Halide Group

$\bar{\nu}$ (cm^{-1})	Intensity	Mode Description
1810–1800	Very strong	C=O stretch, acid chlorides
1415–1405	Strong	CH$_2$ symmetric bend, alpha to —COCl carbonyl

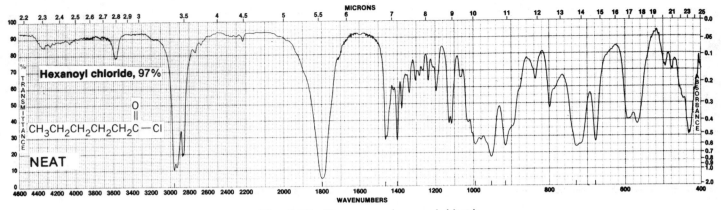

Fig. 7.23 *IR spectrum: hexanoyl chloride.*

HEXANOIC ACID

Hexanoic acid is obtained from hexane by substituting an oxygen and an —OH group for the terminal hydrogens. The acid possesses a very intense band with a width at half peak height of ~1000 cm^{-1}, which covers the region 3500–2200 cm^{-1} (Fig. 7.24 and Table 7.18 or B.13). This absorption is characteristic of very strongly H-bonded carboxylic acid groups. The relatively weak C—H stretching absorption of the aliphatic chain is superimposed on the O—H stretch between 3000 and 2800 cm^{-1}. Also occurring along this broad absorption are a characteristic set of weak overtone and combination bands running from ~2800 to 2200 cm^{-1}. The spectrum of the pure material will be that of the H-bonded dimer, which, as noted earlier (see p. 352), has a center of symmetry. The carbonyl mode (out-of-phase stretch) is found at 1709 cm^{-1}. Two rather broad and intense bands located between 1450–1400 and 1300–1200 cm^{-1} are associated with the in-plane O—H bend and the antisymmetric —CH$_2$—CO—O— stretch. These modes show evidence of considerable mixing, which is quite a different situation from the case of alcohols. A strong broad absorption band at 930 cm^{-1} is assigned to an out-of-plane bending mode of the H-bonded dimer ring. Thus, this band is present only when detectable concentrations of the dimers exist. In dilute solution the dimer band vanishes.

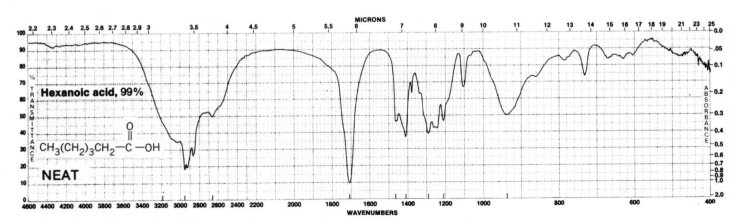

Fig. 7.24 *IR spectrum: hexanoic acid.*

Table 7.18 Normal Modes of the Carboxylic Acid Group

$\bar{\nu}$ (cm^{-1})	Intensity	Mode Description
3500–2500	Very, very strong	O—H stretch intensified by H-bonding
2800–2200	Very weak	Overtone & sum tones
1725–1710	Very strong	C=O, antisymmetric H-bonded dimer stretch
1450–1400	Strong	CH$_2$—CO—O antisymmetric stretch mixed with O—H bend
1300–1200	Strong	CH$_2$—CO—O antisymmetric stretch mixed with O—H bend
950–920	Medium	Out-of-plane O—H bend, acid dimer

This low wavenumber absorption is termed the "acid dimer" band. It is worth noting that even in carboxylic acids in which the normal modes of a large, highly polarized functional group dominate the spectrum, the group frequencies of the aliphatic molecular backbone are still identifiable.

HEXYL ANHYDRIDE Hexyl anhydride can be formed from two molecules of hexanoic acid by removing the elements of water. Coupling of the carbonyls through the ether oxygen splits the carbonyls ($\bar{\nu}_{C=O}$ = 1831, 1761 cm^{-1}) by ~70 cm^{-1} (see Fig. 7.25 and Table 7.19 and B.14). In this instance the higher frequency mode is the in-phase vibration. Strong bands occur in aliphatic anhydrides, near 1050 cm^{-1}, which are directly related to C—O stretching modes. The scissoring deformation of the —CH$_2$— groups alpha to the carbonyls is assigned to the band near 1415 cm^{-1}.

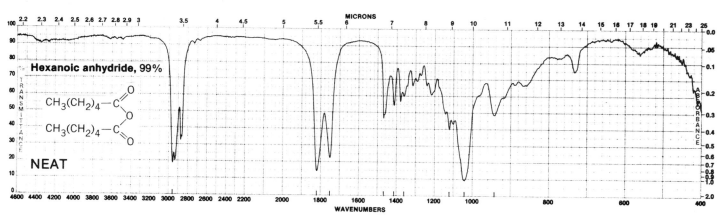

Fig. 7.25 IR spectrum: hexanoic anhydride.

Table 7.19 Normal Modes of the Anhydride Group (Open Chain)

$\bar{\nu}$ (cm^{-1})	Intensity	Mode Description
1840–1810	Very strong	C=O, in-phase stretch
1770–1740	Very strong	C=O, out-of-phase stretch
1420–1410	Strong	CH$_2$ symmetric bend alpha to carbonyls
1100–1000	Very strong	C—O stretch mixed modes

DIHEXYL ETHER

Dihexyl ether is obtained by removing the elements of water from two molecules of 1-hexanol. Long-chain aliphatic ethers have many physical properties similar to the hydrocarbons, and the infrared spectrum of dihexyl ether (Fig. 7.26) is not very different from that of tridecane. The single major departure in the spectrum is the presence of a very strong band near 1100 cm^{-1}. In dihexyl ether this absorption is found at 1130 cm^{-1}. The vibration responsible for this absorption band must involve a considerable amount of antisymmetric C—O—C stretch. Heavy coupling, however, is also involved in this level. Because of the extensive mechanical coupling with the chain carbons, substitution adjacent to the ether linkage can rather significantly shift this frequency. Fortunately, the large intensity associated with this fundamental relative to the other bands occurring in this region make it, in most cases, possible to assign with confidence the antisymmetric C—O—C stretching mode (see Table 7.20 or B.15).

1-HEXYLAMINE

If the units of NH are inserted across a terminal C—H bond of *n*-hexane, we obtain 1-hexylamine. Compare the infrared spectrum of 1-hexylamine (Fig. 7.27) with that of 1-hexanol (Fig. 7.19) and *n*-hexane (Fig. 7.4). The difference in the region above 3310 cm^{-1} is quite remarkable. While *n*-hexane is essentially devoid of absorption in this region and 1-hexanol exhibits a very strong band, 1-hexylamine possesses two bands ($\bar{\nu}_{N—H}$ = 3380, 3290 cm^{-1}) of medium to weak intensity. These latter bands are the antisymmetric and symmetric N—H stretching modes, respectively, of the primary amino group (Table 7.21 or B.16). H-bond intensification of the N—H group in simple primary amines does not equal that of the hydroxyl system. A band of medium intensity just above 1600 cm^{-1} is assigned to the symmetric (scissoring) deformation of the amino group. This vibration is directly related to the methylene mode near 1450 cm^{-1}. The occurrence of this fundamental requires that the amino group be unsubstituted, as the two hydrogen atoms undergo major displacements during the vibration. A second bending mode of primary amino groups can sometimes be observed. This vibration resembles the methylene wagging motion in which the hydrogen atoms are displaced more or less parallel to the molecular axis. As is

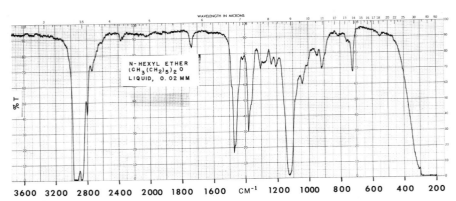

Fig. 7.26   *IR spectrum: dihexyl ether. (Courtesy of Bowdoin College)*

Table 7.20   Normal Modes of the Ether Group

$\bar{\nu}$ (cm^{-1})	Intensity	Mode Description
1150–1050	Strong	C—O—C antisymmetric stretch, mixed mode

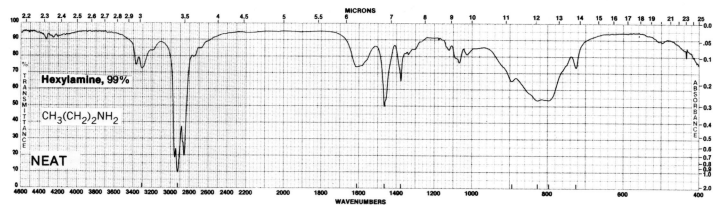

Fig. 7.27 *IR spectrum: hexylamine.*

Table 7.21 Normal Modes of the Primary Amine Group

$\bar{\nu}$ (cm^{-1})	Intensity	Mode Description
3400–3200	Weak-medium	NH$_2$ stretch (antisymmetric & symmetric)
1630–1600	Medium	NH$_2$ symmetric bend
820–780	Medium	NH$_2$ wag

usual for H-bonded bending fundamentals, the mode occurs as a fairly broad and quite strong band near 800 cm^{-1}. Frequencies related to the normal modes of single-bonded C—N systems are highly coupled to the surrounding C—C oscillators and cannot be assigned with any confidence.

HEXANENITRILE   Replacement of three terminal hydrogens of hexane by a nitrogen gives hex-anenitrile. The nitrile group is a very simple two-atom oscillator. The very strong triple bond (as in the case of the alkynes) contributes to an unusually high stretching frequency (Fig. 7.28 and Table 7.22 or B.17), and the polar character of the group gives rise to very strong bands. These two factors allow for easy distinction of nitrile bands from acetylenic absorption. The stretching funda-mental of saturated nitriles falls in the region 2260–2240 cm^{-1}. As expected, conjugation lowers this fundamental (2240–2210 cm^{-1}). In hexanenitrile the

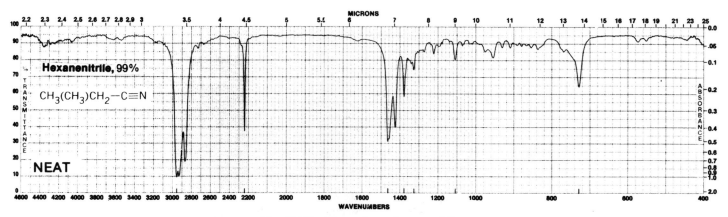

Fig. 7.28 *IR spectrum: hexanenitrile.*

Table 7.22   Normal Modes of the Nitrile Group

$\bar{\nu}$ (cm^{-1})	Intensity	Mode Description
2260–2240	Strong	C—N stretch, aliphatic
2240–2210	Strong	C—N stretch, conjugated

normal mode occurs at 2255 cm^{-1}. Primary nitrile groups can interact through hyperconjugation with the adjacent methylene groups so that the symmetric deformation (scissoring) mode is lowered into the 1425 cm^{-1} region. In hexanenitrile this bending vibration is assigned to a band at 1430 cm^{-1}. The remaining absorption bands are the aliphatic group frequencies.

HEXANAMIDE   Hexanamide is obtained by replacing the terminal hydrogens of hexane with the elements ONH_2. The highly polar amide group leads to very strong H-bonding, which in turn leads to greatly intensified N—H antisymmetric and symmetric stretching modes ($\bar{\nu}_{N—H} = 3375, 3200$ cm^{-1}, Fig. 7.29 and Table 7.23 and B.18). The carbonyl stretch occurs at low values (1675 cm^{-1}) for a saturated system substituted with an electronegative atom. Resonance between the carbonyl and the nitrogen lone-pair, plus strong hydrogen bonding, appears to overcome the inductive effect. The carbonyl band is accidently degenerate (two fundamentals occurring at the same frequency by chance rather than being required to have the same frequency by symmetry restrictions), but does not interact to any appreciable extent with the symmetric bending mode of the —NH_2 group in hexanamide (symmetry constraints restrict the interaction in this case). In some cases both bands can be resolved, but they will often occur as a single band. The wagging vibration of the amino group in which the hydrogen atoms are displaced parallel to the chain axis is found in the 700 cm^{-1}

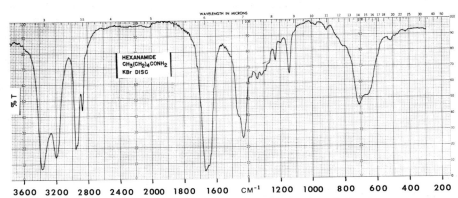

Fig. 7.29   *IR spectrum: hexanamide. (Courtesy of Bowdoin College)*

Table 7.23   Normal Modes of the Primary Amide Group

$\bar{\nu}$ (cm^{-1})	Intensity	Mode Description
3400–3150	Very strong	NH_2 antisymmetric & symmetric stretch-H bond
1680–1650	Very strong	C=O stretch, H-bonded
1660–1620	Strong	NH_2 symmetric bend (overlap with C=O stretch)
1430–1410	Strong	CH_2 symmetric bend alpha to amide carbonyl
750–650	Medium	NH_2 wag

region as a strong broad band. The symmetric deformation of the methylene group adjacent to the amide carbonyl undergoes the conventional frequency drop to the 1425 cm^{-1} region. The perturbed scissoring fundamental, however, gains considerable intensity from the interaction with the highly polarized amide carbonyl. Because of the presence of very strong intermolecular H-bonds, the overall spectra of solid or pure liquid amides exhibit broad, rather ill-defined, absorption bands.

N-METHYLHEXANAMIDE

If we substitute on the amide group by replacing a hydrogen with a methyl group, we obtain the secondary amide, **N**-methylhexanamide. The single N—H group gives rise to a very strong band at ~3300 cm^{-1}, which is very indicative of strong H-bonding (Fig. 7.30). A medium-intensity band near 3100 cm^{-1} is the overtone of the N—H bending mode identified at 1570 cm^{-1} in Fermi resonance with the N—H stretching fundamental. The overtone does not match the fundamental particularly well, but it is close enough to acquire substantial intensity enhancement. The carbonyl stretching mode at 1650 cm^{-1} exhibits the very intense and broad characteristics of the amide C=O system. The N—H in-plane bend of the single oscillator occurs near 1570 cm^{-1}. The drop in frequency from that of the primary scissoring mode near 1600 cm^{-1} allows for confident assignment of the substitution on amide groups (Table 7.24 or B.19). The band at 1570 cm^{-1}, while often referred to as the N—H bending mode, is in reality a heavily mixed mode. The pure in-plane N—H bend naturally falls near 1450 cm^{-1}. Resonance between the carbonyl and the nitrogen lone-pair results in a stiffening of the C—N bond. The resulting C—N stretch is raised into the 1400 cm^{-1} region. Mechanical coupling between the N—H bend and the C—N stretch results in first-order coupling and a splitting of the levels. The upper level occurs near 1570 cm^{-1} while the lower level often can be identified near 1300 cm^{-1} as a weak to medium band. The out-of-plane bend of the N—H group is identified as a broad, medium-intensity band centered near 700 cm^{-1}. The symmetric deformation of the methylene group alpha to the carbonyl is located at 1410 cm^{-1}. The unusual intensity of this mode results from interaction with the heavily polarized carbonyl system.

Studies of amide carbonyl frequencies in dilute nonpolar solution indicate that H-bonding effects are largely responsible for the low frequencies observed with primary and secondary amides, but play no role in tertiary amides.

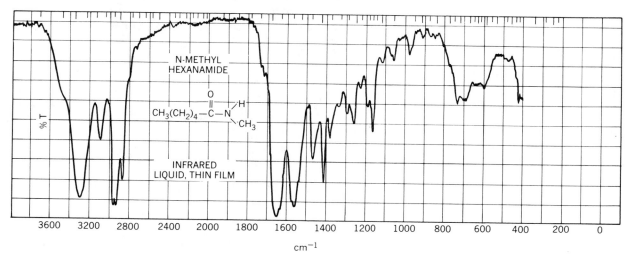

Fig. 7.30 *IR spectrum: N-methylhexanamide. (Courtesy of Bowdoin College)*

Table 7.24   Normal Modes of the Secondary Amide Group

$\bar{\nu}$ (cm^{-1})	Intensity	Mode Description
3350–3250	Strong	N—H stretch, intensified by H-bonding
3125–3075	Medium	Overtone N—H bend in Fermi resonance with N—H stretch
1670–1645	Very strong	C=O stretch, H-bonded
1580–1550	Strong	N—H in-plane bend mixed with C—N stretch
1415–1405	Strong	CH$_2$ symmetric bend alpha to amide carbonyl
1325–1275	Medium	C—N stretch mixed with N—H in-plane bend
725–680	Medium	N—H out-of-plane bend

Table 7.25   Amide Carbonyl: Solution and Solid Phase Data

Amide	Dilute solution (cm^{-1})	Solid (cm^{-1})
R—CO—NH$_2$	~1730	~1690–1650
R—CO—NHR	~1700	~1670–1630
R—CO—NR$_2$	~1650	~1650

The data (Table 7.25) indicate that when H-bonding effects are removed in primary amides, the inductive effect of the nitrogen dominates over the influence of conjugation, but not as much as in the case of esters. This is consistent with the relative electronegativities involved in esters and amides. In secondary amides with an electron-releasing N-alkyl group replacing a hydrogen, conjugation involving the nitrogen lone-pair with the carbonyl begins to overcome the inductive effect. In tertiary amides with two N-alkyl substituents present, conjugation now dominates the inductive effect. Under these conditions polarized resonance forms make large contributions to the character of the carbonyl and the C=O frequency decreases.

1-HEXYL ISOCYANATE   1-Hexyl isocyanate is obtained by replacing a terminal hydrogen atom with an —N=C=O group. The out-of-phase stretching mode of the isocyanate group attached to the hexyl chain occurs at 2275 cm^{-1} as a broad and very strong band (Fig. 7.31). This functional group is representative of a number of cumulated double bond systems that possess vibrations mechanically identical to that of carbon dioxide ($\bar{\nu} = 2350$ cm^{-1}). The range of stretching frequencies observed for alkyl-substituted isocyanates is very narrow, $\bar{\nu} = 2280$–2260 cm^{-1},

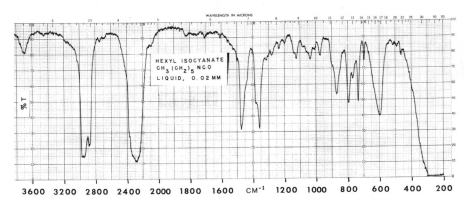

HEXYL ISOCYANATE
CH$_3$(CH$_2$)$_5$ NCO
LIQUID, 0.02 MM

Fig. 7.31   *IR spectrum: hexyl isocyanate. (Courtesy of Bowdoin College)*

Table 7.26   Normal Modes of the Isocyanate Group

$\bar{\nu}$ (cm^{-1})	Intensity	Mode Description
2280–2260	Very strong	—N=C=O antisymmetric stretch

which implies little coupling to the rest of the system (Table 7.26 and B.20). Interestingly, conjugation appears not to have any significant effect on the mode. The symmetric stretching fundamental is not easily observed in the infrared as it is a weak band occurring in the fingerprint region. The remaining group frequencies in the spectrum of 1-hexyl isocyanate are those of the alkyl group.

1-HEXANETHIOL   Inserting a sulfur atom in a terminal C—H bond of *n*-hexane gives 1-hexanethiol. The spectrum of this material resembles that of hexane itself except for small changes in the fingerprint region and a weak band near 2570 cm^{-1} (Fig. 7.32). The latter absorption is assigned to the S—H stretching fundamental (Table 7.27 or B.21). Although this mode is quite weak, it is not involved in any significant coupling, and it occurs in a region of the spectrum sparsely populated by other absorption bands. The S—H stretch, therefore, can be considered a reliable group frequency. The S—H bending and C—S stretching modes also are weak, and as they fall in the fingerprint region they are not useful as group frequencies. The remaining bands of 1-hexanethiol that can be assigned belong to the alkyl portion of the molecule.

1-CHLOROHEXANE   Replacement of a terminal hydrogen atom of *n*-hexane by a chlorine atom gives 1-chlorohexane. The massive chlorine atom is connected to the alkyl section by a fairly weak, but highly polarized, bond that dictates that the C—Cl stretching frequency appears as an intense band at low frequencies (Fig. 7.33 and Table 7.28 or B.22). The spectrum of 1-chlorohexane does possess a number of moderately intense absorption bands in the low-frequency region (800–600 cm^{-1}). Some coupling to the main structure adjacent to the C—Cl bond is expected

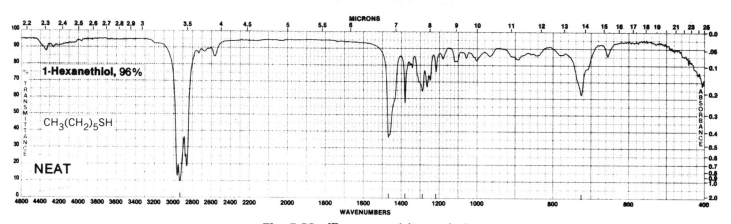

Fig. 7.32   *IR spectrum: 1-hexanethiol.*

Table 7.27   Normal Modes of the Thiol Group

$\bar{\nu}$ (cm^{-1})	Intensity	Mode Description
2580–2560	Weak	S—H stretch

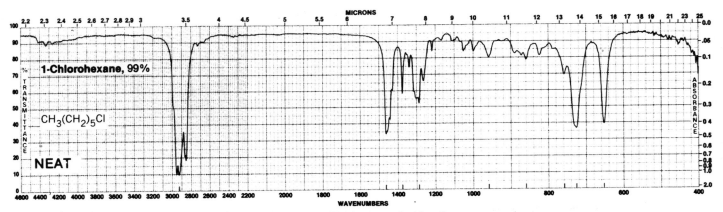

Fig. 7.33 *IR spectrum: 1-chlorohexane.*

Table 7.28 Normal Modes of the Chlorine Group

$\bar{\nu}$ (cm^{-1})	Intensity	Mode Description
750–650	Strong	C—Cl stretch, rotamers, and mixed modes occur

since the carbon atom will be carrying out the majority of the displacement. Reliable assignment of the halogen stretching mode, therefore, is not easy, because the surrounding C—C modes will pick up intensity from the polar C—Cl bond. In the case of 1-chlorohexane, it is possible to assign two C—Cl stretching modes ($\bar{\nu}_{\text{C—Cl}}$ = 731, 658 cm^{-1}), based on Raman spectral data. The presence of two modes is attributed to the presence of rotamers. The higher frequency is assigned to the anti (trans) conformer, and the lower frequency to the gauche conformer. Note that the stretching frequency of the anti isomer, 731 cm^{-1}, falls at the same frequency as the methylene rocking vibration of the hexyl chain. Thus, without additional data it would have been difficult to assign the C—Cl stretching modes even in these fairly simple systems. The carbon-halogen stretching vibration must be employed with care as a group frequency!

The bending modes of the halogens usually occur at such low frequencies as to be of little use as conventional group frequencies. The remaining bands are related to the hydrocarbon portion of the molecule.

```
      R                 R                 R
  H  ╱ ╲  H        H  ╱ ╲  Cl       Cl ╱ ╲  H
   ╲     ╱          ╲     ╱          ╲     ╱
  H       H        H       H        H       H
      Cl                H                 H

 Anti conformer        Gauche conformer
```

CHLOROBENZENE   The final spectrum to be considered in this section is the spectrum of the C$_6$-compound chlorobenzene (Fig. 7.34). Here we have introduced a new hydrocarbon platform bearing the functional group. Chlorobenzene is a material in which the oscillators are tightly bound together as a single conformer. The spectrum contains many needle-sharp bands as compared, for example, to that

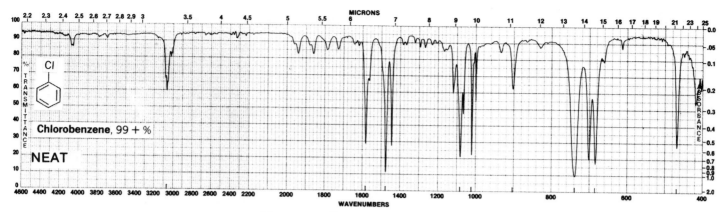

Fig. 7.34 *IR spectrum: chlorobenzene.*

Table 7.29 Group Frequency Assignments for Chlorobenzene

$\bar{\nu}$ (cm^{-1})	Intensity	Mode Description
3080	Medium	C—H stretching, C—H bonded to sp^2 carbon
1585	Strong	ν_{8a} ring stretching
1575	Weak	ν_{8b} ring stretching
1475	Strong	ν_{19a} ring stretching
1450	Strong	ν_{19b} ring stretching
747	Strong	C—H all in-phase, out-of-plane bend
688	Strong	Ring deformation
1945, 1865, 1788, 1733	All weak	Sum tones, out-of-plane C—H bends, pattern matches monosubstitution of ring

of 1-chlorohexane. Thus, it is often possible to identify the presence of either aliphatic or aromatic systems based simply on the overall appearance of the spectrum. In this spectrum the group frequencies of the hydrocarbon portion can be assigned as in Table 7.29 and B.23.

PART D: Strategies for Interpreting Infrared Spectra

1. Divide the spectrum at 1350 cm^{-1}.
2. Above 1350 cm^{-1}, absorption bands have a high probability of being good group frequencies. The interpretation is usually reliable and free from ambiguities. We can be much more confident of our assignments in this region.
3. Because of the reliability of the high wavenumber region, we always begin the interpretation of a spectrum at this end.
4. Bands below 1350 cm^{-1} may be either group frequencies or fingerprint frequencies. (The fingerprint region is often considered to begin at slightly higher wavenumber values, close to 1500 cm^{-1}, but for interpretation purposes we will consider the region to be that lying below 1350 cm^{-1}.)
5. Below 1350 cm^{-1} group frequencies are less easily assigned. In addition, even if a reliable group frequency occurs in this region, absorption at that frequency is not necessarily a result of that mode.
6. In order to make more confident assignments below 1350 cm^{-1}, it is helpful to be able to associate a secondary property such as band shape

with the particular mode. For example, the band is very intense, broad, sharp, occurs as a characteristic doublet, gives the correct frequency shift on isotopic substitution, or the like.

7. A good rule to remember is that in the fingerprint region the *absence* of a band is more important than the presence of a band.

8. Before beginning the interpretation establish the sampling conditions and as much other information about the sample as possible (such as molecular weight, mp, bp, color, odor, elemental analysis, solubility, refractive index, etc.).

9. In the interpretation try to assign the most intense bands first. These bands very often will be associated with a polar functional group.

10. Do not try to assign all the bands in the spectrum. Fingerprint bands are unique to a particular system. Occasionally intense bands will be fingerprint-type absorptions; these bands, generally, will be ignored in the interpretation.

11. Try to utilize the so-called *macro group frequency* approach. That is, if the functionality requires the presence of more than a single group frequency mode, make sure that all modes are correctly represented.

12. The use of the correlation chart (back endpaper) can act as a helpful quick aid for checking potential assignments. It is *not* a substitute for understanding the theory and operation of group frequency logic. *The use of the correlation chart without a good knowledge of group frequencies is the shortest path to disaster!*

QUESTIONS   **7-13.** The form of the C—H out-of-plane bending vibrations of the vinyl group are shown below:

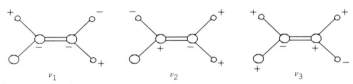

The first two modes give rise to excellent group frequencies, while the third fundamental does not lend itself to these correlations.

a. Explain the factors that lead to the third mode being such a poor group frequency.

b. Predict the location in the spectrum of the third fundamental.

7-14. In the figure below the mass of the terminal hydrogen atoms on acetylene is hypothetically varied from zero to infinity. The response of the C—H symmetric stretching (3374 cm^{-1}) and triple bond stretching (1974 cm^{-1}) modes to the change in mass is shown.

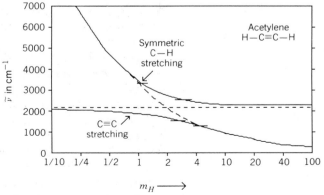

Variation of C—H and C≡C stretching frequencies of C$_2$H$_2$ with change in mass of end atoms.

a. Calculate the expected D-isotopic shift for the C—H symmetric stretching. Is the hypothetical value close to the calculated value? Explain.

b. Explain why the triple bond stretching frequency is approximately 100 cm^{-1} higher for high-mass terminal isotopes as compared to the low-mass terminal isotopes.

7-15. Acetylene has two C—H groups. It will have two C—H stretching frequencies, the in-phase and out-of-phase stretching modes. The in-phase (symmetric) stretch occurs at 3374 cm^{-1} and the out-of-phase stretch at 3333 cm^{-1}. Explain why the in-phase vibration is located at a higher frequency than the out-of-phase stretch in the case of acetylene.

7-16. The carbonyl stretching frequencies of a series of benzoyl derivatives are listed below:

	$\nu_{C=O}(CCl_4)$
x = 2	1677
3	1686
4	1687
5	1686

If we consider the $\bar\nu_{C=O}$ of acetone at 1715 cm^{-1} as a reference frequency, then identify the factors affecting $\bar\nu_{C=O}$ in the series of compounds listed.

7-17. Explain how mass inductive effects act to lower the carbonyl frequency, as well as how hyperconjugation effects act to raise the carbonyl frequency of aldehydes relative to ketones.

7-18. The carbonyl stretching frequency of aliphatic carboxylic acids in dilute solution is located near 1770 cm^{-1}. This frequency is much higher than the carbonyl frequency of these substances when measured neat (~1720 cm^{-1}). Also, it is considerably higher than the corresponding simple aliphatic ester (1745 cm^{-1}) value. Explain.

7-19. In a number of cases dipolar interactions control the frequency shifts found in carbonyl stretching vibrations. The following series lists wavenumber shifts in going from neat to dilute nonpolar solutions.

Table 7.10 Carbonyl Dipolar Interactions^a

Compound	$\Delta\bar\nu_{C=O}$ (cm^{-1})
Acetyl chloride	15
Phosgene	13
Acetone	21
Acetaldehyde	23
Dimethylformamide	50

^{a}Shift measured between dilute nonpolar solution and neat sample.

Explain the observed values.

7-20. The antisymmetric —CH$_2$—CO—O— stretching vibration in carboxylic acids is heavily mixed with the in-plane bending mode of the O—H group. In alcohols these two vibrations seldom show evidence of mechanical coupling. Explain.

7-21. Conjugation of the functional group in alkyl isocyanates has little impact on the antisymmetric —N=C=O stretching vibration located near 2270 cm;-^{1}. Explain.

7.22. In the infrared spectrum of 2-aminoanthraquinone (I) two carbonyl stretching frequencies are observed at 1673.5 and 1625 cm^{-1}. (The spectrum is obtained from a sample mixed with Nujol. Nujol is a complex mixture of saturated hydrocarbons. The Nujol mull is an alternative infrared sampling procedure to employing KBr. This method avoids placing the sample in a highly ionic matrix. The mulling agent, however, does introduce the saturated hydrocarbon group frequencies, which mask a number of regions of the spectrum.)

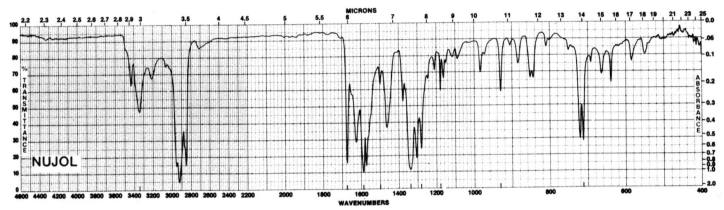

2-Aminoanthraquinone
I

1-Hydroxyanthraquinone
II

a. Assign carbonyl bands in the infrared spectrum to the carbonyl groups in structure I and explain your reasoning.

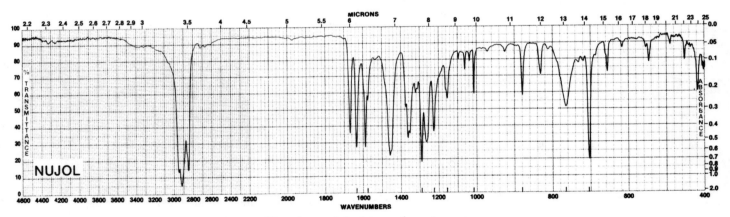

IR spectrum: 2-aminoanthraquinone.

b. The infrared spectrum of 1-hydroxyanthraquinone (II) also exhibits two carbonyl frequencies, which are located at 1675 and 1637 cm^{-1}. Assign the carbonyl groups to the related absorption bands. Explain your reasoning.

IR spectrum: 1-hydroxyanthraquinone.

c. The spectrum of 2-hydroxyanthraquinone exhibits a single carbonyl stretching frequency near 1673 cm^{-1}. Explain why a single carbonyl band would be expected in this system and why the vibration is located at 1673 cm^{-1}.

7-23. Suggest a possible structure for the hydrocarbon C_6H_{14}, which has the following infrared spectrum:
Is there more than one correct structure?

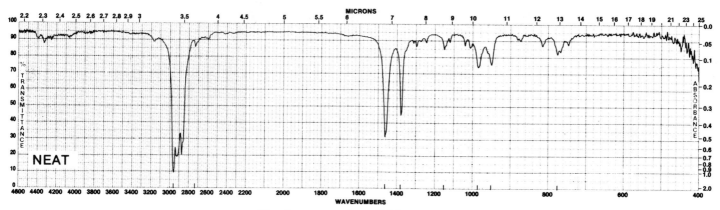

IR unknown spectrum C_6H_{14}.

7-24. The hydroxylamine I can be oxidized by MnO_2 to the amide oxohaemanthidine (II). In dilute solution the carbonyl absorption band of II occurs at 1702 cm^{-1}. Explain this observation.

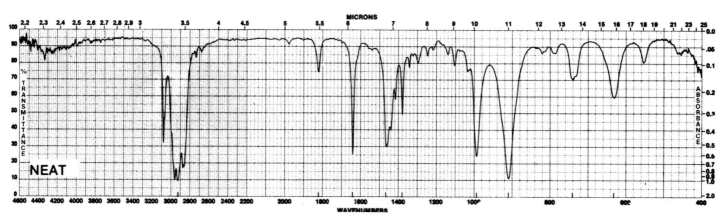

7-25. Identify the following olefinic hydrocarbons. All samples were obtained from distillation cuts in the C_6 boiling range.

IR unknown spectrum (a).

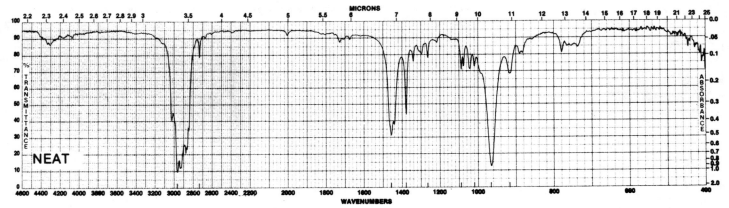

IR unknown spectrum (b).

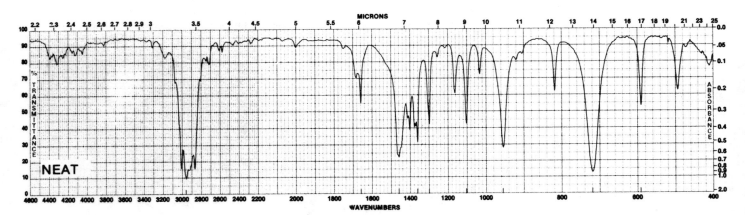

IR unknown spectrum (c).

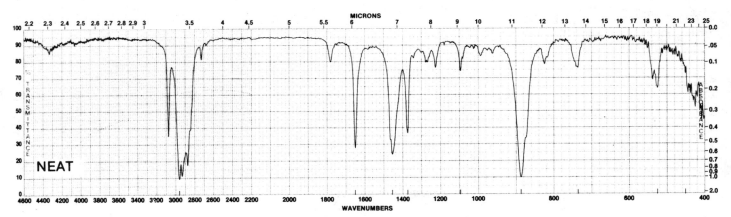

IR unknown spectrum (d).

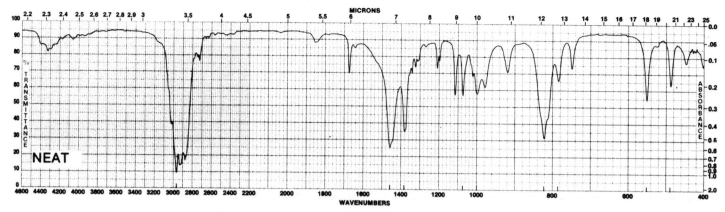

IR unknown spectrum (e).

7-26. The infrared spectra of the xylene isomers, and an additional aromatic hydro-carbon, are given below. Assign the spectra and suggest a potential structure for the remaining unknown substance.

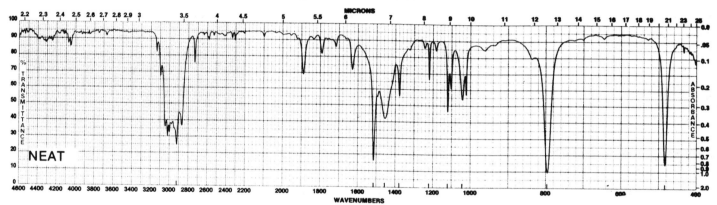

IR unknown spectrum (a).

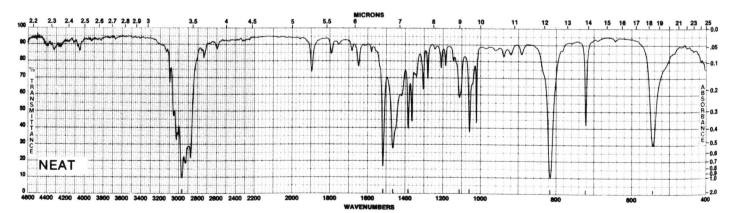

IR unknown spectrum (b).

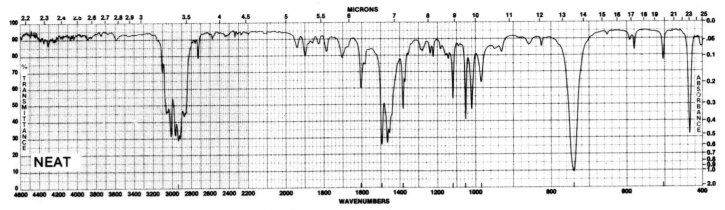

IR unknown spectrum (c).

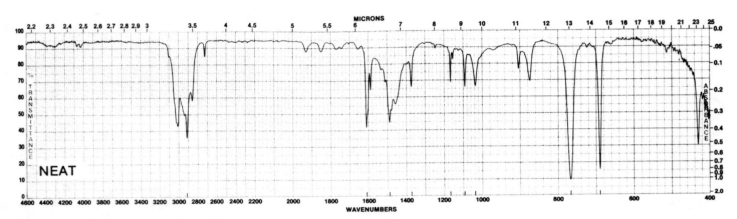

IR unknown spectrum (d).

7-27. The C—H stretching mode of chloroform, $CHCl_3$, which occurs at 3022 cm^{-1}, is one the rare exceptions to the 3000 cm^{-1} rule. What is the rule? Suggest an explanation for this exception.

III. Nuclear Magnetic Resonance

This section will provide a brief introduction to the use of nuclear magnetic resonance (nmr) spectroscopy for structure determination. A description of the theory underlying this method is given in Appendix C.

INTERPRETATION OF UNKNOWN SPECTRA

Sample Preparation

Spectra may be obtained on quantities of the order of 0.1 mmol in most cases. Some instruments and some compounds will yield good spectra on a few micromoles of sample. Spectra are usually obtained on liquids or solutions. The sample tube (generally 5-mm diameter thin-walled glass) is filled to a depth of about 2 cm, which corresponds to the active portion of the magnetic field. Filling the tube to this depth maximizes the number of protons in the active part of the instrument and hence the strength of the signal. Adding more material just wastes sample. Solvents may include aprotic liquids [CCl_4, $CDCl_3$, D_2O, $(CD_3)_2SO$, etc.] or protonated solvents [$CHCl_3$, $(CH_3)_2CO$, etc.] if the protonated solvent does not have a resonance in the area of interest. Another consideration in the choice of solvent is its ability to dissolve, but not otherwise interfere with, the compound of interest.

Since sampling details are highly dependent on the compound of interest and on the nmr facility available, further details will be provided by your instructor. An example of nmr sample preparation techniques is provided in Experiment 7.

Observed Chemical Shifts

Figure 7.35 summarizes the chemical shifts of protons in a large range of chemical environments. It is, however, a bit dangerous to use figures such as this one, without understanding some of the factors that underlie shielding and the chemical shift. To give some flavor of the factors that determine chemical shifts and the range of values observed, we will briefly examine chemical shifts in methyl groups and chemical shifts for protons on sp^2 carbons.

Methyl groups bonded to an sp^3 carbon generally have chemical shifts in the range 0.9–1.9 ppm as long as there is no more than one electron-withdrawing group attached to the carbon. The shifts generally increase as the strength of the electron-withdrawing group increases or as more electron-withdrawing groups are added. Groups that inductively withdraw electrons reduce the electron density near the methyl group protons. This results in less shielding and a downfield shift of the methyl resonance. This effect is clearly seen in the spectra of 1-nitropropane and 1,1-dibromoethane (Appendix C, p. 406 and 405, respectively). The chemical shifts for methyl groups bonded to sp^2 carbons fall in the range 1.6–2.7 ppm.

In the case of a proton bonded to an sp^2 carbon, the location of the proton relative to the π cloud plays an important role in determining the chemical shift. In unconjugated olefins the chemical shifts fall in the range 5–6 ppm. Where more than one proton is bonded to a double bond system, complex second-order spectra can be expected for low operating frequencies since the coupling constants are usually fairly large (Appendix C). In oximes, **RCH=NOH**, the

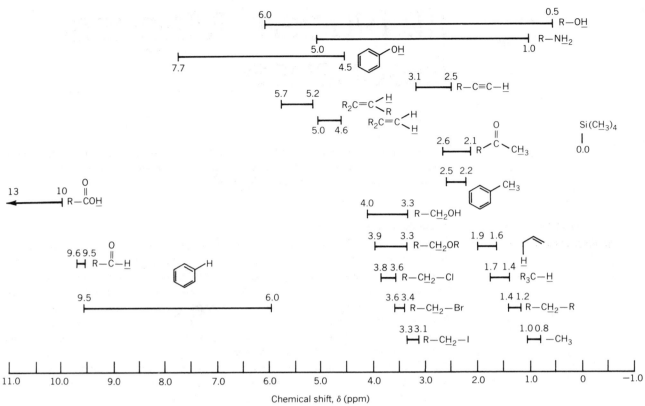

Fig. 7.35 *NMR H¹ chemical shifts. (From Zubrick, J. W. "The Organic Lab Survival Manual"; 1984. Courtesy of John Wiley & Sons, New York, NY.)*

increased electronegativity of the nitrogen increases the deshielding and the chemical shift falls in the range 6.5–7.5 ppm. This trend continues with aldehydes, which have chemical shifts in the range 9.5–10.5 ppm.

The chemical shift in an aromatic system is generally greater than that for olefins. For example, the chemical shift in benzene is 7.37 ppm, substantially greater than the 4.56 ppm for ethylene. Much of this difference results from the "ring current" effect and the orientation of the proton relative to the aromatic π electrons. If the ring substituents are not strongly electron withdrawing, the chemical shift for ring protons may be shifted to slightly higher fields. Furthermore, these substituents generate only small chemical shift differences among the ring protons. Thus, the 60-MHz spectra for toluene and *ortho*- and *para*-xylene appear to have a single resonance in the aromatic region at about 7.1 ppm. If, on the other hand, the substituents are electron withdrawing, the ring protons will be somewhat deshielded relative to benzene. Electron-withdrawing groups will also cause differentiation of the ring proton chemical shifts, and complex second-order spectra are likely to be seen in the range 6.5–8.5 ppm.

Coupling Constants

The sign of the coupling constant (usually symbolized as J) may be positive or negative. However, first-order spectra are not sensitive to the sign of the coupling constant. In second-order cases, the sign of J may be determined by a careful analysis of the spectrum.

Proton–proton coupling involves an effect extending over at least two bonds. The first case we shall consider is that of geminal coupling in which two protons are connected to the same atom, as in H—C—H. When the central atom is an sp^3 carbon atom, the coupling constant ranges from -4 to -20 Hz and is sensitive to the electronegativity and the valence orbitals of the other substituents.

Geminal coupling constants fall in the range -3 to $+2$ for protons bonded to the same sp^2 carbon atom.

Coupling extending over three bonds, as in H—C—C—H, is very sensitive to the angle of rotation about the central bond. In cases where the central bond is an sp^3 C—C bond, internal rotation can occur on a time scale that is very short relative to the nmr time scale. In these cases the effect of internal rotation is completely blurred as far as nmr is concerned and only an average coupling constant is observed. Vicinal coupling constants in ethyl groups are usually observed in the $+6.5$–8 Hz range.

When the central C—C bond is a double bond, internal rotation is restricted and separate coupling constants for cis and trans protons may be observed. Cis coupling constants fall in the range $+5$ to $+12$ Hz, whereas the trans coupling constants range from $+12$ to $+20$ Hz. As a result of these large coupling constants, second-order effects will be quite noticeable in substituted ethylenes.

Longer range coupling involving four or more bonds is common in allylic systems and in aromatic rings and other conjugated π bond systems. These coupling constants are generally smaller than the values considered above (less than 3 Hz).

The nmr user should be alert to the presence of other nuclei that may affect proton spectra. [19]F may be significant because its coupling constant can be quite large. [13]C is present only in about 1% in natural abundance but can have a large effect because it can be directly bonded to the proton. In cases where there are several equivalent carbon atoms (as in tertiary butyl groups), the probability of occurrence of a ^{13}C atom is increased and the coupling may be seen in the presence of weak resonances on either side of a stronger resonance of the ^{12}C species. [31]P is not common among compounds considered in this course, but the coupling constant for a proton bonded to phosphorus can be a few hundred hertz. [14]N has $I = 1$ and is influenced by quadrupole effects (Appendix C). Its presence when bonded to a proton is often reflected in a broadening of the proton resonance.

QUESTIONS   Several 60-MHz spectra are given below.[12] You should be able to account for at least one acceptable structure and for all of the observed resonances.

7.28. C_4H_8O. Spectrum (*a*).

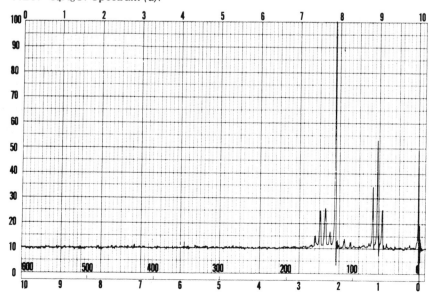

NMR unknown spectrum (a).

[12]From Pouchert, C. J. "The Aldrich Library of NMR Spectra"; Aldrich Chemical Company; Milwaukee, WI, 1983.

7.29. $C_3H_6O_2$. Spectra *(b)* and *(c)*. Two compounds with the same empirical formula.

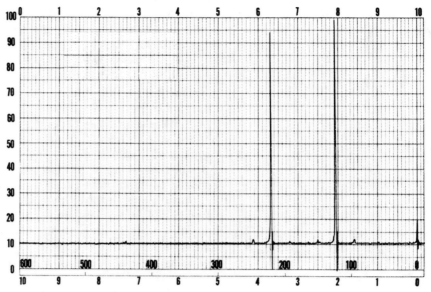

NMR unknown spectrum (b).

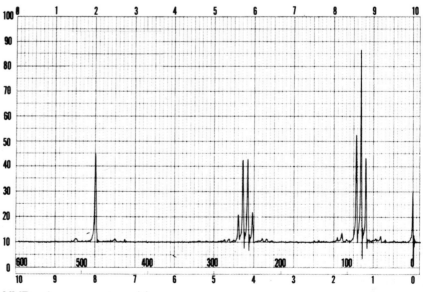

NMR unknown spectrum (c).

7.30. C$_4$H$_8$O again, spectrum *(d)*. Also give some thought to the weak resonances at 0.5 and 1.8 ppm.

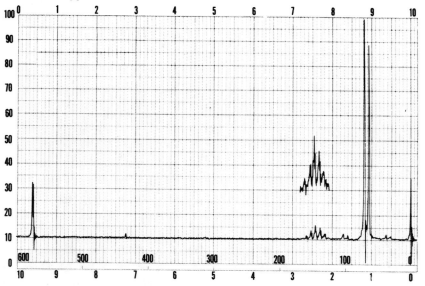

NMR unknown spectrum (d).

7.31. C$_7$H$_7$Cl. Spectrum *(e)*.

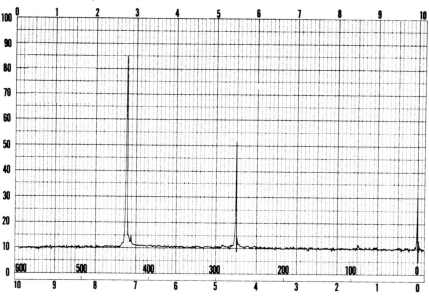

NMR unknown spectrum (e).

7.32. $C_8H_{10}O$. Spectrum *(f)*.

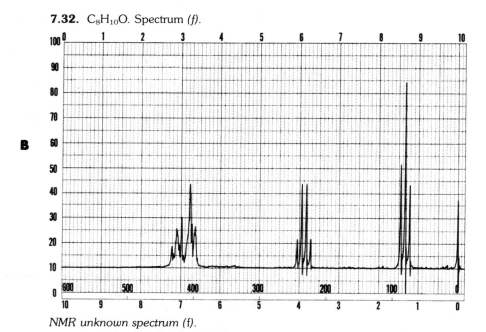

NMR unknown spectrum (f).

Appendix A

Estimation of Concentrations in Laboratory Air[1]

A simple method is presented for estimating concentrations of vapors in the laboratory. This approach is based on models often used to estimate concentrations in enclosed spaces [see for example Turk, 1963; Wadden and Scheff, 1983; Meyer (1983), and Esmen (1978)]. This model is intended for guidance only. It should not be used as a substitute for exposures determined with personal monitors [see for example National Research Council (1981) and Freifeld (1982)], especially for those cases in which the uncertainties are great and the estimates approach the threshold limit values (TLV).

The modeled concentration factors into two components. One of these, the emission factor, represents the vapors emitted by each student for a given experiment. This factor should be transferable from one institution to another, although it may have to be modified for special circumstances. The second factor, the dilution factor, describes the amount of dilution to be expected in a given laboratory. This factor is unique to each laboratory, but should be transferable from one experiment to another.

A suggested emission factor is given in the experiment section for any potentially hazardous materials involved. Where possible, emission estimates are based on measurements of similar processes. For many cases measurements are not available, and we have taken the conservative approach of assuming that the entire quantity is emitted. In some cases there will be a great deal of uncertainty in this emission factor, and some institutions may wish to develop their own values. For many laboratories the dilution factor will be the same from day to day, and we suggest that the instructor work out this factor for the benefit of each laboratory section.

VENTILATION MODEL

The concentration of a pollutant in an indoor space is often expressed by the following equation for the case in which the concentration is uniform throughout the room.

$$\frac{V dc}{dt} = G - Qc \tag{A.1}$$

In this equation the term on the left expresses the overall rate of change of the total amount of material in the room. V is the volume of the room (in cubic meters) and c is the concentration (in milligrams per cubic meter). The first term on the right, G, is the rate at which material is added to the room (in milligrams

---

[1]Much of the material in this appendix has been previously published: Butcher, S.S.; Pike, R.M.; Mayo, D.W.; Hebert, S.M. *J. Chem. Educ.* **1985**, *62*, A-238.

per minute). The second term on the right is the rate at which material is removed. Q is the ventilation rate for the room (in cubic meters per minute). Equation (A.1) will not be used in this form, but is introduced to give a sense of the mass balance principle, which is also the basis of refined models.

Using Eq. (A.1) as a springboard, three other important factors will be added. The first, and most important, is an empirical factor that describes the degree of incomplete mixing. This is called a mixing factor and is symbolized here by k. The second factor accounts for the possibility that the pollutant may also be present at a concentration c_0 in the air brought into the building. The third factor allows for recycling of air back into the laboratory at a rate Q_r, with a loss of pollutant in the building air-handling system (by filters or sorbents) described by an efficiency factor E.

The mass balance equation now takes the form given in Eq. (A.2) (see Esmen, 1978).

$$\frac{V dc}{dt} = G - kQc - kQ_rEc + kQc_0 \qquad (A.2)$$

This equation is included for completeness here, but it will take a much simpler form for most laboratory situations. We will make use of two approximations here that may not apply in all circumstances. Where these approximations do not apply, it will be necessary to have additional information on the ventilation equipment and on ambient air quality in order to use Eq. (A.2). The first approximation we will make is that there is no recycling of laboratory air. That is, Q_r and thus the third term on the right side of Eq. (A.2), is zero. Second, we assume that the concentration of the pollutant in the outdoor air is zero and that the last term on the right-hand side of Eq. (A.2) is also zero.

The next step in the evaluation of the concentration is to determine the steady-state concentration. This is determined by setting $dc/dt = 0$ and solving for c to obtain Eq. (A.3).

$$c = \frac{G}{kQ} \qquad (A.3)$$

It should be noted that the concentration in the laboratory will probably seldom reach a steady state. That is, emissions may occur for a short period of time and then cease. It may be shown, however, that in many cases Eq. (A.3) does well represent the *average concentration* over the laboratory period if G represents the emission rate averaged over the entire laboratory period.

We want to make one more modification of this equation. The average emission rate over the lab period, G, will be expressed by nm/t_0. The number of students in the lab section is given by n; m is the milligrams of a particular substance emitted by each student; and t_0 is the length of the lab period in minutes. The steady-state equation may then be written

$$c = \frac{m}{\dfrac{kQt_0}{n}} \qquad (A.4)$$

The numerator on the right-hand side is called the emission factor; it will have the units milligrams per student. The denominator, kQt_0/n, represents the volume of air available to dilute the emissions of each student. We will refer to this quantity as the dilution factor. It will have the units cubic meters per student.

The effect of multiple exposures may be evaluated by considering the equivalent exposure, E.

$$E = \frac{c_1}{L_1} + \frac{c_2}{L_2} + \dots \frac{c_n}{L_n} \qquad (A.5)$$

c_k is the concentration of the kth vapor and L_k is the TLV of that vapor. For multiple exposures E should be less than one. This assumption does not consider any effects of synergism from different vapors, and it does not allow for cases where different vapors have completely independent effects. Equations (A.4) and (A.5) may be combined to obtain an expression that relates the equivalent exposure to the modeled concentration.

$$\frac{kQt_0}{n} > \frac{m_1}{L_1} + \frac{m_2}{L_2} + \cdots \frac{m_n}{L_n} \qquad (A.6)$$

The terms in expression (A.6) are all volumes. The left-hand side represents the total volume of ventilation *available* to dilute the emissions of each student. The individual terms on the right-hand side represent the volumes of air *required* to reduce the concentration resulting from the emission of a mass m_k to the TLV concentration.

Values for m_k and L_k are given for each experiment. (kQt_0/n) needs to be obtained only once for each laboratory section. The determination of the dilution factor and other parameters are discussed below.

Mixing Factor

If a receptor is very close to the source, a small fraction of the total amount of ventilation air will be available to dilute the emissions before they reach the receptor and k will be less than one. For receptors on the opposite side of the room from the source, a large amount of air will be able to dilute the emissions and the mixing factors will be greater than one.

For a given student receptor in a real laboratory, k will be less than one for their own emissions and greater than one for emissions on the opposite side of the lab. Lab instructors and teaching assistants move about a great deal in the lab and to the extent that they are "where the action is," they may often be close to sources of emission. For present purposes, we assume a value of 0.3 for k. This value is close to a worst case value for a variety of studies (Ishizu, 1980, Drivas et al., 1972) for cases in which the receptor is close to the source. This value may lead to a conservative estimate of the concentration expected for the student (one that errs on the side of the protecting health), but we feel that this factor is justified when considering exposures to lab instructors who move about in the lab, will be exposed to a variety of chemicals, and may also be exposed for more than one laboratory session each week. It is likely that the value of k may be refined as more experience is gained about actual exposures from known emissions.

Emission Factor

The amount of material emitted by each student will be described for each individual experiment. (See Experimental section, Chapters 5 and 6.) In many cases experiments will be modified or only rough estimates of the m values will be available. Also, individual institutions may wish to measure emission factors for specific cases involving their own variations of these experiments. The emission factor will depend on the apparatus, procedure followed, and volatility of the substance used.

Dilution Factor, kQt_0/n

Of the terms in this factor, k has been discussed above (rather dogmatically); and t_0 and n may be easily evaluated for each lab section. Q may be estimated by a number of methods; three will be discussed here.

Direct Measurement of Air Flow

Where either the air inlet or the exhaust for the laboratory occurs at well-defined points, the volume rate of air flow may be obtained as the product of an air

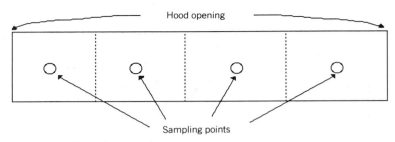

Fig. A.1 *Sampling points.*

velocity and an area. For the general case the reader is referred to the Committee on Industrial Ventilation (1980). In many laboratories, all the air exhausted is taken out through the hoods. Flow rates through hood faces may be conveniently measured with any of several air-velocity instruments. For these purposes an accuracy of 20% is adequate. Velocities may not be the same everywhere, and it is important to measure at several points. The hood opening may be set to a height that will give an air velocity in the readable range of the instrument being used. Then if the velocity readings represent equal areas, the area-averaged velocity is simply the numerical average of these readings. As an example, if one wishes to take four readings across the face of the hood, the hood opening is divided into four equal areas and a velocity reading is taken at the center of each of these areas. Figure A.1 shows the locations of the sampling points for the case of four points and a narrow hood opening.

For a wider hood opening one might use eight readings as two rows of four readings. The number of points required depends on the degree to which the velocity varies in the hood opening.

The volume flow through the hood is then the product of the area-averaged velocity and the area of the hood opening. Many velocity-measuring instruments and equipment design specifications are expressed in feet per minute or cubic feet per minute. It may be convenient to get the flow rate in cubic feet per minute and then make the conversion to cubic meters per minute, since the engineering staff responsible for hood maintenance is more likely to be familiar with cubic feet notation (1 cubic foot = 0.0283 cubic meters). After the volume flow rate for each hood or exhaust port has been measured, the sum of these rates is the room ventilation rate, Q.

Measurement of Clearance Time

The concept of clearance time may be examined by going back to Eq. (A.1). A tracer is injected into the room at $t = 0$ and the concentration of the tracer is measured as a function of time as the ventilation equipment clears the room. The concentration of the tracer in the room will be given by

$$c = c_0 \exp\left(-Qt/V\right) = c_0 \exp\left(-t/t_v\right) \tag{A.7}$$

In this case c_0 is the initial concentration of tracer (at $t = 0$); t is the time at which the measurements of c are taken; and Q and V retain their previous meanings. The clearance time, or well-mixed ventilation time, is also expressed by t_v ($= V/Q$). The reciprocal of t_v is the *air exchange rate* (usually expressed in ACH, or air changes per hour).

One may obtain a value for t_v by plotting $\ln(c)$ as a function of t. The slope of this curve is then $-1/t_v$. It is important for present purposes that the room remain well-mixed during these measurements. Large ventilating fans have worked well in our studies. This does *not* mean that the fans must be used during the laboratory periods. The absence of mixing during the laboratory period is accounted for by the mixing factor, k.

Bayer (1982) has measured t_v by using a nontoxic hydrocarbon tracer and determining the concentration by direct injection of air samples into a gas chromatograph. We have used an integrating nephelometer to measure the light scattering from MgO smoke produced by a flash powder as a function of time.

The method used should have the capability of acquiring at least three to four measurements of concentration in time t_v. Poorly ventilated laboratories will have a t_v in excess of 20 minutes, whereas t_v will be less than 10 minutes for well-ventilated spaces. Once t_v is available, Q is obtained as V/t_v in Eq. (A.7). The reason for mentioning t_v and the air-exchange rate separately is that these parameters provide a measure of the performance of ventilation equipment that may be useful for comparative purposes.

Ventilation Rate from Equipment Design

In some cases, physical plant personnel are able to provide an estimate of the ventilation rate from knowledge of the design characteristics of the air-handling equipment. Such an estimate may provide a useful approximate value for Q, but care should be exercised in taking this number very far. These flow rates may depend on the accessibility of makeup air, the conditions of filters, and how clean blower vanes are. Of perhaps greater use may be a comparison of the design ventilation rate with the value measured by either of the methods described above. A large disparity may indicate a major problem with the ventilation equipment.

Data Sheet for Ventilation Parameters

It is not necessary to fill in all of the blanks, but enough of them should be filled in to enable a calculation of the *bottom line*.

Laboratory section _____

Room number _____

Number of students, n _____

Length of laboratory session, t_0 _____ minutes

Volume of room, V _____ cubic meters

Ventilation rate, Q _____ cubic meters per minute

Air exchange rate, Q/V or $1/t_v$ _____ per hour

Notes on determination _____

_____

Bottom Line

Dilution factor $k = 0.3$, kQt_0/n _____ cubic meters per student

REFERENCES Wadden, R. A.; Scheff, P. A. ''Indoor Pollution''; Wiley Interscience: New York, 1983.
Meyer, B. ''Indoor Air Quality''; Addison-Wesley: Reading, MA, 1983.
National Research Council ''Indoor Pollutants''; National Academy Press: Washington, 1981.
Committee on Industrial Ventilation ''Industrial Ventilation''; American Council of Governmental Industrial Hygienists: Lansing, MI, 1980.
Esmen, N. A. *Environ. Sci. Technol.* **1978,** *12(3),* 337.
Turk, A. *ASHRAE J.* **1963,** *5(10),* 55.

Bayer, R. *J. Chem. Educ.* **1982**, *59(12)*, A385.
Freifeld, M. *J. Chem. Educ.* **1982**, *59(11)*, A351.
Ishizu, Y. *Environ. Sci. Technol.* **1980**, *14(10)*, 1254.
Drivas, P. J.; Simmonds, P. G.; Shair, F. H. *Environ. Sci. Technol.* **1972**, *6(7)*, 609.

Appendix B

Introduction to the Theory of Infrared Spectroscopy
I. Interpretation of Infrared Spectra

INTRODUCTION TO THEORY OF EFFECT

The wavelike character of electromagnetic radiation can be expressed in terms of the velocity v, frequency ν, and wavelength λ of sinusoidally oscillating electric and magnetic vectors traveling through space (Fig. B.1). Frequency is defined as the number of waves passing a reference point per unit time, usually expressed as cycles per second, $\sec^{-1}$, or Hertz, Hz. The velocity of the wave, therefore, equals the product of frequency and wavelength.

$$v = \nu\lambda$$

If the wavelength (the distance between each wave maxima or alternate nodes) is measured in centimeters, v is expressed in centimeters per second (cm/sec). For radiation traveling in a vacuum, v becomes a constant c ($c \sim 3 \times 10^{10}$ cm/sec) for all wavelengths. When electromagnetic radiation traverses other media, however, the velocity changes. The ratio of the speed in vacuum c, to the matrix velocity v, is termed **refractive index** n, of the material

$$n = c/v$$

Since n is frequency dependent, the frequency at which the refractive index is measured must be specified. Frequency, however, has been shown to be independent of the medium and, therefore, remains constant. Wavelength thus varies inversely with n.

$$\lambda = c/n\nu$$

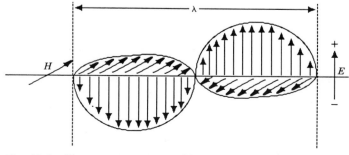

Fig. B.1 *Electromagnetic wave. H magnetic field, E electric field, λ wavelength.*

Since the velocity of electromagnetic radiation in a vacuum is normally greater than that in any other medium, n will generally be greater than one at all frequencies. Thus, the wavelength must become shorter for a particular frequency when measured in any matrix.

Frequency can be considered to be a more fundamental property of radiation because it is independent of the medium. This property also requires that the energy E associated with the radiation be matrix independent because E is directly proportional to frequency by

$$E = h\nu$$

where E equals the energy of a photon, which is related to frequency ν by Planck's constant (6.6×10^{-27} erg-sec or 6.6×10^{-34} joule-sec).

The vibrational states present in molecules can be excited by absorption of photons. The nuclear masses and bond force constants determine the separation of these states and, therefore, the energies of the photons involved in the absorption process. The corresponding radiation frequencies fall predominantly in the infrared region (10^{14}–10^{12} Hz) of the electromagnetic spectrum.

The infrared spectrum currently is measured in **wavenumbers** $\bar{\nu}$, which are units proportional to frequency and energy. The wavenumber is defined as

$$\bar{\nu} = \nu/c = E/hc$$

and as

$$\nu = c/n\lambda \qquad \text{then in air} \qquad \bar{\nu} = \sim 1/\lambda$$

The wavenumber, as expressed in units of cm^{-1} (the number of waves per centimeter), offers several advantages.

1. Wavenumbers are directly proportional to frequency, and wavenumbers are expressed in much more convenient numbers (in this region of the spectrum), 5000 to 500 cm^{-1}.
2. As shown above, wavenumbers are easily converted to wavelength values. The reciprocal of $\bar{\nu}$ and conversion of centimeters to wavelength units is all that is required (this is particularly handy since a large amount of early infrared data is recorded linear in wavelength). The wavelength units employed in most of these spectra are microns μ. The micron unit has now been replaced by units expressed in meters, the micrometer, μm ($1 \; \mu$m $= 1 \times 10^{-6}$ m).
3. Because the wavenumber is directly proportional to frequency and energy, the use of wavenumbers allows spectra to be displayed linear in energy. This is a distinct aid in sorting out related vibrational transitions.

Note. *It should also be pointed out the cm^{-1} is not a unit of frequency. Wavenumbers are only proportional to frequency. Thus, it is not correct to refer to a vibrational absorption band as having a frequency of 3000 cm^{-1} or that the vibration of the C—H bond possesses a frequency of 3000 cm^{-1}. The C—H oscillator, however, can be said to absorb radiation with an energy of 3000 cm^{-1}, or that the C—H bond vibrates with a frequency of 9×10^{13} Hz.*

MOLECULAR ENERGY   The total molecular energy W may be expressed as the sum of the molecular translational, rotational, vibrational, and electronic energies:

$$W_{\mathrm{mol}} = W_{\mathrm{trans}} + W_{\mathrm{rot}} + W_{\mathrm{vib}} + W_{\mathrm{elec}}$$

In this approximation, W_{vib} is assumed to be independent of the other types of molecular energy. Translational and rotational motion involve much smaller energies, having little influence on the spectra under observation. W_{elec} is the energy of the electrons. The energies of these latter transitions are very much larger than vibrational spacings, and their energy changes fall outside the infrared. As a result the infrared region is active mainly to vibrational energy changes of ground electronic state molecules.

MOLECULAR VIBRATIONS

Molecules can be characterized as being in constant vibrational motion. If we are to describe this motion for nuclei of a polyatomic system, we can utilize the Cartesian coordinates x_m, y_m, z_m for nucleus m referred to a fixed coordinate system. Then for n nuclei, we would generate $3n$ coordinates ($3n$ **degrees of freedom**) to describe the motion of all the atoms. Three of these coordinates, however, may be used to locate the center of mass of the system in space. These three coordinates define the translation of the entire system through space. Because translational energies have a small impact on vibrational spectra, the three coordinates of the center of mass can be dropped from the total required to determine the vibrational degrees of freedom. Therefore, $3n - 3$ coordinates are sufficient to determine the positions of the n nuclei with respect to the center of mass. However, the molecular system is still free to rotate about the center of mass. For nonlinear molecules three additional coordinates are required to fully describe rotational motion about the center of mass. For linear molecules only two coordinates are necessary to define rotation, as all the nuclei lie along one of the principal axes and are considered to be point groups. Thus, for nonlinear molecules $3n - 6$ coordinates fully define the vibrational motion of the nuclei. These coordinates are often referred to as the **vibrational degrees of freedom**. In linear systems one rotational degree of freedom can be considered to have been transformed into a vibrational degree of freedom ($3n - 5$).

The number of vibrational degrees of freedom is directly related to the number of fundamental vibrational frequencies possessed by the molecular system. These fundamental frequencies are often referred to as the **"normal modes"** of vibration.

To get a feel for the function of the normal modes in the vibrational pattern of a molecular system, let us consider a very simple arrangement of a single nucleus vibrating in two dimensions.[1] The nucleus of mass m is held by a rigid but elastic rectangular bar or rod. The mass m can vibrate only in the plane perpendicular to the bar (Fig. B.2). If the nucleus is displaced along the x axis and then released, the system will oscillate with simple harmonic motion at a frequency given by

$$\nu_x = \frac{1}{2\pi} \sqrt{\frac{k_x}{m}}$$

where k_x = force constant of the bar in the x direction where the restoring force $F = -k_x X$ for displacement = X.

If displacement and release are carried out in the y direction, a similar type of oscillation will occur with a frequency given by

$$\nu_y = \frac{1}{2\pi} \sqrt{\frac{k_y}{m}}$$

where k_y = the force constant in the y direction.

The frequency ν_x that results from displacement in the x direction is different from the natural frequency ν_y, which results from displacement in the y direction,

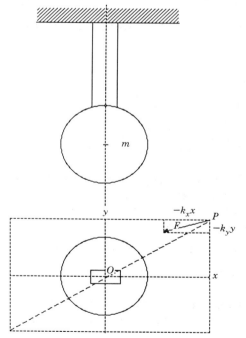

Fig. B.2 *Harmonic vibration in two dimensions.*

[1]Herzberg, G. "Molecular Spectra and Molecular Structure"; D. van Nostrand: New York, 1945; Vol. 2, p. 62.

because the rectangular bar will possess different force constants k_x and k_y. If the rectangular bar is replaced with a bar of square cross section, then the force constants that result from displacement in the x and y directions are equal. The frequencies of x and y motion also will be equal. In this case the bar of rectangular dimensions can be referred to as having **degenerated** to a bar of square dimensions.

We now consider the displacement and release of the nucleus in a direction not along a principal axis. For example, the corner position P. Now on release, the motion performed by the nucleus is no longer simple harmonic. The restoring force F will have components $-k_x X$ and $-k_y Y$, which are unequal and not directed toward the origin.

The complex motion of the nucleus in the x-y plane, however, will still contain components that are simple harmonic in nature. The motion of the nucleus can be represented as the sum of these "normal modes" of vibration, which are perpendicular to each other. Thus, the position of the nucleus at any point in time following release can be expressed by the two coordinates:

$$x = x_0 \cos 2\pi \nu_x t$$

$$y = y_0 \cos 2\pi \nu_y t$$

where x_0 and y_0 are the coordinates of the initial position P and t is the time lapse from release.

The complicated pattern of motion performed by the nucleus on release from position P is termed Lissajous motion. This type of motion is the superposition of two simple harmonic motions of differing frequency that are normal to each other. These are termed the "normal modes" or fundamental frequencies of the Lissajous motion of the nucleus of mass m. The x and y coordinates, thus, become the "normal coordinates."

In the case of a diatomic molecule the frequency (in wavenumbers) is given by

$$\bar{\nu} = \frac{1}{2\pi c} \sqrt{\frac{k}{\mu}}$$

where k is the force constant and μ the reduced mass.

$$\frac{1}{\mu} = \frac{1}{m_1} + \frac{1}{m_2} \quad \text{or } \mu = \frac{m_1 m_2}{(m_1 + m_2)}$$

QUANTIZED VIBRATIONAL ENERGY   Nature has been kind in distributing vibrational energy in molecules! The vibrational states associated with a particular normal mode are not influenced, to a first approximation, by the energies of adjacent states. The portion of W_{vib} contributed by a particular normal mode ν_1, is given by

$$W_{vib} = (v_i + \tfrac{1}{2})h\nu_i$$

v_i is the vibrational quantum number for the normal mode and takes the values 0, 1, 2. . . .

Each normal mode will possess a similar energy–quantum number relationship and the total vibrational energy scheme can be obtained by summing over all $3n - 6$ fundamental vibrations

$$W_{vib} = \sum_{n=1}^{3n-6} (v_i + \tfrac{1}{2})h\nu_i$$

The characteristic energy-level pattern for a normal mode as determined by the quantum relationship dictates that the level spacings will be equal with a value of $h\nu_i$. In the upper states, however, the potential energy curve begins to

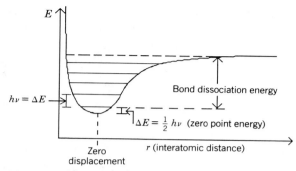

$h\nu = \Delta E$

Bond dissociation energy

$\Delta E = \frac{1}{2} h\nu$ (zero point energy)

Zero
displacement

r (interatomic distance)

Fig. B.3 *Vibrational energy level diagram.*

depart from the harmonic values and the vibration becomes anharmonic. In general, anharmonicity results in lower energy transitions or a contraction in the level spacing (Fig. B.3). In rare cases the potential energy well develops steeper sides (quartic terms become important), and the spacing actually becomes greater at higher levels. We will see a few of the more well-known departures of this type.

In addition to the equal spacing of the energy levels associated with each normal mode, the quantization of the vibrational energy requires that the lowest or zero vibrational (v_0) level does not occur at zero energy, but at $\frac{1}{2}h\nu$. Thus, the molecule retains, even at absolute zero, some small amount of vibrational energy. This is termed the "zero point" energy, and its origin lies with the Uncertainty Principle.

SELECTION RULES The range of vibrational energy level transitions associated with the fundamental frequencies runs from somewhat under 5000 to approximately 100 cm^{-1}. Thus, the study of the absorption spectra of the normal modes appears in the infrared region of the spectrum. While most molecules possess a large set of normal modes, the equal spacing of the vibrational levels and the operation of molecular selection rules greatly simplify what would otherwise be a very complex absorption pattern.

For infrared spectra the selection rules define the changes in the vibrational and rotational quantum numbers. The changes for the $3n - 6$ vibrational quantum numbers are given below:

1. Only one quantum number can change during a transition.
2. The change Δv is restricted to $+1$, -1.
3. For certain vibrations, Δv must always be zero.

In a very large percentage of cases, an absorbed photon excites only a single normal mode (this rule also holds for emission spectra). Thus, each frequency observed in the spectrum corresponds to a normal mode present in the molecule. The maximum number of frequencies observed corresponds directly to the $3n - 6$ (or $3n - 5$ for linear molecules) vibrational degrees of freedom.

In normal modes where $\Delta v = 0$, an incomplete set of frequencies will be observed. The symmetry elements present in the molecule largely determine whether $\Delta v = 0$ for a particular normal mode. This results from the further requirement that there be a change in molecular dipole moment during the vibration for the absorption of a photon to occur.

We can see why a variation in the magnitude of the dipole moment is essential to the absorption process by considering hydrogen and hydrogen chloride molecules placed between condenser plates (Fig. B.4).

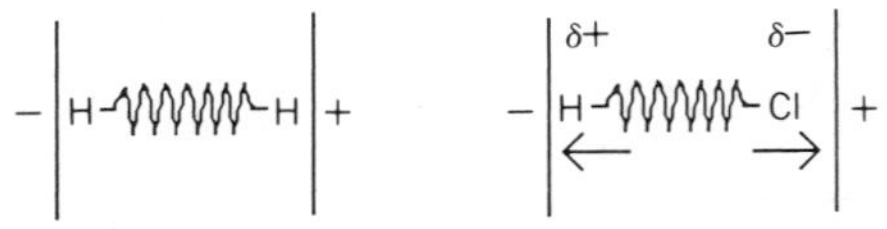

Fig. B.4 *H_2 and HCl oscillators.*

In the case of HCl, a permanent dipole moment exists along the molecular axis with the negative pole closer to the chlorine atom and the positive pole closer to the hydrogen atom. When placed between condenser plates, as shown in the figure, the molecule will experience attractive forces at both ends that will exert a stretching action. If the charge on the condenser plates is quickly reversed, the molecule will then experience replusive forces at each end and be compressed. If it were possible to alternate the charge on the plates fast enough to match the natural frequency of the HCl molecule, the system would resonate. In the resonance or "tuned" condition, the oscillator will absorb energy from the condenser and expand its vibrational displacements. (If enough energy were absorbed, the molecule would dissociate.) The frequency remains constant, but the amplitude of the vibration will increase (much the same as the input of periodic energy with the correct phase into a child's swing increases the amplitude of the swing, but leaves the frequency constant). Within the band of infrared radiation (5000 to 100 cm^{-1}), oscillating electric fields of the radiation will act on molecules in a fashion similar to alternating condenser fields. The frequencies in this spectral region correspond to the natural vibrational frequencies present in the molecular systems. Thus, at the particular radiation frequency that matches the vibrational frequency of the HCl molecule, resonance will occur and the photon of corresponding energy will be absorbed as the molecule moves to the next higher vibrational state. In the case of the hydrogen molecule, there is no vibrating electric dipole present because of the symmetry of the system. Thus, no interaction with the oscillating electric vector of the radiation at the natural frequency can occur (this would correspond to no interaction with the condenser plate fields). In this case the normal mode for the hydrogen molecule is not observed in the infrared spectrum, and the selection rule $\Delta v = 0$ applies.

Let us examine two simple examples to illustrate the above discussion.

The Case of Water

Water is a nonlinear three-atom molecule. It has, therefore, three normal modes of vibration or fundamental frequencies ($3n - 6$, where $n = 3$). The displacements of the normal modes can be derived in much the same fashion as the two-dimensional case and are as shown in Fig. B.5.

All three vibrations are active in the infrared, and three absorption bands are observed. The arrows represent the relative atomic displacements involved in one phase of the vibration. The atoms all move in-phase in simple harmonic motion in each fundamental mode. The high-frequency vibration ν_3 at 3756 cm^{-1} involves predominantly hydrogen motion with one bond contracting while the other is stretching. This vibration is designated as the "antisymmetric stretching" mode. (Modes of this type are often incorrectly referred to as "asymmetric" vibrations. The vibration is not asymmetric, without symmetry, but a vibration of antisymmetric character, opposite symmetry or opposed symmetry.) The other stretching vibration designated ν_1 at 3652 cm^{-1} involves in-phase and identical displacements of the two O—H bonds. It is termed the "symmetric stretching" mode. Finally, the low-frequency mode ν_2 at 1595 cm^{-1} corresponds to a bending of the molecule about the H—O—H bond angle. It occurs at lower frequencies than the two stretching modes since it takes less energy to bend a bond than it does to stretch one. Thus, the force constant k_{bend} is considerably

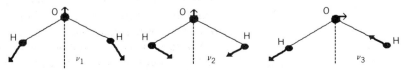

Fig. B.5 *Normal modes of water.*

smaller than $k_{stretch}$. This bending vibration is often termed the scissoring vibration because the symmetric bending motion involved is similar to the action of scissors. The complex molecular vibrational pattern of the water molecule can be resolved into three simple harmonic components or normal modes that correspond directly to three absorption bands in the infrared spectrum. The atomic displacements of these normal modes can be used to characterize the type of fundamental vibration giving rise to the infrared absorption. In the water molecule we have two O—H stretching vibrations and one O—H bending mode. These modes represent the three vibrational degrees of freedom present in the water molecule.

The Case of Carbon Dioxide

Carbon dioxide is a three-atom linear molecule. It will possess four vibrational degrees of freedom (one more than water). The displacements of the normal modes are given in Fig. B.6.

In the case of carbon dioxide only two absorption bands are observed in the infrared spectrum even though the molecule has four vibrational degrees of freedom. The high-frequency antisymmetric stretching mode ν_3 occurs at 2350 cm^{-1}. The form of this mode is close to that of the antisymmetric stretching vibration found in water. The symmetric stretching vibration ν_1 of carbon dioxide is similar to its counterpart in water; however, this vibration does not give rise to a change of dipole moment during the vibration. The arrangement of the atoms in the carbon dioxide molecule places a center of symmetry on the carbon atom. This symmetry element remains intact during the symmetric stretching vibration, and as a consequence no change in the dipole moment occurs. Thus, $\Delta v = 0$ and no absorption occurs at this frequency in the infrared. The normal mode still exists in the molecule, but it is infrared inactive. This fundamental has been located near 1334 cm^{-1} in the Raman effect (The Raman effect is another spectroscopic technique used to observe vibrational energy-level transitions. It does not depend on a change in dipole moment). There are two bending frequencies present in carbon dioxide, ν_{2a} and ν_{2b}. These two modes are identical except that one is rotated by 90 degrees with respect to the other. The bending motion does produce a change in dipole moment, and as a result it will be infrared active. Since the two modes will have identical frequencies, however, only one absorption band will be observed at 667 cm^{-1}. Carbon dioxide can bend in two mutually perpendicular planes, and therefore two bending vibrations are required to fully characterize the vibrational motion of the molecule. The symmetry of the system dictates that the modes will be identical. Vibrations possessing these characteristics are termed **degenerate vibrations** (remember the *degeneration* of the rectangular bar to a square that gave rise to two mutually perpendicular and identical modes). Thus, although carbon dioxide possesses four vibrational degrees of freedom, the observation of two absorption bands in the infrared spectrum can be satisfactorily explained. Indeed, if we examine the bending vibration present in water, we see that water can bend only in a single plane, and "bending" the molecule out of that plane results in rotation. It is this rotational degree of freedom of water that is translated into a vibrational degree of freedom in carbon dioxide as the nonlinear three-atom system is converted into a linear molecule.

If we place the carbon dioxide molecule between the plates of a condenser (Fig. B.7), we can see why some of the CO_2 vibrations are infrared inactive, and other vibrations are active. As carbon dioxide does not possess a permanent dipole moment, it is important to note that the antisymmetric stretching mode and the degenerate bending vibration will develop dipole moments during these vibrations as a result of the nuclear displacements involved. These dipole moments obviously undergo changes during the different phases of the vibrations,

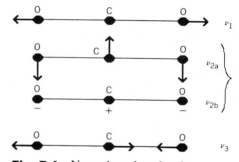

Fig. B.6   *Normal modes of carbon dioxide.*

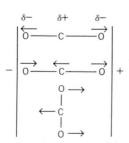

Fig. B.7   *Carbon dioxide oscillator.*

and therefore these modes fit the requirements of the vibrational selection rules for infrared activity. In the symmetric stretching vibration, no dipole moment is developed during the vibration, and we have a case similar to that of the hydrogen molecule. Thus, no absorption band is observed in the infrared spectrum of CO_2 that corresponds to the symmetric stretching mode.

COUPLED OSCILLATORS

Water and carbon dioxide are valuable examples of simple systems in which mechanical coupling between two oscillators is amply demonstrated. If we consider water to be constructed of two O—H oscillators, the individual diatomic systems would be expected to have identical frequencies. When welded together, however, the vibrations of one O—H oscillator interfere with the vibrations of the other O—H oscillator. The coupled oscillators generate two new vibrations, one at higher frequencies and one at lower frequencies (much the same as the resonance that develops between two coupled identical pendulums can be considered to involve two beat frequencies, one higher and one lower than the natural frequency of the pendulum). The coupling interaction is frequency dependent. The closer the frequencies of the two oscillators, the stronger the interaction. In the case of identical frequencies and a direct mechanical connection, the coupling effect will be maximized. Under the conditions of strong interaction, the form of the new vibrations can be quite different from that of the isolated oscillators. Coupling is also angle dependent. Oscillators normal to each other couple poorly, whereas colinear oscillators will undergo maximum coupling. [Note the wavenumber separation in water (bent system) of $104 \, cm^{-1}$, as compared to carbon dioxide $1016 \, cm^{-1}$ (back-to-back C=O oscillators)].

SECOND-ORDER COUPLING

The selection rules break down occasionally, particularly in condensed phases, to give overtone bands (Δv greater than 1). These departures may result from anharmonicities. The overtone frequencies are usually somewhat less than double that of the fundamental mode. The drop in expected frequency results from the compression of upper levels on the potential energy curve (Fig. B.3). The absorption bands that result from these transitions are usually very weak, as the mode is formally forbidden.

One of the most spectacular of the second-order events is Fermi resonance. When the first overtone ($\Delta v = 2$) of a fundamental possesses very nearly the same energy as the $\Delta v = 1$ level of another normal mode, an interaction may occur in which the two close-lying levels are split into two new levels, one higher and one lower in frequency than the original modes. As a result of this mixing, the overtone often undergoes a dramatic intensity gain at the expense of the fundamental. The resulting doublet may even possess components of approximately equal intensity. The intensity distribution is dependent to a large extent on the value of the original frequency match. The classic example of the effect is the symmetric stretching frequency (Raman active only, see above) in carbon dioxide, which should occur near $1334 \, cm^{-1}$, but which, in fact, exists as a doublet (1388 and $1286 \, cm^{-1}$). The perturbation was explained by Fermi as the interaction of the overtone of the bending fundamental at $667 \, cm^{-1}$ with the first excited state of the symmetric stretching mode. For Fermi resonance to occur (1) the oscillators involved must be so arranged that the anharmonic terms can interact (mechanical interaction can occur), and in addition, (2) the modes must meet certain symmetry restrictions. The large majority of all complex organic substances are of such low symmetry that the latter condition usually can be assumed to have been met.

In a few cases the overtone of a fundamental, while weak, will occur with higher than usual intensity, and in a region uncluttered by other absorptions.

These bands can be utilized as confirmatory evidence in making assignments of fundamentals. In even rarer cases the first overtone will be observed to occur at slightly higher than double the fundamental values. These systems are considered to possess "negative anharmonicity" (see quantized vibrational energy, p. 390).

Another second-order effect is the "sum tone" or combination band. While forbidden in the harmonic approximation, occasionally a photon will be absorbed of the appropriate energy to simultaneously excite two normal modes. Combination bands occur as weak absorption bands that possess frequencies near the sum of the two fundamentals. If sum tones occur in regions open to observation, occasionally they can be of importance in group frequency interpretations (see out-of-plane C—H bending modes on aromatic rings; see also Fig. 7.18).

II. Tables of Group Frequencies

Table B.1 Alkane Normal Modes

C—H Vibrational Mode	$\bar{\nu} \pm 10$ (cm^{-1})
Methyl groups	
Antisymmetric (degenerate) stretch	2960
Symmetric stretch	2870
Antisymmetric (degenerate) deformation	1460
Symmetric (umbrella) deformation	1375
Methylene groups	
Antisymmetric stretch	2925
Symmetric stretch	2850
Symmetric deformation (scissor)	1450
Rocking mode (all-in-phase)	720

Table B.2 Substitution Classification of C=C Stretching Frequencies

C=C Normal Mode	$\bar{\nu}$ (cm^{-1})
trans-, tri-, tetra-Substituted	1680–1665
cis-, vinylidene- (terminal-1,1-), vinyl-Substituted	1660–1620

Table B.3 Alkene Normal Modes

C—H Vibrational Mode	$\bar{\nu} \pm 10$ (cm^{-1})
Stretching modes	
Antisymmetric stretch (=CH$_2$)	3080
Symmetric stretch (=CH$_2$)	3020
Uncoupled stretch (=CH)	3030
Out-of-plane bending modes	
Vinyl group	
Trans hydrogens (in-phase)	990
Terminal-hydrogens (wag)	910
Vinylidene group	
Terminal (wag)	890
Trans group	
Trans hydrogens (in-phase)	965
Cis group	
Cis hydrogens (in-phase)	~700
Trisubstituted group	
Uncoupled hydrogen	820
Tetrasubstituted group: no modes	

Table B.4 Alkyne Normal Modes

C≡C, C—H Normal Mode	$\bar{\nu} \pm 10$ (cm^{-1})
Triple bond stretch (monosubstituted)	2120
Triple bond stretch (disubstituted)	2225
C—H bond stretch (monosubstituted)	3300

Table B.5 Arene Out-of-Ring Plane C—H Deformation Modes

Number of Adjacent H's	cm^{-1} range
5	770–730
4	770–735
3	810–750
2	860–800
1	900–860

Table B.6 Arene Group Frequencies

Arene Fundamental	$\bar{\nu}$ (cm^{-1})
C—H stretch	3100–3000
C=C ring stretch (ν_{8a})	1600 ± 10
C=C ring stretch (ν_{8b})	1580 ± 10
C=C ring stretch (ν_{19a})	1500 ± 10
C=C ring stretch (ν_{19b})	1450 ± 10
C—H out-of-plane bend (1H)	900–860
C—H out-of-plane bend (2H)	860–800
C—H out-of-plane bend (3H)	810–750
C—H out-of-plane bend (4H)	770–735
C—H out-of-plane bend (5H)	770–730
C—C ring out-of-plane bend (1; 1,3; 1,3,5)	690 ± 10
C—H out-of-plane bend sum tones	2000–1650

Table B.7 Carbonyl Frequencies

Functional Group	$\bar{\nu}_{C=O}$ (cm^{-1})
Ketones, aliphatic, open chain ($\mathbf{R}_2$CO)	1725–1700
Ketones, conjugated	1700–1675
Ketones, ring	(see Table 7.8)
Acid halides	>1800
Esters, aliphatic	1755–1735
Esters, conjugated	1735–1720
Esters (conjugated to oxygen)	1780–1760
Lactones	(see Table 7.8)
Anhydrides, aliphatic, open chain	1840–1810 and 1770–1740
Acids, aliphatic	1725–1710
Amides	(see Chap. 7, Part C)
Lactams	(see Table 7.8)
Aldehydes	1735–1720

Table B.8   Normal Modes of the Hydroxyl Group

$\bar{\nu}$ (cm^{-1})	Intensity	Mode Description
3500–3200	Very strong	O—H stretch (only strong when H-bonded)
1500–1300	Medium-strong	O—H in-plane bend (overlaps CH$_2$ & CH$_3$ bend)
1260–1000	Strong	C—C—O antisymmetric stretch
650	Medium	O—H out-of-plane bend

Table B.9   Normal Modes of the Aldehyde Group

$\bar{\nu}$ (cm^{-1})	Intensity	Mode Description
2750–2720	Weak-medium	C—H stretch in Fermi resonance with C—H bend
1735–1720	Very strong	C=O stretch
1420–1405	Medium	CH$_2$ symmetric bend, —CH$_2$— alpha to —CHO carbonyl
1405–1385	Medium	C—H in-plane bend

Table B.10   Normal Modes of the Ketone Group

$\bar{\nu}$ (cm^{-1})	Intensity	Mode Description
3430–3410	Very weak	Not fundamental, overtone of carbonyl stretch
1725–1700	Very strong	C=O stretch
1430–1415	Medium	CH$_2$ symmetric bend, —CH$_2$— alpha to ketone carbonyl

Table B.11   Normal Modes of the Ester Group

$\bar{\nu}$ (cm^{-1})	Intensity	Mode Description
1755–1735	Very strong	C=O stretch
1370–1360	Medium	CH$_3$ symmetric bend alpha to ester carbonyl
1260–1230	Very strong	C—CO—O antisymmetric stretch—acetates
1220–1160	Very strong	C—CO—O antisymmetric stretch—higher esters
1060–1030	Very strong	O—CH$_2$—C antisymmetric stretch 1° acetates
1100–980	Very strong	O—CH$_2$—C antisymmetric stretch—higher esters (may overlap with upper band)

Table B.12   Normal Modes of the Acid Halide Group

$\bar{\nu}$ (cm^{-1})	Intensity	Mode Description
1810–1800	Very strong	C=O stretch, acid chlorides
1415–1405	Strong	CH$_2$ symmetric bend, alpha to —COCl carbonyl

Table B.13 Normal Modes of the Carboxylic Acid Group

$\bar{\nu}$ (cm^{-1})	Intensity	Mode Description
3500–2500	Very, very strong	O—H stretch intensified by H-bonding
2800–2200	Very weak	Overtone & sum tones
1725–1710	Very strong	C=O, antisymmetric H-bonded dimer stretch
1450–1400	Strong	CH$_2$—CO—O antisymmetric stretch mixed with O—H bend
1300–1200	Strong	CH$_2$—CO—O antisymmetric stretch mixed with O—H bend
950–920	Medium	Out-of-plane O—H bend, acid dimer

Table B.14 Normal Modes of the Anhydride Group (Open Chain)

$\bar{\nu}$ (cm^{-1})	Intensity	Mode Description
1840–1810	Very strong	C=O, in-phase stretch
1770–1740	Very strong	C=O, out-of-phase stretch
1420–1410	Strong	CH$_2$ symmetric bend alpha to carbonyls
1100–1000	Very strong	C—O stretch mixed modes

Table B.15 Normal Modes of the Ether Group

$\bar{\nu}$ (cm^{-1})	Intensity	Mode Description
1150–1050	Strong	C—O—C antisymmetric stretch, mixed mode

Table B.16 Normal Modes of the Primary Amine Group

$\bar{\nu}$ (cm^{-1})	Intensity	Mode Description
3400–3200	Weak-medium	NH$_2$ stretch (antisymmetric & symmetric)
1630–1600	Medium	NH$_2$ symmetric bend
820–780	Medium	NH$_2$ wag

Table B.17 Normal Modes of the Nitrile Group

$\bar{\nu}$ (cm^{-1})	Intensity	Mode Description
2260–2240	Strong	C—N stretch, aliphatic
2240–2210	Strong	C—N stretch, conjugated

Table B.18 Normal Modes of the Primary Amide Group

$\bar{\nu}$ (cm^{-1})	Intensity	Mode Description
3400–3150	Very strong	NH$_2$ antisymmetric & symmetric stretch, H-bonded
1680–1650	Very strong	C=O stretch, H-bonded
1660–1620	Strong	NH$_2$ symmetric bend (overlap with C=O stretch)
1430–1410	Strong	CH$_2$ symmetric bend alpha to amide carbonyl
750–650	Medium	NH$_2$ wag

Table B.19   Normal Modes of the Secondary Amide Group

$\bar{\nu}$ (cm^{-1})	Intensity	Mode Description
3350–3250	Strong	N—H stretch, intensified by H-bonding
3125–3075	Medium	Overtone N—H bend in Fermi resonance with N—H stretch
1670–1645	Very strong	C=O stretch, H-bonded
1580–1550	Strong	N—H in-plane bend mixed with C—N stretch
1415–1405	Strong	CH$_2$ symmetric bend alpha to amide carbonyl
1325–1275	Medium	C—N stretch mixed with N—H in-plane bend
725–680	Medium	N—H out-of-plane bend

Table B.20   Normal Modes of the Isocyanate Group

$\bar{\nu}$ (cm^{-1})	Intensity	Mode Description
2280–2260	Very strong	—N=C=O antisymmetric stretch

Table B.21   Normal Modes of the Thiol Group

$\bar{\nu}$ (cm^{-1})	Intensity	Mode Description
2580–2560	Weak	S—H stretch

Table B.22   Normal Modes of the Chlorine Group

$\bar{\nu}$ (cm^{-1})	Intensity	Mode Description
750–650	Strong	C—Cl stretch, rotamers and mixed modes occur

Table B.23   Group Frequency Assignments for Chlorobenzene

$\bar{\nu}$ (cm^{-1})	Intensity	Mode Description
3080	Medium	C—H stretching, C—H bonded to sp^2 carbon
1585	Strong	ν_{8a} ring stretching
1575	Weak	ν_{8b} ring stretching
1475	Strong	ν_{19a} ring stretching
1450	Strong	ν_{19b} ring stretching
747	Strong	C—H all in-phase, out-of-plane bend
688	Strong	Ring deformation
1945, 1865, 1788, 1733	(All weak)	Sum tones, out-of-plane C—H bends, pattern matches monosubstitution of ring

Appendix C

Nuclear Magnetic Resonance

THEORY AND EXAMPLE CASES

Nuclear Spin

Nuclear spin is a property intrinsic to a nucleus and analogous to the electron spin that plays such an important role in determining electron configurations. Nuclear spins are represented by I and are different for different nuclei. Values range from 0 through $\frac{7}{2}$ in multiples of $\frac{1}{2}$. For nuclei of greatest interest to organic chemists, the [^{1}H], [^{13}C], [^{19}F], and [^{31}P] nuclei have spins of $\frac{1}{2}$; the [^{12}C], [^{16}O], and [^{32}S] nuclei have spins of 0; the [^{14}N] nucleus has a spin of 1.

For nuclei, as in the case for electrons, the spin may have different orientations with respect to an external magnetic field. These orientations are referred to as the z component of the nuclear spin, m_z. For a nucleus with a spin of $\frac{1}{2}$ the z component of the nuclear spin may be $+\frac{1}{2}$ or $-\frac{1}{2}$. In general for a nucleus of spin I, the z component takes all values from $-I$, $-I + 1$, ... $I - 1$, I, or $2I + 1$ different values in all. For most of this discussion we will limit ourselves to nuclei with spin $\frac{1}{2}$.

In the absence of a magnetic field, the difference in energy of the nuclei with various values of the z component are small enough to be neglected for most chemical purposes. On the other hand, when a sample of protons is placed in a magnetic field the energies of the $m_z = +\frac{1}{2}$ and $-\frac{1}{2}$ states are separated.

The amount of separation of the two states is proportional to the magnetic field and is given by the following expression.

$$E = \mu m_z H_0 / I \tag{C.1}$$

H_0 is the strength of the magnetic field at the nucleus, and μ is a characteristic of the nucleus known as the magnetic moment. When protons in the magnetic field are exposed to radiation of the proper frequency, transitions between the two states are stimulated. This resonance occurs when the frequency and the energy difference are related by the Planck relation $E = h\nu$. The study of these energy changes is known as *nuclear magnetic resonance* spectrometry, nmr among the cognoscenti.

Chemical Shift

The magnetic field at the nucleus depends not only on H_0, the field generated by the instrument (the external field), but also on the electrons near the nucleus. These electrons are also influenced by the external field in such a way that they reduce the actual field at the nucleus. This reduction is very small (relative to the external field) and is of the order of 0.001%, 10 ppm, for most protons. This reduction of the external field is known as *shielding* and it gives rise to differences in the energy separation for protons in different chemical environments in a molecule. The differences in the energy separation are known as *chemical shifts*.

The magnitude of the chemical shift depends on the nature of the valence and inner electrons of the nucleus and also on electrons that are not directly associated with the nucleus. Chemical shifts are influenced by inductive effects, which reduce the electron density near the nucleus and reduce the shielding. The orientation of the nucleus relative to π electrons also plays an important role in determining the chemical shift. A nucleus located outside the π electron system (as in the case of the ring protons on substituted benzenes) will be significantly deshielded. In many molecules the chemical shift is determined by a combination of these factors.

A spectrum may be obtained by either changing the magnetic field while holding the frequency constant or by changing the frequency at constant magnetic field. The magnetic field is usually provided by a large permanent magnet or electromagnet. In the case of protons a field of about 14,000 oersted is required to produce an energy change corresponding to 60 MHz. The 14,000 oersted field may then be tuned over a range of perhaps 0.14 oersted to change the field by 10 ppm. This small change in the magnetic field is usually provided by small coils near the sample.

The observation of a spectrum will be described for the case of a constant frequency instrument. As the sample is irradiated with a constant frequency, the small field coil is adjusted as the instrument electronics detect the absorption of 60-MHz photons (or whatever the operating frequency of the system may be). This condition is illustrated in Fig. C.1, where the energy change is shown for increasing magnetic field strength. In this example we illustrate the case with two different protons, A and X. Since A and X have different chemical shifts, their energies change at different rates as the magnetic field increases.

The actual spectrum in this case would be displayed as shown in the Fig. C.2. The difference in the resonances is known as a *chemical shift* and is often expressed in parts per million (ppm). Since it is not convenient to measure

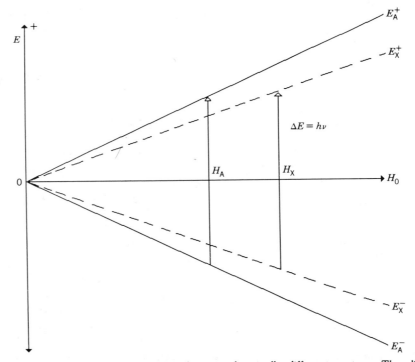

Fig. C.1 *The energy splitting for two chemically different protons. The differences between the A energy levels (solid lines) and the X levels (dashed lines) have been amplified for illustrative purposes. The 60-MHz resonance condition for nucleus A occurs at field H_A and that for X occurs at H_X. Nucleus X is said to be more strongly shielded than A. The resonance for X is said to occur upfield of that for A.*

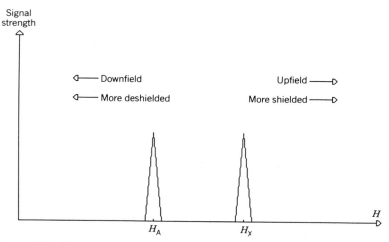

Fig. C.2 *The spectrum for the system in Fig. C.1 as it would be displayed. It is conventional to display the spectrum with magnetic field increasing to the right so that upfield (and more strongly shielded) is toward the right and downfield (and deshielded) is toward the left.*

chemical shifts relative to protons that are not in molecules, chemical shifts are often measured relative to reference molecules that have only one type of proton, such as tetramethyl silane (TMS) or chloroform. The chemical shift relative to TMS is symbolized by δ, which is defined below. TMS is used as a reference substance for a number of reasons. It is more strongly shielded than most other protons and its resonance is thus well removed from other areas of interest. TMS is also relatively inert and its 12 protons per molecule provide a strong signal per mol of TMS.

$$\delta = \frac{10^6 (H_B - H_A)}{H_B}$$

Since most of the early nmr spectrometers operated in the constant frequency mode, it is common to display the spectra as shown in Fig. C.2 even if the instrument operates in the constant magnetic field mode. Thus, the same terms, upfield and downfield, are applied when describing spectra obtained by constant field instruments. The chemical shift may also be expressed as an equivalent difference in hertz (Hz) (or cycles). For a system operating at 60 MHz this difference is

$$\text{Chemical shift (Hz)} = \text{Chemical shift (ppm)} \times 60 \text{ MHz}/10^6$$

Thus, a shift of 3 ppm would correspond to 180 Hz for a 60-MHz spectrometer. It is important to note that the chemical shift in hertz is dependent on the frequency of the system, and since different operating frequencies are in common use, this notation is seldom used unless one is analyzing second-order spectra (see below).

Spin–Spin Coupling

In a molecule with several protons, the energy change when one proton changes its spin state depends not only on the chemical environment of that proton, but also on the chemical environment and the spin states of neighboring protons. The effect of one proton on another is known as *coupling*. Coupling is a term used to describe the interaction between two properties we would otherwise expect to act independently. The horizontal and vertical motions of a sphere might be such independent properties that would be analyzed in calculating the trajectory of a thrown ball. The "curve" in a thrown baseball is caused by the

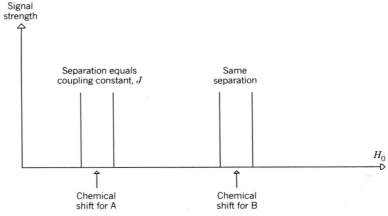

Fig. C.3 *Spectrum of two chemically different protons that are coupled.*

coupling of horizontal and vertical motion resulting from the ball's rotation and stitches.

The spectra resulting from spin–spin coupling depend on the types of nuclei, the distance of separation, the electronic environment, and the total number of spin states possible. This may be illustrated by looking at the spectrum of an imaginary compound that has protons A and X represented in Fig. C.3. In the first approximation we would expect one resonance for A and one resonance for X, and the spectrum would resemble that shown in Fig. C.2. In the presence of coupling the resonance for A splits into two signals, one of which corresponds to X having $m_X = +\frac{1}{2}$ and the other to $m_X = -\frac{1}{2}$. The coupling effect is symmetric in that the X resonance also splits into two resonances, one for each spin state of A. The magnitude of the separation of the A pair (a doublet) or the X pair is known as the coupling constant. It is usually expressed in frequency units (Hz) since this splitting (to a first order) is independent of the field strength.

The splitting becomes more interesting when there are several nuclei of one type. 1,1-Dibromoethane, CH_3CHBr_2, has three equivalent protons in the methyl group and one proton on the 1-carbon. The chemical shift for the 1-carbon proton is 5.86 ppm and that for the methyl protons is 2.47 ppm. (Here we can see an example of decreased shielding resulting from the presence of electronegative substituents.) The methyl group in this case exhibits rapid internal rotation so that its three protons are equivalent. Equivalent protons do not affect one another (this is an important rule in interpreting spectra), but the methyl protons will affect the proton on the 1-carbon.

To analyze the splitting pattern we need to consider the component of spin parallel to the magnetic field for all three methyl protons. Since each proton may have two spin states, there are $2^3 = 8$ spin states in all for the methyl protons. The component of the total spin directed along the magnetic field axis may have only four different values, as shown in Fig. C.4. The symbol $+$ is used to represent $m_z = +\frac{1}{2}$ for a single proton. Thus $(+)(+)(-)$ means that protons 1 and 2 have $m_z = +\frac{1}{2}$, while proton 3 has $m_z = -\frac{1}{2}$.

―――― $(+)(+)(+)$	$m_z = +3/2$
―――― $(+)(+)(-)$ or $(+)(-)(+)$ or $(-)(+)(+)$	$m_z = +1/2$
―――― $(+)(-)(-)$ or $(-)(+)(-)$ or $(-)(-)(+)$	$m_z = -1/2$
―――― $(-)(-)(-)$	$m_z = -3/2$

Fig. C.4 *m_z values for the entire methyl group are shown on the right and the different spin states of the individual protons on the left.*

The number of different m_z states is $2I + 1$, where I is the total spin of the equivalent nuclei ($\frac{3}{2}$ in the case of three protons on the methyl group). In addition to the four different states, we should note that there is only one way to achieve the $+\frac{3}{2}$ and $-\frac{3}{2}$ state and that there are three ways to obtain the $+\frac{1}{2}$ and $-\frac{1}{2}$ states. The number of different ways of having the same m_z value is known as the *degeneracy* of that state.

In the case of 1,1-dibromoethane without coupling, we can arbitrarily represent the 1-carbon proton resonance as having a strength of 8 units. The methyl resonance will then have a strength of 24, three times as great (since there are three times as many methyl protons). This is shown schematically in Fig. C.5a.

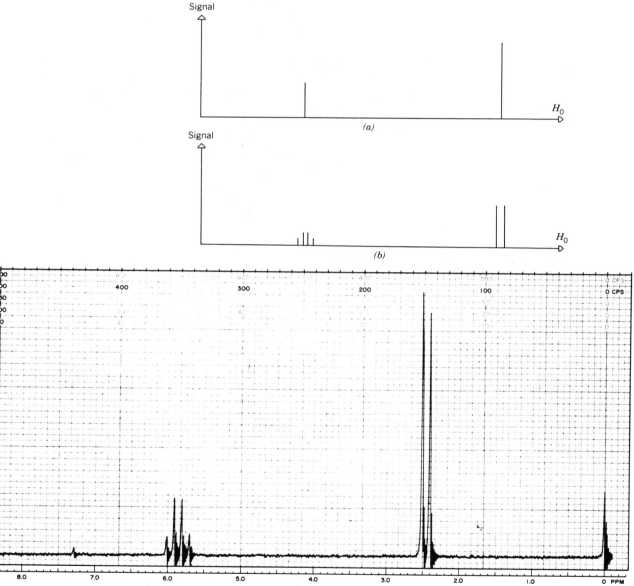

Fig. C.5a–c *Analysis of the spectrum of 1,1-dibromoethane. (Courtesy of Varian Associates, Palo Alto, CA.) (a) The spectrum without any spin–spin coupling. The resonance for the proton on the 1-carbon is shifted downfield. (b) A "stick figure" spectrum that indicates the expected intensities. The actual 60-MHz spectrum is shown in (c). The TMS resonance at 0 ppm is seen as well as a weak resonance at 7.3 ppm, which is not due to this molecule. This spectrum also shows the "ringing," which is a characteristic of a well-tuned spectrometer.*

The separation of each of the 1-carbon proton signals (known as a quartet) will equal the separation of the methyl signal. In the presence of coupling, the four m_z values for the methyl group will cause the 1-carbon proton resonance to split into four signals with relative strengths 1:3:3:1. (Note that they add up to eight.) The methyl resonance will split into two signals (one for each m_z value of the 1-carbon proton). These two resonances will have the same intensity (12 in this case). The proposed spectrum is shown in Fig. C.5b. The coupling constant in this case is about 7 Hz. The 60 MHz 1,1-dibromoethane spectrum is shown in Fig. C.5c. The "N + 1 Rule" is used to describe the number of resonances generated by coupling with N equivalent protons. The total spin of a group of N protons is $N/2$ and the number of m_z values is $2I + 1$. Therefore we can represent the number of m_z values as $N + 1$. This is also the number of resonances generated in neighboring protons as a result of coupling with the N equivalent protons.

A resonance may be affected by more than one group of nuclei. The spectrum of 1-nitropropane, $CH_3CH_2CH_2NO_2$, is shown in Fig. C.6. Can you sort out the splitting pattern? Perhaps the first thing to do is to consider the possible m_z states and their degeneracies for the two equivalent methylene protons. The most difficult part of the analysis is sorting out the methylene protons on the 2-carbon. These are first split into a 1:2:1 triplet as a result of coupling with the 1-carbon protons. Further interaction with the methyl group splits each member of the triplet into a 1:3:3:1 quartet. Of course, in many cases the separate lines will not be completely resolved or may be too weak to be seen.

Nuclei with a spin of zero have only one spin state and do not produce spin–spin splittings. Nuclei with spins of 1 or more might be expected to exhibit

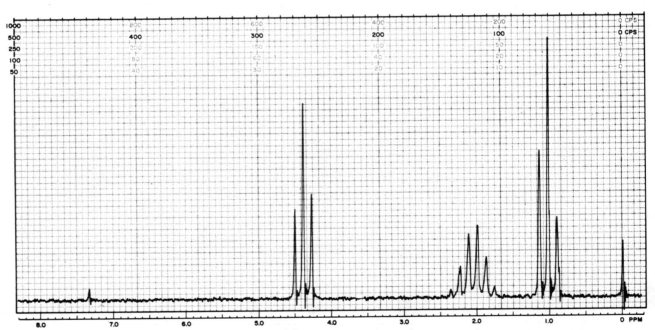

Fig. C.6 *The 60-MHz spectrum of 1-nitropropane.* (Courtesy of Varian Associates, Palo Alto, CA.) *Starting from the right, the TMS resonance at 0 ppm is seen. Next is a 1:2:1 triplet at 1.03 ppm. This results from the protons on the 3-carbon and their coupling with the two protons on the 2-carbon. Next is the pattern of at least six resonances centered at 2.07 ppm. These signals result from the protons on the 2-carbon and their coupling with the protons on the 1-carbon and the 3-carbon. Finally we have the signal from the protons closest to the nitro group centered at 4.38 ppm. They appear as a 1:2:1 triplet due to coupling with the 2-carbon protons.*

complex spin–spin effects. In fact, these nuclei usually interchange very rapidly between different spin states because of their quadrupole moments. As a result, no effects of the spins of Cl, Br, or I are observed in nuclear magnetic resonance. The effects of [^{2}H] and [^{14}N] are small, but observable in some cases.

Intensities

The intensity of a resonance is often represented as the *integral* over a resonance or a group of resonances. As mentioned above, the resonance of a proton that is coupled with a methyl group will exhibit a characteristic 1:3:3:1 intensity pattern. In more complex spectra the intensities are still useful as a measure of the number of protons of a given type. For instance, in the above case the integral over both members of the methyl group doublet will be three times the integral over the quartet of the proton on the 1-carbon. Most nmr spectrometers have the capability of measuring the integral. This can often provide an additional handle on determining the identity of a compound.

Second-Order Effects

The examples so far considered have all consisted of first-order spectra. These highly symmetric and fairly simple spectra are generally observed when the chemical shift differences (expressed as a frequency) are much greater than the coupling constant. Second-order effects occur when the coupling constants become comparable to or greater than the chemical shift differences.

Second-order effects may be understood in qualitative terms by considering the limiting cases. Let us consider the hypothetical disubstituted ethylene shown in Fig. C.7 where R and M are substituents that might be identical or may have very different effects on the olefinic protons. In Figure C.7a the spectrum is shown for the case in which R and M have very different effects. Perhaps M is a strongly electronegative group and R actually contributes some electron density. In this case we will observe a first-order spectrum consisting of two doublets. The coupling constant is the separation in the doublets and the chemical shift of each nucleus is the midpoint of each doublet.

In Fig. 7b, groups R and M are identical. H_A and H_B are identical in this case and only a single resonance is observed. (There is still coupling between the nuclei, but it is not observed in the spectra of equivalent nuclei.)

In Fig. 7c the difference in the chemical environment of H_A and H_B is very slight. The spectrum shown may be seen as intermediate between the limiting cases in Figs. C.7a and C.7b. It may be seen that there is a "leaning in" of the doublets as the central members increase in intensity at the expense of the outer members. For example, see the spectra of substituted cinnamic acids. A full continuum of behavior may be expected with cases observed in which the outer members are lost in the noise and the central members take the appearance of a doublet. This is one example of a class of spectra known as "deceptively simple spectra." It should also be noted that in second-order spectra the coupling constants and the chemical shift differences may not be obtained as simple differences.

The second-order spectra of systems with more than two protons are difficult to describe even in qualitative terms. It is partly for this reason that spectra obtained at high frequencies (and magnetic fields) are useful. As the operating frequency of the instrument is increased, the chemical shift differences (in energy terms) increase while the spin–spin coupling remains constant. Thus, the complicating second effects are likely to be less noticeable in high field spectra. The reader is referred to more extensive treatments of nmr for a discussion of second-order cases.

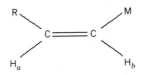

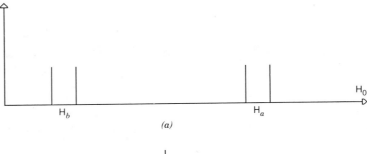

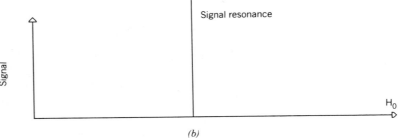

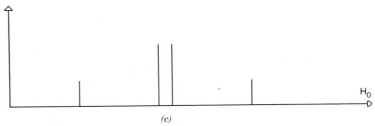

Fig. C.7a–c *Illustration of second-order effects. In (a) the chemical shift difference is much larger than the coupling constant and a first-order spectrum is observed. In (b) protons A and B are equivalent and a single resonance is observed. In (c) the chemical shift difference is the same order of magnitude or less than the coupling constant. The "leaning in" of this spectrum relative to 7a is also seen in second order spectra systems which have more than two protons.*

REFERENCES

1. American Petroleum Institute Research Project 44, "Selected Nuclear Magnetic Resonance Spectral Data."
2. Abraham, R. J. "The Analysis of High Resolution NMR Spectra"; Elsevier: New York, 1971.
3. Bhacca, N. S.; Hollis, D. P.; Johnson, L. F.; Pier, E. A.; Shoolery, J. N. "NMR Spectra Catalog"; Varian Associates: Palo Alto, CA, 1963.
4. Cooper, J. W. "Spectroscopic Techniques for Organic Chemists"; Wiley: New York, 1980.
5. McFarlane, W. In "Elucidation of Organic Structures by Physical and Chemical Methods"; Bentley, K. W.; Kirby, G. W., Eds.; Wiley: New York, 1972; Part I, Chapter IV.

Appendix D

Tables of Derivatives

Table D.1 Derivatives of Carboxylic Acids (Liquids)

Acid	bp (°C)	Melting Point of Derivative (°C)[a]		
		Amide	Anilide	p-Toluidide
Methanoic (formic)	101	—	50	53
Ethanoic (acetic)	118	82	114	153
Propenoic (acrylic)	141	84	104	141
Propanoic	141	81	106	126
2-Methylpropanoic (isobutyric)	155	128	105	109
Butanoic (butyric)	163	115	96	75
2-Methylpentenoic (methacrylic)	163	102	87	—
Pyruvic	165	124	104	109
3-Methylbutanoic	177	135	110	106
Pentanoic (valeric)	186	106	63	74
2-Methylpentanoic	186	79	95	80
2,2-Dichloroethanoic	194	98	118	153
Hexanoic (caproic)	205	100	94	74
Heptanoic (enanthic)	223	96	65; 70	81
Octanoic (caprylic)	239	106; 110	57	70
Nonanoic (pelargonic)	254	99	57	84

[a]Two values are given for those derivatives which may exist in polymorphic forms.

Table D.2 Derivatives of Carboxylic Acids (Solids)

Acid	mp (°C)	Melting Point of Derivative (°C)[a]		
		Amide	Anilide	p-Toluidide
Decanoic	31–32	108	70	78
Lauric	43–45	87	78	100
Myristic	54	103	84	93
Trichloroacetic	54–58	141	97	113
Chloroacetic	61	121	137	162
Palmitic	62	106	90	98
Octadecanoic (stearic)	70	109	95	102
Crotonic	72	158	118	—
3,3-Dimethyl acrylic	69	107	126	—
Phenylethanoic	77	156	65	117
2-Benzoylbenzoic	128	165	195	—
Pentandioic (glutaric)	97	175	223	218
Ethanedioic (oxalic)	101	219	148	169
2-Methylbenzoic (o-toluic)	105	143	125	144
3-Methylbenzoic (m-toluic)	112	94	126	118
Benzoic	122.4	130	160	158
Sebacic	131–134	170 (mono) 210 (di)	122 (mono) 200 (di)	201
trans-Cinnamic	133	147	153	168
2-Acetoxybenzoic (aspirin)	135	138	136	—
cis-Butenedioic (maleic)	137	172	198 (mono) 187 (di)	142 (di)
Malonic	137	—	132 (mono) 230 (di)	86 (mono) 253 (di)
2-Chlorobenzoic	140	—	118	131
3-Nitrobenzoic	140	143	154	162
2-Nitrobenzoic	146	176	155	—
Diphenylacetic	148	168	180	172
2-Bromobenzoic	150	155	141	—
Benzilic	150	153	175	190
Hexanedioic (adipic)	153	125 (mono) 230 (di)	151 (mono) 241 (di)	—
2-Hydroxybenzoic (salicylic)	158	142	136	156
2-Iodobenzoic	162	110	141	—
4-Methylbenzoic (p-toluic)	179	160	144	160; 165
4-Methoxybenzoic (p-anisic)	185	167	170	186
2-Naphthoic	186	192	171	192
Succinic	190	157 (mono) 260 (di)	143 (mono) 230 (di)	180 (mono) 255 (di)
Phthalic	211	149 (mono) 220 (di)	170 (mono) 254 (di)	150 (mono) 201 (di)
3,5-Dinitrobenzoic	205	183	234	—
4-Nitrobenzoic	241	198	204; 211	192; 204

[a]Two values are given for those derivatives which may exist in polymorphic forms.

Table D.3 Derivatives of Alcohols

Alcohol	bp (°C)	Melting Point of Derivative (°C)		
		Phenyl-urethan	α-Naphthyl-urethan	3,5-Dinitro-benzoate
Methyl- (methanol)	65	47	124	108
Ethyl- (ethanol)	78	52	79	93
Isopropyl- (2-propanol)	82	88	106	122
t-Butyl- (t-butanol)	83	136	101	142
Allyl-	97	70	109	49
n-Propyl- (1-propanol)	97	51	80	74
s-Butyl- (2-butanol)	99	65	97	76
t-Pentyl- (2-methyl-2-butanol)	102	42	71	116
Isobutyl- (2-methyl-1-propanol)	108	86	104	87
3-Pentanol	116	48	71	101
n-Butyl- (1-butanol)	118	63	71	64
2,3-Dimethyl-2-butanol	118	—	—	111
2-Pentanol	119	—	76	61
2-Methyl-2-pentanol	121	239	—	72
3-Methyl-3-pentanol	123	50	—	97
2-Methoxyethanol	125	—	113	—
2-Methyl-1-butanol	129	—	—	70
2-Chloroethanol	131	51	101	95
4-Methyl-2-pentanol	132	143	88	65
3-Methyl-1-butanol	132	55	—	61
2-Ethoxyethanol	135	—	67	75
3-Hexanol	136	—	—	77
2,2-Dimethyl-1-butanol	137	—	—	51
1-Pentanol	138	46	68	46
2-Hexanol	139	—	—	39
2,4-Dimethyl-3-pentanol	140	—	—	—
Cyclopentanol	141	132	118	115
2-Ethyl-1-butanol	148	—	—	52
2-Methyl-1-pentanol	148	—	—	51
4-Heptanol	156	—	80	64
1-Hexanol	158	42	59	58
2-Heptanol	159	—	54	49
Cyclohexanol	161	82	128	113
2-Furfuryl-	172	45	129	81
1-Heptanol	177	68	—	47
Tetrahydrofurfuryl-	178	61	—	84
2-Octanol	179	114	—	32
1-Octanol	195	74	—	61
Benzyl-	205	78	—	113
2-Phenylethanol	221	79	—	108
1-Decanol	231	59	—	57
Cinnamyl-	(mp 35)	90	—	121
Benzohydrol	(mp 67)	139	—	141
Cholesterol	(mp 147)	168	—	—

Table D.4   Derivatives of Aldehydes

Aldehyde	bp (°C)	Melting Point of Derivative (°C)[a]	
		Semi-carbazone	2,4-Dinitrophenyl-hydrazone
Acetaldehyde	21	162	168
Propionaldehyde	50	89 (154)	154
Isobutyraldehyde	64	125	187 (183)
n-Butyraldehyde	74	104	123
Isovaleraldehyde	92	107	123
n-Valeraldehyde	103	108	107
Crotonaldehyde	104	199	190
n-Hexaldehyde	131	106	104; 107
n-Heptaldehyde	153	109	108
2-Furaldehyde	161	202	212 (230)
Benzaldehyde	179	222	237
Salicylaldehyde	197	231	252 dec
p-Tolualdehyde	204	221	239
2-Chlorobenzaldehyde	215	146 (229)	213
Citral	228	164	116
4-Anisaldehyde	248	210	253
trans-Cinnamaldehyde	252	215	255
4-Chlorobenzaldehyde	(mp 47)	230	254

[a]Two values are given for those derivatives which may exist in polymorphic forms or as syn and anti geometrical isomers.

Table D.5   Derivatives of Ketones

Ketone	bp (°C)	Melting Point of Derivative (°C)[a]	
		Semi-carbazone	2,4-Dinitro-phenylhydrazone
Acetone	56	187	126
2-Butanone	80	146	117
3-Methyl-2-butanone	94	113	120
2-Pentanone	102	112	143
3-Pentanone	102	139	156
3,3-Dimethyl-2-butanone	106	157	125
4-Methyl-2-pentanone	119	134	95
2,4-Dimethyl-3-pentanone	124	160	88 (94)
2-Hexanone	129	122	106
Cyclopentanone	131	205	142
4-Heptanone	145	133	75
3-Heptanone	149	101	—
2-Heptanone	151	127	89
Cyclohexanone	155	166	162
2-Octanone	173	122	58
Acetophenone	200	198	240
Benzalacetone	(mp 41)	187	223
Benzophenone	(mp 48)	164	239
Benzalacetophenone	(mp 58)	168; 180	245
Benzil	(mp 95)	175 (182)	189
Benzoin	(mp 133)	206 (dec)	245

[a]Two values are given for those derivatives which may exist in polymorphic forms or as syn and anti geometrical isomers.

Table D.6 Derivatives of Primary and Secondary Amines

| Amine | bp (°C) | Melting Point of Derivative (°C)[a] | | |
		Acetamide	Benzamide	Picrate
Methylamine	−6	—	80	—
Ethylamine	17	—	71	—
Isopropylamine	33	—	71	165
t-Butylamine	45	98	134	198
n-Propylamine	49	47	84	135
Allylamine	53	—	—	140
Diethylamine	55	—	42	155
sec-Butylamine	63	—	76	139
Isobutylamine	69	—	57	150
n-Butylamine	77	—	42	—
Di-*iso*propylamine	84	—	—	140
Di-*n*-propylamine	109	—	—	75
Piperidine	106	—	48	152
Ethylenediamine	116	172 (di)	244 (di)	233
Cyclohexylamine	134	104	149	—
Di-*iso*butylamine	139	86	—	121
Di-*n*-butylamine	159	—	—	59
Benzylamine	185	60	105	199
Aniline	185	114	163	198
N-Methylaniline	196	102	67	145
2-Methylaniline	199	110	144	213
4-Methylaniline	200 (mp 45)	148	158	—
3-Methylaniline	203	65	125	200
N-Ethylaniline	205	111	147	194
2-Chloroaniline	208	87	99	134
2,5-Dimethylaniline	215	139	140	171
2,6-Dimethylaniline	216	177	168	180
2,4-Dimethylaniline	217	133	192	209
N-Ethyl-3-methylaniline	221	—	72	—
2-Methoxyaniline	225	85	60 (84)	200
4-Chloroaniline	232 (mp 70)	179	192	178
4-Methoxyaniline	243 (mp 57)	130	154	170
2-Ethoxyaniline	229	79	104	—
4-Ethoxyaniline	254	135	173	—
Diphenylamine	(mp 54)	101	180	182
3-Nitroaniline	(mp 114)	152	155	—
4-Nitroaniline	(mp 147)	—	199	—

[a]Two values are given for those derivatives which may exist in polymorphic forms.

Table D.7   Derivatives of Tertiary Amines

Tertiary Amine	bp (°C)	Melting Point of Derivative (°C)[a] Picrate
Trimethylamine	3	216
Triethylamine	89	173
Pyridine	116	167
2-Methylpyridine (2-picoline)	129	169
2,6-Dimethylpyridine (2,6-lutidine)	142	168 (161)
3-Methylpyridine (3-picoline)	143	150
4-Methylpyridine (4-picoline)	143	167
Tripropylamine	157	116
N,N-Dimethylaniline	193	163
Tributylamine	216	105
N,N-Diethylaniline	216	142
Quinoline	237	203
Tri-isopentylamine	245	125

[a]Two values are given for those derivatives which may exist in polymorphic forms.

Table D.8   Derivatives of Acid Chlorides and Anhydrides

Acid Chloride or Anhydride	bp (°C)	mp (°C)	Melting Point of Derivative (°C) Amide
Acetyl chloride	52	—	82
Propionyl chloride	77–79	—	81
Butyryl chloride	102	—	115
Acetic anhydride	138–140	—	82
Propionic anhydride	167	—	81
Butyric anhydride	198–199	—	115
Benzoyl chloride	198	—	130
3-Chlorobenzoyl chloride	225	—	134
2-Chlorobenzoyl chloride	238	—	142
cis-1,2-Cyclohexanedicarboxylic anhydride	—	32	192d (acid)
Benzoic anhydride	—	39–40	130
Maleic anhydride	—	54–56	181 (mono) 266 (di)
4-Nitrobenzoyl chloride	—	72–74	201
Succinic anhydride	—	119–120	157 (mono) 260 (di)
Phthalic anhydride	—	131–133	149 (mono) 220 (di)

Table D.9 Derivatives of Aromatic Hydrocarbons

			Melting Point of Derivative (°C)[a]
Aromatic Hydrocarbon	bp (°C)	mp (°C)	Picrate
Benzene	80	—	84
Toluene	111	—	88
Ethylbenzene	136	—	96
p-Xylene	138	—	90
m-Xylene	138–139	—	91
o-Xylene	143–145	—	88
Mesitylene	163–166	—	97
1,2,4-Trimethylbenzene	168	—	97
1,2,3,4-Tetramethylbenzene	205	—	92
1-Methylnaphthalene	242	—	142
2-Methylnaphthalene	—	35	116
Pentamethylbenzene	—	51	131
Naphthalene	—	81	149
Acenaphthene	—	94	161
Phenanthrene	—	100	144 (133)
Anthracene	—	216	138

[a]Two values are given for those derivatives which may exist in polymorphic forms.

Table D.10 Derivatives of Phenols

		Melting Point of Derivative (°C)	
Phenol	mp (°C)	Bromo	α-Naphthylurethan
2-Chloro-	7 (bp 175)	48 (mono)	120
		76 (di)	
Phenol	42	95 (tri)	133
4-Methyl- (p-cresol)	35	49 (di)	146
		108 (tetra)	
3-Methyl- (m-cresol)	203 (bp)	84 (tri)	128
3,4-Dimethyl-	229 (bp)	171 (tri)	141
2-Methyl- (o-cresol)	33	56 (di)	142
4-Ethyl-	45	—	128
2-Nitro-	45	117 (di)	113
2,6-Dimethyl-	48	79	176
2-Isopropyl-5-methyl- (thymol)	50	55	160
3,4-Dimethyl-	63	117 (tri)	—
3,5-Dimethyl-	64	166 (tri)	—
4-Bromo-	66	95 (tri)	168
2,5-Dimethyl-	73	178 (tri)	173
1-Naphthol	95	105 (di)	152
3-Nitro-	96	91 (di)	—
4-t-Butyl-	98	50 (mono)	110
		67 (di)	
1,2-Dihydroxy- (catechol)	105	192 (tetra)	175
1,3-Dihydroxy- (resorcinol)	110	112 (tri)	275
4-Nitro-	112	142 (di)	150
2-Naphthol	123	84	157
Pyrogallol	134	158 (di)	173
1,4-Dihydroxy- (hydroquinone)	171	186 (di)	—

Table D.11   Hydrocarbons

Compound	bp (°C)	Compound	bp (°C)
Alkanes			
Pentane	36	2,2,4-trimethyl-pentane	99
Cyclopentane	49		
2,2-Dimethylbutane	50	*trans*-1,4-Dimethyl-cyclohexane	119
2,3-Dimethylbutane	58		
2-Methylpentane	60	Octane	126
3-Methylpentane	63	Nonane	151
Hexane	69	Decane	174
Cyclohexane	81	Eicosane	343 (mp 37)
Heptane	98	Norbornane	(mp 87, subl.)
		Adamantane	(mp 268, sealed)
Alkenes and Alkynes			
1-Pentene	30	3-Hexene	82
2-Methyl-1,3-butadiene (isoprene)	34	Cyclohexene	84
		2-Hexyne	84
trans-2-Pentene	36	1-Heptene	94
cis-2-Pentene	37	1-Heptyne	100
2-Methyl-2-butene	39	2,4,4-Trimethyl-1-pentene	102
Cyclopentadiene	41		
1,3-Pentadiene (piperylene)	41	2,4,4-Trimethyl-2-pentene	104
3,3-Dimethyl-1-butene	41	1-Octene	123
1-Hexene	63	Cyclooctene	146
cis-3-Hexene	66	1,5-Cyclooctadiene	150
trans-3-Hexene	67	*d,l*-α-Pinene	155
1-Hexyne	71	$(-)$-β-Pinene	167
1,3-Cyclohexadiene	80	Limonene	176
		1-Decene	181

Table D.12   Halogenated Hydrocarbons

Compound	bp (°C)	Compound	bp (°C)
Alkyl Halides			
Chlorides		*Bromides*	
n-Propyl-	47	Ethyl-	38
t-Butyl-	51	Isopropyl-	60
sec-Butyl-	68	Propyl-	71
Isobutyl-	69	*t*-Butyl-	72
n-Butyl-	78	Isobutyl-	91
Neopentyl-	85	*sec*-Butyl-	91
t-Pentyl-	86	Butyl-	101
Cyclohexyl-	143	*t*-Pentyl-	108
Hexachloroethane	185 (mp 187, subl.)	Neopentyl-	109
Triphenylmethyl-	(mp 113)	1-Bromoheptane	174; 180
		Iodides	
		Methyl-	43
		Ethyl-	72
		Isopropyl-	90
		Propyl-	102

Compound	bp (°C)	mp (°C)
Aryl Halides		
Chlorobenzene	132	—
Bromobenzene	156	—
2-Chlorotoluene	157–159	—
4-Chlorotoluene	162	—
1,3-Dichlorobenzene	172–173	—
1,2-Dichlorobenzene	178	—
2,4-Dichlorotoluene	196–203	—
3,4-Dichlorotoluene	201	—
1,2,4-Trichlorobenzene	214	—
1-Bromonaphthalene	279–281	—
1,2,3-Trichlorobenzene	—	51–53
1,4-Dichlorobenzene	—	54–56
1,4-Bromochlorobenzene	—	66–68
1,4-Dibromobenzene	—	87–89
1,2,4,5-Tetrachlorobenzene	—	138–140

Table D.13   Nitriles

Compound	bp (°C)	Compound	mp (°C)
Acrylonitrile	77	4-Chlorobenzylcyanide	30.5
Acetonitrile	81	Malononitrile	34
Propionitrile	97	Stearonitrile	40
Isobutyronitrile	108	2-Chlorobenzonitrile	41
n-Butyronitrile	117	Succinonitrile	48
Benzonitrile	191	Diphenylacetonitrile	75
2-Methylbenzonitrile	205	4-Cyanopyridine	80
3-Methylbenzonitrile	212		
4-Methylbenzonitrile	217		
Benzylcyanide	234		
Adiponitrile	295		

Table D.14   Amides[a]

Compound	bp (°C)	mp (°C)
N,N-Dimethylformamide	153	—
N,N-Diethylformamide	176	—
N-Methylformamide	185	—
N-Formylpiperidine	222	—
N,N-Dimethylbenzamide	—	41
N-Benzoylpiperidine	—	48
N-Propylacetanilide	—	50
N-Benzylacetamide	—	54
N-Ethylacetanilide	—	54
N,N-Diphenylformamide	—	73
N-Methyl-4-acetotoluidide	—	83
N,N-Diphenylacetamide	—	101
N-Methylacetanilide	—	102
Acetanilide	—	114
N-Ethyl-4-nitroacetanilide	—	118
N-Phenylsuccinimide	—	156
N-Phenylphthalimide	—	205

[a]Also see Table 6.1 for amides prepared as derivatives of carboxylic acids.

Table D.15 Nitro Compounds

Compound	bp (°C)	mp (°C)
Nitrobenzene	211	—
2-Nitrotoluene	225	—
2-Nitro-*m*-xylene	225	—
3-Nitrotoluene	231	—
3-Nitro-*o*-xylene	245	—
4-Ethylnitrobenzene	246	—
2-Chloro-6-nitrotoluene	—	36
4-Chloro-2-nitrotoluene	—	38
3,4-Dichloronitrobenzene	—	42
1-Chloro-2,4-dinitrobenzene	—	50
4-Nitrotoluene	—	54
1-Nitronaphthalene	—	56
1-Chloro-4-nitrobenzene	—	84
m-Dinitrobenzene	—	90

Table D.16 Ethers

Compound	bp (°C)	mp (°C)
Furan	32	—
Ethyl vinyl ether	33	—
Tetrahydrofuran	67	—
n-Butyl vinyl ether	94	—
Anisole	154	—
4-Methylanisole	174	—
3-Methylanisole	176	—
4-Chloroanisole	203	—
1,2-Dimethoxybenzene	207	—
4-Bromoanisole	215	—
Anethole	234–237	—
Diphenyl ether	259	—
2-Nitroanisole	273	—
Dibenzyl ether	298	—
4-Nitroanisole	—	50–52
1,4-Dimethoxybenzene	—	56–60
2-Methoxynaphthalene	—	73–75

Table D.17 Esters

Compound	bp (°C)	Compound	bp (°C)
Liquids			
Methyl formate	32	Pentyl formate	132
Ethyl formate	54	Ethyl 3-methylbutanoate	135
Methyl acetate	57	Isobutyl propanoate	137
Isopropyl formate	68	Isopentyl acetate	142
Ethyl acetate	77	Propyl butanoate	143
Methyl propanoate	80	Ethyl pentanoate	146
Methyl propenoate	80	Butyl propanoate	147
Propyl formate	81	Pentyl acetate	149
Isopropyl acetate	91	Isobutyl 2-methylpropanoate	149
Methyl 2-methylpropanoate	93	Methyl hexanoate	151
sec-Butyl formate	97	Isopentyl propanoate	160
t-Butyl acetate	98	Butyl butanoate	165
Ethyl propanoate	99	Propyl pentanoate	167
Propyl acetate	101	Ethyl hexanoate	168
Methyl butanoate	102	Cyclohexyl acetate	175
Allyl acetate	104	Isopentyl butanoate	178
Ethyl 2-methylpropanoate	110	Pentyl butanoate	185
sec-Butyl acetate	112	Propyl hexanoate	186
Methyl 3-methylbutanoate	117	Butyl pentanoate	186
Isobutyl acetate	117	Ethyl heptanoate	189
Ethyl butanoate	122	Isopentyl 3-methylbutanoate	190
Propyl propanoate	122	Ethylene glycol diacetate	190
Butyl acetate	126	Tetrahydrofurfuryl acetate	194
Diethyl carbonate	127	Methyl octanoate	195
Methyl pentanoate	128	Methyl benzoate	200
Isopropyl butanoate	128	Ethyl benzoate	213

Compound	mp (°C)	Compound	mp (°C)
Solids			
d-Bornyl acetate	29 (bp 221)	Ethyl 3,5-dinitrobenzoate	93
Ethyl 2-nitrobenzoate	30	Methyl 4-nitrobenzoate	96
Ethyl octadecanoate	33	2-Naphthyl benzoate	107
Methyl cinnamate	36 (bp 261)	Isopropyl 4-nitrobenzoate	111
Methyl 4-chlorobenzoate	44	Cyclohexyl 3,5-dinitrobenzoate	112
1-Naphthyl acetate	49	Cholesteryl acetate	114
Ethyl 4-nitrobenzoate	56	Ethyl 4-nitrobenzoate	116
2-Naphthyl acetate	71	t-Butyl 4-nitrobenzoate	116
Ethylene glycol dibenzoate	73	Hydroquinone diacetate	124
Propyl 3,5-dinitrobenzoate	74	t-Butyl 3,5-dinitrobenzoate	142
Methyl 4-bromobenzoate	81	Hydroquinone dibenzoate	204

Index

About the Authors

Dana W. Mayo holds the Charles Weston Pickard Professor of Chemistry Chair at Bowdoin College. A former Fellow of the School for Advanced Study at MIT and a Special Fellow of the National Institute of Health at the University of Maryland, he received his PhD in Chemistry from Indiana University. Professor Mayo is Director of the Bowdoin College Summer Course in Infrared Spectroscopy. His research interests include the application of vibrational spectroscopy to molecular structure determination, natural products chemistry, and environmental studies of oil pollution.

Samuel S. Butcher is Professor of Chemistry at Bowdoin College. His research interests lie in the areas of atmospheric chemistry and air pollution. He is the coauthor of a textbook on air chemistry and has written several papers on air pollution and effects of small wood-burning stoves. He has taught a wide range of undergraduate chemistry and science courses for nonspecialists. Professor Butcher received his PhD in Chemistry from Harvard University.

Ronald M. Pike is Professor of Chemistry at Merrimack College. He received his PhD from MIT. His main research interests involve the synthesis of organofunctional silanes and related silicone polymers. He is the author of numerous papers and patents in this area. Professor Pike was previously associated with Union Carbide Corporation and the Lowell Technological Institute. He was a recent Visiting Charles Weston Pickard Professor of Chemistry at Bowdoin College.

THE USE OF THE CORRELATION CHART OF INFRARED GROUP FREQUENCIES

This chart is by no means complete. It is scarcely possible to crowd onto one piece of paper of reasonable size all that might possibly be desired; further, as infrared continues to spread in application, new group frequencies are still being discovered. The user of such charts is urged to place his own notations for newfound correlations on the chart.

The chart has been organized into chemical group types, whose designations appear along the left-hand edge. Across the top of the chart have been indicated the various classes of molecular motions that form usable group frequencies.

The short heavy horizontal line under each group symbol indicates the extremes of the frequency region in which such groups are known to have a characteristic absorption. (Often in the past, such regions have had to be extended when a particular group is placed in molecules with less familiar groups in the vicinity, with the result that its characteristic frequencies have strayed.)

The thickness of the line is a very rough index of the intensity of this absorption. A line of tapering thickness indicates the intensity for this group to be quite variable. Often these intensity variations can be correlated with structure, but no clever way to represent this on the limited space of a chart was at hand.

An open, cross-hatched line represents a region in which there is usually more than one absorption characteristic of the particular group. For example, halogenated aliphatic hydrocarbons often have several strong, sharp absorptions in the region 950–1300 cm^{-1} to which it is difficult to ascribe specific vibrational motions, but nonetheless are characteristic of that class of molecules.

The chemical symbols are those of standard organic nomenclature. A few, perhaps, should be amplified: X = halogen *except* fluorine; M = metal; N$^+$—H = hydrogen attached to a positively charged nitrogen atom, as in amine salts of acids; (Σ) = a "summation" band—i.e., combination or overtone—*not* a fundamental; ϕ = phenyl ring; CJ = conjugated.

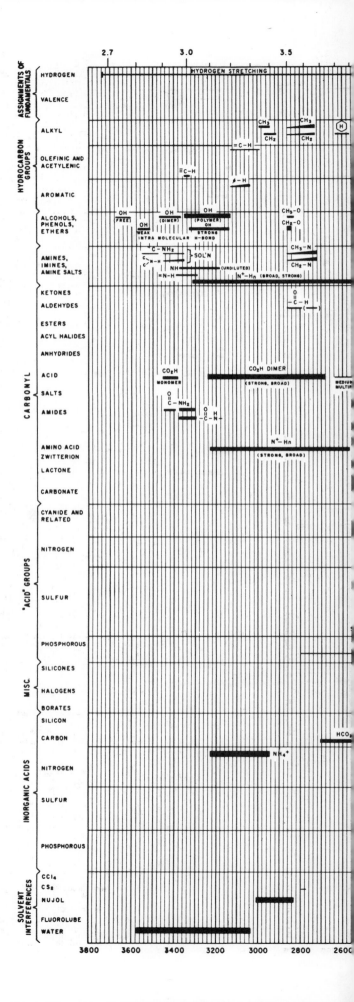